Jaguar E Type Owners Workshop Manual

by J H Haynes
Associate Member of the Guild of Motoring Writers
and Bill Harper

Models covered

UK and USA:
   Roadster. 3781 cc (230 cu in) & 4235 cc (258 cu in). Series 1, 2 & 3
   Coupe 2 seater. 3781 cc (230 cu in) & 4235 cc (258 cu in). Series 1 & 2
   Coupe 2 + 2. 4235 cc (258 cu in). Series 2 & 3

ISBN 0 85696 140 X

© J H Haynes and Company Limited 1974

ABCDE 140
FGHIJ
KLMNO
PQRST

All rights reserved. No part of this book may be reproduced or transmitted in any form or by any means, electronic or mechanical, including photocopying, recording or by any information storage or retrieval system, without permission in writing from the copyright holder.

Printed in England

**HAYNES PUBLISHING GROUP
SPARKFORD YEOVIL SOMERSET ENGLAND**
distributed in the USA by
**HAYNES PUBLICATIONS INC
861 LAWRENCE DRIVE
NEWBURY PARK
CALIFORNIA 91320
USA**

Acknowledgements

Our first thanks must go to Jaguar Cars British Leyland UK Limited of Coventry for allowing us to use some of their illustrations and for being most forthcoming with certain technical information.

Castrol Limited have been helpful with lubrication details.

Brian Horsfall stripped the car with his usual dexterity and Les Brazier took photographs at many angles.

Ted Frenchum has page edited this manual and our thanks must go to him.

Special note: The supply of illustrations and material by Jaguar Cars British Leyland UK Limited does not imply that Jaguar Cars has approved the contents of this book or is in any way responsible for the accuracy of any information printed. The copyright in illustrations and other technical material provided by Jaguar Cars British Leyland UK Limited remains vested in that company.

About this manual

The aim of this book is to help you get the best value from your car. It can do so in two ways. First, it can help you decide what work must be done, even should you choose to get it done by a garage; the routine maintenance and the diagnosis and course of action when random faults occur. But it is hoped that you will also use the second and fuller purpose by tackling the work yourself. This can give you the satisfaction of doing the job yourself. On the simpler jobs it may even be quicker than booking the car into a garage and going there twice, to leave and collect it. Perhaps most important, much money can be saved by avoiding the costs a garage must charge to cover their labour and overheads.

The book has drawings and descriptions to show the function of the various components so that their layout can be understood. Then the tasks are described and photographed in a step by step sequence so that even a novice can cope with complicated work. Such a person is the very one to buy a car needing repair yet be unable to afford garage costs.

The jobs are described assuming only normal spanners are available, and not special tools. But a reasonable outfit of tools will be a worthwhile investment. Many special workshop tools produced by the makers merely speed the work, and in these cases guidance is given as to how to do the job without them, the oft quoted example being the use of a large hose clip to compress the piston rings for insertion in the cylinder. But on a very few occasions the special tool is essential to prevent damage to components, then their use is described. Though it might be possible to borrow the tool, such work may have to be entrusted to the official agent.

To avoid labour costs a garage will often give a cheaper repair by fitting a reconditioned assembly. The home mechanic can be helped by this book to diagnose the fault and make a repair using only a minor spare part. The classic case is repairing a non-charging dynamo by fitting new brushes.

The manufacturer's official workshop manuals are written for their trained staff, and so assume special knowledge; detail is left out. This book is written for the owner, and so goes into detail.

The book is divided into twelve Chapters. Each Chapter is divided into numbered sections which are headed in bold type between horizontal lines. Each section consists of serially numbered paragraphs.

There are two types of illustration: (1) Figures which are numbered according to Chapter and sequence of occurrence in that Chapter. (2) Photographs which have a reference number on their caption. All photographs apply to the Chapter in which they occur so that the reference figure pinpoints the pertinent section and paragraph number.

Procedures, once described in the text, are not normally repeated. If it is necessary to refer to another Chapter the reference will be given in Chapter number and section number thus: Chapter 1/16.

If it is considered necessary to refer to a particular paragraph in another Chapter the reference is eg, 'Chapter 1/5:5'. Cross references given without use of the word 'Chapter' apply to sections and/or paragraphs in the same Chapter, eg, 'see Section 8' means also 'in this Chapter'.

When the left or right side of the car is mentioned it is as if looking forward from the drivers seat.

Great effort has been made to ensure that this book is complete and up to date. The manufacturers continually modify their cars, even in retrospect.

Whilst every care is taken to ensure that the information in this manual is correct no liability can be accepted by the authors or publishers for loss, damage or injury caused by any errors in, or omissions from, the information given.

Contents

Chapter	Section	Page	Section	Page
Introductory sections	Buying spare parts	7	Recommended lubricants	17
	Introduction	2	Routine drawings 'RM Nos.'	11-15
	Lubrication chart	16	Routine maintenance	8
1 Engine	Camshaft bearings	48	Re-assembly - general	53
	Decarbonisation	49	Removal	25
	Dismantling - general	30	Tappets & valve adjusting pads	49
	Examination and renovation	43	Valve clearance adjustment	58
	Exhaust manifold	61	Valve timing	59
2 Cooling system	Anti-freeze	74	Radiator	69
	Cooling fan	71	Thermostat	71
	Draining, flushing & refilling	68	Water pump belt	72
	Header tank	69	Water pump - general	72
3 Fuel system - carburation and exhaust emission control systems	Air cleaner & filter element	99	Fuel gauge unit	84
	Carbon canister	97	Fuel pump	79
	Carburettors	89	Fuel tank	84
	Exhaust emission control	96	Gulp valve	100
4 Ignition system	Coil	104	Distributor	106
	Condenser	106	Spark plugs and leads	109
	Contact points	105	Timing	107
5 Clutch	Bleeding	113	Refitting	122
	Hydraulic system	123	Removal	117
	Master cylinder	123	Slave cylinder	124
6 Gearbox and automatic transmission	Manual gearbox:-		Fluid level	144
	Dismantling	128	Manual linkage adjustment	145
	Examination	137	Oil pan - removal	145
	Re-assembly	141	Shift speeds	146
	Automatic transmission:-		Throttle/kickdown cable	144
	Band adjustment front & rear	145		
7 Propeller shaft	Propeller shaft) lubrication	150	Universal joints) dismantling	151
	) removal	150	) reassembly	152
	Sliding joint - dismantling	151		
8 Rear axle	Axle unit	155	Final drive	161
	Differential	161	Rear hubs	158
	Drive shafts	160	Universal joints	158
9 Braking system	Bellows type vacuum servo	174	Handbrake - general	185
	Bleeding the hydraulics	168	Lockheed dual-line servo	177
	Brake calipers front & rear	171	Master cylinders	174
	Brake/clutch box	181	Remote servo	179
	Brake overhaul	173	Vacuum reservoir	176
	Friction pads	168	Vacuum servo unit	176
10 Electrical and instruments	Alternator	198	Revolution counter	213
	Battery	191	Speedometer	214
	Dynamo	192	Starter motor	205
	Electric clock	212	Switches	212
	Flasher unit	201	Voltage regulator	196
	Fuses units	201	Windscreen washer	209
	Horns	202	Windscreen wiper	210
	Lamps	203	Wiring diagrams	218
11 Suspension and steering	Anti-roll bar	229	Rear wheel camber angle	238
	Front suspension	223	Road spring and damper	233
	Front wheel alignment	245	Steering wheel	239
	Rack and pinion assembly	243	Torsion bar	229
	Rear suspension	233	Wishbone	236
12 Bodywork	Air conditioning	263	Hardtop	260
	Body repairs major or minor	251	Heater	263
	Bonnet	255	Maintenance	247
	Bumpers front and rear	257	Tyre inflation pressures	265
	Door rattles	252	Wheel alignment	266
	Exhaust system	262	Wire spoke wheels	266

NB. Specifications and general descriptions are given in each Chapter immediately after the 'list of contents'. Where applicable Fault diagnosis is given at the end of each appropriate Chapter.

List of 'line drawings'	268-271
Metric conversion tables	272-273
Use of English	274
Grades of Castrol lubricants	18
Index of contents	275-277

Jaguar 'E' Type Series I Roadster 3.8

Jaguar 'E' Type Series I FH Coupe' 3.8

Jaguar 'E' Type Series I 2 + 2 Coupe' 4.2

Jaguar 'E' Type Series II FH Coupe' 4.2

Jaguar 'E' Type Series II FH Coupe' 4.2 (North American specification)

Jaguar 'E' Type Series III 2 + 2 Coupe' 4.2

Buying spare parts and vehicle identification numbers

Spare parts are available from many sources, for example: Jaguar garages, other garages and accessory shops, and motor factors. Our advice regarding spare part sources is as follows:

Officially appointed Jaguar garages - This is the best source of parts which are peculiar to your car and are otherwise not generally available (eg complete cylinder heads, internal gearbox components, badges, interior trim etc). It is also the only place at which you should buy parts if your car is still under warranty - non-Jaguar components may invalidate the warranty. To be sure of obtaining the correct parts it will always be necessary to give the storeman your car's engine and chassis number, and if possible, to take the 'old' part along for positive identification. Remember that many parts are available on a factory exchange scheme - any parts returned should always be clean! It obviously make good sense to go straight to the specialists on your car for this type of part for they are best equipped to supply you.

Other garages and accessory shops - These are often very good places to buy materials and components needed for the maintenance of your car (e.g. oil filters, spark plugs, bulbs, fan belts, oils and greases, touch-up paint, filler paste etc). They also sell general accessories, usually have convenient opening hours, charge lower prices and can often be found not far from home.

Motor factors - Good factors will stock all of the more important components which wear out relatively quickly (e.g. clutch components, pistons, valves exhaust systems, brake cylinders/pipes/seals/shoes and pads etc). Motor factors will often provide new or reconditioned components on a part exchange basis - this can save a considerable amount of money.

Vehicle identification numbers

When ordering new parts it is essential that you give full information about your particular model of Jaguar otherwise it cannot be guaranteed that you will be supplied with the correct item and there is nothing more frustrating than to find that a part (for which you might have had to wait some time due to supply difficulties) will not fit.

It is imperative, therefore, that the car and engine number together with any prefix or suffix letters are quoted when ordering parts. Look at the item you are replacing, it may have a part number stamped on it, if so, quote this number when ordering or better still take the part along with you for correct identification.

The car number is stamped on the right-hand frame/cross member above the hydraulic damper mounting.

The engine number, of early models, is stamped on the right-hand side of the cylinder block above the oil filter and is also marked at the front of the cylinder head casting, however, in case the cylinder head has been renewed at some time we advise use of the cylinder block marking. The engine number of later models is stamped on the crankcase bellhousing flange on the left hand side of the engine adjacent to the oil dipstick. The figure 8 or 9 following the engine number denotes the compression ratio.

The gearbox number is stamped on a shoulder at the left-hand rear corner of the gearbox casing and also on the top cover of the gearbox.

Identification of automatic transmission units will be found on a plate attached to the unit.

The car body number is shown on a plate attached to the right hand side of the scuttle.

To comply with Federal Safety Standards, cars exported to the United States of America and Canada will have a Certification plate attached to the left hand door shut face which will show the month/year of manufacture and the commission number of the car.

If you intend to use the Jaguar exchange scheme, make sure that the component you wish to exchange is clean and is also complete to the standard of the exchange item before taking it to the stores; needless to say you should have removed those items which are not supplied with the exchange item.

Routine maintenance

Periodic servicing of your car should be looked upon as an essential, not only for the purpose of obtaining economy and the best performance from the vehicle, but also for ensuring safety and for finding defects at an early date before anything serious, and probably expensive, occurs. You will find that by far the largest element of the maintenance routine is a purely visual examination which will take up very little time.

The maintenance instructions which follow are those recommended by the manufacturer and they are supplemented by additional tasks which we have found, from practical experience, need to be carried out as a purely preventative measure.

The service periodicity recommended by Jaguar for early models is tied to a 2500 mile (4000 km) cycle but for later models it is 3000 miles (5000 km). In view of the similarity of components between models we feel that this extension can be applied to all "E" type cars and consequently the following servicing procedures are written to a 3000 mile cycle.

Daily before use

1 Remove the dipstick and check the engine oil level which should be up to the "MAX" mark. Top up as necessary.
2 Remove the radiator filler cap, **when the engine is cold**, and top up the coolant as necessary.
3 Top up the windscreen washer bottle.
4 Look at the level of brake fluid in the reservoir. The level of the fluid will be readily seen at a quick glance by noting the position of the float needle. Investigate the cause if frequent topping up is necessary.
5 Check correct operation of services-horn, windscreen wipers, windscreen washer and lights.

Weekly

1 Check tyre pressures. See Chapter 12.

Monthly

1 Check condition of tyres for compliance with legal limits of wear and condition. But, depending on mileage covered and road conditions, this check may be necessary at more frequent intervals.
2 Check the level of electrolyte in the battery. Check the battery connections for security and cleanliness. See Chapter 10.

Every 3000 miles (5000 km)

Engine

1 Drain the oil from the engine sump by removing the plug at the right-hand rear corner of the sump as shown in Fig.RM.1. Do this work when the engine is warm so that the oil will flow more freely.
2 Clean and replace the sump drain plug and refill the sump to the correct level with engine oil.
Note: Where the car is used for low speed driving, stop/start driving particularly in cold weather where the choke is used more than is normal, or long periods of use in dusty conditions the oil should be changed at least every 1000 miles (1600 km). Oil changes for 4.2 litre Service 2 cars may be deferred to the 6000 mile servicing if desired.

Distributor

1 Remove the moulded cap at the top of the distributor, lift off the rotor arm and apply a few drops of oil round the screw "A" shown in Fig.RM.2. Do not remove the screw, it has clearance to permit the passage of oil.
2 Apply one drop of oil to the post "B" on which the contact breaker pivots.
3 Lightly smear the cam "C" with grease.
4 Lubricate the centrifugal advance mechanism by injecting a few drops of oil through the aperture at the edge of the contact breaker base plate.
5 Refer to Fig.RM.3 and check the gap between the contact points, this should be 0.014" - 0.016" (0.36 - 0.41 mm). If the gap is incorrect, slacken the screw (2), insert a screwdriver in the slot (3) and move the contact plate as required. Finally tighten down on the screw and recheck the gap.

Spark plugs

1 Remove the plugs and clean them with a wire brush. A preferable alternative is to have them cleaned on a round blasting machine specially designed for the purpose.
2 Set the plug gaps to 0.025" (0.64 mm).

Carburettors

1 Remove the carburettor piston dampers and top up with SAE 20 engine oil as shown in Fig.RM.4.
2 Refer to Chapter 3 and follow the instructions for adjustment of the idling speed if this is required.

Cooling system

1 Check condition of water hoses and replace as necessary.
2 Check for leaks and rectify.

Gearbox/Automatic transmission

1 Place the car on level ground.
2 Check the level of oil in the gearbox by removing the combined level/filler plug located on the left-hand side of the gearbox. This plug is accessible after lifting the carpet and then removing the rubber bung as shown in Fig.RM.5.
3 Top up the gearbox as necessary.
4 To check the level of fluid in the automatic transmission

Routine maintenance

unit: Run the engine (with "P" selected) until it attains its normal operating temperature and is running at the correct idling speed; remove the dipstick (Fig.RM.6), wipe it clean, replace it in the filler tube (in its correct position) and then withdraw it immediately. Check the level of fluid.
5 Top up with Castrol TQF automatic transmission fluid as necessary. The difference between the "FULL" and "LOW" marks on the dipstick represents approximately 1½ pints (2 US pints or 0.75 litres). Do not overfill the unit.

Final drive unit
1 Place the car on level ground and then check the level of oil in final drive unit by removing the combined filler/level plug (shown in Fig.RM.7).
2 Top up as necessary to bring the oil level to the bottom of the filler hole.

Handbrake
Refer to Chapter 9 and adjust as required.

Ignition timing
For those cars fitted with an exhaust emission control system, refer to Chapter 4, and check and reset the ignition timing in the manner described.

Air conditioning equipment (where fitted)
1 Check the compressor drive belt for wear - renew if necessary.
2 Check the compressor drive belt for correct tension; this is best done, with an approved belt tensioning gauge, the deflection should be 0.15" (4 mm) at 10 lb (4.5 kg) load.
3 The belt is adjusted by means of two nuts on the adjuster which is accessible from underneath the car. To tighten the belt, slacken the outer nut and tighten the inner nut - vice-versa to relieve tension on the belt.

Clutch
Remove the cap from the clutch fluid reservoir (Figs.RM.8 and 9) and check the level of fluid. Top up as necessary using the recommended type of fluid.

Every 6000 miles (10,000 km)

Carry out the 3000 mile (5000 km) servicing, plus:

Engine
1 Refer to Chapter 1. Remove and discard the oil filter element after first draining the oil from the canister by taking out the drain plug "B" (shown in Fig.RM.10). Take out the central bolt "A" and then remove the canister and element.
2 Remove the rubber sealing ring in the filter head and replace it with a new item which will be supplied with the replacement element.
3 Thoroughly clean the canister, spring, pressure plate and the central rod.
4 Assemble the new element and refit the canister to the car.
5 Check for oil leaks at the filter assembly when the engine is running and the oil hot - remedy as may be required. The most common fault which arises is poor seating of the rubber seal in the head of the filter.

Water pump and dynamo/alternator drive belt
1 Examine for wear and replace if necessary.
2 The belt is removed by slackening the dynamo/alternator towards the engine as far as possible. Now press inwards on the jockey pulley to relieve all tension on the belt and then lift it off the pulleys. Refit the belt or fit a new belt in the reverse order; but, make sure that it is seating properly in the 'vee' of the pulleys before releasing the jockey pulley.

SU carburettors
1 Remove the bolt securing the banjo union to each float chamber; take out the bolt, and move the banjo union to one side - do not lose the fibre washers each side of the union.
2 Remove the filter and its spring from the float chamber (see Fig.RM.11).
3 Clean the filter in petrol - do not use a cloth as fibre particles will stick to the gauze.
4 Refit the filter with the spring leading.
5 Refit the banjo union not forgetting the fibre washers and then secure with the bolt.
6 Remove the glass bowl of the fuel feed line filter (Fig.RM.12) by slackening the finger nut and then swinging the retaining clip to one side.
7 Clean the filter gauze and the bowl by washing in petrol.
8 Examine the sealing washer and renew if necessary.

Note: Although we suggest that this work is done at the 6000 miles (10,000 kms) service it may be that (due to prevailing conditions) more frequent cleaning is required.

Steering and wheel swivels
1 Use a grease gun to lubricate the rack and pinion housing. The grease nipple (shown in Fig.RM.13) is accessible from underneath the car. Do not pump in grease to the extent that the rubber bellows at the end of the housing become distended.
2 Check that the clips at the ends of the bellows are fully tightened otherwise grease will escape from the housing.
3 Use a grease gun to lubricate the balljoints of the two steering tie-rods. The location of the grease nipple is shown in Fig.RM.14. A bleed hole is provided at each balljoint and this is covered by a nylon washer which will lift under pressure to indicate when sufficient lubricant has been applied.
4 Examine the rubber seals at the bottom of the ball housing to see if they have become displaced, or are split. Reposition or replace them as necessary; this is important, because any dirt or water which enters the joint will cause premature wear.
5 Use a grease gun to lubricate the wheel swivels (two each side of the car (see Fig.RM.15). The grease nipples are accessible from underneath the car at the front and the joint, as in the case of the steering tie-rods, has a bleed hole covered by a nylon washer.

Front wheel alignment
The alignment of the front wheels should be checked. This (as stated in Chapter 11) is a task which should be left to your local Jaguar agent.

Propeller shaft
The propeller shaft of early model cars is provided with grease nipples for lubrication of the universal joints and of the sliding spline. Later cars are fitted with "sealed-for-life" universal joints and sliding spline and no lubrication is required.
1 Raise the rear of the car to allow one rear wheel to be turned to bring the grease nipples to an accessible position.
2 Clean the grease nipples and then apply a grease gun to lubricate the joints.
3 The grease nipple for the rear universal joint is accessible from underneath the car. The grease nipple for the front universal joint and for the sliding spline is approached, via, an access hole in the gearbox cowl trim panel covered by a metal or rubber bung.

Halfshafts - universal joints
Some early model cars are fitted with nipples for lubrication of the halfshaft universal joints. These were omitted later but have been re-introduced with effect from the following chassis numbers on 4.2 litre models:

	RH Drive	LH Drive
Roadster	1E.1926	1E.16721
Fixed Head Coupe	1E.21669	1E.34851
2 + 2	1E.51067	1E.77705

1 Raise the rear of the car to bring one rear wheel clear of the ground or (if the car can be placed over a pit) locate it so that it

can be pushed forwards or backwards to bring the nipples to an accessible position for application of a grease gun.
2   Clean the nipples and then lubricate the joints using a grease gun.

Rear suspension
1   Use a grease gun to lubricate the outer pivot bearings, via, the grease nipple provided. A bleed hole is located opposite the grease nipple to indicate when sufficient grease has been applied - make sure that this hole is clear before commencing work. The location of the grease nipple is shown in Fig.RM.16.
2   Fig.RM.16 also shows the location of two grease nipples at either end of the wishbone fork pivot bearing. Use a grease gun to lubricate sparingly.

Brake friction pads
Refer to the instructions given in Chapter 9. Examine the brake friction pads for wear, and renew them, if they have worn down to a thickness of ¼" (7 mm).
This work may be necessary at more frequent intervals if a loss in braking efficiency is noticed.

Bodywork
Clear the drain holes in the bottom of the doors using a piece of stiff wire. This task should be done at more frequent intervals if conditions warrant it.

Oil can lubrication
Lubricate the following points:
Seat runner and adjusting mechanism
Handbrake lever ratchet
Door locks
Luggage compartment lid hinges and catches
Bonnet hinges and catches
Windscreen wiper arms
Accelerator and carburettor linkage
Fuel filler cover hinge
Dynamo bearing (see Chapter 10).

Every 12,000 miles (20,000 km)

Carry out the 6000 mile (10,000 km) servicing plus:-
Engine
1   Remove and clean each camshaft cover.
2   If the timing chain has proved noisy in use - adjust it in the manner described in Chapter 1.
3   Check and adjust the valve clearances as described in Chapter 1.
4   Slacken the clip securing the pipe at the crankcase breather union and disconnect the pipe from the breather.
5   Remove the domed nuts securing the breather to the timing cover and remove the breather. Collect the two gaskets and the gauze filter (see Fig.RM.17).
6   Clean the assembly in petrol; paying particular attention to the filter gauze.
7   Refit the crankcase breather. New gasket should be used for those cars fitted with exhaust emission control systems but on other models the gaskets can be re-used if they are in very good condition.
8   Check tightness of the inlet manifold securing nuts and (on cars fitted with an exhaust emission control system) also check for tightness the secondary throttle housing nuts and the nuts securing the primary mixture pipe to the secondary throttle housing.
9   Starting at the connections on the manifold; check through the exhaust system for leaks. This is especially important for those cars required to comply with exhaust emission control regulations.

Air cleaner
1   Renew the paper element.
2   To gain access to the element (shown at Fig.RM.18) release the three spring clips which retain the top cover to the air box and lift out the element cover, remove the serrated nut and retainer plate from the base of the unit and withdraw and discard the element.
3   Fit the new element in the reverse order to the above.
Note: Depending on prevailing conditions, more frequent renewal of the paper element may be necessary.

Fuel tank filter
The fuel tank filter is located within the drain plug tube on the underside of the fuel tank.
1   Remove the drain plug tube (preferably when the tank is low on fuel) and allow the fuel to drain away into a clean container.
2   Remove the filter from the fuel pump inlet pipe by slackening the union nut and at the same time, holding the filter. Now unscrew the filter.
3   Clean the filter assembly in petrol. Do not use cloth as fibre particles may stick to the gauze.
4   Examine the drain plug tube sealing washer and renew if its condition is doubtful.
5   Refit the filter in the reverse manner to the above.
6   Filter the fuel before pouring it back into the tank.

Spark plugs
Remove and replace with new plugs set to the correct gap. Make sure that the gasket seatings in the cylinder head are clean before fitting the plugs.

Distributor
1   Refer to Chapter 4 and follow the instructions for removal of the contact breaker.
2   Examine the contact breaker points and clean them in the manner described or, if they are badly burnt and pitted, fit a new set.
3   Replace the contact breaker and set the gap in the manner described in the 3000 mile servicing procedures.
4   Whilst the plugs are removed (see previous operation); check the ignition timing, and reset as described in Chapter 3 (exhaust emission control system) and Chapter 4.

SU carburettors
1   Remove and dismantle the carburettors. Instructions for this are given in Chapter 3.
2   Thoroughly clean each carburettor paying particular attention to the bottom of each float chamber where you will find a considerable amount of sediment.
3   Follow the procedures given in Chapter 3 for re-assembly and refitment of the carburettors to the car and then adjust them for mixture and slow running in the manner described.

Stromberg carburettors with exhaust emission control system
1   Remove and dismantle the carburettors in the manner described in Chapter 3.
2   Thoroughly clean each carburettor.
3   Obtain Emission Pack (Part No 11549) and fit the parts provided, when re-assembling the carburettors.
4   Refit the carburettors to the car, and then tune, and adjust them, in accordance with the instructions given in Chapter 3. An analysis of the exhaust gas CO content will now have to be made and for this the car will have to be taken to your local Jaguar agent.

Gearbox
1   Change the oil in the gearbox by removing the drain plug located at the front end of the gearbox casing (shown in Fig.RM.19). This work is best done when the oil is hot so that it will run freely.
2   After all oil has drained away, clean the drain plug and refit it.
3   Refill the gearbox with the recommended grade of oil through the combined filler/level hole depicted in Fig.RM.5.

Final drive unit
1   Change the oil in the unit by removing the drain plug shown

Routine maintenance

in Fig.RM.20. This is best done after a run when the oil is hot so that it will run freely.
2   After all oil has drained away, clean the drain plug and refit it.
3   Remove the combined filler/level plug (shown in Fig.RM.7) and refill the unit with the recommended grade of oil. The level of the oil should be to the bottom of the hole when the car is standing on level ground.

Front wheel bearings
1   Check for end float (see Chapter 11).
2   Removal of the wheels will expose a grease nipple in the wheel bearing hubs (Fig.RM.21).
3   Clean the grease nipple and, using a grease gun, lubricate sparingly with the recommended grade of grease. An indication that sufficient grease has been applied will be given by the escape of grease past the outer hub bearing - this can be observed, through the bore of the splined hub.

Rear wheel bearings
1   Check for end float (see Chapter 11).
2   Removal of the rear wheels will expose a hole in each hub bearing housing for lubrication of the wheel bearings (Fig.RM.22). The hole is sealed by a dust cap.
3   Clean all dirt from the area of the dust cap and then prise it off.
4   Inject grease of the recommended grade into the hole until no more will enter. If you use a pressure gun, be careful not to build up excess pressure in the hub as this will cause grease to escape past the oil seal.
5   Clean off excess grease and refit the dust cap.

General
1   Check and tighten all chassis and body nuts, bolts and screws.
2   Check all steering connections, balljoints etc; for security and wear.
3   Arrange to have the headlamp alignment checked and adjusted as necessary.
4   Examine all flexible brake connections for chafing or other deterioration, and renew as necessary.

Every 21,000 miles (35,000 km)

Carry out the 3,000 mile (5,000 km) servicing, plus:

Automatic transmission
1   Remove the drain plug (Fig.RM.23) from the oil pan and allow the oil to drain into a container. Do this work after a run when the oil is hot, so that it will run freely. However, as the unit runs at a very high temperature take care that you do not scald yourself when taking out the drain plug.
2   Refer to the instructions given in Chapter 6: Remove the oil pan and adjust the front and rear brake bands in the manner described. However, as special tools are required to do this work it is advisable to leave it to your local Jaguar agent.

Every 24,000 miles (40,000 km)

Carry out the 12,000 mile (20,000 km) servicing, plus:

Brakes
1   Slacken all bleed screws and drain off the old fluid by pumping on the brake pedal.
2   Carefully examine all brake pipes and unions for damage - renew as necessary.
3   Flush out the brake system using clean hydraulic fluid of the correct specification.
4   Tighten all bleed screws, refill the system with hydraulic fluid and then carry out the procedure described in Chapter 9 to expel air from the sytem.

Stromberg carburettors
1   Remove the carburettors from the car as described in Chapter 3.
2   Dismantle and clean the carburettors.
3   Obtain (for each carburettor) a Red Emission Pack (Part No 11791). The following items will also be required:
   4 secondary throttle housing/carburettor gaskets and spacers
   1 secondary throttle housing/manifold gasket.
   1 primary mixture pipe/secondary throttle housing gasket.
4   Re-assemble the carburettor using the replacement parts supplied in the emission pack.
5   Refit the carburettors to the car using the new gaskets listed at paragraph 3.
6   Have your local Jaguar agent carry out for you an engine and ignition diagnostic check and an analysis of the exhaust gas CO content.

Periodic servicing

Every 12 months

Air conditioning equipment (where fitted)
The compressor oil level should be checked but as this will entail the use of special equipment you are advised to take the car to your local Jaguar agent to have the work done. Neglect to the extent of allowing the level of the oil to fall below the "minimum" will result in damage to the equipment.

Bodywork
It is a good idea to have the underbody of the car steam cleaned once a year; preferably, in the summer. The body should then be examined and any deterioration in the paintwork, metalwork, or underseal, made good.

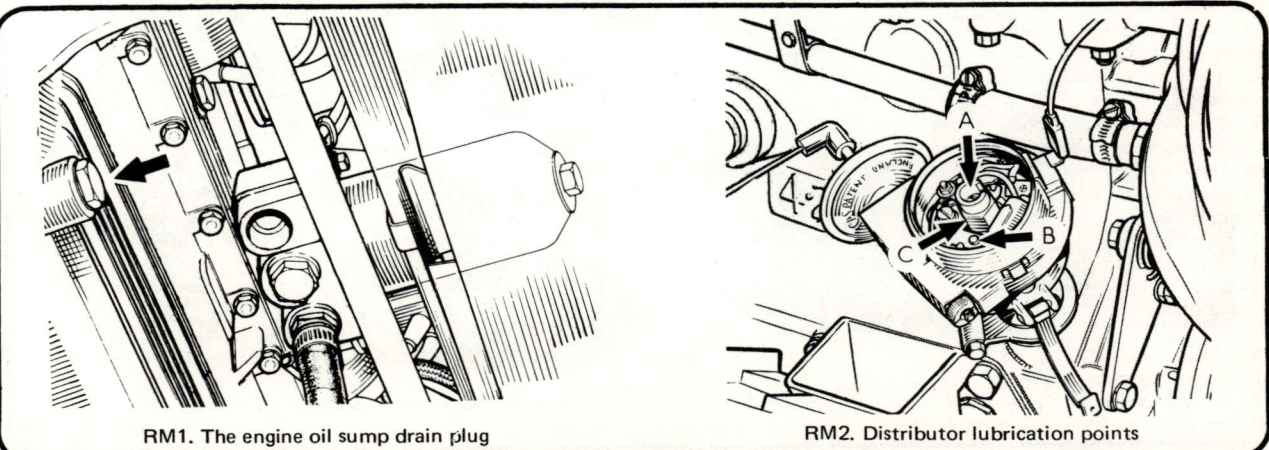

RM1. The engine oil sump drain plug   RM2. Distributor lubrication points

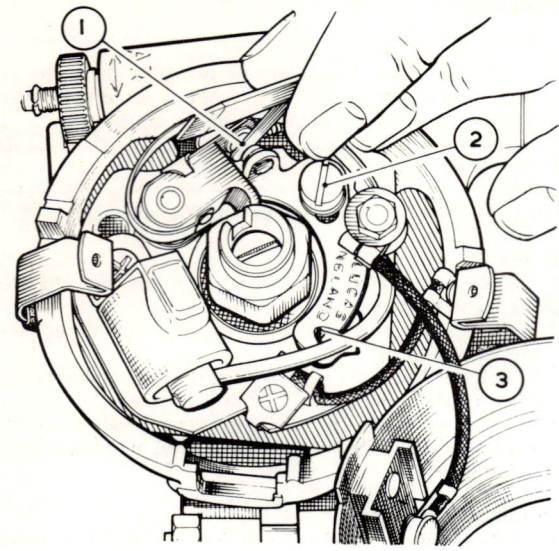

RM3. Checking the gap between the contact points

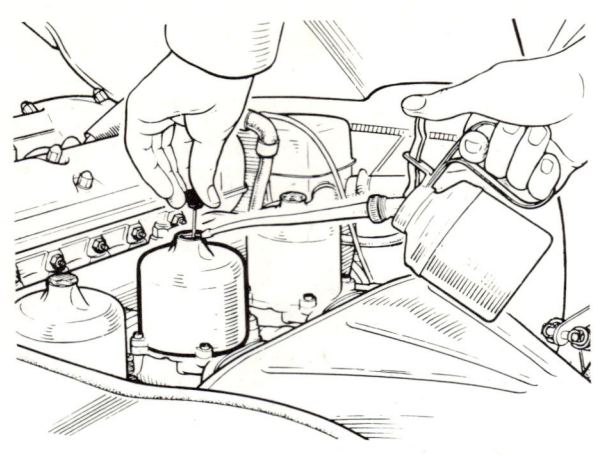

RM4. Topping up the piston damper

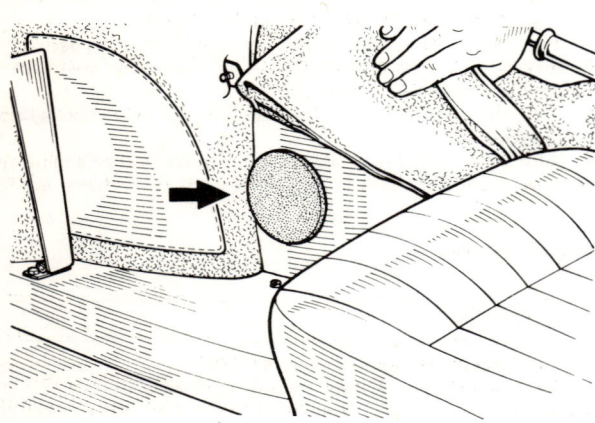

RM5. Access point to gearbox filler/level plug

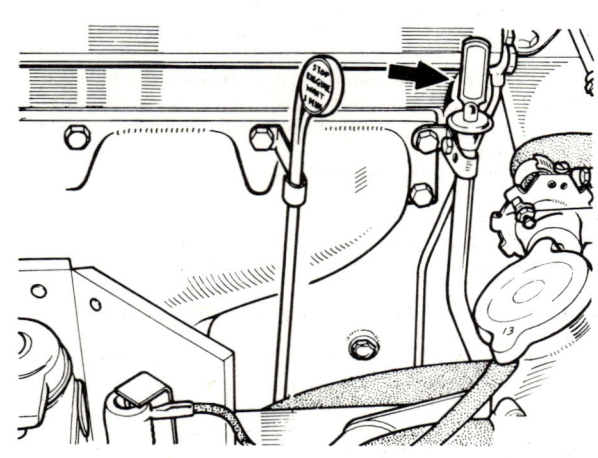

RM6. Automatic transmission dipstick and filler point

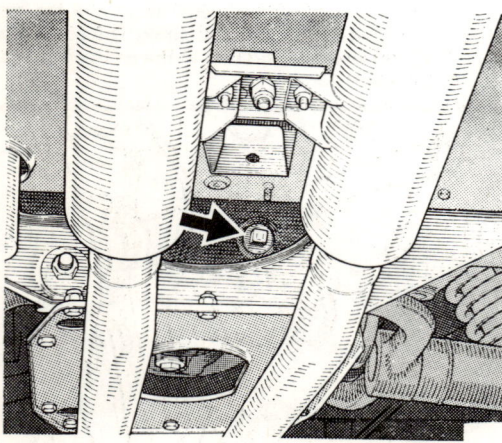

RM7. Rear axle oil filler/level plug

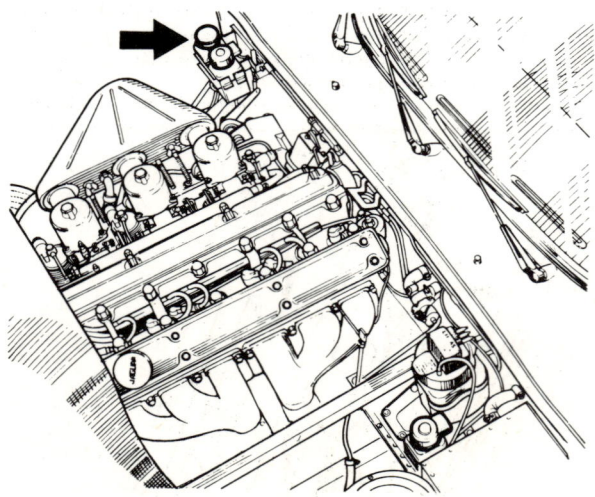

RM8. Clutch fluid reservoir RH drive

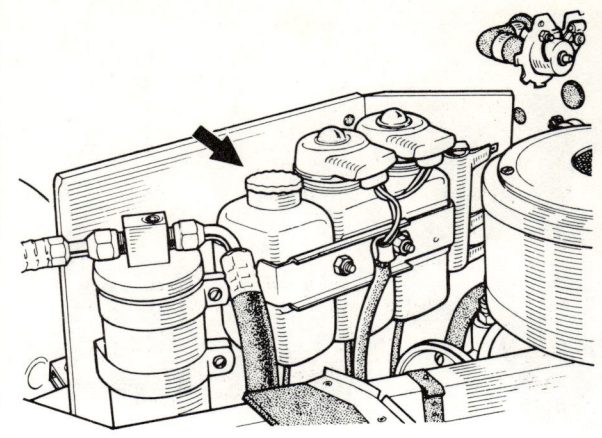
RM9. Clutch fluid reservoir LH drive

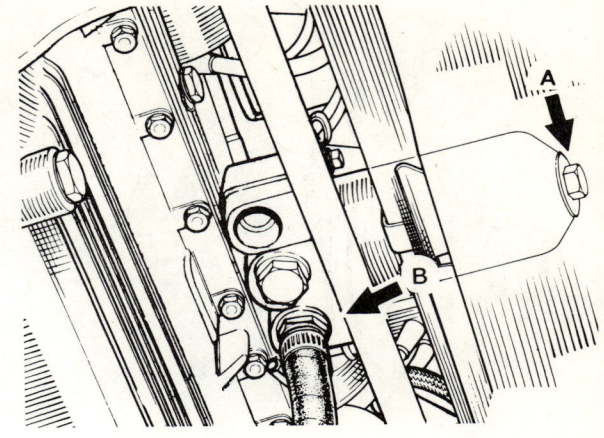
RM10. The engine oil filter

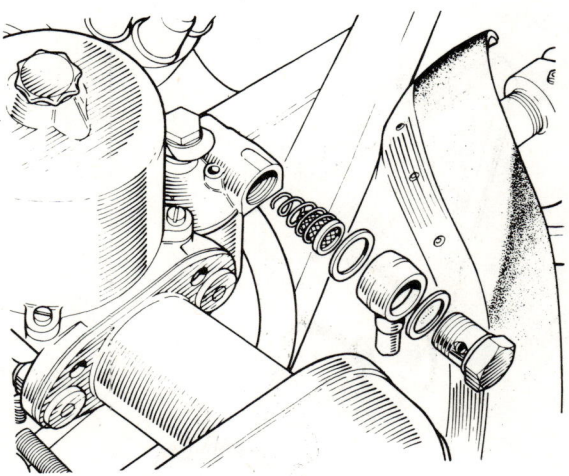
RM11. Float chamber filter

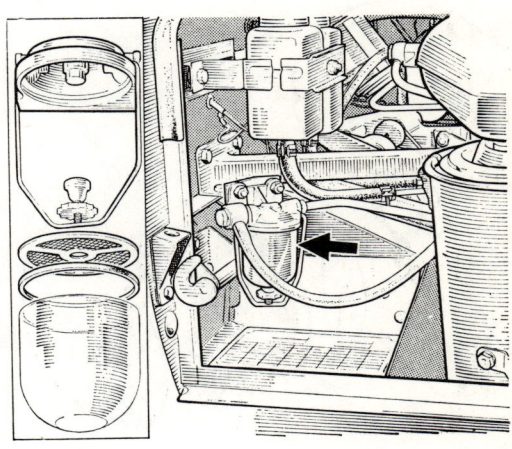
RM12. The fuel feed line filter

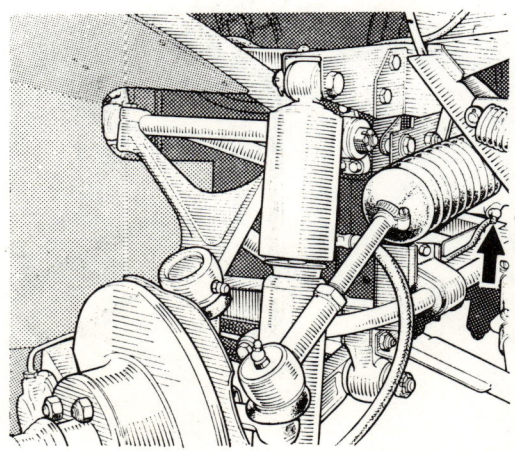
RM13. The steering housing grease nipple

RM14. A steering tie rod grease nipple

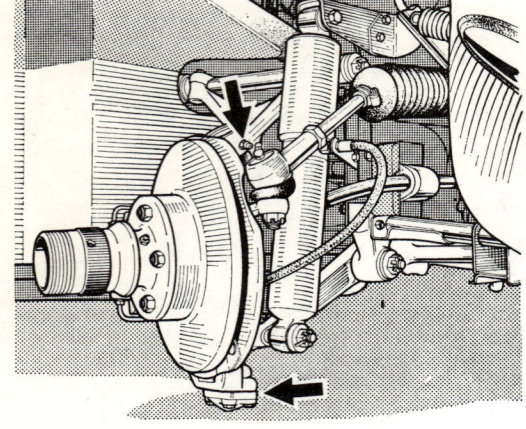

RM15. Steering swivel grease nipples

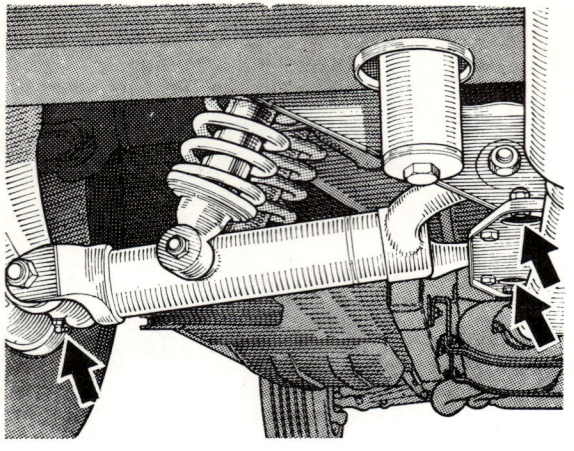

RM16. Outer and inner pivot bearing grease nipples

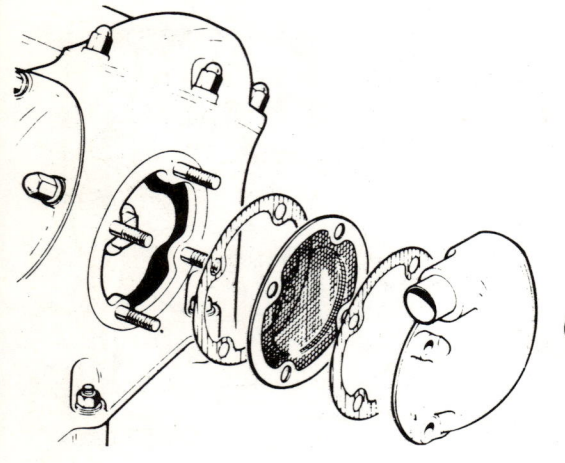

RM17. The crankcase breather

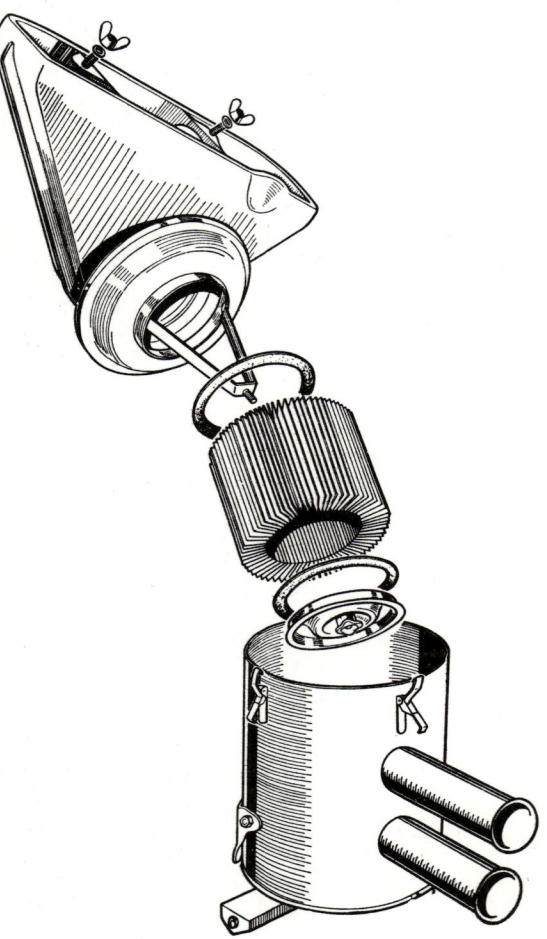

RM18. The air cleaner

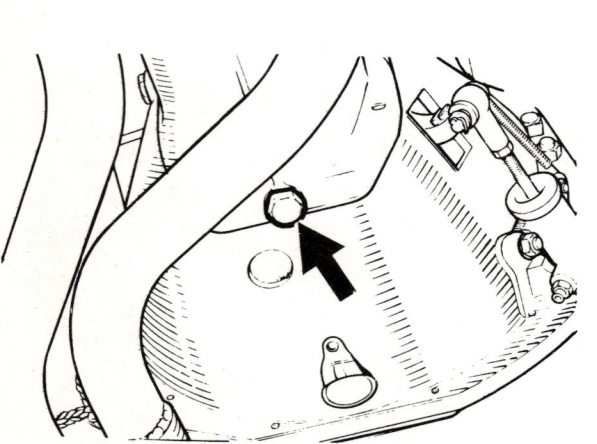

RM19. The gearbox drain plug

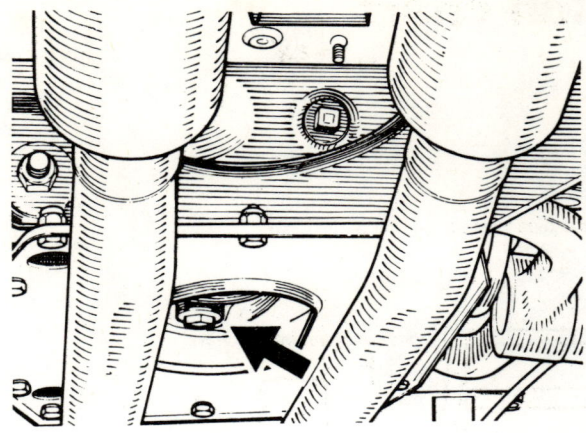
RM20. Final drive unit drain plug

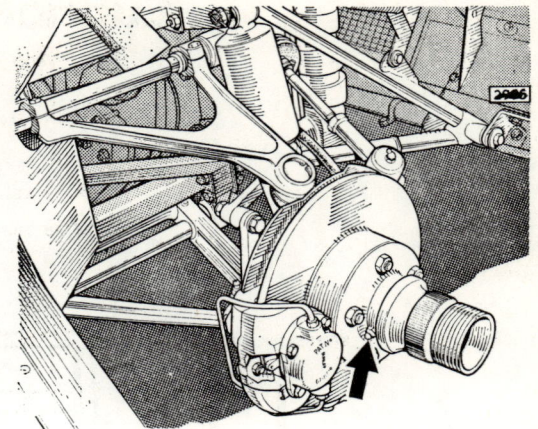
RM21. Front wheel hub bearing grease nipple

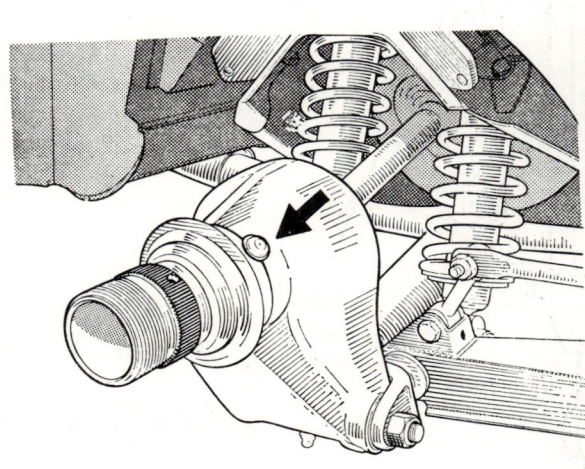
RM22. Rear wheel hub bearing grease cap

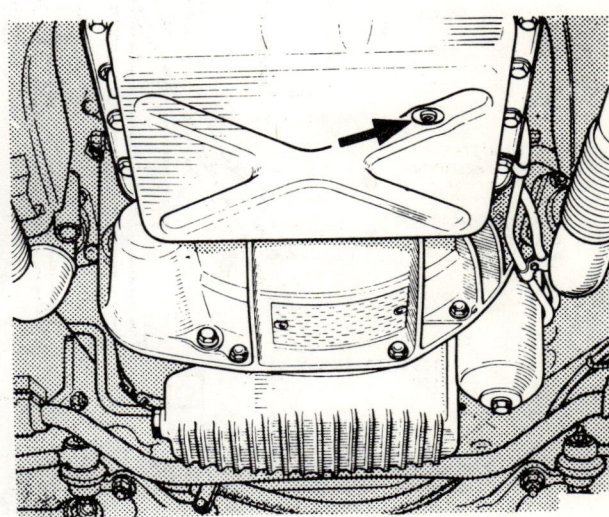
RM23. The automatic transmission drain plug

Lubrication chart

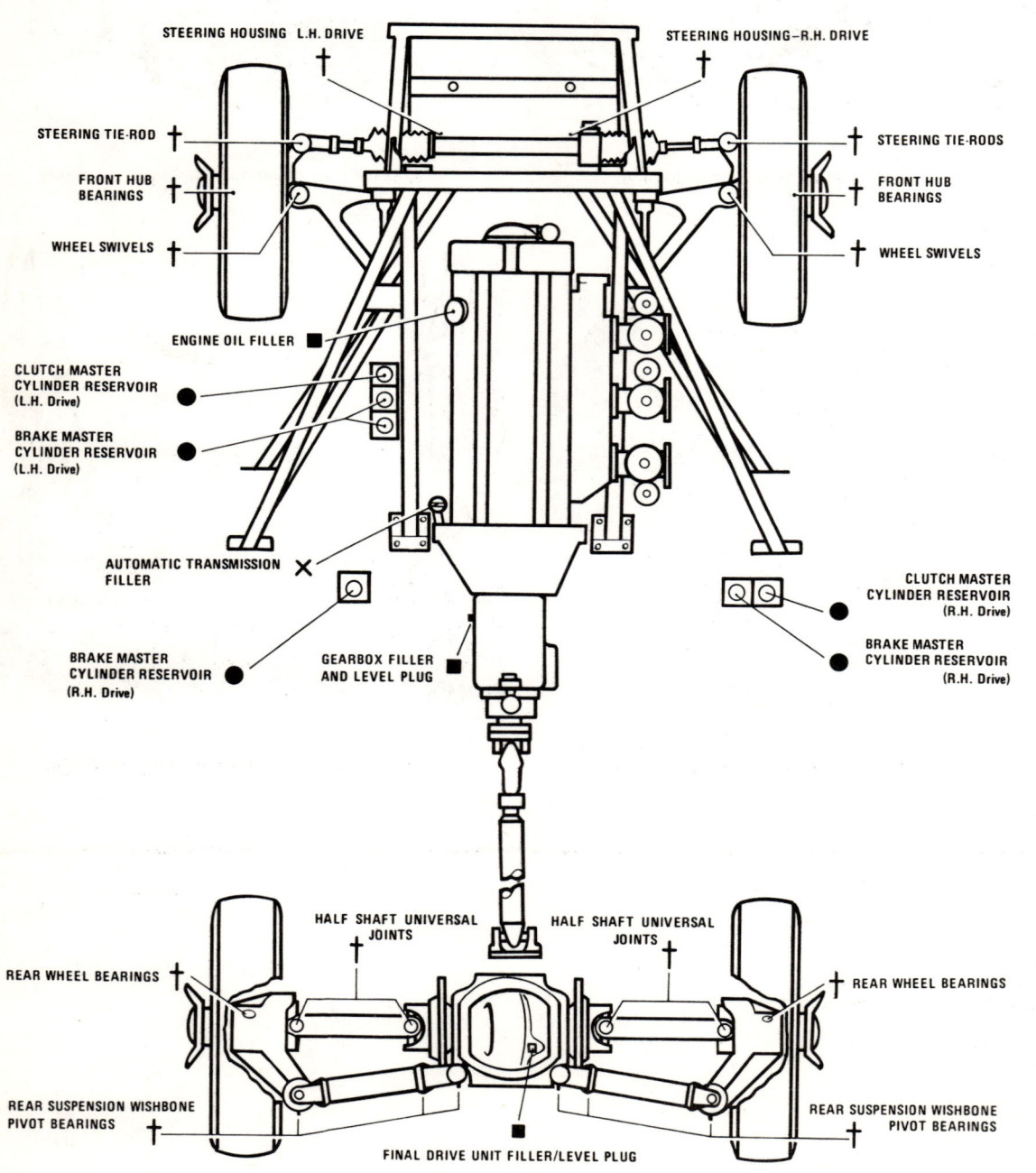

Recommended lubricants

The following table gives details of the recommended lubricants.
Almost all lubricating oils contain additives and although it is permissible to mix different brands of lubricant it is an undesirable practice. If you wish to change from one brand to another it is advisable to wait until the sump or gearbox are drained and then to follow the Oil Company's recommendations in regard to flushing procedures before refilling with the different make of oil. In cases where the grade or make of oil in your engine or gearbox is not known, our advice is that you drain off and refill with a known grade and make of oil rather than run the risk of sludge formation and gumming up which might result from mixing.

Component	Lubricant
ENGINE	Castrol GTX
UPPER CYLINDER LUBRICATION	Castrollo
DISTRIBUTOR OIL CAN POINTS	Castrol GTX
OIL CAN LUBRICATION	Castrol GTX
GEARBOX	Castrol Hypoy
FINAL DRIVE UNIT (Not 'Powr-Lok')	Castrol Hypoy
FRONT WHEEL BEARINGS	Castrol LM Grease
REAR WHEEL BEARINGS	Castrol LM Grease
DISTRIBUTOR CAM	Castrol LM Grease
FINAL DRIVE HALF-SHAFTS	Castrol LM Grease
STEERING TIE-RODS	Castrol LM Grease
WHEEL SWIVELS	Castrol LM Grease
DOOR HINGES	Castrol LM Grease
STEERING HOUSING	Castrol LM Grease
AUTOMATIC TRANSMISSION UNIT	Castrol TQF
POWER STEERING SYSTEM	Castrol TQF

Recommended hydraulic fluid: The manufacturer's latest recommendation for clutch and brake systems is for the use of Castrol Girling Universal Brake and Clutch Fluid conforming to Specification J1703A modified for additonal safety to give a higher boiling point. Where this fluid is not available; only that guaranteed to conform to Specification J1703A may be used as an alternative after fully draining and flushing the system.

Interior layout of the Series II Roadster 4.2

Chapter 1 Engine

Contents

Air cleaner - general	47
Ancillary engine components - removal	9
Big end and main bearings - examination	27
Camshaft and camshaft bearing - examination and renovation	33
Camshaft - removal	12
Compression pressures - general	38
Connecting rods - examination and renovation	32
Connecting rod to crankshaft - reassembly	53
Crankshaft damper assembly cone and distance piece - refitting	74
Crankshaft - examination and renovation	25
Crankshaft gear and sprocket - reassembly	54
Crankshaft pulley and damper - examination and renovation	26
Crankshaft - removal	21
Crankshaft - replacement	49
Cylinder block - examination and renovation	29
Cylinder head and bore - decarbonisation, examination and renovation	37
Cylinder head oil feed pipe - refitting	71
Cylinder head - reassembly	65
Cylinder head - refitting	67
Cylinder head removal - engine in car	11
Cylinder head removal - engine on bench	10
Distributor and oil pump drive gear - reassembly	55
Distributor drive - removal	20
Distributor - refitting	64
Engine and gearbox - separation (manual)	6
Engine and transmission unit - separation (automatic)	7
Engine breather assembly - refitting	73
Engine dismantling - general	8
Engine examination and renovation - general	24
Engine - initial start up after overhaul and major repair	77
Engine - major operations (with engine in place)	2
Engine - major operations (with engine removed)	3
Engine mountings - examination and renovation	46
Engine mountings - general	45
Engine reassembly - general	48
Engine removal	5
Engine removal - method	4
Engine replacement	76
Engine stabiliser - examination and renovation	44
Exhaust manifolds - refitting	70
Fault diagnosis - engine	78
Flywheel and clutch - refitting	63
Flywheel - examination and renovation	28
General description	1
Gearbox and clutch housing - refitting	75
Gudgeon pin - removal	17
Inlet manifold - refitting	69
Lubrication system - description	22
Oil filter - removal and replacement	23
Oil pump and pipes - reassembly	56
Oil pump assembly - removal	19
Oil pump - examination and renovation	40
Oil sump - cleaning and examination	41
Oil sump - refitting	62
Piston and connecting rod - reassembly	50
Pistons and piston rings - general	30
Pistons, piston rings and gudgeon pin - examination and renovation	31
Piston - replacement	52
Piston ring - removal	18
Piston ring - replacement	51
Sump, piston, connecting rod and big end bearing - removal	15
Tappets and valve adjusting pads - examination and renovation	36
Timing chain tensioners (bottom) - adjustment	60
Timing chain tensioner (bottom) - examination and renovation	42
Timing chain tensioner (bottom) - reassembly	59
Timing cover - refitting	61
Timing gear and chain - examination and renovation	39
Timing gear - assembly	57
Timing gear - reassembly to engine	58
Timing gear - removal	16
Valves and seats - examination and renovation	34
Valve clearance adjustment	66
Valve guide and tappet guide - removal	14
Valve removal	13
Valve springs - examination and test	35
Valve timing	68
Water pump belt tensioner (automatic) - examination and renovation	43
Water pump - refitting	72

Specifications

Camshaft

Number of journals ...	Four per shaft
Journal diameter ...	1.00 in. - 0.0005 in. - 0.001 in. (25.4 mm - 0.013 mm - 0.025 mm)
Thrust taken ...	Front end
Number of bearings ...	Four per shaft (eight half bearings)

Chapter 1/Engine

Type of bearing	White metal steel backed shell
Diameter clearance	0.0005 in. to 0.002 in. (0.13 to 0.05 mm)
Permissable end float	3.8 litre 4.2 litre
	0.0045 in. to 0.008 in. 0.004 in. to 0.006 in.
	(0.11 to 0.20 mm) (0.10 to 0.15 mm)
Tightening torque - bearing cap nuts	15 lb. f. ft (2.0 kg. f. m)

Connecting rods

Length centre to centre	7.75 in. (19.68 cm)
Big end bearing shell type	Lead bronze, steel backed
Bore for big end bearings	2.233 in. to 2.335 in. (56.72 to 56.73 mm)
Big end width	1.1875 in. - 0.006 in. - 0.008 in.
Big end - diameter clearance	3.8 litre 4.2 litre
	0.0015 in. to 0.0033 in. 0.0015 in. to 0.0033 in.
	(0.04 to 0.08 mm) (0.037 to 0.083 mm)
- side clearance	0.0058 in. to 0.0087 in. (0.15 to 0.22 mm)
Bore for small end bush	1.00 in $\pm$ 0.0005 in. (25.4 mm $\pm$ 0.13 mm)
Small end bush type	Phosphor bronze - steel backed
Small end width	1.0781 in. (27.4 mm)
Small end bush bore diameter	0.875 in. + 0.0002 in. — 0.0000 in.
	(22.22 mm + 0.005 mm — 0.000)
Tightening torque - connecting rod bolts	37 lb. f. ft (5.1 kg. f.m)

Crankshaft

Number of main bearings	Seven
Main bearing shell type	Lead bronze - steel backed
Journal diameter	2.750 in. to 2.7505 in. (69.85 mm to 69.86 mm)
	Intermediate 3.8 litre 2.7495 in. to 2.750 in.
	(69.84 to 69.85 mm)
Journal length:	**3.8 litre** **4.2 litre**
Front	1.6875 in. $\pm$ 0.005 in. 1.5625 in. (39.06 mm)
	(42.86 mm $\pm$ 0.13 mm)
Centre	1.75 in. + 0.0005 in. 1.375 in. + 0.001 in.
	+ 0.001 in. — 0.0005 in.
	(44.45 mm + 0.013 mm (34.37 mm + 0.025 mm
	+ 0.025 mm) — 0.0215 mm)
Rear	1.875 in. (47.63 mm) 1.6875 in. (42.86 mm)
Intermediate	1.2188 in. $\pm$ 0.002 in. 1.219 in. $\pm$ 0.002 in. (30.96 mm
	(30.96 mm $\pm$ 0.05 mm) $\pm$ 0.05 mm)
Thrust taken	Centre bearing thrust washers
Thrust washer thickness	0.092 in. $\pm$ 0.001 in. and 0.096 in. $\pm$ 0.001 in.
	(2.33 mm $\pm$ 0.025 mm and 2.43 mm $\pm$ 0.025 mm)
End clearance	0.004 in. to 0.006 in. (0.10 to 0.15 mm)
Main bearing:	
Length:	
Front)	
Centre)	1.5 in. $\pm$ 0.005 in. (38.1 mm $\pm$ 0.13 mm)
Rear)	
Intermediate	1.00 in. $\pm$ 0.005 in. (25.4 mm $\pm$ 0.13 mm)
Diameter clearance	0.0025 in. to 0.0042 in. (0.063 to 1.06 mm)
Crankpin - diameter	2.086 in. + 0.0006 in. — 0.000 in.
	(52.98 mm + 0.015 mm — 0.000 mm)
- length	1.1875 in. + 0.007 in — 0.0002 in.
	(30.16 mm + 0.18 mm — 0.006 mm)
- regrind undersize	0.10 in., 0.020 in., 0.030 in. and 0.040 in.
	(0.25, 0.51, 0.76 and 1.02 mm)
- minimum diameter for regrind	— 0.040 in. (— 1.02 mm)
Tightening torque main bearing bolts	83 lb. f. ft (11.5 kg.f.m)

Cylinder block (3.8 litre)

Bore size for fitting liners	3.561 in. to 3.562 in. (90.45 to 90.47 mm)
Outside diameter of liner	3.563 in. to 3.566 in. (90.50 to 90.58 mm)
Interference fit	0.001 in. to 0.005 in. (0.025 to 0.125 mm)
Overall length of liner	6.34 in. (17.7 cm)
Outside diameter of lead-in	3.558 in. to 3.560 in. (90.37 to 90.42 mm)
Size of bore honed after assembly in cylinder block -	
nominal	3.4252 in. (87 mm)
Maximum rebore size	+ 0.30 in. (0.76 mm)

Chapter 1/Engine

Cylinder block (4.2 litre)
Bore size for fitting liners	3.761 in. to 3.762 in. (94.03 mm to 94.05 mm)
Outside diameter of liner	3.765 in. to 3.766 in. (94.13 to 94.15 mm)
Interference fit	0.003 in. to 0.005 in. (0.08 to 0.13 mm)
Overall length of liner	6.959 in. to 6.979 in. (17.39 to 17.45 mm)
Outside diameter of lead-in	3.758 in. to 3.760 in. (93.95 to 94.00 mm)
Size of bore honed after assembly - cylinder block - nominal	3.625 in. (92.07 mm)
Maximum rebore size	+ 0.30 in. (0.76 mm)

Cylinder head
Type	Straight port (Gold top)
Material	Aluminium alloy
Valve seat angle:	
inlet	45°
exhaust	45°
Valve throat diameter:	
inlet	1.5 in. (38.1 mm)
exhaust	1.375 in. (34.9 mm)
Tightening torque - cylinder head nuts	54 lb. f. ft (7.5 kg. f. m)
Firing order	1 5 3 6 2 4 (No. 1 cylinder being that at the rear of the engine unit)

Gudgeon pin
Type	Fully floating	
Length	**3.8 litre**	**4.2 litre**
	2.840 in. to 2.845 in. (72.14 to 72.26 mm)	3.00 in. (76.2 mm)
Inside diameter	0.625 in. (15.87 mm)	
Outside diameter	0.8750 in. to 0.8752 in. (22.22 to 22.23 mm)	

Lubrication system
Oil pressure (hot)	40 lbs per sq. in. at 3000 r.p.m.
Oil pump:	
Type	Eccentric rotor
Clearance at end of lobes	0.006 in. (max) (0.15 mm)
End clearance	0.0025 in. (max) (0.06 mm)
Clearance between outer rotor and body	0.010 in. (max) (0.25 mm)

Pistons and piston rings (3.8 litre)
Make	Brico
Type	Semi-split skirt
Piston skirt clearance (measured at bottom of skirt at 90° to gudgeon pin axis)	0.0011 in. to 0.0017 in. (0.028 to 0.043 mm)
Piston rings - number:	
Compression	2
Oil control	1
Piston rings - width:	
Compression	0.777 in. to 0.787 in. (1.97 to 2 mm)
Oil control	0.155 in. to 0.156 in. (3.94 to 3.96 mm)
Piston rings - thickness:	
Compression	0.124 in. to 0.130 in. (3.15 to 3.30 mm)
Oil control	0.119 in. to 0.127 in. (3.02 to 3.23 mm)
Piston rings side clearance in groove:	
Compression and oil control	0.001 in. to 0.003 in. (0.02 to 0.07 mm)
Piston rings - gap when fitted to cylinder bore:	
Compression (both)	0.015 in. to 0.020 in. (0.38 to 0.51 mm)
Oil control	0.011 in. to 0.016 in. (0.28 to 0.41 mm)

Pistons and piston rings (4.2 litre)
	4.2 Early models	**4.2 Later models**
Make	Brico	Hepworth and Grandage
Type	Semi-split skirt	Solid skirt
Piston skirt clearance (measured at bottom of skirt at 90° to gudgeon pin axis)	0.0011 in. to 0.0017 in. (0.028 to 0.043 mm)	0.0007 in. to 0.0013 in. (0.018 to 0.03 mm)

Piston rings - number:
   Compression 2
   Oil control 1

Piston rings - width:
   Compression ... ... 0.0770 in. to 0.0780 in. (1.97 to 2.00 mm)
   Oil control ... ... Self expanding (Maxiflex)

Piston rings thickness ... ... 0.151 in. to 0.158 in. (3.775 to 3.95 mm)

Piston rings side clearance in groove:
   Compression and oil control ... ... 0.001 in. to 0.003 in. (0.02 to 0.07 mm)

Piston rings - gap when fitted to cylinder bore:

Compression (top)	0.015 in. to 0.020 in. (0.38 to 0.51 mm)	0.015 in. to 0.020 in. (0.38 to 0.51 mm)
Compression (lower)	0.015 in. to 0.020 in. (0.38 to 0.51 mm)	0.010 in. to 0.015 in. (0.254 to 0.38 mm)
Oil control	0.015 in. to 0.033 in. (0.38 to 0.82 mm)	0.015 in. to 0.045 in. (0.38 to 1.143 mm)

Tappets and tappet guides
   Tappet - material ... ... Chilled cast iron
      - outside diameter ... ... 1.3738 in. to 1.3742 in. (34.89 to 34.90 mm)
      - diameter clearance ... ... 0.0008 in. to 0.0019 in. (0.02 to 0.048 mm)
   Tappet guide - material ... ... Austenitic iron
      - inside diameter before reaming ... ... 1.353 in. to 1.357 in. (34.37 to 34.48 mm)
      - reaming size when fitted to cylinder head 1.375 in. + 0.0007 in. − 0.000 in. (34.925 + 0.018 mm − 0.000 mm)
      - interference (shrink) fit in head ... ... 0.003 in. (0.07 mm)

Timing chains and sprockets
   Type ... ... Duplex
   Pitch ... ... 0.375 in. (9.5 mm)
   Number of pitches - top chain 100
      - bottom chain 82
   Crankshaft sprocket - teeth 21
   Intermediate sprocket - outer teeth 28
   Intermediate sprocket - inner teeth 20
   Camshaft sprocket - teeth 30
   Idler sprocket - teeth 21

Valve timing
   Inlet valve opens ... ... 15º BTDC) With valve clearances set at 0.010 in.
   Inlet valve closes ... ... 57º ABDC) (0.25 mm)
   Exhaust valve opens ... ... 57º BBDC)
   Exhaust valve closes ... ... 15º ATDC)

Valves and valve springs
   Valves - material:
      inlet ... ... Silicone chrome steel
      exhaust ... ... Austenitic steel
   Valve head diameter - inlet ... ... 1.75 ± 0.002 in. (44.45 ± 0.05 mm)
      - exhaust ... ... 1.625 in. ± 0.002 in. (41.27 ± 0.05 mm)
   Valve stem diameter, inlet and exhaust ... ... 0.3125 in. − 0.0025 in. − 0.0035 in. (7.95 − 0.06 − 0.09 mm)
   Valve lift ... ... 0.375 in. (9.28 mm)

	Touring	Racing
Valve clearance (cold) - inlet	0.004 in. (0.010 mm)	0.006 in. (0.15 mm)
exhaust	0.006 in. (0.15 mm)	0.010 in. (0.25 mm)
Later cars - inlet and exhaust	0.012 in. − 0.014 in. (0.304 − 0.355 mm)	

   Valve seat angle - inlet 45º
      - exhaust 45º
   Valve spring free length - inner ... ... 1.657 in. (42 mm)
      - outer ... ... 1.937 in. (49.2 mm)
   Valve spring fitted length - inner ... ... 1.218 in. (30.96 mm)
      - outer ... ... 1.312 in. (33.34 mm)
   Valve spring - fitted load - inner ... ... 30.33 lbs (13.76 kg)
      - outer ... ... 48.375 lbs (21.94 kg)
   Valve spring solid length (max) - inner ... ... 0.810 in. (20.57 mm)
      - outer ... ... 0.880 in. (22.35 mm)

Number of free coils	- inner	...	...	...	...	...	6
	- outer	...	...	...	...	...	5
Diameter of wire	- inner	...	...	...	...	...	12 SWG (0.104 in.) (2.64 mm)
	- outer	...	...	...	...	...	10 SWG (0.128 in.) (3.25 mm)

Valve guide and valve seat insert

Valve guides - material ... ... Cast iron
Valve guide - length:
   inlet ... ... 1.8125 in. (46.04 mm)
   exhaust ... ... 1.9375 in. (49.21 mm)
Valve guide inside diameter:
   inlet ... ... 0.3125 in. − 9.0005 in. − 0.0015 in.
      (7.94 mm − 0.013 − 0.038 mm)
   exhaust ... ... 0.3125 in. ± 0.0005 in. (7.94 ± 0.01 mm)
Interference fit in head ... ... 0.0005 in. to 0.0022 in. (0.013 to 0.055 mm)
Valve seat inserts - material ... ... Cast iron (centrifugally cast)
           - inside diameter:
   inlet ... ... 1.5 in. + 0.003 − 0.001 in.
      (38.1 + 0.076 − 0.025 mm)
   exhaust ... ... 1.379 in. to 1.383 in. (35.03 to 35.13 mm)
           - interference (shrink) fit in head ... 0.003 in. (0.076 mm)

Capacity

Engine refill with lubricant ... ... 15 Imperial pints
      18 U.S. pints
      8.5 litres

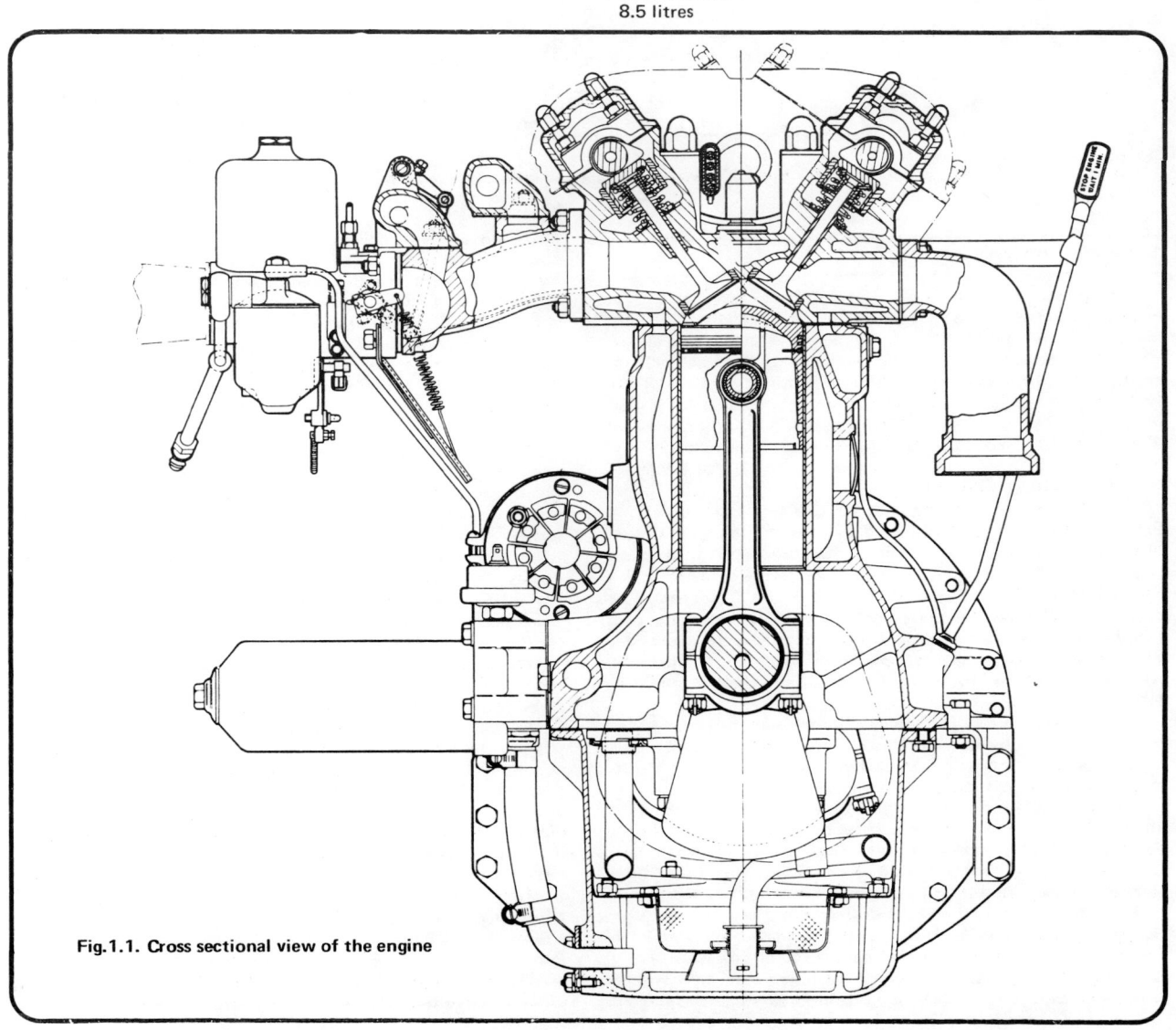

Fig.1.1. Cross sectional view of the engine

Chapter 1/Engine

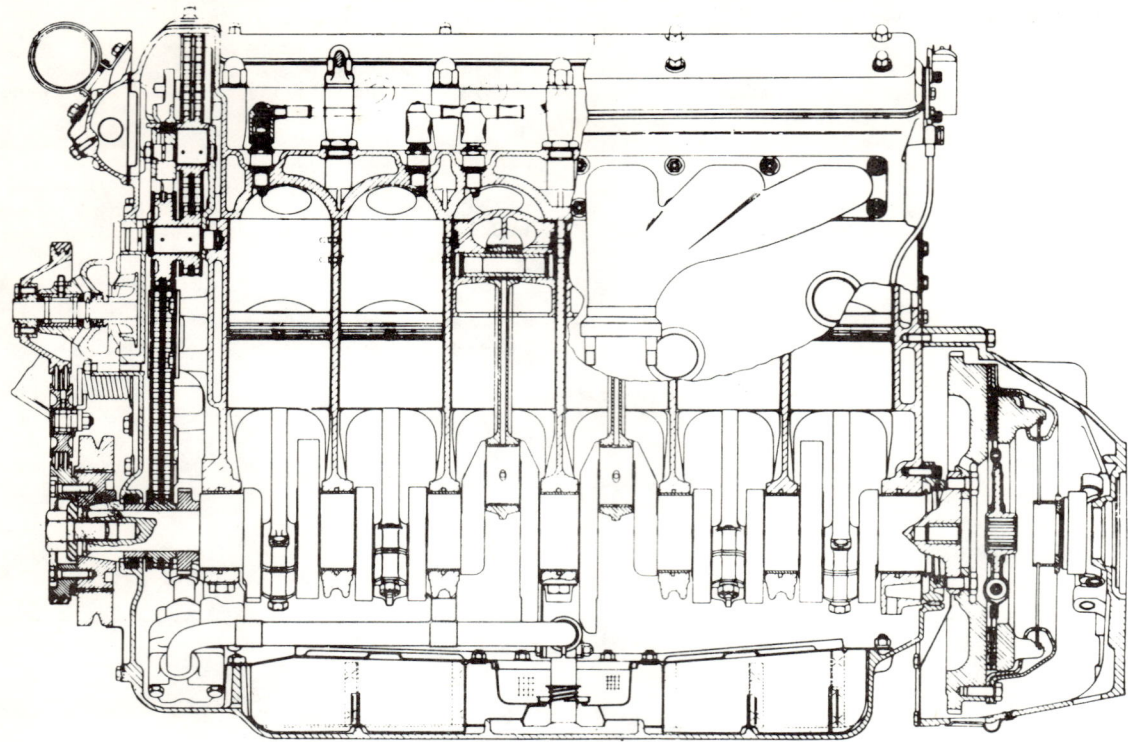

Fig.1.2. Longitudinal section view of the engine

1 General description

The engines fitted to the 3.8 and 4.2 litre range of E type cars are identical except for certain dimensional differences (particularly in the bore of the cylinders) details of which are given under Specifications at the beginning of this Chapter. The engine is a six cylinder, twin overhead camshaft Jaguar XK unit with, in the case of the 3.8 litre version, a bore of 3.4252" (87 mm) and stroke of 4.1732 (106 mm) and a cubic capacity of 3781 cc whilst the 4.2 litre unit has the same stroke but a bore of 3.625" (92.07 mm) and a cubic capacity of 4235 cc Triple SU HD.8 carburettors are fitted.

The engines are available with compression ratios of 8:1 or 9:1 the differences being obtained by varying the crown design of the piston as described later in this Chapter. The compression ratio of an engine is indicated by either /8 or /9 following the engine number. Power output (9:1) is 265 bhp at 5400 rpm. The cylinder block and the upper half of the crankcase are cast together and a pressed steel oil sump is bolted to the underside. The cylinders are 'Brivadium' dry liners which are pressed into position and which have a maximum oversize limit of +0.030" (+0.76 mm).

The cylinder head is made of aluminium alloy and has machined hemispherical combustion chambers. The cast iron valve inserts, the tappet guides and valve guides are shrunk into the cylinder head casting.

The pistons are manufactured in low expansion aluminium alloy and are of the semi-split skirt type in 3.8 and early 4.2 litre engines, later 4.2 litre units are fitted with Hepworth and Grandage pistons having a solid skirt. The pistons have three rings each, two compression and one oil control. The top piston ring is chromium plated and is also Cargraph treated on its outer diameter to assist in bedding down the plated surface; both compression rings have a tapered periphery. The oil control ring is of the Maxiflex type and consists of two steel rails with a spacer between. The gudgeon pin is fully floating and is retained in the piston by a circlip at each end. The camshafts are made of cast iron and each shaft is supported in four white metal steel backed bearings; end float of the shafts is taken on flanges formed at each side of the front bearing. The shafts are driven by a Duplex roller chain from a sprocket on the crankshaft and micrometer adjustment for radial positioning of the shafts is made at matching sprockets attached to the front end of each shaft. The right hand camshaft, that is the shaft for the inlet valves, drives an electric generator to operate the revolution counter.

The counter balanced manganese molybdenum steel crankshaft is supported in seven precision shell bearings with end thrust being taken on two semi-circular white metal faced steel thrust washers which fit in recesses in the centre main bearing cap. A torsional vibration damper is keyed to a taper at the front end of the shaft whilst a flange at the rear of the shaft carries the flywheel to which is attached the clutch unit. Initially, the crankshaft is balanced both statically and dynamically and is then rebalanced as an assembly with the flywheel and clutch unit attached.

The air cleaner of the paper element type is mounted on the right side of the engine compartment and is connected to the carburettors by means of an elbow trumpet plate.

The oil pump is of the eccentric rotor type, it is driven by a gear on the crankshaft and feeds oil, under a pressure of 40 lbs per sq in. at 3000 rpm, to the main and big end bearings through drillings, and to the camshafts via an external oil pipe and an external oil filter of the full flow type fitted with a replaceable element. A pressure relief valve is provided in the head of the filter and oil passing this valve is returned to the sump by an external rubber hose. A balance valve is also fitted in the filter head which will open at a pressure differential of 10 to 15 lbs per sq. in. (0.7 to 1.1 kg/cm^2) and so provide a safeguard against the possibility of the filter element becoming so choked that oil is prevented from reaching the bearings.

The maximum permissable rpm for the engine is 5500 and under no circumstances should this figure be exceeded.

2 Major operations with the engine in place

The following major operations can be carried out with the engine installed in the car:
1 Removal and replacement of the cylinder head assembly.
2 Removal and replacement of the sump.
3 Removal and replacement of the pistons, connecting rods and big end bearings.
4 Removal and replacement of the timing gear assembly.
5 Removal and replacement of the camshafts.
6 Removal and replacement of the oil pump.

3 Major operations with the engine removed

The following operations will necessitate removal of the engine from the car:
1 Removal and replacement of the crankshaft and main bearings.
2 Removal and replacement of the flywheel.
3 Removal and replacement of the clutch assembly.
4 Removal and replacement of the gearbox.

4 Methods of engine removal

The engine and gearbox, or automatic transmission unit, must be removed from the car as an assembly and although this is a major task it is not quite so difficult as it may at first appear, provided, you have adequate lifting equipment and approach the work in a methodical manner. No special tools are required and the only equipment you need is a wheeled jack (preferably) and a hoist capable of supporting a weight of eight hundredweights and also capable of lifting that weight to a height of about three feet. Needless to say the lifting tackle must be in good condition. As the engine has to be tilted to an angle of about 45° to clear the front subframe when lifting it out, you will find that the rear of the car, depending on your method of working, will have to be raised to allow clearance fo the gearbox. Therefore axle stands or blocks on which to support the rear end will be required, as an alternative, the front of the car can be lowered. If a pit is available, this will greatly facilitate working underneath the car.

The engine is lifted out from above the car, the method you adopt will depend on the facilities you have available. The photographs which follow later depict the removal of an engine from a 4.2 litre model but work on a 3.8 litre car will be identical. The method we used was to make all the disconnections, as described later, inside the car and in the engine compartment with the exception of the engine mountings. The car was then pushed over a pit and all work underneath completed with the exception of removal of the rear engine mounting. The next step was to position the car to give a firm base for a wheeled jack to take the weight of the gearbox but, before taking the weight, the rear of the car was raised about one foot and the rear wheels chocked and supported on blocks. The weight of the engine was then taken on the hoist and the weight of the gearbox was taken on the jack but we were careful to ensure that no undue strain was placed on the mounting. The front and rear mountings were now disconnected and the engine was inched forward horizontally as far as possible (about two inches). Lifting then commenced and at the same time the jack under the gearbox was lowered until the engine assumed the attitude depicted to enable it to be lifted clear of the car. We repeated the foregoing removal procedure but this time instead of raising the rear of the car we removed the front wheels and then lowered the front of the car to rest the stub axles on wooden blocks. Neither method of giving clearance for the gearbox had any real advantage over the other.

We feel that the work of removing the engine will be made much easier if it its divided into four separate stages and if you complete the work in each stage before starting the next, i.e.

Stage - Work inside the car
Stage 2 - Work in the engine compartment
Stage 3 - Work underneath the car
Stage 4 - Lifting out the engine

5 Engine removal

The various operations entailed in removing the engine are depicted in the photographs - which can be related to the text by their Section/paragraph numbers.
1 Refer to Chapter 2 and drain the coolant from the sytem in the manner described.
2 Place a container below the sump, remove the drain plug (Fig. 1.3) and drain off the engine oil.

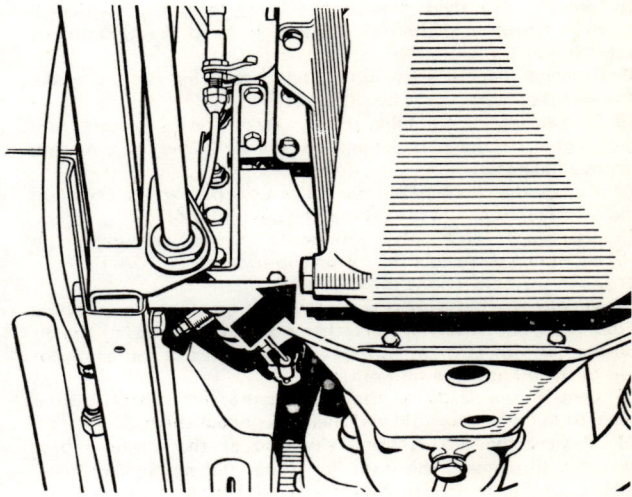

Fig.1.3. The sump drain plug

3 Open the bonnet and disconnect the battery.
4 On manual transmission cars, follow the instructions given in paragraphs 5 - 13 inclusive.
5 Unscrew the chrome nut at each side of the centre console and then lift the console rearwards (photo). Disconnect the input and the aerial leads (if applicable) and lift the console clear of the car.
6 Slacken the conical locknut at the gear lever knob. Unscrew the knob and then remove the locknut (photo).
7 The next task is to remove the cover panel from the gearbox tunnel. On 2 + 2 cars, remove the seats by lifting out the cushions and then remove the four nuts and washers securing each seat pan to the runners and lift off the seat. Now raise the central arm rest, lift out the bottom panel, withdraw the five self tapping screws and remove the central arm rest. On other models the seats can be left in position: unclip the seat belts and now unscrew and remove each seat belt anchorage (photo). For all models, make sure that the handbrake lever is in the fully applied position and then lift the cover panel up at the rear and at the same time tilt it to the left to clear the handbrake lever (photo), once the panel is clear of the handbrake lever lift it up over the gear lever and remove it from the car.
8 Undo the self tapping screws securing the gearbox cover to the floor and then remove the cover (photo).
9 Remove the self locking nut from the gear lever pivot pin (photo), collect the flat washer and spring and lift out the gear lever assembly.
10 Identify the leads to the reverse light switch on the top cover of the gearbox and then disconnect them (photo).
11 Slacken the clamps securing the gearbox breather pipe (these can be seen in photograph 5.10) and remove the pipe.

12 Undo the union nut and disconnect the speedometer drive (photo).
13 Using a centre punch, mark the relative positions of the gearbox and propeller shaft flanges. Remove the nuts from the four bolts securing the propeller shaft to the gearbox flange (photo) and then separate the propeller shaft by pushing to the rear on the sliding joint; collect the four bolts. It will be necessary to move the car in order to bring all the flange securing nuts into a convenient position for removal.
14 When removing an engine with an automatic transmission unit, follow the instructions given in paragraphs 7 and 8 for removal of the gearbox tunnel panel and for removal of the cover plate on the transmission tunnel. The propeller shaft is now disconnected in the manner described in paragraph 13.
15 Refer to Chapter 12 and follow the instructions for removal of the bonnet.
16 Identify and then remove the leads from the revolution counter generator at the rear of the right hand (inlet) camshaft cover (photo).
17 Remove the butterfly nuts (photo) securing the air intake box and then remove the box (photo).
18 Slacken the clips holding the two heater hoses running from the engine to the connections on the bulkhead (photo) and remove the hoses.
19 Slacken the clip securing the brake servo pipe to the union on the inlet manifold and disconnect the pipe (photo).
20 Disconnect the throttle linkage at the rear carburettor by pulling outwards against the spring clip (photo).
21 Disconnect the choke cable (photo).
22 We suggest that you now remove the breather pipe running from the front cover and breather housing assembly to the air cleaner (photo) in order to improve access to the starter motor and to the oil pressure transmitter.
23 Identify the leads to the oil pressure transmitter located beneath the inlet manifold and then disconnect them.
24 Unscrew the nut securing the lead at the starter motor terminal, disconnect the lead. For those later model cars fitted with a Pre-engaged Starter Motor you will have to disconnect the two leads to the actuating solenoid. (photo)
25 Disconnect the petrol feed pipe from below the centre carburettor using two spanners, one to hold the bottom union nut and the other to unscrew the top nut.
26 Unscrew the nut securing the electric lead to the water temperature gauge element located on the water rail on the right hand side of the engine and remove the lead from the terminal (photo).
27 The header tank and radiator can be removed as an assembly, slacken the clips and remove the right and left hand water hoses running from the header tank to the top of the radiator.
28 Slacken the clips and remove the water hose running from the header tank to the thermostat outlet.
29 Slacken the clips and remove the bypass pipe running from the radiator to the water pump.
30 Disconnect the leads from the thermostat in the front of the header tank.
31 Remove the two nuts securing the header tank to the subframe. Collect the flat washers and the rubber mounting washers.
32 Move below the radiator. Slacken the clips holding the rubber junctions of the water pipe running from the engine to the bottom of the radiator and then remove the pipe.
33 Remove the nuts from e two mounting studs at the bottom of the radiator collect the flat washers and the rubber mounting washers.
34 Working from above the radiator, lift it and the header tank upwards and outwards clear of the car taking care not to damage the cooling fan or to allow the fan to damage the radiator matirx (photo).
35 Remove the fan motor and the cooling fan.
36 Slacken the alternator, or dynamo, mounting bolts including the adjuster link bracket (photo).
37 Slacken the automatic belt tensioner pivot belt and then push the tensioner out of the way clear of the belt.

38 Remove the water pump belt.
39 Remove the cables from the alternator by removing two cable nuts (identify the leads to these terminals) and one pull-off connector (photo).
40 Disconnect the HT lead at the ignition coil. Identify the other two leads and then remove them from the coil by pulling off from the Lucar tags.
41 Remove the nuts from the coil mounting bracket and then lift off the coil (photo).
42 Remove the four nuts (remember when reassembling that these are special nuts) holding the exhaust pipe flanges to the exhaust manifold. You will find that there is not much space in which to use a spanner so you may find it easier, if you have a long socket extension, to remove these nuts when you come to work underneath the car (photo).
43 Remove the two rearmost nuts from the exhaust manifold and disconnect the gearbox breather pipe which is held by them (photo).
44 Move to the front of the engine and take out the two setscrews holding the ignition timing pointer (photo). Clearances are so small that the heads of these setscrews will catch on the sub-frame when the engine is being lifted out.
45 Knock back the tabs locking the two setscrews holding the crankshaft damper bolt locking plate. Remove the setscrews and the locking plate (photo).
46 Remove the large centre bolt holding the crankshaft damper (photo).
47 Place two screwdrivers behind the crankshaft damper and try to lever the damper off the cone. You may have to tap on the end of the cone to release the grip of the taper and, if so, use a copper or brass drift to prevent damaging the metal. Remove the damper (photo).
48 Now move underneath the car. The ideal, of course is to position the car over a pit but if you have to raise the car to give access to the underside, do ensure that it is safely supported before crawling underneath it.
49 The first job is to clear the exhaust system out of the way and we suggest you remove this as a complete assembly. If you have not already disconnected the exhaust pipes from the manifold do this first (see photograph 5.41).
50 Remove the nut and bolt holding the tail pipes to the rear bracket (photo).
51 Now take the nuts from the bolts holding the silencers to their mounting brackets. The bolts (two for each silencer) will remain in position and hold the assembly until you are ready to lift it off (photo).
52 Have an assistant to help take the weight of the assembly and then spring the silencer mounting brackets off the bolts and lift the exhaust assembly clear of the car (photo).
53 Unhook the return spring from the clutch slave cylinder and remove it (photo).
54 Push off the spring clip (photo) securing the clutch operating lever pin and remove the pin.
55 Remove the nuts from the studs holding the clutch slave cylinder and then pull the cylinder clear of the engine.
56 Take out the self tapping screws holding the fairing below the oil filter. Take out the central bolt passing through the oil filter canister and remove the oil filter assembly (photo).
57 Remove the nut and bolt holding the earth connection at the left hand side of the engine (photo).
58 Remove the bolts (one each side) holding the torsion bar reaction plate (photo).
59 Remove the two lower nuts securing the torsion bar reaction plate on each side and then tap the bolts back flush with the face of the plate (photo).
60 With the aid of a helper, place a lever between the head of the bolt just released and the torsion bar. Exert pressure on the bolt just released and the torsion bar. Exert pressure on the bolt head to relieve the tension on the upper bolt, remove the nut and tap this bolt back flush with the plate. Repeat for the other side. Note:- Failure to relieve tension on the upper bolts when tapping them back will result in stripping the threads.
61 Remove the reaction tie plate (photo).

5.5 Remove centre console

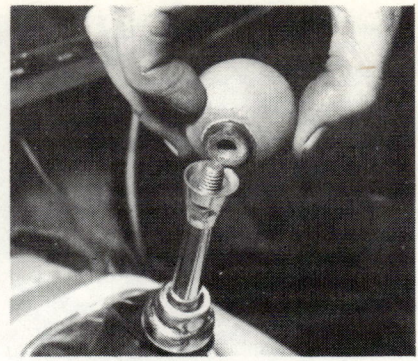
5.6 Remove gear lever knob and locknut

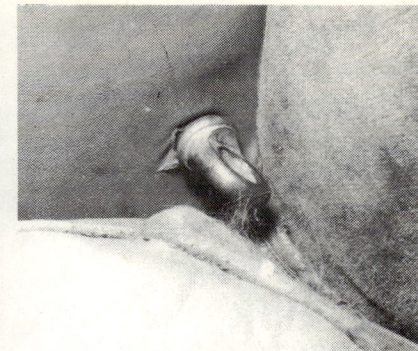
5.7a Seat belt anchorage

5.7b Removing the cover panel

5.8 Removing the gearbox cover

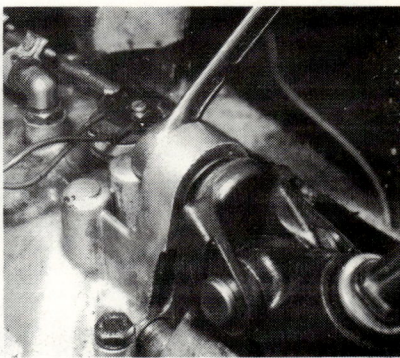
5.9 Remove gear lever assembly

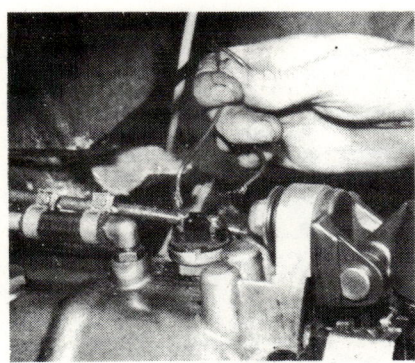
5.10 Leads to the reverse light switch

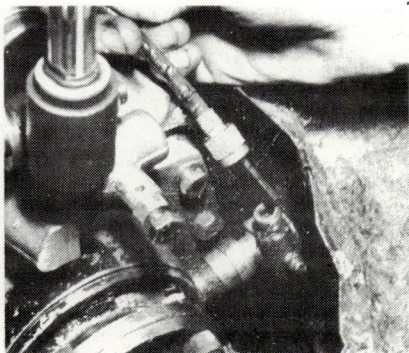
5.12 Speedometer drive disconnected

5.13 Disconnecting gearbox and propellor shaft flanges

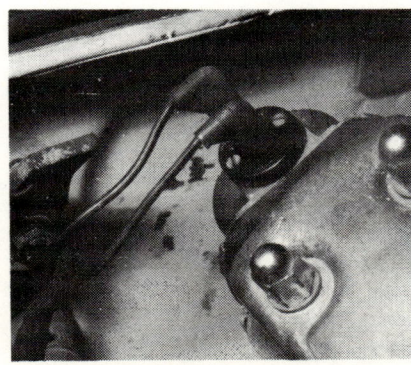
5.16 Leads to revolution counter generator

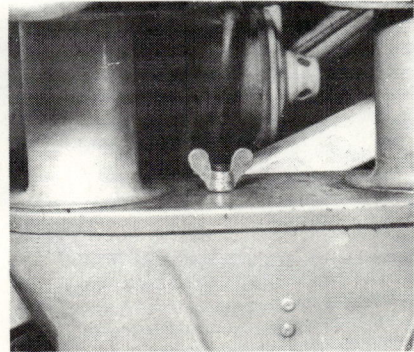
5.17a Removing air intake box (1)

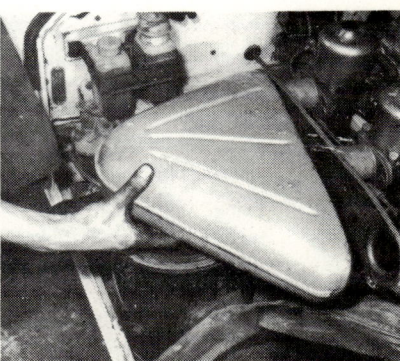
5.17b Removing air intake box (2)

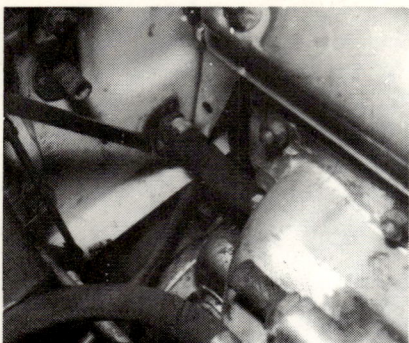
5.18 Remove heater hoses

5.19 The brake servo pipe

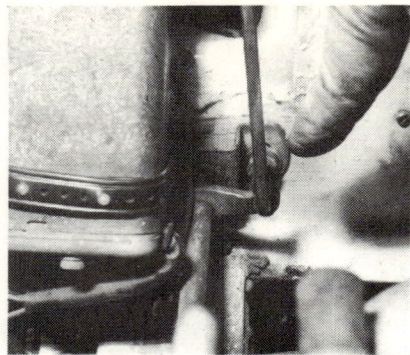
5.20 The throttle linkage

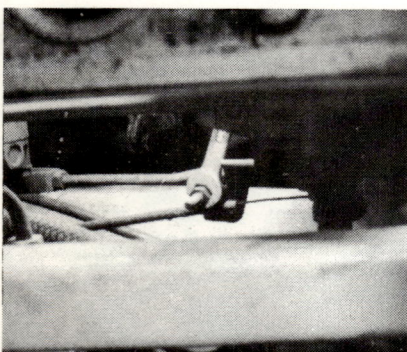
5.21 Disconnect choke cable

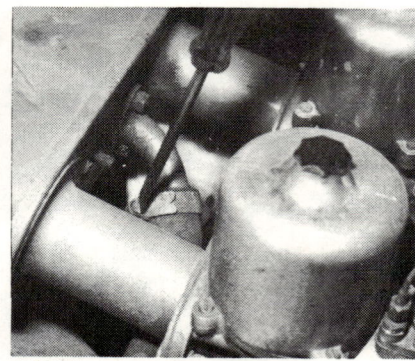
5.22 The engine breather pipe

5.24 Disconnect leads to starter motor. Oil pressure transmitter also shown

5.26 Disconnect lead at water temperature gauge element

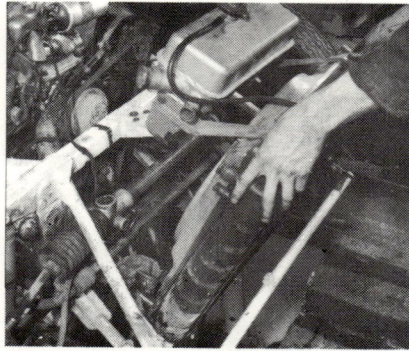
5.34. Remove radiator and header tank

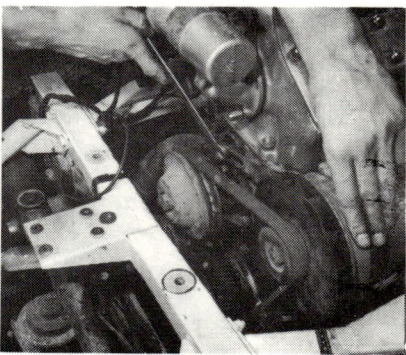
5.36 Slacken the dynamo/alternator mounting bolts

5.39 The alternator connections

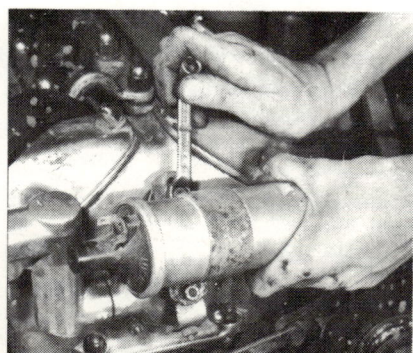
5.41 Removing the coil

5.42 Using long socket extension to remove exhaust pipe flange nuts

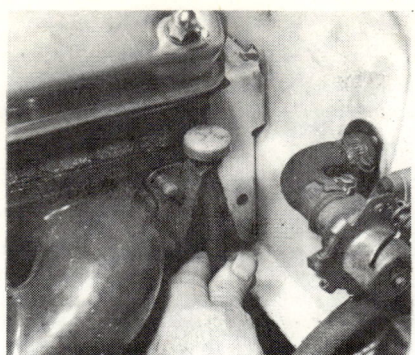
5.43 Disconnect gearbox breather pipe at exhaust manifold

5.44 Remove ignition timing pointer

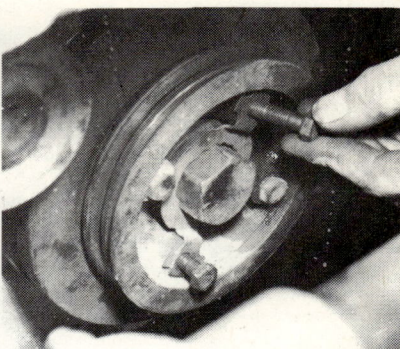
5.45 Remove crankshaft damper setscrews and locking plate

5.46 Centre bolt of crankshaft damper

5.47 Remove crankshaft damper

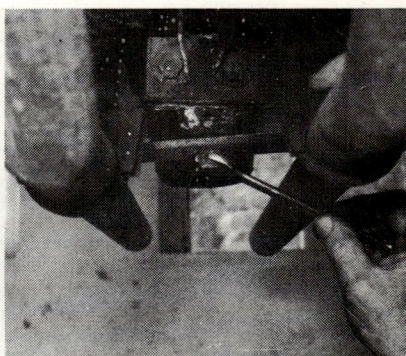
5.50 Exhaust tail pipe mounting bolt

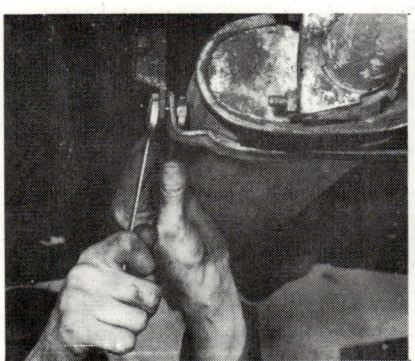
5.51 Remove nut from each silencer mounting bolt

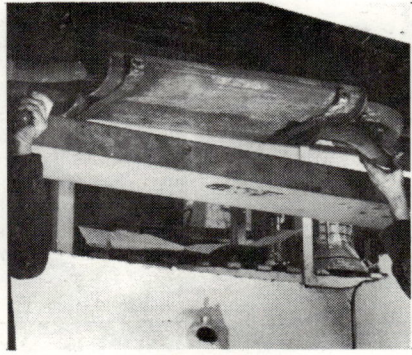
5.52 Lift out exhaust assembly

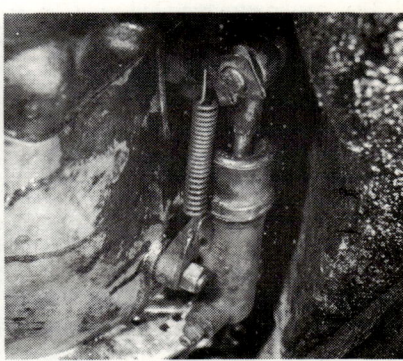
5.53 The clutch slave cylinder

5.54 Spring clip securing operating lever pin

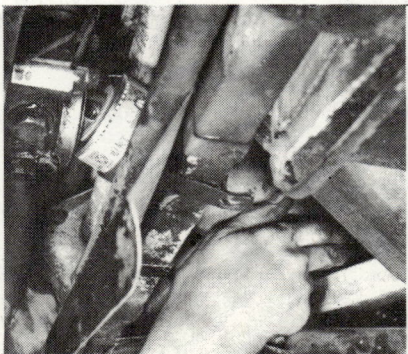
5.56 Remove oil filter

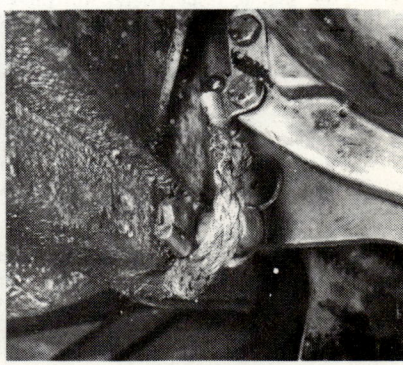
5.57 Disconnect engine earth connection

62 If the gearbox and engine are to be separated after removal, we think it a good plan at this stage to remove the cover plate located between the engine and the clutch bell housing. Remove the four setscrews, collect the spring washers and remove the plate (photo).
63 On cars fitted with automatic transmission, make sure that the selector lever is in 'L' and then, still underneath the car, unscrew the nut securing the selector cable adjustable ball joint to the transmission lever, and release the nut securing the outer cable clamp to the abutment bracket. Now remove the speedometer drive cable from the transmission extension housing.
64 Make a final check underneath the car to ensure that all items between the engine/gearbox/transmission unit (other than the rear mounting) have been disconnected.
65 Working in the engine compartment, unscrew the top self locking nut of the engine stabiliser and then screw the bottom nut downwards away from the bracket. As the stabiliser can prove a nuisance when lifting out the engine, and especially when replacing it, we suggest that it is disconnected from the engine at this stage. Remove the nut from the bolt securing the stabiliser to the engine and then remove the bolt. You will find that, due to restricted access, this is not an easy job but a little patience now will pay dividends later.
66 Make a final check in the engine compartment to ensure that all connections, other than the front mountings, between the engine and the body have been disconnected.
67 Position the car for lifting out the engine; from now onwards the procedures will depend on the equipment and facilities available to you. All we can do is to detail the procedure which we followed and from this you should be able to adopt a satisfactory method of working. However, the following factors must be taken into consideration:-
a) The lifting tackle must be in good condition and must be capable of lifting a weight of at least 8 cwt to a height of 3 ft.
b) The gearbox/transmission unit must be supported and, indeed, lifted slightly during the early stages of removal.
c) Although the assembly will finally take up a position at an angle of about 45° it must first be moved towards the front of the car in a horizontal plane. This means that the lifting tackle must be capable of being moved with the engine slung or that the car must be moved rearwards, but:-
d) If the car is not located over a pit, it will be necessary to either raise the rear of the car about one foot and then support the wheels on blocks or to remove the front wheels and then lower the front of the car and rest the stub axles on wooden blocks so that the discs are just clear of the ground.
68 Remove the second pair of cylinder head nuts from the front of the engine; fit a lifting plate in position over the studs and secure in position with the nuts. The 4.2 litre engine is provided with two lifting plates as standard equipment held in position on the second pair of cylinder head studs from the front, and the rear of the engine, respectively, this pre-supposes the use of a double sling or two sets of lifting tackle and we feel that this may be awkward for engine removal purposes. It is suggested, therefore, that you use the front lifting plate only. A lifting plate - if not available - can be made out of of two pieces of 3" angle iron of about 1½" width and drilled to accept the cylinder head studs and your lifting attachment (photo).
69 If you are not working over a pit, either raise the rear of the car about one foot, support the rear wheels on blocks and then securely chock the front wheels, or, remove the front wheels and then lower the car onto wooden blocks placed under the stub axles so that the discs are clear of the ground.
70 Attach your lifting tackle to the lifting plate and take the weight of the engine; but not to the extent of placing undue strain on the front mountings.
71 Move under the car. Place a trolley jack underneath the gearbox, and take its weight.
72 Now undo the setscrews of the rear mounting a part turn at at time keeping them even against the pressure of the mounting spring. Finally remove the rear mounting.
73 Outside the car again, remove the centre bolt to each front mounting (photo).
74 Raise the engine slightly at the front and then raise the jack under the gearbox until the front of the clutch housing is seen to be clear of the protruding bracket for attachment of the torsion bar reaction tie plate.
75 Move the engine forward as far as you can in a horizontal plane; now start raising at the front and lowering the jack under the gearbox. Carry on like this until the engine is clear of the car but, during the early stages, check that all connections have been broken. The various stages in lifting out the engine are shown in photographs 5.75A, B and C.
76 Lower the engine to the floor and rest it so that the weight is evenly distributed.(photo).
77 Now complete the job, clear any loose nuts, bolts and tools from the engine compartment and the floor and place them where they will not become lost.

6 Separating the engine from the gearbox (manual)

1 Remove the two bolts securing the starter to the clutch housing and remove the starter (photo).
2 If you did not remove the cover plate before taking the engine out of the car, tilt the engine to give access to the plate, undo the securing nuts and bolts and remove the plate.
3 Remove the bolts and setscrews securing the clutch housing to the engine, starting at the bottom and working towards the top. The gearbox must be supported during this operation to avoid straining the constant pinion shaft and the clutch driven plate.
4 Carefully draw the gearbox rearwards clear of the engine (photo).

7 Separating the engine from the transmission unit (automatic)

1 Drain the fluid from the transmission unit oil pan by removing the Allen screw using a ¼" AF key.
2 Remove the dipstick, detach the dipstick tube top securing clip from its anchorage at the rear of the exhaust manifold and now unscrew the tube from the oil pan.
3 Remove the transmission retaining bolts commencing with the two bottom ones. Support the unit during this operation and place a container beneath the unit to catch the fluid from the converter when the transmission unit is withdrawn. **DO NOT** place supports immediately below the sump tray.

8 Dismantling the engine - general

1 It is is best to mount the engine on a dismantling stand but if one is not available, stand the engine on a strong bench at a comfortable working height. Alternatively, the engine can be dismantled on the floor in the position in which it was placed on removal from the car but this will make for very uncomfortable, awkward and inefficient working and, therefore, cannot be recommended.
2 During the dismantling process, the greatest care should be taken to keep the exposed parts free from dirt. To that end, thoroughly clean down the outside of the engine, removing all traces of oil and congealed dirt. Use paraffin or proprietry solvent. The latter will make the job much easier for, after the solvent has been applied and allowed to stand for a time, a vigourous jet of water will wash off the solvent with all the dirt. If the dirt is thickly and deeply embedded, work in the solvent with a stiff brush.
3 Finally wipe down the exterior of the engine with rag and only then when the engine is quite clean, should the dismantling process begin. As the engine is stripped, clean each part in a bath of paraffin or solvent.
4 Never immerse parts with oilways (for example the crankshaft) in the cleaning bath. To clean such items, carefully wipe down with a clean paraffin rag and then wipe dry. Oilways can

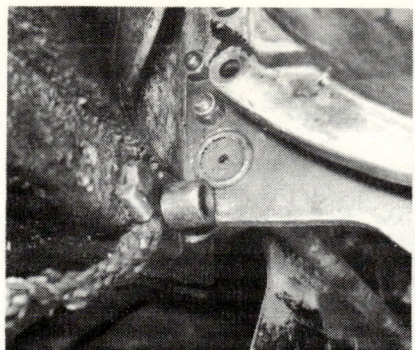
5.58 Remove bolt holding bottom of torsion bar reaction plate

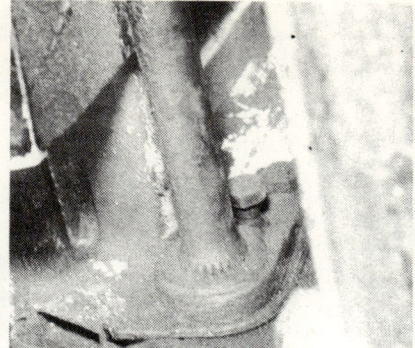
5.59 Tap back the lower bolt

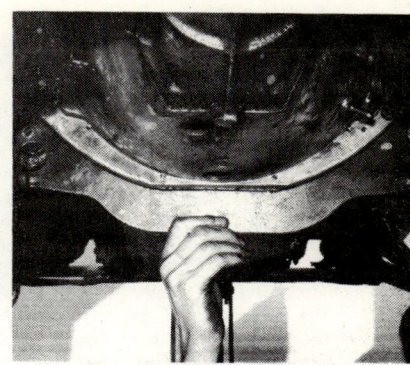
5.61 Remove the reaction tie plate

5.62 Remove the cover plate

5.68 Engine lifting attachment

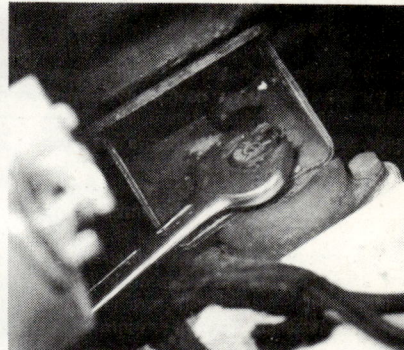
5.73 Remove centre bolt of front mountings

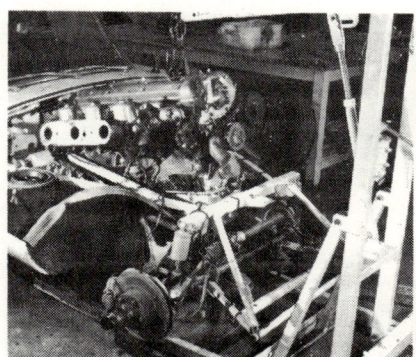
5.75A Lifting engine - Stage 1

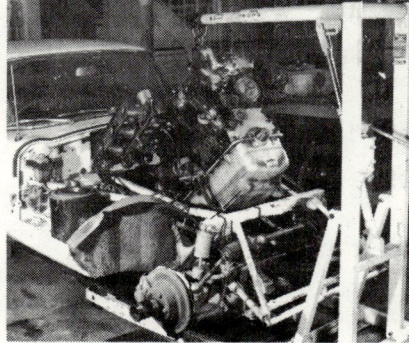
5.75B Lifting engine - Stage 2

5.75C Lifting engine - Stage 3 - engine clear

5.76 Engine out of car

6.1 Remove starter

6.4 Gearbox separated from engine (typical)

be cleaned out with nylon pipe cleaners or blown through with an air blast.

5 Re-use of old engine gaskets, copper washers etc; is false economy and will, in all probability, lead to oil or water leaks. Always use new items throughout.

6 Retain the old gaskets until the job is finished - for it sometimes happens that a replacement is not always immediately available and in such cases the old one will come in handy for use as a template.

7 When stripping the engine it is best to work from the top down - the underside of the crankcase, when supported on wooden blocks, makes a firm base from which to work. We always recommend, therefore, that the sump is removed at an early stage.

8 Whenever possible, replace nuts, bolts and washers finger tight from wherever they were removed. This helps to avoid loss and muddle later; if they cannot be replaced lay them out in such a fashion that it is clear from whence they came.

9 Removing the ancillary engine components

Before basic engine dismantling begins it is necessary to strip it of ancillary components as follows:-

    Clutch   Thermostat
    Flywheel   Water pump
    Distributor   Oil supply pipe to camshafts

1 Remove the clutch by slackening the mounting screws a turn one at a time by diagonal selection until the spring pressure is released. Remove the setscrews (photo) and withdraw the clutch assembly from the flywheel (photo). Note that the clutch and flywheel are balanced as an assembly and correct location should be marked by balance marks 'B' on both the clutch and the flywheel (Fig. 1.4). If the marks are not present, mark the assembly location of the components yourself using a centre punch for the purpose. Store the driven plate assembly so that the faces will not become contaminated by oil or dirt.

2 Knock back the tabs to the locking plate of the flywheel bolts. Undo the bolts (photo) and remove the locking plate. Remove the flywheel from the crankshaft by tapping with a hide faced hammer (photo).

3 Remove the setscrew holding the plug harness clip at the front of the engine, identify the leads to the plugs and then disconnect them from the plugs. Turn the engine to bring No. 6 piston on TDC on the compression stroke and then remove the distributor. Note the position of the offset driving slot (Fig. 1.5).

4 Remove the oil supply pipe to the camshaft at the rear of the engine (photo). Collect the fibre washers under the union nuts.

5 Remove the petrol feed pipe assembly linking the carburettors by undoing the banjo unions. Collect the fibre washers from between the unions.

6 Remove the three nuts and spring washers holding the water outlet pipe on the inlet manifold. Remove the outlet pipe and lift out the thermostat which is now uncovered.

7 Remove the bolts securing the water pump to the front cover and then remove the pump and the gasket between the pump and the cylinder block. (photo)

8 Remove the split cone (over which the crankshaft damper fits), the distance piece and the oil thrower behind it (photo). Note the Woodruffe keys locating these items on the crankshaft; remove them and store in a safe place.

9 Remove the inlet and exhaust manifolds. The engine is now stripped of all ancillary items and is ready for major dismantling to begin.

10 Cylinder head removal - engine on bench

1 Remove the eleven dome nuts securing each camshaft cover and remove the covers.

2 Remove the four nuts securing the engine breather at the front of the cylinder head. Remove the breather, the two gaskets and the gauze filter (photo).

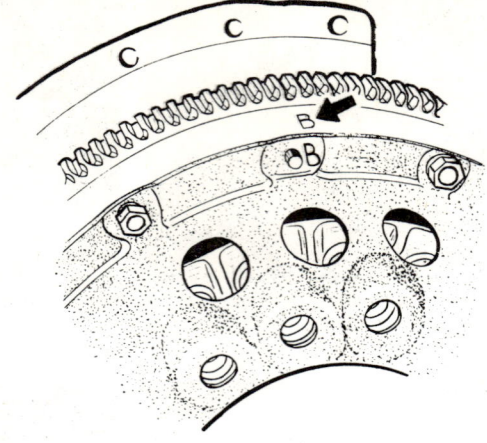

Fig. 1.4. Balance marks on clutch and flywheel

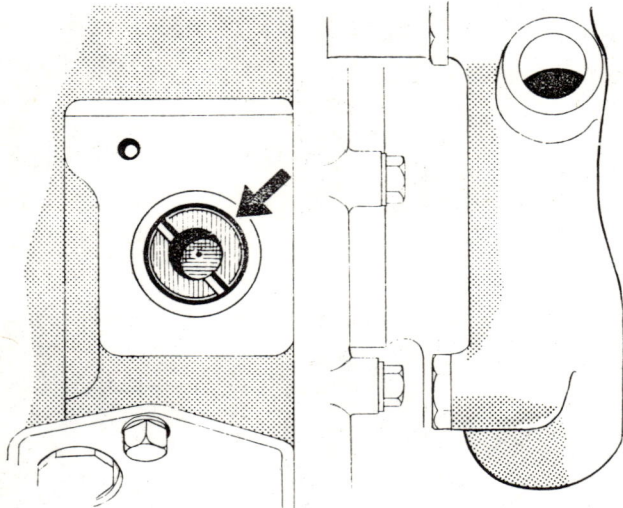

Fig. 1.5. Position of distributor drive shaft offset when No.6 piston is at TDC

9.1a Removing the clutch assembly

9.1b Clutch and driven plate assembly removed

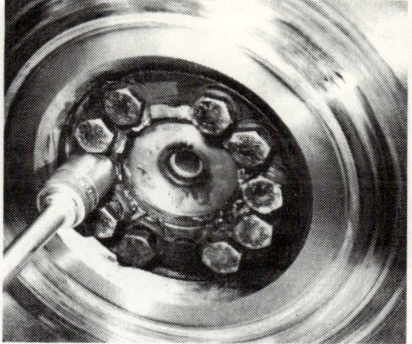
9.2a Flywheel securing bolts

9.2b Removing the flywheel

9.4 The camshaft oil supply pipe

9.7 Removing the water pump (typical)

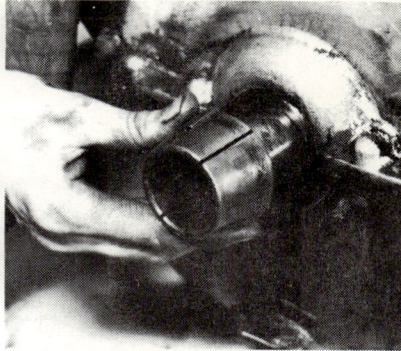
9.8 Split cone for hub of crankshaft damper

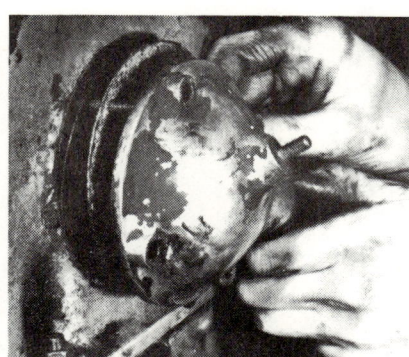
10.2 Removal of engine breather (typical)

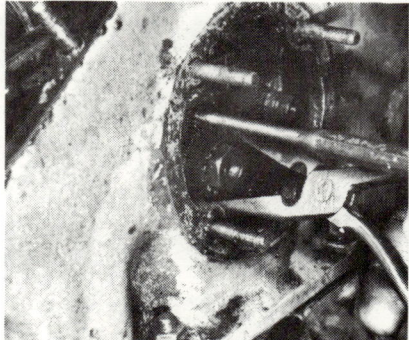
10.3 Use of round nosed pliers to rotate timing chain tensioner plate

10.4 Remove wire locking the sprocket securing bolts

10.5 Fit thin nut to sprocket shaft

10.6 Remove sprocket securing bolts

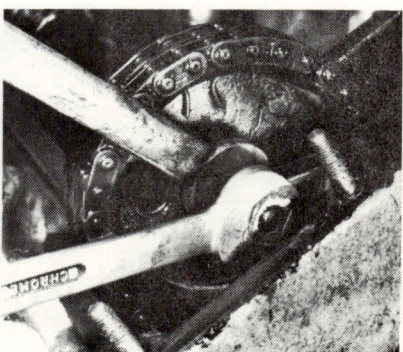
10.7 Tighten down on thin nut

3 In the space covered by the engine breather will be seen a serrated plate secured by a bolt locked by a tab washer. Knock up the tab washer to the bolt and then slacken the bolt. Depress the spring loaded plunger locating the plate and then rotate the plate in a clockwise direction (a pair of round nosed pliers entered in the holes in the plate is a handy tool for this (photo)) and this will relieve some of the tension on the top timing chain. Conversely, rotating the plate in an anti-clockwise direction will tighten the chain.

4 Break the locking wire to the two bolts securing the top sprocket to each camshaft (photo).

5 It is a good tip at this stage to fit a thin nut (7/16" A/F) to the threaded end of the sprocket shaft outside of the support slide (photo), a nut cut in half will suffice. The object of this is to retain the sprocket and chain when disconnected from the camshaft and so prevent them from falling into the sump.

6 Remove the sprocket securing bolts (photo).

7 Tighten down on the nut you fitted to the sprocket shaft (photo) and this will pull the sprocket away from the camshaft.

8 Push the sprocket up the support slide to slacken the chain as much as possible and then lock the sprocket in this position with the nut, hold the sprocket shaft with a spanner on the flats behind the support slide. (photo).

9 Repeat the above for the other sprocket. Both sprockets will now be in position shown in photograph 10.8.

10 Refer to Fig. 1.7 and slacken the cylinder head nuts a part turn at a time in the sequence shown (note that nuts 15 to 20 are below the head and above the front timing cover).

11 With the help of an assistant, lift off the cylinder head. If the gasket is partly stuck to the head, clear it off because it will catch on the cylinder head studs and prevent the head being lifted. Also watch the nut you have fitted to the sprocket shaft as this may foul the front of the head if it is too thick, or if the sprocket shaft is protruding too far.

12 Place the cylinder head on wooden blocks on the bench to avoid damage to the valves which will be protruding clear of the head at this stage.

10.8 Sprocket locked at end of support shaft

Fig.1.6. Exploded view of the cylinder head assembly

1	Cylinder head	29	Oil thrower	57	Gasket
2	Stud	30	Setscrew	58	Dome nut
3	Ring dowel	31	Copper washer	59	Spring washer
4	'D' washer	32	Sealing ring	60	Flexible pipe
5	Plug	33	Sealing plug	61	Clip
6	Copper washer	34	Seal	62	Clip
7	Valve guide	35	Adaptor	63	Exhaust manifold
8	Valve insert	36	Driving dog	64	Exhaust manifold
9	Tappet guide	37	Circlip	65	Gasket
10	Gasket	38	Generator	66	Clip
11	Stud	39	Sealing ring	67	Stud
12	Stud	40	Screw	68	Sealing ring
13	Stud	41	Plate washer	69	Inlet manifold
14	Stud	42	Lock washer	70	Inlet manifold
15	Stud	43	Inlet camshaft cover	71	Inlet manifold
16	Stud	44	Exhaust camshaft cover	72	Gasket
17	Inlet valve	45	Gasket	73	Air balance pipe
18	Exhaust valve	46	Gasket	74	Gasket
19	Valve spring	47	Dome nut	75	Stud
20	Valve spring	48	Copper washer	76	Adaptor
21	Seat	49	Filler cap	77	Gasket
22	Collar	50	Fibre washer	78	Water pipe
23	Cotter	51	Oil pipe	79	Gasket
24	Tappet	52	Banjo bolt	80	Thermostat
25	Adjusting pad	53	Copper washer	81	Plate
26	Inlet camshaft	54	Breather housing	82	Gasket
27	Exhaust camshaft	55	Pipe	83	Elbow
28	Bearing	56	Baffle	84	Gasket

35

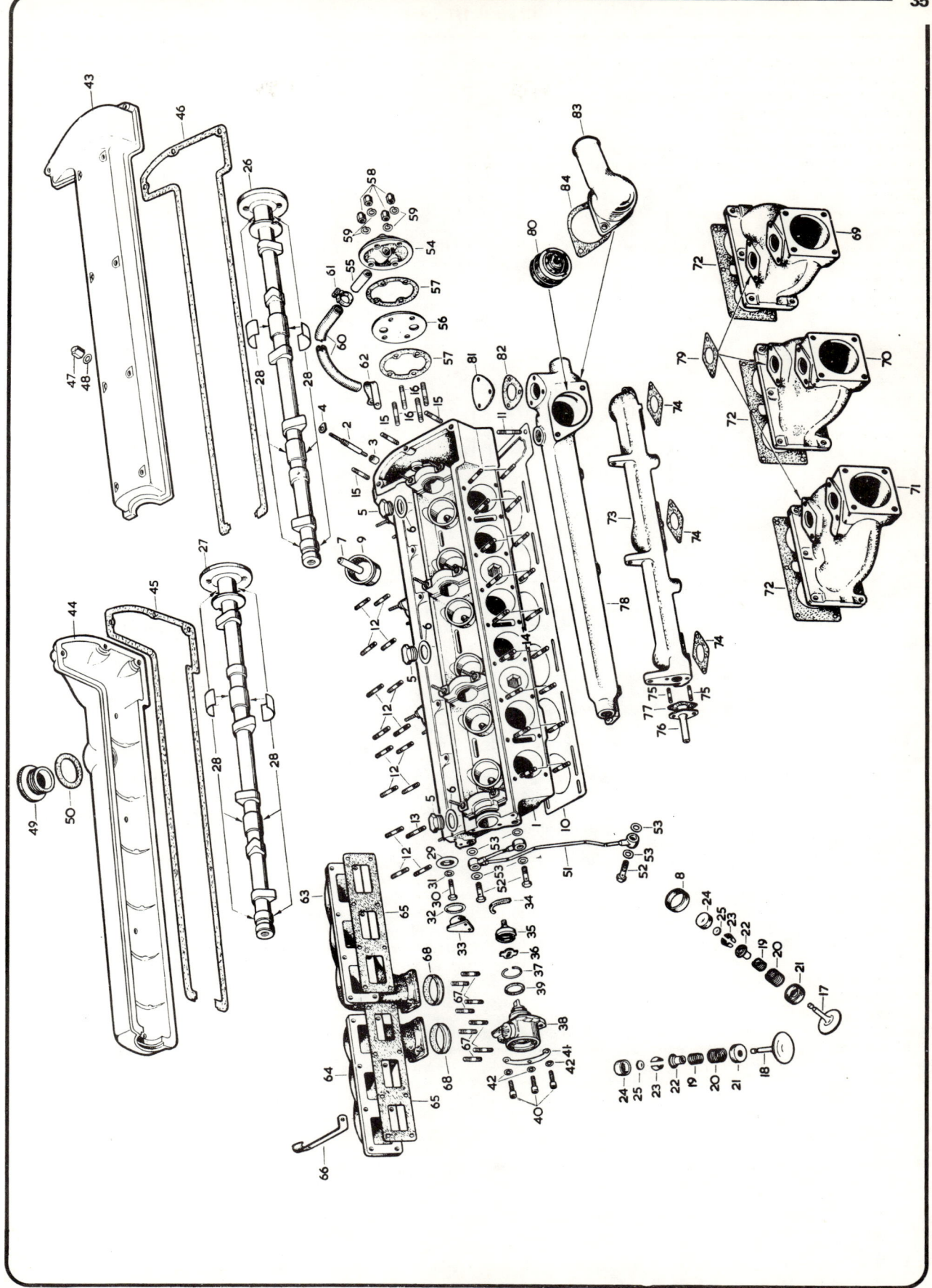

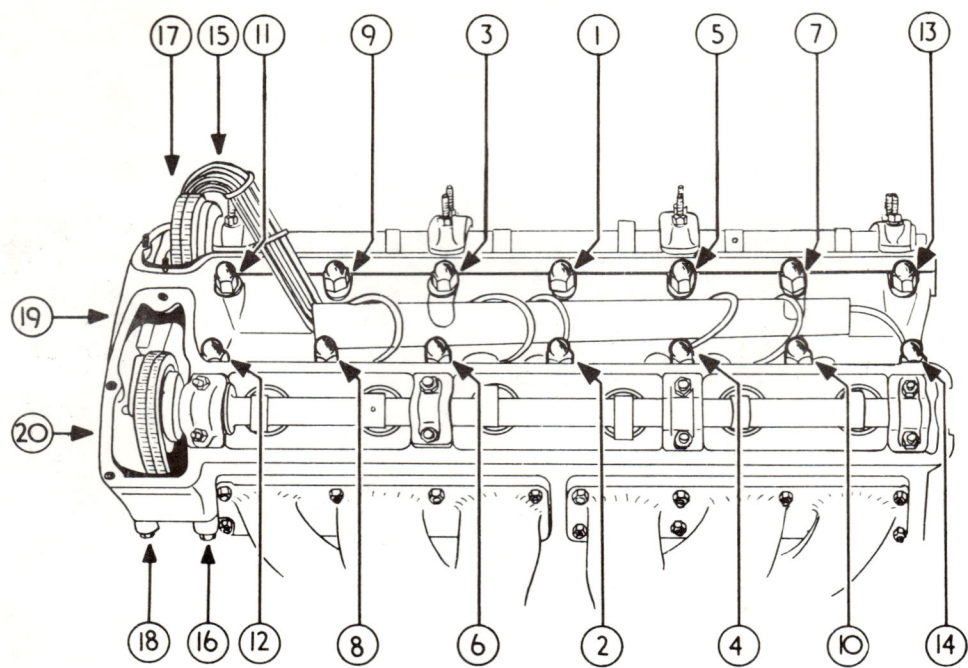

Fig.1.7. Loosening and tightening sequence of cylinder head nuts

11 Cylinder head removal - engine in the car

1 Drain coolant from the system in the manner described in Chapter 2.
2 Disconnect the battery.
3 Follow the instructions given in Chapter 12 for removal of the bonnet.
4 Identify and then disconnect the leads to the revolution counter generator at the rear of the inlet camshaft.
5 Remove the butterfly nuts securing the air intake box and then remove the box.
6 Slacken the clips holding the two heater hoses running from the engine to the connections on the bulkhead and remove the hoses.
7 Slacken the clip securing the brake servo pipe to the union on the inlet manifold and disconnect the pipe.
8 Disconnect the throttle linkage at the rear carburettor by pulling outwards against the spring clip.
9 Disconnect the choke cable.
10 Disconnect the petrol feed pipe from below the centre carburettor using two spanners, one to hold the bottom union nut and the other to unscrew the top nut.
11 Unscrew the nut securing the electric lead to the water temperature gauge element to the water rail on the right-hand side of the engine and remove the lead from the terminal.
12 Slacken the clips and remove the left and right-hand water hoses running from the header tank to the top of the radiator.
13 Slacken the clips and remove the water hose running from the header tank to the thermostat outlet.
14 Identify and disconnect the leads from the thermostat in the front of the header tank.
15 Remove the two nuts securing the header tank to the sub-frame, collect the flat washers and the rubber mounting washers.
16 Remove the setscrews attaching the header tank to the radiator, unclip the overflow pipe and then lift off the header tank.
17 Disconnect the HT lead at the ignition coil. Identify the other two leads and then remove them from the coil by pulling them off Lucar tags.
18 Remove the nuts from the coil mounting bracket and lift off the coil.
19 Remove the setscrew holding the plug harness clip at the front of the engine. Identify the plug leads and then disconnect them from the plugs.
20 Remove the four nuts holding each exhaust pipe flange to the exhaust manifold (remember when re-assembling that these are special nuts) and then disengage the exhaust pipes from the manifold.
21 Remove the two rearmost nuts holding the exhaust manifold and disengage the gearbox breather pipe (and the automatic transmission dipstick tube if applicable).
22 Undo the banjo unions to disconnect the oil supply pipe to each camshaft. Collect the fibre washers under the unions.
23 Now follow the instructions given in Section 10 to remove the cylinder head.

12 Camshaft removal

1 The camshafts can be removed with the cylinder head in-situ provided you follow instructions given in Sections 10 and 11 for disconnection of the lead to the revolution counter generator; the removal of the camshaft covers and the disconnection of the sprockets.
2 Unscrew the three Allen setscrews attaching the revolution counter generator to the right hand (inlet) camshaft (photo) and the three setscrews attaching the sealing plug on the rear of the exhaust camshaft. Note the copper washers under the heads of the setscrews. Remove the circular rubber sealing rings.
3 Release the nuts securing the bearing caps a part turn at a time. Remove the nuts and 'D' washers from the bearing studs.
4 Remove the bearing caps noting that the caps and the cylinder head are marked with corresponding numbers as illustrated in Fig. 1.8 Also note that the bearing caps are located to the lower bearing housings by hollow dowels.
5 Lift out the camshaft.
6 Remove the camshaft bearing shells from the cylinder head and keep them with their counterparts for correct reassembly in their original positions if you are not fitting new items.
Important do not rotate either a camshaft if the other shaft is still fitted as the valves will foul each other.

Chapter 1/Engine

12.2 Removing the revolution counter generator

3 Remove the tappets (Fig. 1.10). A valve grinding suction tool applied to the head of the tappet makes this an easier task.
4 Remove the valve adjusting pad (photo) if it is has not adhered to the tappet. Keep it and the tappet together.
5 Place the wooden valve support block beneath the cylinder head to support the valves.
6 Compress the valve spring and have an assistant remove the cotters as they are released. We have found that an easy way approximately ¾" internal diameter and 9" in length, over the end of the spring, strike the tube a sharp blow with a hammer (photo) and the cotters will fall out of their own accord.
7 Remove the valve collar (photo) springs (photo), valve and the valve spring seat and keep them as a set for replacement in the position from which they were removed.
8 Repeat the above for the removal of all the valves.

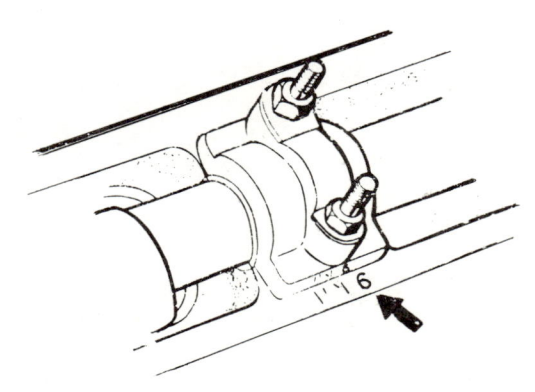

Fig.1.8. Showing corresponding numbers on the bearing cap and the cylinder head

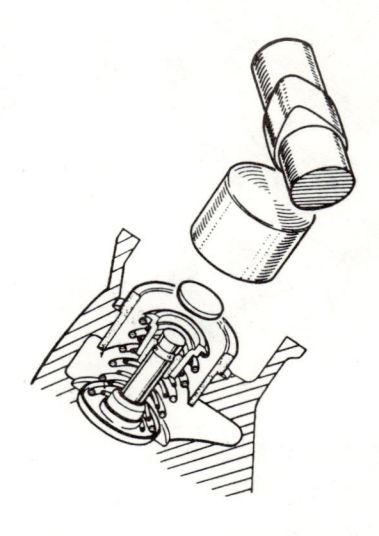

Fig.1.10. Valve tappet and adjusting pad

13 Valve removal

1 Make some arrangement to keep the valves, springs, tappets and adjusting pads related to each other and to their position in the cylinder head. A board (as illustrated in photograph 13.1) is ideal for this purpose.
2 Make up a wooden block (as illustrated in Fig. 1.9) to support the valves when compressing the springs.

14 Valve guide and tappet guide - removal

The valve and tappet guides are shrunk into the cylinder head and although their removal is a fairly simple task it is recommended that you do not tackle it - as their replacement is too difficult to do accurately. It is far better to leave this task to a Jaguar garage which will be fully equipped to work within the accurate limits required.

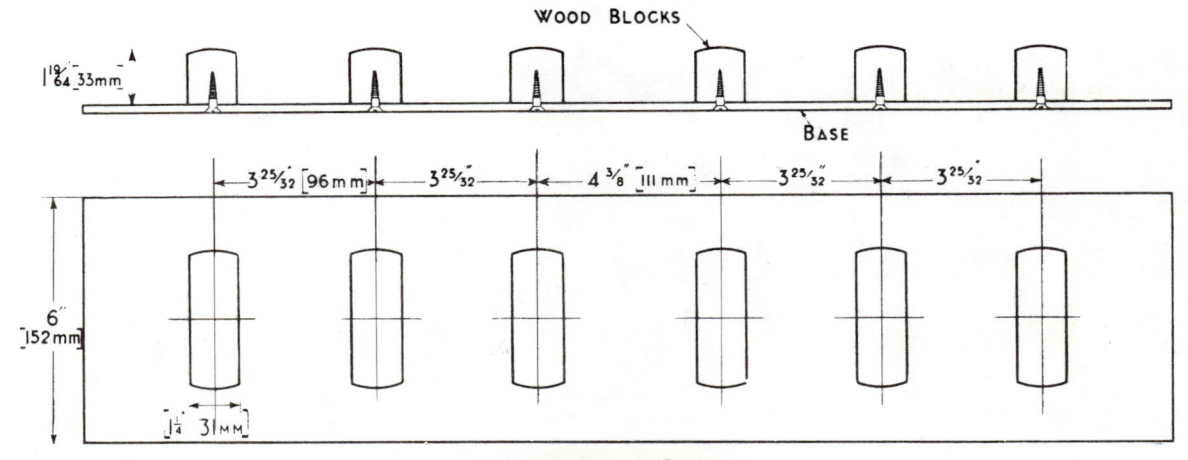

Fig.1.9. Valve support block

13.1 Suggested valve holder

13.4 Remove valve adjusting pad

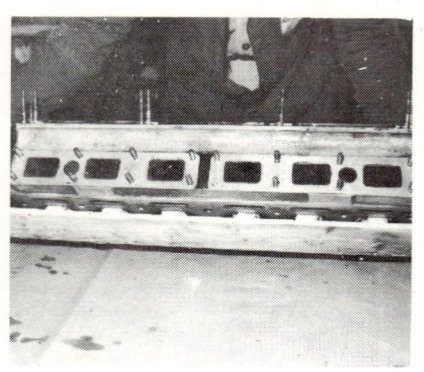

13.5 Supporting cylinder head on wooden block

13.6 Removing the valve cotters

13.7a The valve collar

13.7b The valve springs

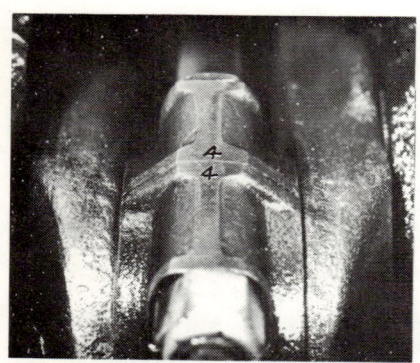

15.6a The connecting rod and big end bearing cap

15.6b The connecting rod and big end bearing cap separated

15.8 Removing the piston and connecting rod

16.3 The bottom timing chain tensioner

16.4 The serrated adjuster plate and locking plunger

16.8 The timing gear sprocket on crankshaft

Chapter 1/Engine

15 Sump, piston, connecting rod and big end bearing - removal

The sump, pistons and connecting rods can be removed with the engine in the car. The pistons and connecting rods are drawn up out of the top of the cylinder bore so the cylinder head must be removed and for this task follow the instructions given in Sections 10 or 11 as applicable.
1 If the engine is in-situ, the first job is to remove the sump drain plug and allow the oil to drain into a container.
2 Refer to Section 5 and follow the instructions given in paragraphs 45 - 47 for removal of the crankshaft damper.
3 In both cases: engine installed; or on the bench, slacken the clips to the oil return hose running from the sump to the head of the oil filter.
4 If the engine is on the bench, lay it on its side and in both cases remove the twenty-six setscrews and the fournuts securing the sump to the crankcase. Note that a short setscrew is fitted at the right hand front corner of the sump.
5 Remove the sump. If the engine is installed you may find that the engine stabiliser will have to be slackened to allow the engine to be raised to give clearance for removing the sump.
6 Remove the nuts (may be split pinned or self locking) to the big end bearing caps (photo) and lift off the cap and shell bearing (photo). Note that the big end bearing cap and the connecting rod are marked with the number of the cylinder to which they belong and that they are assembled so that the numbers on the rod and cap are together, if they are not so makred, or if there is any doubt, stamp them or mark them in some way for correct reassembly.
7 If the bearings ar not to be changed, keep them with their respective caps and connecting rods.
8 Push the connecting rod, and the piston, up the bore. Withdraw them from the top of the cylinder (photo) and lay them out in their correct order ofr replacement in the same bore. Refit the caps and bearings to the connecting rods and replace the nuts finger tight to keep them in position. The piston and connecting rod assembly is illustrated in Fig. 1.11.

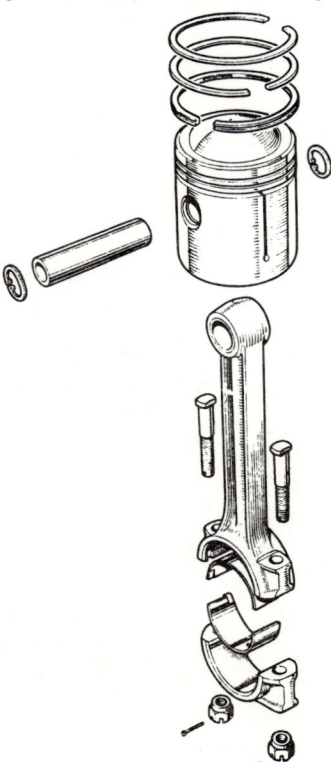

Fig.1.11. The piston and connecting rod assembly

16 Timing gear - removal

1 If the engine is installed, the following items must be removed to give access to the timing gear. Instructions for the work have already been given in previous Sections.
    Bonnet   Crankshaft damper
    Radiator and header Sump
    tank
    Water pump
2 Now remove the timing cover by undoing the set bolts. Note that the cover is located by two dowels, check that these are a tight fit in the cylinder block and if they are loose, remove them to prevent loss.
3 Remove the bottom timing chain tensioner (photo) by knocking up the tab washer and undoing the two mounting bolts. Withdraw the bolts and take off the tensioner together with the backing plate and shim. Note the conical gauge filter fitted in the tensioner oil feed hole in the cylinder block, this should be removed for subsequent cleaning.
4 Withdraw the bolt holding the serrated adjuster plate. Remove the plate and the spring loaded plunger (photo).
5 Unscrew the four set bolts securing the front mounting bracket of the timing gear to the cylinder block. Remove the two screwdriver slotted setscrews securing the rear mounting bracket, these setscrews also secure the intermediate damper bracket.
6 Lift out the right and left hand upper chain damper assembly and distance pieces and the vibration damper for the lower chain which have now been released by removal of the mounting bolts.
7 Lift the timing gear assembly away from the cylinder block.
8 Remove the timing gear sprocket from the crankshaft. Note the Woodruffe key (photo).

17 Gudgeon pin - removal

1 The fully floating gudgeon pins are a finger push fit in the piston at normal room temperature and are retained by a circlip at each end.
2 Remove the circlips using a pair of circlip pliers and discard them.
3 Apply finger pressure to the gudgeon pin, if it does not move try from the other end.
4 If the gudgeon pin cannot be moved, immerse the piston in a bath of hot oil and after a few minutes you will find that the pin will move quite easily.
5 Push out the pin far enough to clear the small end of the connecting rod. Separate the piston and the connecting rod. We advise that you do not push the pin right out of the piston unless this is absolutely necessary, if you do take it out of the piston make sure that it is replaced in its original position.
6 The pistons should be marked with the number of the cylinder to which they belong. If they are not marked, ensure that they are correctly identified because it is most important that they are reassembled to the bore from which they were removed.

18 Piston ring - removal

1 To remove the piston rings, slide them over the top of the piston taking care not to scratch the surface of the piston and not to distort the rings. Never slide them off the bottom of the piston skirt. Piston rings are very brittle and are easily broken if they are pulled off roughly.
2 It is helpful to use an old feeler gauge to facilitate the removal of the rings. Lift one end of the piston ring to be removed out of its groove and insert the end of the feeler gauge under it. Turn the feeler gauge slowly round the piston andas the ring comes out of its groove it will rest on the land above. It can then be eased off the piston with the feeler gauge stopping it

Chapter 1/Engine

from entering an empty groove if it is any but the top ring that is being removed.

19 Oil pump assembly - removal

1 Remove the sump.
2 Remove the nut and bolt securing the oil pump inlet pipe clip to the bracket on the main bearing cap. (photo)
3 Tap back the tab washers and unscrew the two set bolts securing the oil feed pipe flange to the bottom of the crankcase.
4 Open the tab washers to the three bolts securing the oil pump to the front main bearing cap, remove the bolts.
5 The oil pump can now be withdrawn (photo).

20 Distributor drive - removal

1 Tap back the tab washer securing the distributor drive gear nut and remove the nut and washer.
2 Tap the squared end of the distributor drive shaft through the gear. Note that the gear is keyed to the shaft.
3 Remove the gear and thrust washer (photo) and withdraw the shaft.
4 Remove the distributor/oil pump helical drive gear (photo). Remove the key locking it to the shaft.

21 Crankshaft - removal

1 Knock back the tab washers securing the fourteen main bearing cap bolts.
2 Note the corresponding numbers stamped on the caps and on the bottom face of the crankcase. The caps must be correctly identified if they are not marked.
3 Undo the bolts and remove the main bearing caps. If a main bearing shell does not come away with its cap, remove it from the crankshaft and keep it with its cap.
4 Note the thrust washers fitted in the recess at each side of the centre main bearing cap. (Fig. 1.12)
5 Detach the bottom of the oil return thread cover by removing the two Allen securing screws (photo). Note that the two halves are located by hollow dowels.
6 The crankshaft can now be lifted away from the crankcase.
7 Collect the remaining halves of the main bearing shells and, if they are to be refitted, identify them with the position from which they were removed.

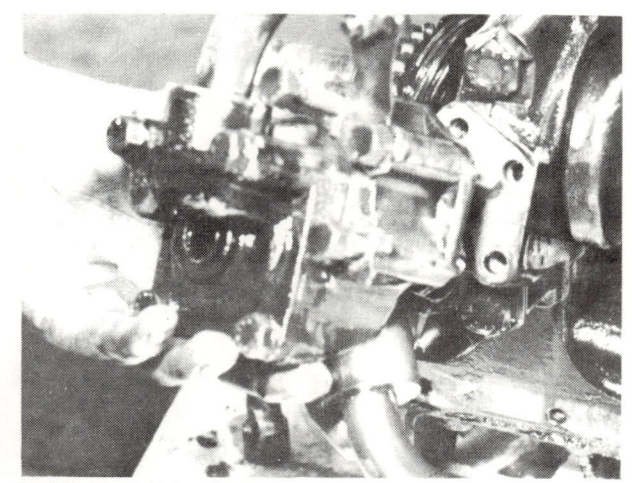

19.2 Clip securing oil pump inlet pipe

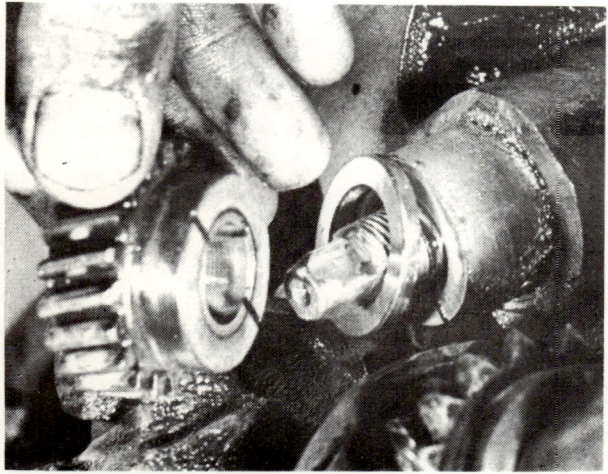

19.5 Removing the oil pump

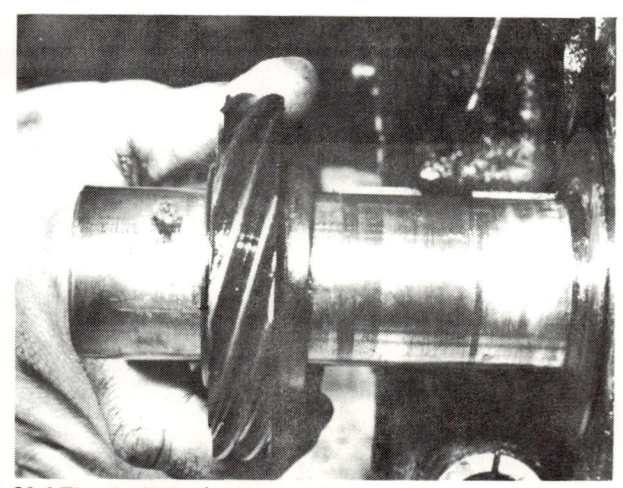

20.3 The distributor drive gear and thrust washer

20.4 The distributor/oil pump helical drive gear

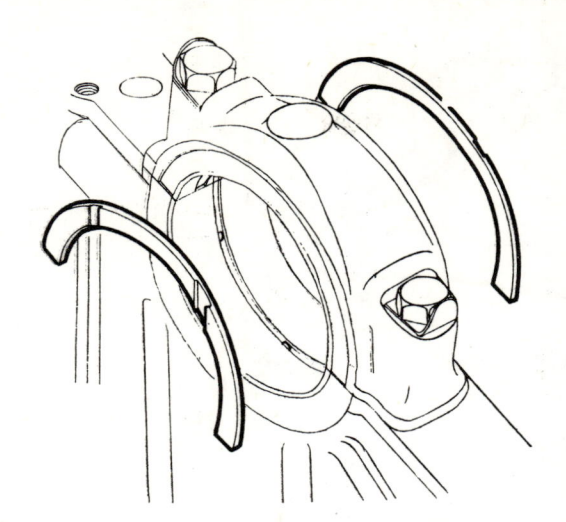

Chapter 1/Engine

connecting rod throws a small jet of oil to the cylinder wall with each revolution. The oil filter (Fig.1.15) is of the full flow type with a replaceable element. The filter head assembly incorporates a removable oil pressure relief valve and a balance valve which provides a safeguard against the possibility of the filter element becoming so choked as to prevent oil reaching the bearings. Oil which passes the oil pressure relief valve is returned to the engine sump by an external rubber hose.

23 Oil filter - removal and replacement

1 The external oil filter which is of the disposable cartridge type and which is located at the right hand rear side of the engine as illustrated in Fig. 1.1. It is most important to renew the filter element at the recommended periods (5000 mile servicing) to avoid the possibility of it becoming choked with impurities.
2 Place the car over a pit or raise the car to give access to the filter, Make sure that the car is properly supported before crawling underneath it.
3 Take out the self tapping screws from the fairing below the filter and remove the fairing.
4 Place a container below the filter to catch any oil which may escape when it is removed.
5 Remove the central bolt securing the canister to the head of the filter and remove the canister complete with the filter element.
6 Remove and discard the old element and allow the oil in the canister to drain into the container.
7 Remove the rubber sealing ring from the head of the filter (photo).
8 Withdraw the canister retaining bolt and note the order of assembly of the spring clip, the pressure plate (and which way up it faces), felt washer, plain washer and spring.
9 Thoroughly clean all parts, especially the interior of the canister where it will be found that sediment has collected in the base and this will have to be removed using paraffin and a brush.
10 Reassemble the filter in the reverse order to the above but fitting a new element, rubber sealing ring in the head and a new felt and rubber washer on the canister securing bolt. You will find that these items are supplied with the replacement element.
11 Check after a short run, when the oil is hot, that there are no leaks. The usual source of trouble is poor seating between the canister and the sealing rubber in the head.

Fig.1.12. The crankshaft thrust washers

21.5 Removing the bottom oil return thread cover

22 Lubrication system - description

A force feed system of lubrication is employed with oil being circulated round the engine from the sump below the cylinder block. The level of the oil in the sump is indicated by the dipstick located on the left hand rear side of the engine, High and Low level of oil is indicated by marks on the dipstick and ideally the level of oil should not be above the high mark and should never be allowed to fall below the low mark. Oil is replenished via the filler cap in the left hand (exhaust) camshaft cover. The oil is circulated round the engine by an eccentric rotor type oil pump which consists of five main parts; the body, the driving spindle with the inner rotor pinned to it, the outer rotor and the cover which is secured to the main body by four bolts. The pump is illustrated in Fig. 1.14.

Oil is drawn from the sump and is then passed under pressure by the pump to the filter on the right hand exterior of the crankcase and thence through drillings to the big end and main bearings; the camshaft bearings are fed via an external oil pipe. A longitudinal drilling through the connecting rod feeds the small end and the gudgeon pin with oil and a small 'hold' in each

23.7 The rubber sealing ring in the head of the oil filter

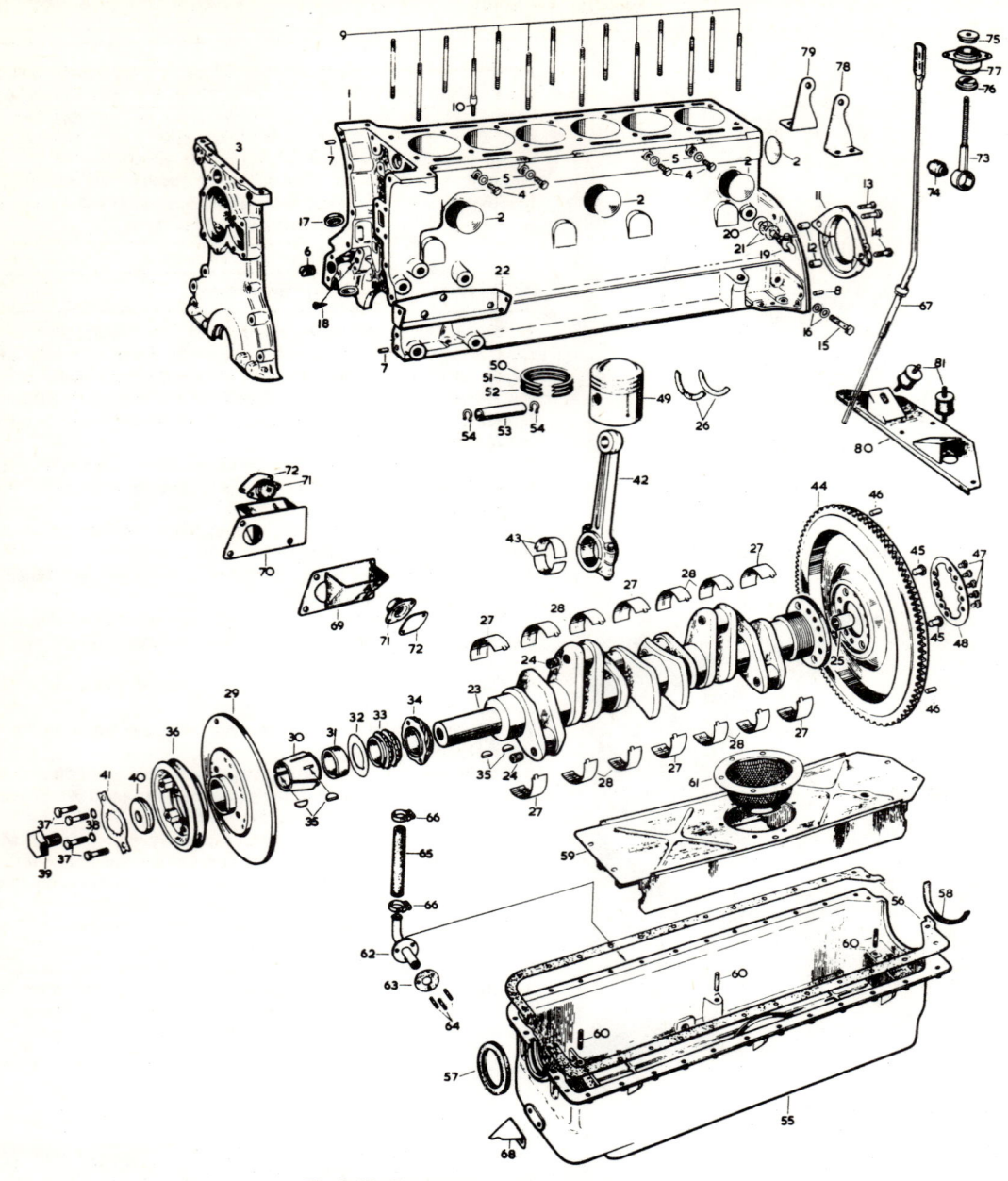

Fig.1.13. Exploded view of the cylinder block assembly

1	Cylinder block	22	Mounting bracket	42	Connecting rod	62	Adaptor
2	Core plug	23	Crankshaft	43	Bearings	63	Gasket
3	Timing cover	24	Plug	44	Flywheel	64	Stud
4	Setscrew	25	Bush	45	Dowel	65	Hose
5	Copper washer	26	Thrust washer	46	Dowel	66	Clip
6	Plug	27	Main bearing	47	Setscrew	67	Dipstick
7	Dowel	28	Main bearing	48	Locking plate	68	Pointer
8	Dowel	29	Crankshaft damper	49	Piston	69	Bracket
9	Stud	30	Cone	50	Compression ring	70	Bracket
10	Dowel stud	31	Distance piece	51	Compression ring	71	Engine mounting
11	Cover	32	Oil thrower	52	Scraper ring	72	Plate
12	Ring dowel	33	Sprocket	53	Gudgeon pin	73	Link
13	Setscrew	34	Gear	54	Circlip	74	Bush
14	Bolt	35	Key	55	Oil sump	75	Stepped washer
15	Banjo bolt	36	Pulley	56	Gasket	76	Stepped washer
16	Copper washer	37	Bolt	57	Seal	77	Rubber mounting
17	Sealing ring	38	Locking washer	58	Cork seal	78	Bracket
18	Gauze filter	39	Bolt	59	Baffle plate	79	Bracket
19	Drain tap	40	Washer	60	Stud	80	Support bracket
20	Copper washer	41	Tab washer	61	Filter basket	81	Rubber mounting
21	Fibre washer						

Chapter 1/Engine

24 Engine - examination and renovation - general

With the engine stripped and all parts thoroughly cleaned, every component should be examined for wear. The following items should be checked, and where necessary, renewed or renovated as described later.

25 Crankshaft - examination and renovation

1 Examine the crankpin and main journal surfaces for signs of scoring or scratches; a typical example of unserviceability due to this cause is given in photograph 25.1 If the bearings have failed it is possible that the journals will have white metal adhereing to them, this can be removed by very light rubbing with fine crocus paper.

2 If the journals are scored or scratched it is a waste of time and money to fit new bearings as these will fail rapidly. The crankshaft should be reground or a factory reconditioned item fitted. This can be obtained on an exchange basis provided the old shaft is fit for reconditioning.

3 Clean the journals and crankpins and measure their diameter at different positions with a micrometer. Regrinding is generally

Fig.1.14. Exploded view of the oil pump

1 Body
2 Rotor assembly
3 Cover
4 Setscrew
5 Setscrew
6 Washer
7 'O' ring
8 Drive shaft
9 Bush
10 Washer
11 Helical gear
12 Key
13 Nut
14 Tab washer
15 Coupling
16 Dowel bolt
17 Tab washer
18 Oil delivery pipe
19 Gasket
20 Oil suction pipe
21 Clip
22 Strut
23 Strut
24 Plate
25 Spring
26 Split pin

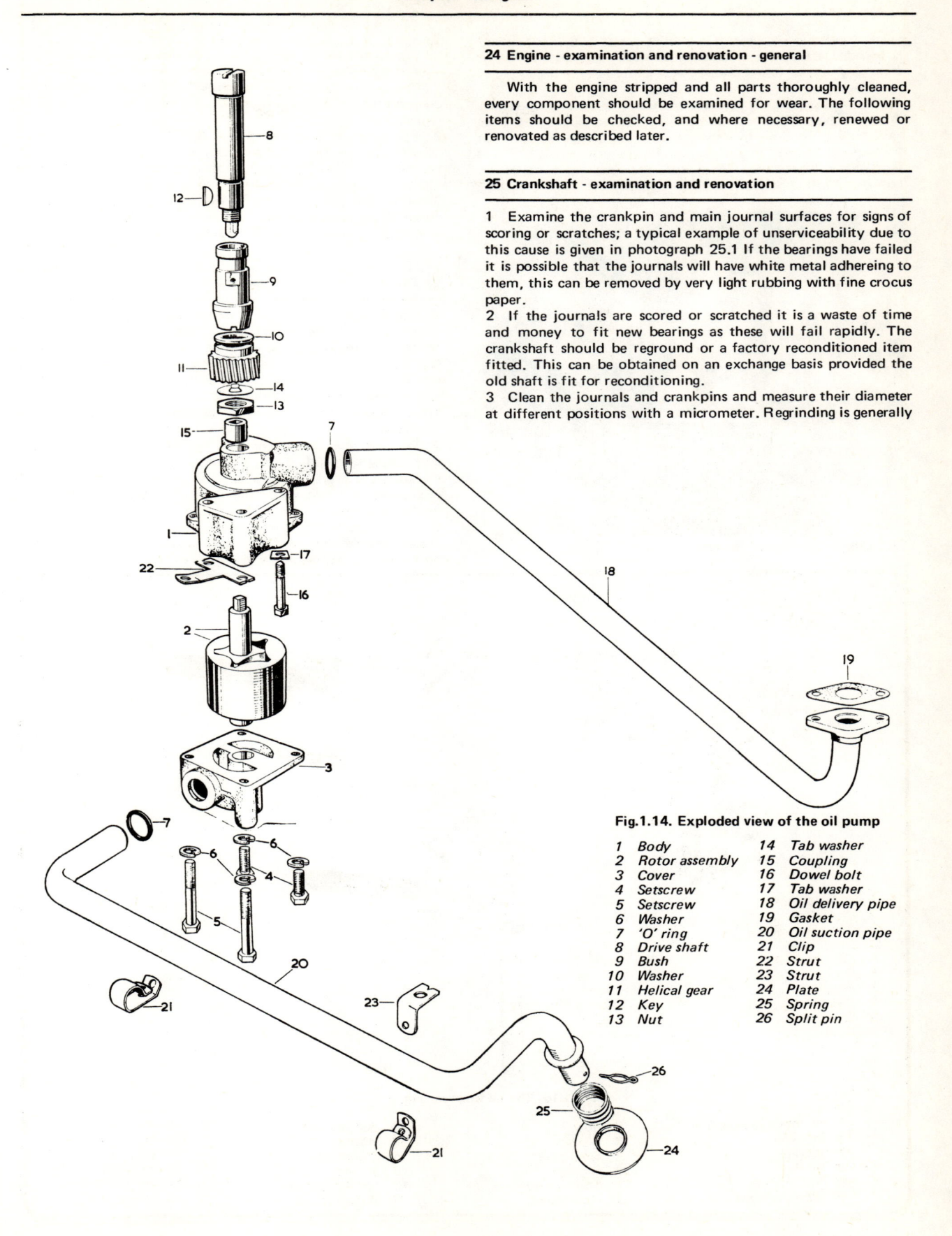

recommended when wear or ovality in excess 0.003" (0.08 mm) is found. Details of the standard diameter of the journals and crankpins will be found under Specifications at the beginning of this Chapter.
4 Ensure that the oil passages are clear. If the original crankshaft is to be refitted, remove the Allen headed plugs in the webs and thoroughly clean out any accumulated sludge using a high pressure jet followed by blowing out by compressed air.
5 Replace the plugs and fix in position by staking with a centre punch or blunt chisel.

25.1 Crankshaft journals scored

26 Crankshaft pulley and damper - examination and renovation

1 The rubber portion of the damper should be examined for deterioration. If the rubber appears to be perished it should be replaced.
2 The drive on the pulley should be taken on sides of the 'V'. Check that the belt does not bottom in the 'V'. If it does, recheck with a new belt and if this does not bottom discard the old belt. If the new belt bottoms, renew the pulley.
3 Note that the damper and the pulley are balanced as an assembly at production and if they are to be separated they should be marked for correct reassembly.

27 Big end and main bearings - examination

1 Big end bearing failure is usually accompanied by a noisy knocking from the engine and a drop in oil pressure. Main bearing failure gives rise to vibration which can be quite severe as the engine speed increases and reduces, a drop in oil pressure will also be noticed. However, if engine vibration is experienced, do not immediately jump to the conclusion that the main bearings failed because there are quite a number of other factors which can cause this.
2 We feel that having gone to the trouble of removing and completely dismantling the engine, your aim will be to restore the engine as nearly as possible to an 'as new' condtiion. To this end we suggest that you fit new main and big end bearing as a matter of course.
3 If it is decided to re-use the old bearings; proceed as follows: Inspect the bearings, and the thrust washers, for signs of general wear, scoring, pitting and scratches. The bearings should be matt

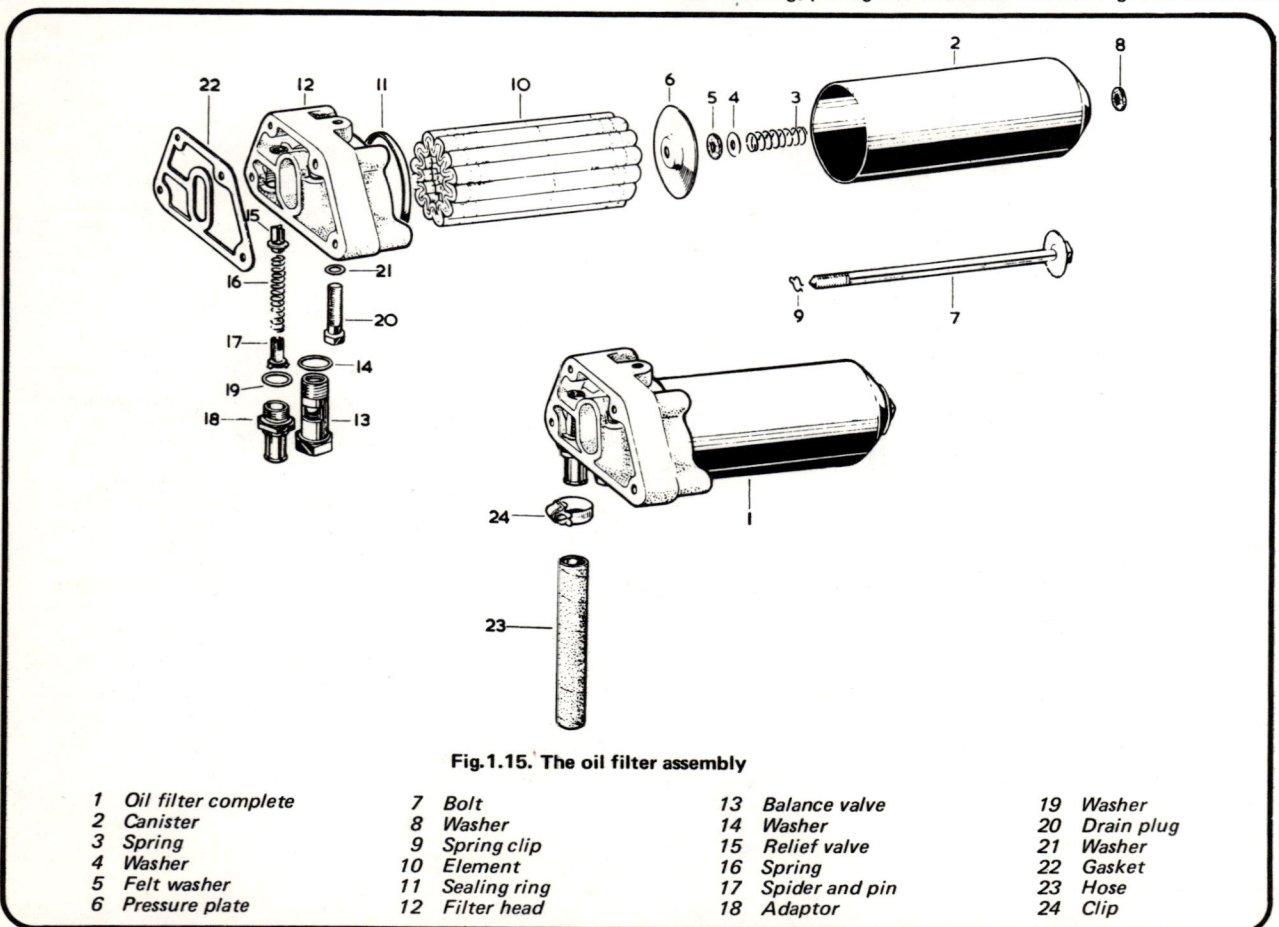

Fig.1.15. The oil filter assembly

1 Oil filter complete	7 Bolt	13 Balance valve	19 Washer
2 Canister	8 Washer	14 Washer	20 Drain plug
3 Spring	9 Spring clip	15 Relief valve	21 Washer
4 Washer	10 Element	16 Spring	22 Gasket
5 Felt washer	11 Sealing ring	17 Spider and pin	23 Hose
6 Pressure plate	12 Filter head	18 Adaptor	24 Clip

Chapter 1/Engine

grey in colour; should any trace of copper or bronze be seen it means that the bearings are badly worn, for the lead bearing has worn away to expose the underlay. Renew the bearings if they are in this condition or if there is any sign of pitting or scoring. Typical examples of bearing failure and failure of the thrust washers are shown in photographs 27.3a and 27.3b.

4 The undersizes available are designed to correspond with regrind sizes in steps of -0.010". The bearings are in fact slightly larger than the stated undersize as running clearances have been allowed for during their manufacture.

5 Bearing shells must be changed in pairs. It is no use fitting a new half bearing to one that has been in use.

6 Very long engine life can sometimes be obtained by changing big end bearings at 30,000 and main bearings at 50,000 miles respectively irrespective of the visual condition of the bearings. Normally, crankshaft wear is infinitesimal and regular changes of bearings may ensure mileages of between 100,000 and 120,000 miles before regrinding becomes necessary. Crankshaft wear and scoring is usually the result of bearing failure.

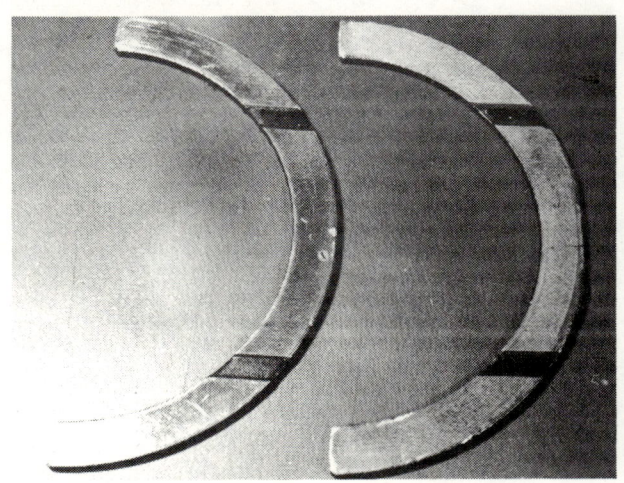

27.3a Worn big end bearings

27.3b Worn thrust washers

28 Flywheel - examination and renovation

1 Examine the starter teeth for wear or damage which, if found, will necessitate replacement of the flywheel.

2 Look at the front face of the flywheel and check for radial scoring and for signs of oil having been burnt onto the surface. A typical example of both fauls is shown in photograph 28.2.

3 If the front face is scored, not only will rapid wear of the clutch driven plate result but other clutch troubles can be anticipated. In this case, the flywheel should be replaced.

4 Burnt oil can be removed with paraffin and rubbing with fine emery cloth. Make sure that all traces of paraffin is removed before reassembly.

5 The flywheel and clutch are balanced as an assembly and should be marked accordingly as shown in Fig. 1.4. If a new flywheel is brought into use the items should be balanced as follows.

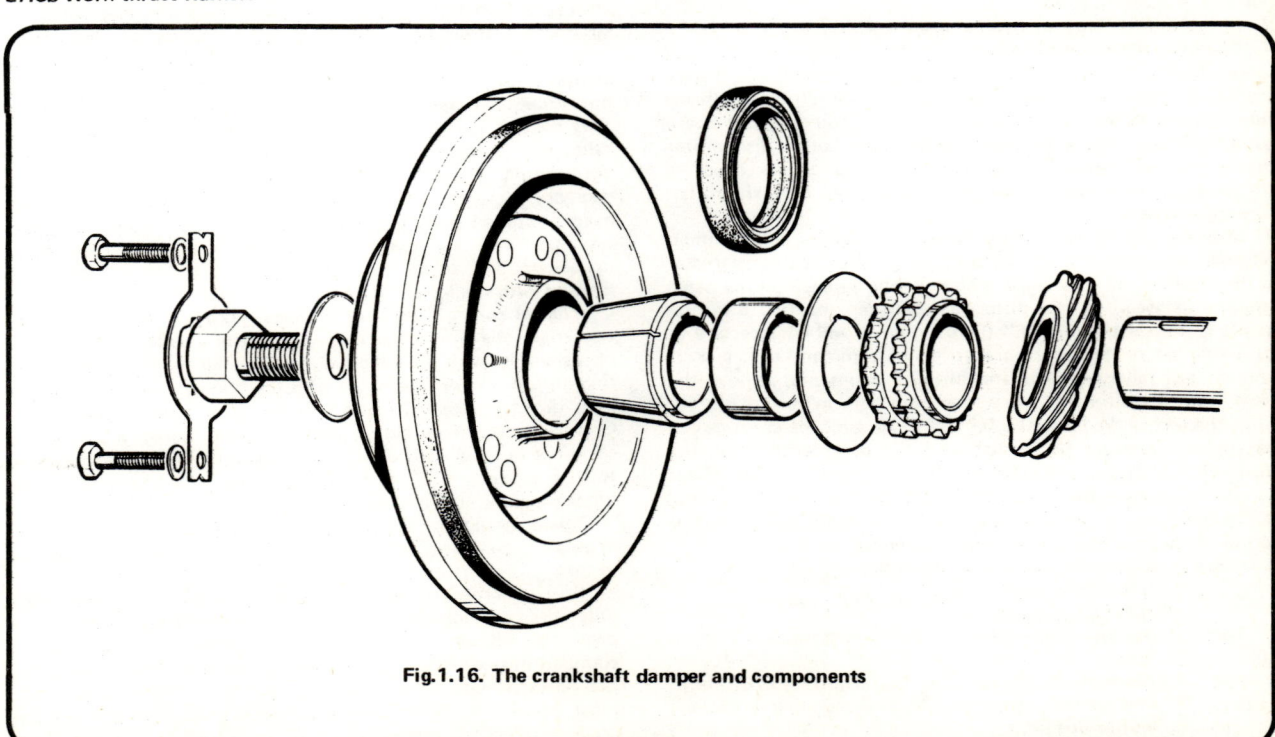

Fig.1.16. The crankshaft damper and components

6 Mount the assembled flywheel and clutch on a mandrel and set them up on parallel knife edges. If they are in balance, mark the relative position of the clutch and the flywheel.
7 If necessary to obtain balance, remove the clutch and drill 0.375" (9.5 mm) holes not more than ½" (12.7 mm) deep at a distance of 0.375" (9.5 mm) from the edge of the flywheel. However it will be appreciated that this is a task best left to a Jaguar garage having the necessary equipment and experience.

7 If the bores are slightly worn but not so badly as to justify reboring them, special oil control rings can be fitted which will restore compression and stop the engine burning oil. Several different types are available and the manufacturers instructions concerning their fitment must be followed closely. However, fitting special rings is a comparatively short term remedy; if the engine is out of the car and is completely stripped it seems false economy not to return the engine to an 'as new' condition by reboring.

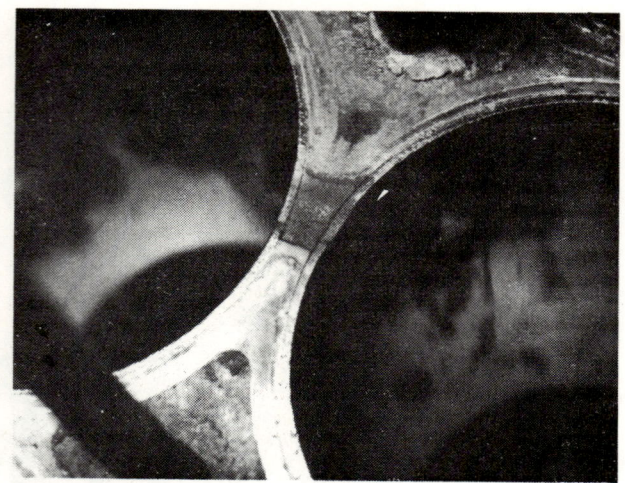

28.2 Flywheel scored and oil fouled

29.1 Cracked and burnt cylinder block

29 Cylinder block - examination and renovation

1 Thoroughly clean the top face of the cylinder block and examine it for damage. Pay particular attention to the condition of the face between the webs of the cylinder bores, look for burning or cracks (photo).
2 Check the top face for truth using a straightedge.
3 Examine the cylinder bores for taper, ovality, scratches and scores. Start by carefully examining the tops of the bores, if they are worn fractionally a very slight ridge will be felt on the thrust side, this marks the top of the piston travel. You will have a very good indication of the condition of the bores before dismantling the engine or removing the cylinder head as excessive oil consumption accompanied by blue smoke from the exhaust is a sure sign of wear.
4 Measure the diameter of the bore just under the ridge with an internal micrometer or vernier and compare it with the diameter at the bottom of the bore which is not subject to the same amount of wear. If the difference between the two measurements is greater than 0.006" (0.152 mm) it will be necessary to fit a ring set or to rebore and fit oversize pistons and rings. If you do not have precision measuring instruments, remove the rings from a piston and place it in each bore in turn about ¾" from the top of the bore. If a 0.010" (0.25 mm) feeler gauge can be entered between the piston and the cylinder wall on the thrust side of the bore then remedial action must be taken. Refer to specifications at the beginning of this Chapter for rebore sizes, the present size of the bore (i.e 0.010" + 0.020" etc) should be found marked on the cylinder block alongside each bore.
5 Oversize pistons are available in the following sizes:-
   +0.010" (0.25 mm)   +0.030" (0.76 mm)
   +0.020" (0.51 mm)

There are no selective grades in oversize pistons (see Section 30).
6 The maximum limit for reboring is +0.030" (0.762 mm). Liners and standard size pistons should be fitted when bores will not clean up within this limit.

30 Pistons and piston rings - general

The pistons are made from low expansion aluminium alloy and, in the case of the 3.8 and early 4.2 litre engines are of the semi-split skirt type. Later engines are fitted with Hepworth and Grandage pistons having a solid skirt. The pistons have three rings each, two compression and one oil control. The top compression ring is chromium plated and is Cargraph treated on the outer diameter to assist 'bedding-in' the chromium surface. This coating is red in colour and **should not be removed**; the rings may be wiped down with a clean paraffin dampened rag but should not, in any circumstances, be immersed in a degreasing agent. Both the top and the second compression rings have a tapered periphery and these must be fitted the correct way up. The narrowest part of the ring should be fitted uppermost and, to assist in identification, this part of the ring is marked with the letter 'T' or 'TOP' (Fig. 1.17).

The oil control ring is of the Maxiflex type which consists of two steel rails with a spacer between, the rails are held together on assembly with an adhesive. An expander is fitted inside the ring and should be assembled with the two lugs positioned in the hole directly above the gudgeon pin bore. The Hepworth and Grandage pistons employ a ring of similar construction to the Maxiflex but the ends of the expander ring are butted together. If this expander ring is fitted with the ends overlapping, the outer ring assembly will not seat properly.

There should not be a variation in weight between the pistons of more than 3.5 grammes; replacements are, therefore, supplied in sets.

Five selective grades of piston are available in standard sizes only, they are identified by a letter stamped on crown of the piston as shown in Fig.1.18 and a corresponding letter is stamped also on the top face of the cylinder block adjacent to the bores. If you have to order a set of standard pistons you must, therefore, quote the identification letter of the selective grade you require, these are:-

Chapter 1/Engine

Grade identification letter	To suit cylinder bore size
	3.8 litre
F	3.4248" to 3.4251" (86.990 to 86.997 mm)
G	3.4252" to 3.4255" (87.000 to 87.007 mm)
H	3.4256" to 3.4259" (87.010 to 87.017 mm)
J	3.4260" to 3.4263" (87.020 to 87.027 mm)
K	3.4264" to 3.4267" (87.030 to 87.037 mm)
	4.2 litre
F	3.6250" to 3.6253" (92.075 to 92.0826 mm)
G	3.6254" to 3.6257" (92.0852 to 92.0928 mm)
H	3.6258" to 3.6261" (92.0953 to 92.1029 mm)
J	3.6262" to 3.6265" (92.1055 to 92.1131 mm)
K	3.6266" to 3.6269" (92.1156 to 92.1123 mm)

There are no selective grades in oversize pistons as grading is purely for factory production methods. Oversize pistons are available in the following sizes:-

+0.010" (0.25 mm), +0.020" (0.51 mm), +0.030" (0.76 mm)

Fig 1.19 shows the difference in pistons to give either 8:1 or 9:1 compression ratio.

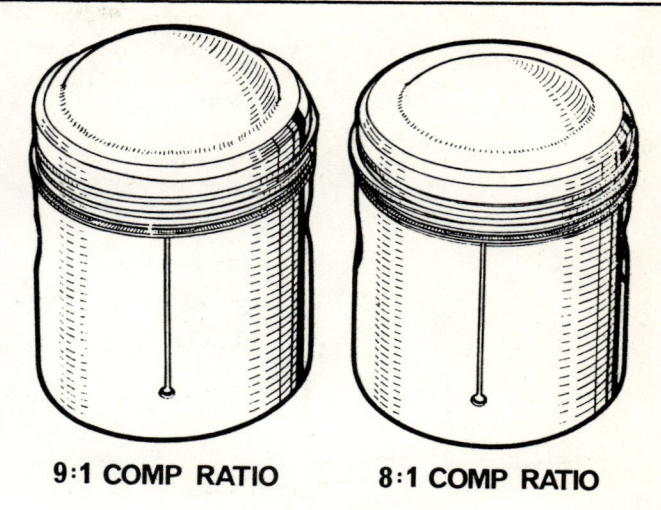

Fig.1.19. 'E' Type pistons

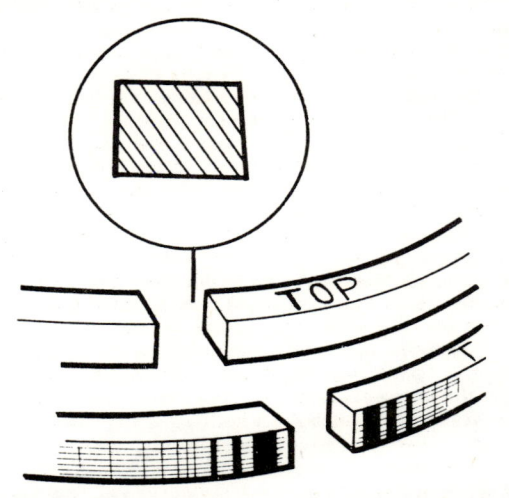

Fig.1.17. Identification marks on compression rings

Fig.1.18. Identification marks on the piston crown

31 Pistons, piston ring and gudgeon pin - examination and renovation

1 The method of removing the gudgeon pin and piston rings has already been described in Sections 17 and 18 respectively.
2 Clean carbon from the head of the piston using worn emery cloth and paraffin. Do not use a scraper or any tool that may score the head.
3 Do not use an abrasive to clean the outside of the piston despite the discolouration that may be present; a wipe with a cloth and paraffin will suffice.
4 Examine the lands for burrs as these may prevent freedom of movement of the ring, rectify as necessary using fine emery cloth.
5 Clean all dirt out of the grooves especially in the corners. A broken piston ring is a handy tool for this job but be careful not to dig in or remove metal.
6 Examine the skirt for fractures at the extremity of the split.
7 When a new piston ring is brought into use its gap, when sprung out in the cylinder bore, must be measured and adjusted as necessary. If the gap is too small seizure will result when the ring expands, if the gap is too great compression pressure will be lost.
8 Push the new ring down the bore as far as possible using a piston; this will ensure that the ring is square in the bore.
9 Refer to Fig. 1.20 and measure the gap using a feeler gauge.

The correct gaps are:

	3.8 litre	4.2 litre
Top compression	0.015" to 0.020" (0.38 to 0.51 mm)	0.015" to 0.020" (0.38 to 0.51 mm)
Lower compression	0.015" to 0.020" (0.38 to 0.51 mm)	0.010" to 0.015" (0.254 to 0.38 mm)
Oil control	0.011" to 0.016" (0.28 to 0.41 mm)	0.015" to 0.045" (0.38 to 1.143 mm)

10 With the rings fitted to the piston, check the side clearance in the grooves, this, for both engines, should be 0.001" to 0.003" (0.025 to 0.076 mm).
11 Apply pressure longitudinally to the gudgeon pin when it, and the connecting rod, are assembled to the piston. Watch for movement of the pin in the piston and if movement is noted the piston and pin must be replaced.
12 Two grades of gudgeon pin are supplied for the 4.2 litre engine, these are:-

Colour coding	Diameter	Clearance in piston
Red	0.8753" to 0.8754" (22.23 to 22.24 mm)	0.0001" to 0.0003" (0.0025 to 0.0076 mm)
Green	0.8752" to 0.8753" (22.22 to 22.23 mm)	As above

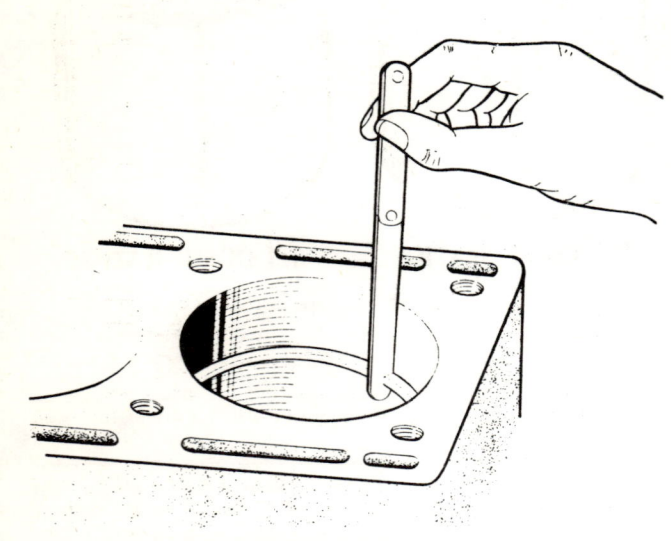

Fig.1.20. Checking the piston ring gap

32 Connecting rods - examination and renovation

1 If the connecting rods have been in use for a very high mileage or if bearing failure has been experienced, it is advisable to renew the affected rods owing to the possibility of fatigue failure.
2 When a new connecting rod is to be fitted, although the small end bush is reamed to the correct dimensions it may be necessary to hone the bush to obtain the correct gudgeon pin fit.
3 Check rock of the connecting rod on the gudgeon pin. If movement is observed and it can be established that this is due to wear at the small end, a new small end bush must be fitted.
4 The small end bush is a press fit in the connecting rod. Force it out using a vice or a press and fit a new item. Ream the new bush after fitting to the diameter quoted under specifications at the beginning of this Chapter or to the dimensions given in Section 30 paragraph 12 as the case may be.
5 Check that the big end caps have not been filed. If they have, there is no alternative but to replace the complete connecting rod.
6 The alignment should be checked and corrected as necessary on an approved connecting rod alignment jig. Arrangements should be made for this to be done at your local Jaguar garage.

33 Camshaft and camshaft bearing - examination and renovation

1 The camshafts and bearings normally give a very long life but there are always exceptions to the rule. Photograph 33.1 shows a camshaft which has failed due to a breakdown in the surface hardening of the cam lobe, it is estimated that the engine from which this shaft was removed had done well over 100,000 miles. This type of fault is not so common as general wear on the lobes, this point can be checked by making a comparison with a new shaft although a good estimate of their condition can be obtained by comparing one lobe with another and with those on the other camshaft.

2 Scoring on the bearing surfaces is a more likely fault to be found. It may be possible to remove slight score marks by gently rubbing down with fine emery cloth or an oilstone but this must not be overdone as undersize bearings are not supplied. Therefore, if the scoring cannot be rectified, or if wear on the lobes is found, the shaft should be scrapped.
3 Examine the shell bearings for scoring, pitting and general signs of wear. It is advisable to fit new bearings if there is any doubt as to their condition.
4 Remember that the camshafts are not interchangeable, the inlet shaft has a dog drive at the rear for the revolution counter generator whereas the exhaust shaft is plugged at this end.

34 Valves and seats - examination and renovation

1 Examine the heads of the valves for pitting and burning, especially the exhaust valves.
2 If the valves appear to be fit re-use after grinding to their seats in the cylinder head, scrape all the carbon away and carefully clean the stem of the valve. Clean the valve guide in the cylinder head and fit the valve to its guide.
3 With the valve about threequarters of its way in the guide, check it for sideways movement. If the movement appears to be excessive, remove the valve and measure the diameter of the stem, this should be not less than 0.309". If the stem diameter is satisfactory it means that the valve guide is worn and remedial action as indicated in Section 37 will have to be taken.
4 If no wear is present, check the valve stem for distortion by moving the valve up and down in its guide and at the same time rotating it, no restriction to movement should be felt.
5 Grinding the valves to their seats is easily carried out. First place the cylinder head upside down on the bench resting on a block of wood at each end to give clearance for the valve stems.
6 Smear a trace of coarse carburundum (photo) on the seat face and apply a suction grinding tool to the head of the valve. With a semi-rotary action, grind the valve to its seat (photo), lifting the valve occasionally to re-distribute the paste. When a continuous ring of dull matt even finish is produced on both the valve seat and the valve, then wipe off the coarse paste and repeat the process with a fine paste, lifting and turning the valve as before. A light spring placed under the head of the valve will assist in the lifting operation. When a smooth unbroken ring of light grey matt is produced on both valve and valve seat faces, the grinding operation is complete. Be very careful during the grinding operation not to get the abrasive paste on the stem of the valve, do not handle the valve stem once you have started to use the paste because it will be transferred from the fingers to the stem and the result will be rapid wear of the valve guide. Trouble is often experienced with the suction tool not gripping the valve head, this can be overcome if the valve head and the tool are kept free of oil and grease at all times.
7 When grinding is completed, thoroughly clean the cylinder head to remove any trace of carburundum as this can cause a lot of damage after first start up of the engine.

35 Valve springs - examination and test

After considerable mileage some deterioration in the valve springs, and consequent reduction in engine effeciency, must be expected. It is considered advisable, therefore, to test the springs when they are removed from the cylinder head to ensure that they are fit for use. We do not consider that measurement of the free length of a spring is a satisfactory method of deciding its suitability and so the following method of test is advocated:
1 Obtain a new inner and outer valve spring.
2 Place the new spring and the one to be tested between the jaws of a vice or under a press with a flat metal plate interposed between the two springs.
3 Apply a load to partly compress the two springs and then measure their lengths whilst under load. If the old spring is obviously shorter than the new one it is a sign that deterioration has set in.

Chapter 1/Engine

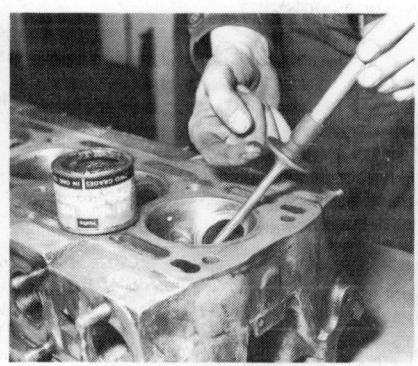

33.1 Unserviceable camshaft

34.6a Coat valve seat with carburundum paste

34.6b Valve grinding

36 Tappets and valve adjusting pads - examination and renovation

1 Examine the bearing surface of the tappet on which the camshaft bears. Any indentation on this surface or any cracks indicate serious wear and the tappet must be renewed.
2 It is unlikely that the sides of the tappet will be worn, but if the tappet can be rocked in its guide in the cylinder head it should be established by measurement which item is at fault; details of dimensions are given under Specifications at the beginning of this Chapter.
3 The tappet should also be checked to see that it moves freely in its guide, the most likely cause of restriction is dirt but rectify as necessary.
4 Clean and examine the valve adjusting pads. After considerable mileage it is most probable that the pads will be indented on the face which bears on the valve stem, if indentation is found the pad should be replaced. Adjusting pads are available rising in 0.001" (0.03 mm) increments from a thickness of 0.085" to 0.110" (2.16 to 2.79 mm) and are etched on the surface with the letter A to Z each letter indicating an increase in size of 0.001".

37 Cylinder head and bore - decarbonisation, examination and renovation

1 This operation can be carried out with the engine either in or out of the car.
2 With the cylinder head removed, carefully clean, with a wire brush and blunt plastic scraper, all traces of carbon deposits from the combustion spaces and ports. Wash the combustion spaces with paraffin and scrape the cylinder head surface free of any foreign matter. Take care not to scratch or damage the cylinder head surface in any way. Do this work with the sparking plugs (preferably an old set) fitted to the head to prevent hard carbon getting into the plug threads. If this happens, and the carbon is not cleaned out, there is risk of damaging the soft threads in the head when the plugs are screwed in.
3 Examine the face of the cylinder head for damage such as score marks, indentations or burning as illustrated in photograph 37.3. Damage of this type will prevent the cylinder head seating properly and will result in failure of the cylinder head gasket and burning of the head as shown, or in water leaks to the cylinders.
4 Examine the valve seat inserts, firstly for burning or pitting to the extent of preventing seating of the valves by grinding and secondly to check the insert for security.
5 Burned or pitted valve seat inserts can be reclaimed by recutting provided the damage is not too serious. Your local garage can probably do this work for you. The angle of the valve seat, both inlet and exhaust, is 45°.

6 The valve seat insert, valve guide and tappet guide are shrunk into the cylinder head and if either of these items are loose or are damaged to the extent of requiring replacement it is advisable to have the work done by a Jaguar agent having the necessary equipment and experience.
7 If the spark plug thread has been damaged to the extent of preventing correct fitment of the plug it is possible to reclaim the head by fitting an insert in the following manner;
a) Refer to Fig. 1.21
b) Bore out the damaged thread to 0.75" (19.05 mm) diameter and tap ½" BSP.
c) Counterbore 57/64" (22.62 mm) diameter to accommodate the larger diameter of the insert.
d) Fit the screwed insert ensuring that it sits firmly at the bottom of the thread.
e) Drill and ream a 1/8" diameter hole 3/16" (4.76 mm) deep between the side of the insert and the cylinder head. Drive in the locking pin and peen over the insert and the locking pin.
8 Clean the pistons and the top of the cylinder bores. If the pistons are still in the bores it is essential that great care is taken to ensure that no carbon gets into the bore for this will scratch the cylinder walls or cause damage to the pistons or rings. To stop this happening, first turn the crankshaft so that two of the pistons are at TDC, then place a clean non-fluffy rag into the other bores or seal them off with paper and masking tape. Seal off all other openings into the cylinder head and into the sump.
9 It is a matter of opinion how much carbon ought to be removed from the piston crown. Some consider that a ring of carbon should be left around the edge of the piston crown and on the cylinder bore wall as an aid to keep oil consumption low. We feel that this is probably true for engines with worn bores; but with an engine in good condition we recommend removal of all carbon.
10 If all traces of carbon are to be removed, press a little grease into the gap between the cylinder walls and the pistons that are to be worked on. With a blunt scraper, carefully scrape away all carbon from the piston crown taking care not to scratch away all carbon from the piston crown taking care not to scratch the surface, Also, scrape away carbon from the surrounding lip of the cylinder wall. When all carbon has been removed, scrape away carbon from the surrounding lip of the cylinder wall. When all carbon has been removed, scrape away the grease which will now be contaminated with carbon particles taking care not to press any into the bores. Polishing the piston crown with metal polish will retard the subsequent build up of carbon but be careful not to allow any polish to run into the bore. Remove the seals from the other cylinders and repeat the foregoing until all pistons have been cleaned.
11 If you decide to leave a ring of carbon around the piston then this can be helped by inserting an old piston ring into the bore to rest on the piston and this will ensure that carbon is not accidentally removed.

12 Check that there are no particles of carbon in the cylinder bores and clean the face of the cylinder block. Decarbonisation is now complete.

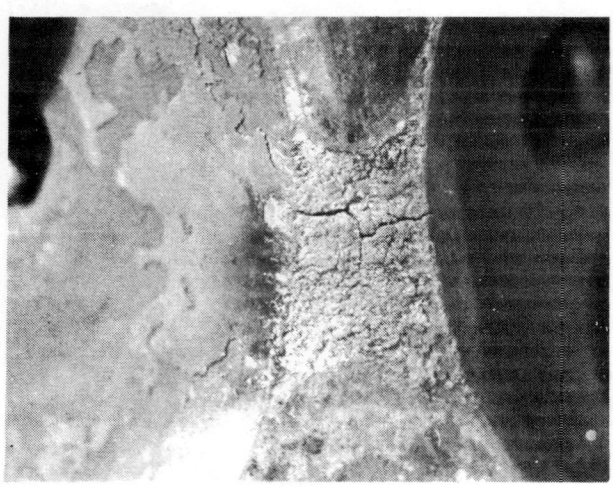

37.3 Burnt and cracked cylinder head

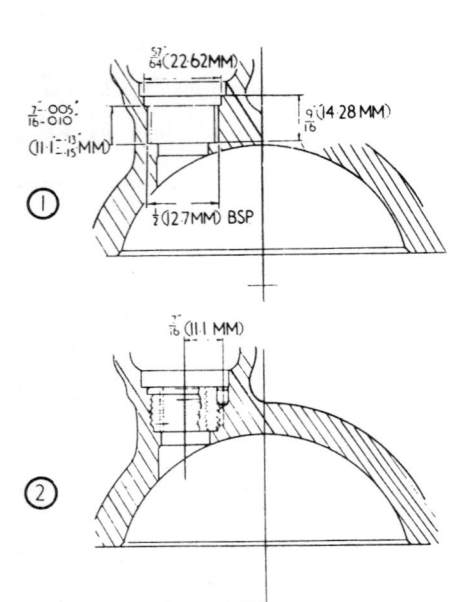

Fig.1.21. Dimensions for fitting a sparking plug insert

38 Compression pressures - general

The compression pressures for all six cylinders should be even and should approximate to the pressures quoted later. If one or more compressions are weak it will be due (most probably) to poor valve seating. In which case the cylinder head will have to be removed and the valves ground in as already described.

Pressures must be taken with all spark plugs removed, carburettor throttles wide open and the engine at its normal operating temperature (70°C approx:). The engine should be rotated by operating the push button on the starter solenoid with the ignition switched off. Models with automatic transmission should have the selector lever in the 'P' (park) position.

Pressures:

8 to 1 compression ratio - 155 lbs per sq. in. (10.90 kg/cm^2)
9 to 1 compression ratio - 180 lbs per sq. in. (12.65 kg/cm^2)

The difference in compression ratios are obtained by varying the crown design of the piston (refer to Fig. 1.19).

39 Timing gear and chain - examination and renovation

1 The timing gear assembly as fitted to 3.8 and 4.2 litre engines us illustrated in Figs. 1.22 and 1.23 respectively and an exploded view (3.8 litre) is given in at Fig. 1.24.
2 Examine the teeth of the sprockets for wear. Each tooth forms an inverted 'V' with the periphery of the sprocket and, if worn, the side of the tooth under tension will be slightly concave in shape when compared with the other side of the tooth. If any wear is present the sprocket should be renewed.
3 Examine the links of the chain for side slackness and renew the chain if any slackness is noticeable when compared to a new chain. It is sensible to replace the chain if the engine is stripped for overhaul and when the cylinder head is removed for decarbonisation or some other purpose, if it is known that the chain has been in use for a considerable time.

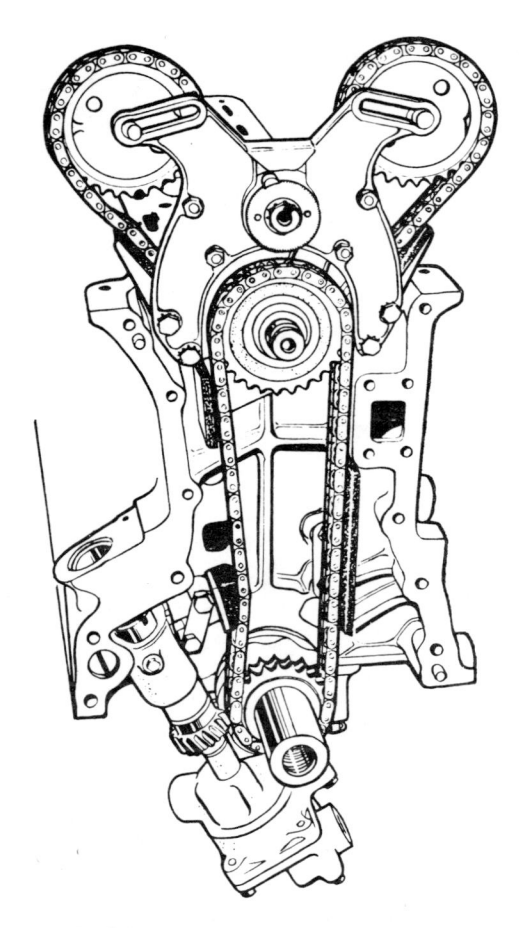

Fig.1.22. 3.8 litre timing gear arrangement

Fig. 1.23. 4.2 litre timing gear arrangement
(letters are referred to in text)

Fig. 1.25. Oil pump - checking clearance between the inner and outer rotors

Fig. 1.26. Oil pump - measuring the clearance between the outer rotor and the pump body

Fig. 1.24. Exploded view of the timing gear assembly (3.8 litre)

1 Camshaft sprocket
2 Adjusting plate
3 Circlip
4 Guide pin
5 Star washer
6 Circlip
7 Timing gear front mounting bracket
8 Timing gear rear mounting bracket
9 Idler sprocket
10 Eccentric shaft
11 Plug
12 Adjustment plate
13 Plunger pin
14 Spring
15 Intermediate sprocket of top timing chain
16 Intermediate sprocket of lower timing chain
17 Key
18 Shaft
19 Circlip
20 Top timing chain
21 Damper for top timing chain (left hand)
22 Damper for top timing chain (right hand)
23 Distance piece
24 Intermediate damper
25 Bottom timing chain
26 Vibration damper
27 Hydraulic chain tensioner
28 Shim
29 Filter gauze
30 Front timing cover
31 Gasket
32 Oil seal

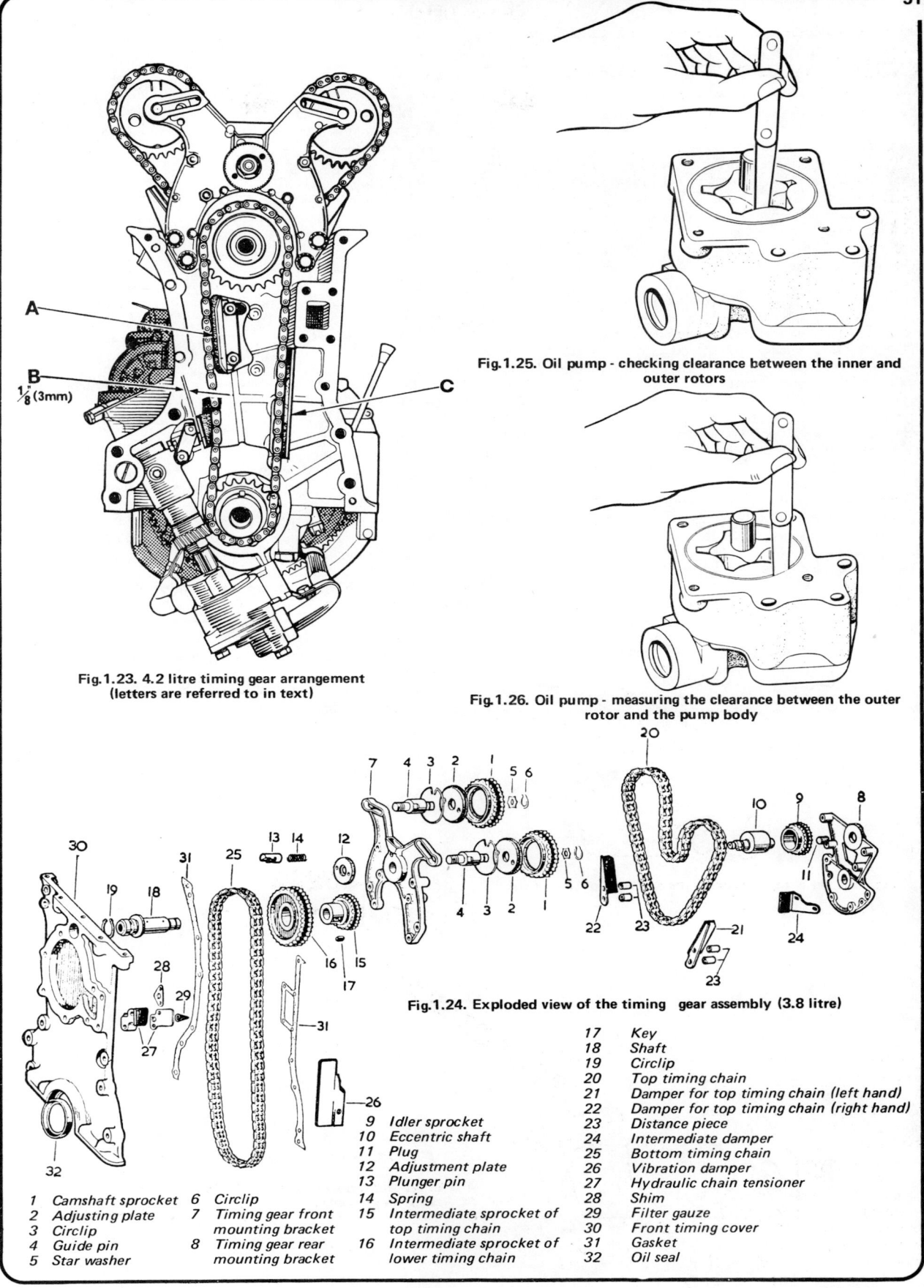

40 Oil pump - examination and renovation

1 An exploded view of the oil pump is given in Fig. 1.14.
2 Unscrew the four bolts and detach the bottom cover from the pump.
3 Withdraw the inner and the outer rotors from the oil pump body. The inner rotor is pinned to the drive shaft and must not be dismantled.
4 Refer to Fig. 1.25 and check the clearance between the inner and outer rotors, this should be 0.006" (0.15 mm) maximum.
5 Refer to Fig. 1.26, check the clearance between the outer rotor and the pump body. This should not exceed 0.010" (0.25 mm).
6 Check the end float of the rotors by placing a straight edge across the joint face of the body and measuring the clearance between the rotors and the straight edge in the manner depicted in Fig. 1.27. The clearance should be 0.0025" (0.06 mm), if outside of this limit it can be restored by lapping the pump body and the outer rotor on a surface plate to suit the inner rotor.
7 Examine the pump body and the bottom cover for signs of scoring and the drive shaft for signs of wear especially in the squared drive recess accommodating the distributor drive shaft.
8 At this point examine also the squared end of the distributor drive shaft and ensure that the flats are not worn.
9 Place the oil pump drive shaft in a vice fitted with soft jaws and check that the inner rotor is tight on the securing pin.
10 Any part failing the above examination should be replaced as its retention can only lead to low oil pressure and possible bearing failure.
    Note that the drive shaft, the inner and outer rotors are supplied only as an assembly.

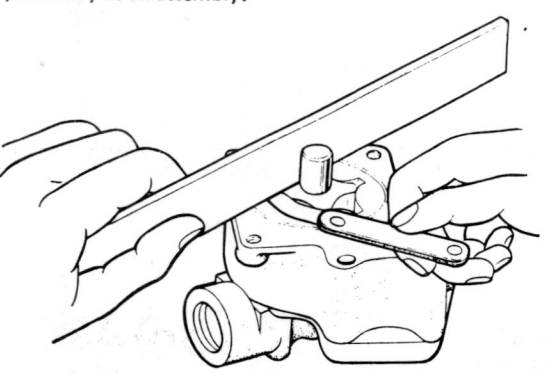

Fig.1.27. Oil pump - checking the end float of the rotors

41 Oil sump - cleaning and examination

1 Clean off any part of the old gasket between the oil sump and the cylinder block which may be adhering to the top face of the sump.
2 Take out, and discard, the cork/rubber seal at the rear of the sump.
3 Remove the six nuts and take out the filter basket.
5 Clean all components in paraffin, the use of a stiff brush may be required to remove sediment from the bottom of the sump casing.
6 Examine all instances of damage to the sump arising from impact of stones or contact with the ground and assess their acceptibility. Check carefully that any such damage has not resulted in fractures.
7 There is no necessity to remove the oil return pipe assembly unless there are signs of leakage at the gasket.

42 Bottom timing chain tensioner - examination and renovation

1 Refer to Fig. 1.28.
2 Examine the rubber slipper for wear or deterioration. Replace if necessary.
3 Check the plunger for freedom of movement in the body and the restraint cylinder for freedom in the plunger. Rectify as necessary.
4 Assemble the restraint cylinder and spring to the plunger. Rotate the restraint cylinder in a clockwise direction and check that it is retained by the limit peg.

43 Automatic water pump belt tensioner - examination and renovation

   Examination of this item is limited to looking for wear on the pulley, freedom of the pulley and satisfactory operation of the spring.

44 Engine stabiliser - examination and renovation

   The only fault that is likely to be found in this item is that the rubber may be perished (photo) after a long period of use. The importance of this item is often overlooked, failure of the rubber can lead to engine vibration and/or fouling of the gearbox in its cowl.

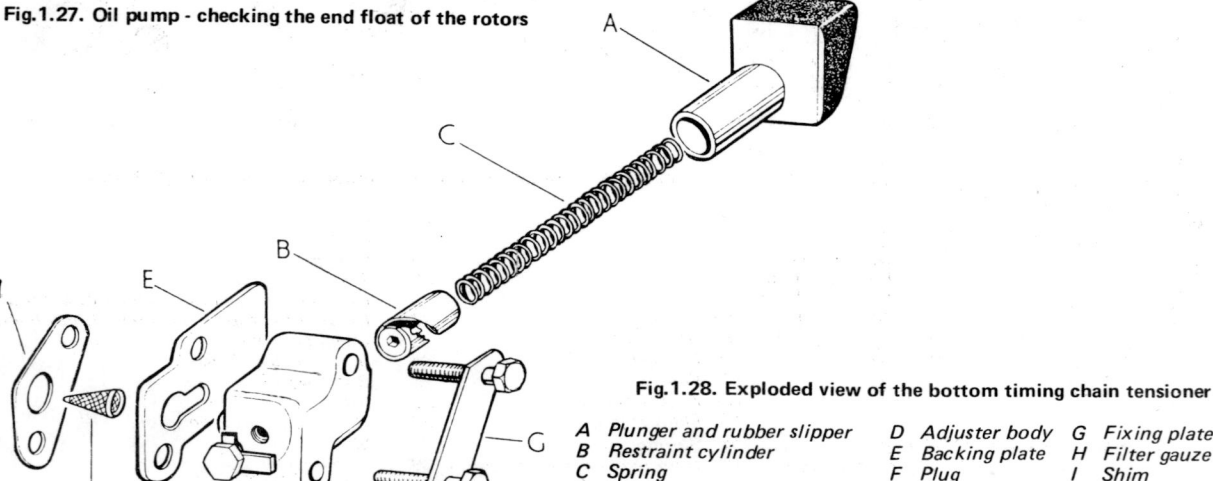

Fig.1.28. Exploded view of the bottom timing chain tensioner

A Plunger and rubber slipper
B Restraint cylinder
C Spring
D Adjuster body
E Backing plate
F Plug
G Fixing plate
H Filter gauze
I Shim

Chapter 1/Engine

45 Engine mountings - general

The engine is supported at the front on two rubber mountings which are attached to brackets on the body underframe. The rear of the power unit, for both automatic and standard transmission models, is supported on a coil spring mounted in a channel support which is bolted to the body floor, this mounting is illustrated in photograph 45 and also in Fig. 6 in Chapter 6.

46 Engine mountings - examination and renovation

With the engine out of the car it is a simple matter to carefully examine the front mountings for sponginess of the rubber and the condition of the adhering metal parts. Change the mountings if there is any doubt as to their condition because it will now be a much easier task than when the engine is installed.

Make sure that the spring of the rear mounting is not fractured and that the rubber components are in good condition. The rubber spring seat should be stuck to the support bracket with adhesive, if it is not loose and appears to be in good condition it is as well not to disturb it.

44 Engine stabiliser - rubber perished

45 The rear engine mounting

47 Air cleaner - general

The air cleaner (Fig. 1.29) is of the paper element type and is mounted on the right hand side of the engine compartment and is connected to the carburettors by means of an elbow trumpet plate.

No maintenance is necessary but the element should be renewed every 10,000 miles (16,000 km) or more frequently in dusty territories. The element is removed by first releasing the three spring clips which retain the top cover to the base. Now remove the two wing nuts attaching the cleaner to the air box and lift out the element and cover. Remove the serrated nut and retainer plate from the base of the unit and withdraw the element.

Fig.1.29. The air cleaner assembly

48 Engine reassembly - general

To ensure maximum life with minimum trouble from a rebuilt engine not only must every part be correctly assembled but everything must be spotlessly clean, all oilways must be clear, locking washers and new tab washers must always be fitted where needed. All bearings and other working surfaces must be thoroughly lubricated during assembly in order to afford initial lubrication of parts when first starting the engine. Before assembly begins, renew any bolts or studs the threads of which are in any way damaged, it is advisable to use new spring washers and new self locking nuts, especially at important locations such as big end bearing caps etc: Never re-use a split pin despite the fact that the removed pin may appear to be in good condition. Use soft iron locking wire, or annealed copper wire, in places where locking wire is called for.

Apart from your normal tools, a good supply of non-fluffy rag, an oil can filled with engine oil, a set of new gaskets and a torque wrench should be collected together.

49 Crankshaft - replacement

Ensure that the crankcase is thoroughly clean and that all oilways are clear. A thin twist drill is handy for cleaning them out. If possible, blow them out with compressed air. Treat the crankshaft in the same fashion and then inject engine oil into the oilways.

1 Fit the top half of the rear oil seal cover assembly and then fit the top half of the oil seal. First prepare the new oil seal by carefully tapping it on its side face to narrow the section. Now fit the seal to the housing and press into the groove, using a hammer handle, until the seal does not protrude from the ends of the housing. DO NOT cut the ends off the seal if they do protrude but continue pressing into the groove until both ends are flush. Using a knife or similar tool, press all loose ends of asbestos into the ends of the groove so that they will not be trapped between the two halves of the housing when assembled. Fit the asbestos seal to the bottom half of the cover assembly in the same manner as described above. (Fig. 1.30).

2 Clean the locations for the half main bearing shells in the crankcase and fit the half bearing shells.

3 Lay the crankshaft in the bearing shells.

4 Fit the bottom half of the oil return thread cover to the top half which is bolted to the cylinder block behind the rear main bearing.

5 Check the clearance between the oil return thread cover and the oil return thread on the crankshaft, this should be 0.0025" to 0.0055" (0.06 to 0.14 mm).

6 Fit the centre main bearing cap with a new thrust washer, (Fig. 1.12), white metal side outwards, in the recess at each side of the cap. Tighten down on the cap and check the crankshaft end float (photo), this should be 0.004" to 0.006" (0.10 to 0.15 mm). The thrust washers are supplied in two thicknesses, standard and 0.004" (0.10 mm) oversize and should be selected to bring the end float within permissable limits. There is no objection to the use of a standard and an oversize washer on the same shaft. The oversize washers are stamped +0.004 (0.010) on the steel face.

7 Fit the main bearing caps and shells to the crankshaft. But do make sure that the numbers stamped on the caps correspond with those stamped on the crankcase (Fig 1.31).

8 Fit the main bearing cap bolts and tab washers and tighten down to a torque of 83 lb.f.ft. The tab washers for the rear main bearing bolts are longer than the remainder and the plain ends should be tapped down round the bolt hole bosses.

9 Test the crankshaft for freedom of rotation.

10 If there is no undue restriction to movement of the crankshaft, knock up the tab washers to secure the bolts.

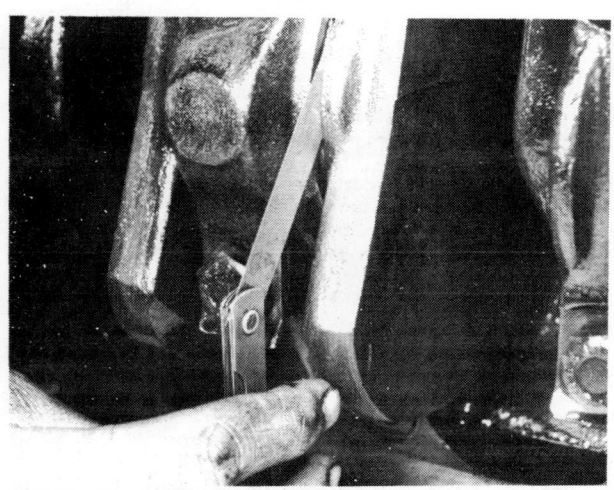

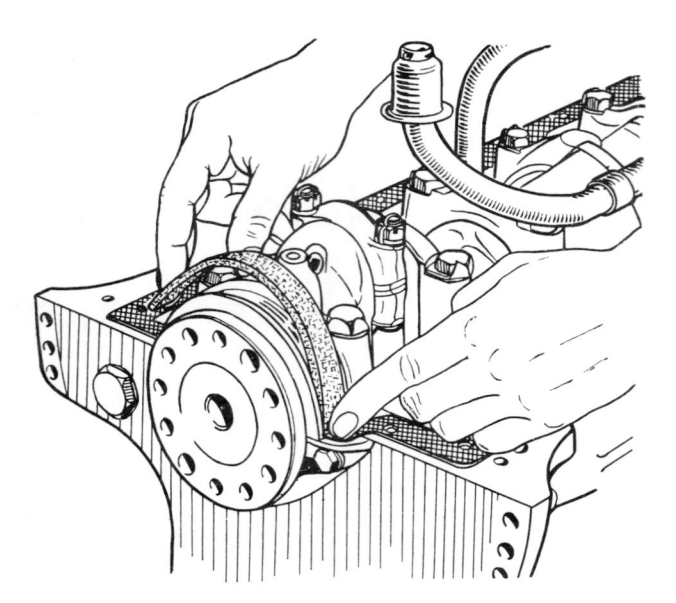

Fig.1.30. Fitting the bottom half of the rear oil seal

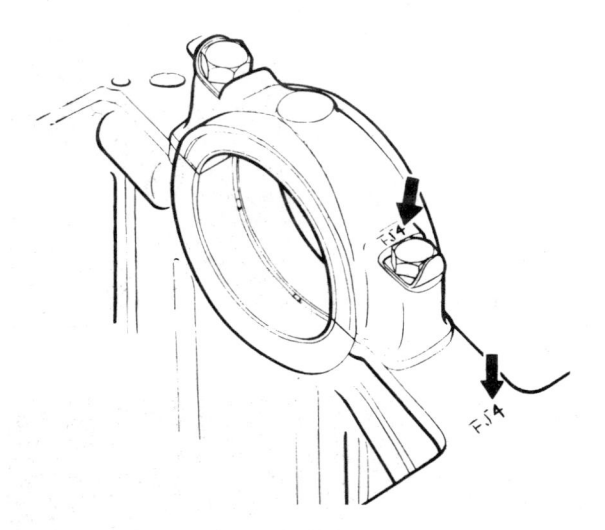

Fig.1.31. Corresponding marks on the main bearing caps and crankcase

49.6 Checking end float of the crankshaft

50 Piston and connecting rod - reassembly

1 If the gudgeon pin was removed during dismantling of the assembly it must be replaced in the piston and connecting rod in the same position before removal.

2 If the gudgeon pin will not enter the piston, it must not be forced. Immerse the piston in a bath of hot oil for a few minutes, remove the piston from the oil and the gudgeon pin should now enter freely under finger pressure. Enter the pin in one half of the piston (if applicable), engage the connecting rod, original way round, and push the gudgeon pin home.

3 Secure the gudgeon pin with a new circlip at each end.

Chapter 1/Engine

51 Piston ring - replacement

1   Check that the piston ring grooves are thoroughly clean and that oilways are not blocked. Piston rings must always be fitted over the head of the piston and never from the bottom.
2   Refitment is the exact opposite to the removal procedure (see Section 18). Make sure that the chromium plated compression ring is fitted to the top groove in the piston and that the narrowest part of the compression rings, which is marked with a letter 'T' or with the word 'TOP' (Fig. 1.17), are indeed uppermost. The oil control ring is not tapered and can be fitted either way up.
3   As each ring is fitted, make sure that it is free in its groove.
4   When new rings are fitted to the piston, the side clearance in the grooves should be checked on assembly. This clearance should be 0.001" to 0.003" (0.025 to 0.076 mm).
5   Finally, set all ring gaps at 90° to each other.

52 Piston - replacement

1   Turn the engine on its side.
2   Wipe the cylinder bores clean with a non-fluffy rag, and then liberally lubricate the walls of each.
3   The pistons, complete with connecting rods, must be fitted to their respective bores from the top of the cylinder block. As each piston is inserted in the bore, make sure that it is the correct assembly for that bore by checking the number stamped on the connecting rod (no. 1 cylinder is at the rear of the engine).
4   Ensure that the pistons are the correct way round in the cylinder, the piston crown is marked 'Front' to aid correct assembly (Fig. 1.18).
5   Check that the piston rin gaps are at 90° to each other.
6   Compress the piston rings in a clamp. Guide the piston into the bore until it reaches the ring compressor. Gently tap the piston into the bore with a wood or hide faced hammer.
7   Do not try to fit the pistons without a ring compressor as the chance of breaking a ring and scoring the bore is very high. If a ring compressor is not available then a suitable jubilee clip is better than nothing but make sure that the clip is not tightened too much.

53 Connecting rod to crankshaft - reassembly

1   Wipe the connecting rod half of the big end bearing location and the underside of the shell bearing clean. Fit the shell bearing in position with its locating tongue engaged with the corresponding groove in the connecting rod (Fig. 1.32).
2   Wipe clean and then generously lubricate the crankpin journals with engine oil. Turn the crankshaft to a handy position for the connecting rod to be drawn onto it and for the connecting rod cap to be fitted.
3   Fit the bearing shell to the connecting rod cap in the same manner as with the connecting rod itself.
4   Generously lubricate the shell bearing and offer up the cap to the connecting rod, ensure that the numbers are mating (Fig. 1.32).
5   Fit the connecting rod bolts (it is advisable to use new bolts), fit the nuts and tighten down to a torque of 37 lb.f.ft (5.1 kf.f.m) Lock the nuts with new split pins if applicable.

54 Crankshaft gear and sprocket - reassembly

1   Fit the Woodruffe key and drive on the helical distributor drive gear with the widest part of the boss to the rear.
2   Fit the Woodruffe key and drive on the crankshaft timing gear sprocket.
3   Fit the oil thrower, washer and distance piece.
4   Turn the engine until No's 1 and 6 pistons are at TDC (see Fig. 1.33).

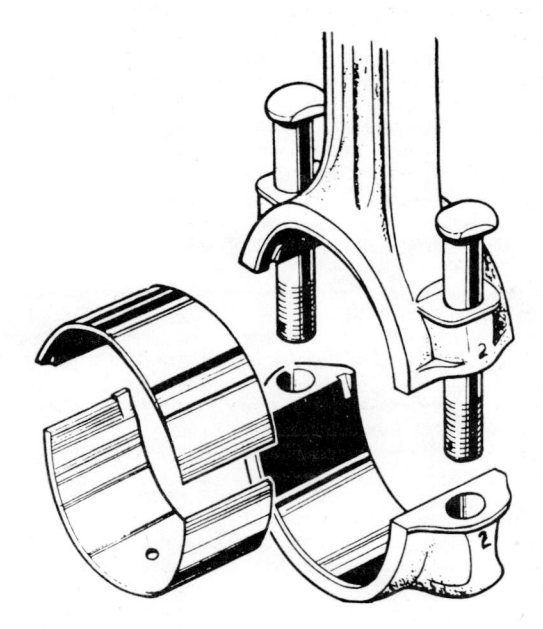

Fig. 1.32. Assembly of big end bearing shells and caps

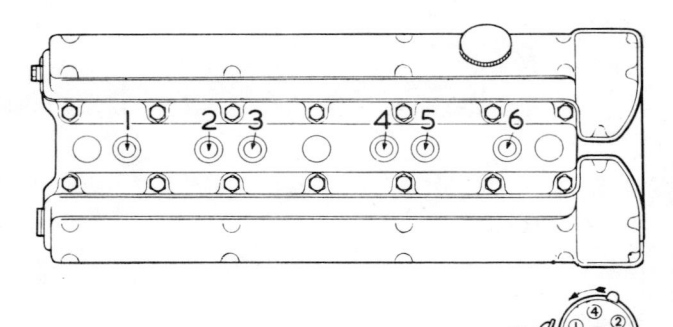

Fig. 1.33. View of the engine showing cylinder numbers and firing order

55 Distributor and oil pump drive gear - reassembly

1   Fit the distributor drive shaft to the bush on the front face of the cylinder block with the offset in the top of the shaft positioned as shown in Fig. 1.5.
2   Fit the thrust washer and drive gear to the shaft, noting that the gear is keyed to the shaft.
3   Fit the pegged tab washer with the peg in the keyway of the drive gear.
4   Engage the retaining nut, fully tighten it and then check the end float of the shaf which should be 0.004" to 0.006" (0.10 to 0.15 mm).

If no clearance exists, fit a new oil pump/distributor driving gear and this should restore the clearance. However, in an emergency it is possible to adjust the clearance by rubbing down the thrust washer on a piece of emery cloth placed on a surface plate to ensure that the face of the washer is kept flat.
5   Secure the tab washer when the clearance is satisfactory.

56 Oil pump and pipes - reassembly

1 Fit the coupling shaft between the squared end of the distributor drive shaft and the driving gear of the oil pump.
2 Secure the oil pump to the front main bearing cap by the three dowel bolts and tab washers. Check that there is appreciable end float of the short coupling shaft.
3 Fit the oil delivery pipe from the oil pump to the bottom face of the crankcase with a new 'O' ring and gasket.
4 Fit the suction pipe with a new 'O' ring at the oil pump end and secure to its clip on the main bearing cap.

57 Timing gear - assembly

1 Fit the eccentric shaft to the hole in the front mounting bracket.
2 Insert the spring and locking plunger for the serrated plate to the hole in the front mounting bracket.
3 Fit the serrated plate and secure with its shake proof washer and nut.
4 Fit the idler sprocket (21 teeth) to the eccentric shaft.
5 Fit the two intermediate sprockets (20 and 28 teeth) to their shaft with the larger sprocket forward and press the shaft through the lower central hole in the rear mounting bracket. Secure with the circlip at the rear of the bracket. The foregoing applies to early model cars; you will probably find that the intermediate sprockets fitted to your car are a one piece casting.
6 Fit the top timing chain (longer chain) to the small intermediate sprocket and the bottom timing chain (shorter chain) to the large intermediate sprocket.
7 Loop the upper timing chain under the idler sprocket and offer up the front mounting bracket to the rear mounting bracket with the two chain dampers interposed between the brackets.
8 Fit the intermediate damper to the bottom of the rear mounting bracket and secure with the two screwdriver slotted screws, do not lock with the tab washers at this stage if the timing gear assembly fitted to your car is similar to that depicted in Fig. 1.23.
9 Pass the four securing bolts through the holes in the brackets, the chain dampers and spacers. Fit shake proof washers under the bolt heads.
10 Secure the two mounting brackets together with the four nuts and shake proof washers.

58 Timing gear - reassembly to engine

1 Fit the lower timing chain damper and bracket to the front face of the cylinder block and secure with the two set bolts and locking plate but do not lock the bolts at this stage if the timing gear assembly fitted to your car is similar to that depicted in Fig 1.23.
2 Offer the timing gear assembly up to the cylinder block. Loop the bottom timing chain over the crankshaft sprocket and secure the mounting brackets to the front face of the cylinder block with the four long securing bolts and the two screwdriver slotted setscrews.
3 Do not fully tighten the two setscrews until the four long securing bolts are tight.

59 Bottom timing chain tensioner - reassembly

1 Place the bottom timing chain tensioner, the backing plate and the filter in position so that the spigot on the tensioner aligns with the hole in the cylinder block.
2 Fit shims as may be necessary between the backing plate and the cylinder block to bring the rubber slipper central on the timing chain (Fig. 1.34).
3 Fit the tab washer and the two securing bolts. Tighten the bolts but do not lock them at this stage if the timing gear assembly fitted to your car is similar to that depicted in Fig. 1.23.
4 It is important that the locking mechanism is not released until the adjuster has been finally mounted on the engine with the timing chain in position.
5 Remove the hexagon headed plug from the end of the body after knocking up the tab washer.
6 Insert an Allen key (0.125" A/F) into the hole until it registers in the end of the cylinder. Turn the key clockwise until the tensioner head moves forward under spring pressure against the chain. DO NOT attempt to force the tensioner head into the chain by external pressure.
7 Refit the plug and lock with its tab washer.

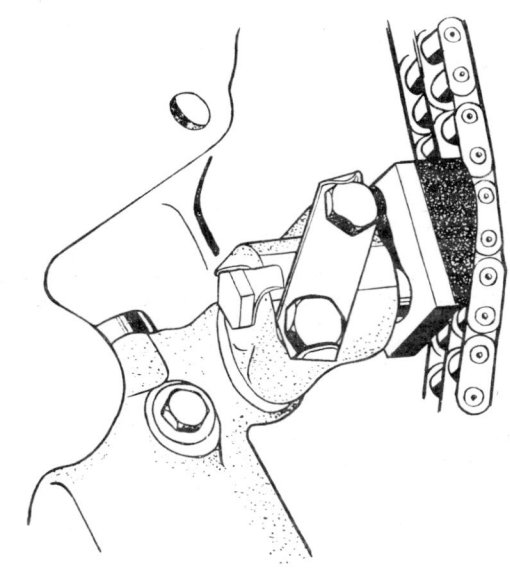

Fig. 1.34. Positioning of the bottom timing chain tensioner

60 Bottom timing chain tensioners - adjustment

1 Refer to Figs. 1.22 and 1.23 and note the differences in damper arrangement.
2 If your car is fitted with the later type of timing gear assembly depicted in Fig. 1.23, the dampers will have to be positioned as follows.
3 Set the intermediate damper (A) in light contact with the chain when there is a 1/8" (3 mm) gap (B) between the rubber slipper and the body of the lower timing chain tensioner. However, this clearance may have to be increased in the case of a worn chain in order to avoid fouling between the chain and the cylinder block. Set the lower damper (C) in light contact with the chain.
4 Having made the necessary adjustments, lock the damper and tensioner securing bolts with their tab washers.

61 Timing cover - refitting

1 Fit a new oil seal to the recess in the bottom face of the timing cover and ensure that the seal is well bedded in its groove.
2 Replace the dowels if you removed them during dis-assembly.
3 Smear the mating faces of the timing cover and the cylinder block with a good quality jointing compound. Fit a new gasket and secure the timing cover to the front face of the cylinder block with the securing bolts.
4 Do not forget to fit the dynamo/alternator adjusting link over its timing cover stud with the distance piece interposed between the link and the timing cover.

Chapter 1/Engine

62 Oil sump - refitting

The oil sump may be refitted at this stage or, if you wish to use the base of the cylinder block on which to rest the engine for further assembly work, it may be left until later. However, to refit the sump:
1 Clean the mating faces of the sump and the crankcase. Although not really necessary, they may be treated with jointing compound if desired.
2 Fit a new gasket to the bottom face of the crankcase.
3 Fit the oil seal to the recess in the rear main bearing cap (Fig. 1.30).
4 Fit the sump to the crankcase and secure with the twenty-six setscrews and the four nuts and washers. Remember that the short setscrew goes at the right hand front corner of the sump.
5 Fit a new gasket over the fixing studs followed by the oil return pipe elbow and secure with the three nuts.
6 Fit the hose running from the oil return pipe to the oil cleaner and tighten down the clip.

63 Flywheel and clutch - refitting

1 Turn the engine upright.
2 Check the crankshaft flanges, the holes for the flywheel bolts and the dowels for freedom from burrs.
3 Check that No. 1 and 6 pistons are at TDC.
4 Fit the flywheel of the crankshaft flange so that the 'B' stamped on the edge of the flywheel is approximately at the BDC position. This will ensure that the balance mark 'B' is in line with the balance point of the crankshaft (this is a group of letters stamped on the crank throw just forward of the rear main journal).
5 Tap the two mushroom headed dowels into position.
6 Fit the locking plate and the flywheel securing setscrews and tighten them to a torque of 67 lb.f.ft (9.2 kg.f.m.) Secure the screws with the locking tabs.
7 It is advisable to fit a new clutch driven plate now that the engine has been completely overhauled. Make sure that the flywheel is clean and free of oil or grease and then assemble the driven plate to the flywheel noting that one side of the plate is marked 'Flywheel side'. Now offer up the clutch assembly so that the 'B' stamped on the flywheel or that the marks you made, when separating the clutch from the flywheel, match up.
8 Centralise the driven plate by means of a dummy shaft which fits the splined bore of the driven plate and the spigot bush in the crankshaft. An old constant pinion shaft may be used for this purpose (photo). Now push the clutch assembly home and secure in postiion with the six setscrews. Tighten the setscrews a turn at a time by diagonal selection.
9 Remove the dummy shaft.

64 Distributor - refitting

1 Check that No. 6 (front) piston is at TDC and check that the distributor drive shaft offset is in the position shown in Fig. 1.5. It may be 180° out in which case rotate the engine through a complete revolution to again bring No.6 piston to TDC, Again check the position of the slot.
2 Fit the cork seal to the recess at the top of the hole for the distributor.
3 Secure the distributor clamping plate to the cylinder block with the setscrew. Slacken the clamping plate bolt.
4 Set the micrometer adjustment in the centre of the scale.
5 Enter the distributor in the cylinder block with the vacuum advance unit connection facing rearwards.
6 Rotate the rotor arm until the driving dog engages with the distributor drive shaft. In this position the rotor arm should be in the approximate postiion of No. 6 cylinder segment in the distributor cap. (Fig. 1.33).

65 Cylinder head - reassembly

1 Turn the head upside down on the bench and rest it on wooden blocks.
2 Generously lubricate the valve stems with engine oil and then fit the valves in their correct positions.
3 Place the wooden block which was made up for valve removal, in position to retain the valves and invert the head allowing it to rest on the block on the bench.
4 Fit each valve with its valve seat.
5 Fit the valve springs and collars.
6 Depress the valve springs and, with the help of an assistant, fit the cotters and make sure that they bed down correctly over the valve stem when pressure on the springs in released. It has been found that a tool as illustrated in Fig. 1.35 makes compression of the springs much easier. It is essentially a ¾" internal diameter tube with a part section cut away and its method of use is to apply the cut away part ot the valve spring and to bear down on the spring until your assistant can enter the cotters through the cut away section (photo). Care must be taken when compressing the springs, whatever method is used, to ensure that side loads are not put on the valve stem as in all probability they would be bent.
7 Replace the tappets and valve adjusting pads in their correct positions.

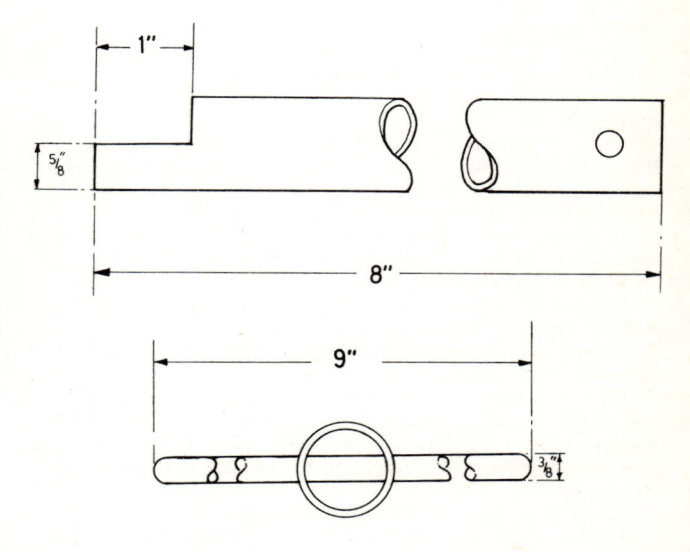

Fig. 1.35. Suggested tool for compressing valve springs

Chapter 1/Engine

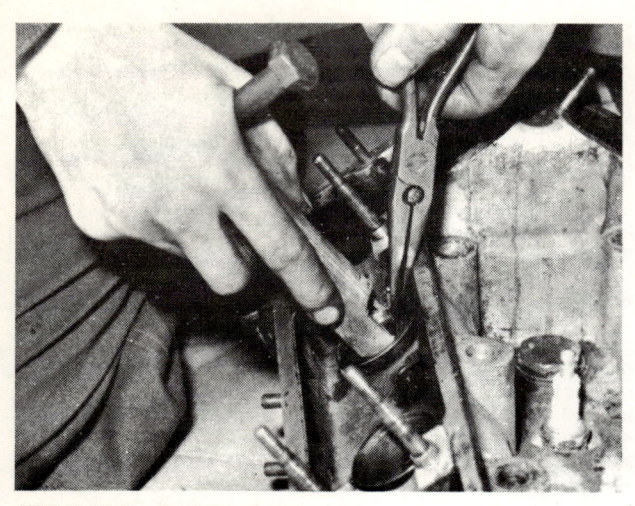

65.6 Refitting the valve cotters

66 Valve clearance adjustment

Valve clearances are:

Normal touring use
 Inlet   0.004" (0.10 mm)
 Exhaust   0.006" (0.15 mm)

Racing
 Inlet   0.006" (0.15 mm)
 Exhaust   0.010" (0.25 mm)

Important

When checking the valve clearances, the camshafts must be fitted **one at a time** as, if one camshaft is rotated when the other shaft is in position, fouling is likely to take place between the inlet and the exhaust valves. When checking the clearance of one set of valves is completed, the shaft must either be removed or the bearing cap nuts must be slackened to the extent of relieving all pressure on the valves.

1 Clean the location in the cylinder head for camshaft bearing shells. Fit the half bearing shells to the head.
2 Clean the bearing location in the camshaft bearing caps and fit the half shell bearings.
3 Fit one camshaft to the cylinder head making sure that the correct shaft is being offered to the correct set of valves. check the numbers on the caps to that shaft and the corresponding numbers on the cylinder head and fit the caps in their correct positions.
   Fit the D washers and the nuts to the bearing studs.
4 Tighten down on the nuts evenly a turn at a time to a torque of 15 lb.f.ft (2.0 kg.f.m).
5 Rotate the camshaft to bring the back of a cam (heel) to the valve tappet. Measure, and record, the clearance between the cam and the tappet (photo). Repeat for all the valves in that bank.
6 Adjusting pads are available rising in 0.001" (0.03 mm) increments from 0.085" to 0.110" (2.16 to 2.79 mm) and are etched on the surface with the letter A to Z each letter indicating an increase in size of 0.001" (0.03 mm).
7 Should any valve clearance require adjustment, remove the camshaft and the affected tappet and adjustment pad.
8 Observe the letter stamped on the adjusting pad and should the recorded clearance for this valve have shown, say 0.002" (0.05 mm) in excess of the correct value, select a new adjusting pad with a letter two lower than the original pad.
9 After all the adjusting pads have been changed as required, reassemble the camshaft and carry out a final check to ensure that the clearances are indeed correct. Remove the camshaft or slacken all the nuts to relieve all pressure on the valves.
10 Repeat the foregoing to check the clearances of the other set of valves.
11 When you are satisfied that the clearances are correct, tighten down on the nuts (paragraph 4) of one camshaft and then turn it so that the square slot in the shaft is at 90° to the camshaft cover face (Fig. 1.36).
12 Fit the other camshaft and position it so that its slot is also at 90° to the camshaft cover face. Fit the D washers and nuts and tighten these down to the correct torque figure (paragraph 4).
13 From this point onwards the camshafts must not be rotated independently of each other.

66.5 Measuring valve clearance

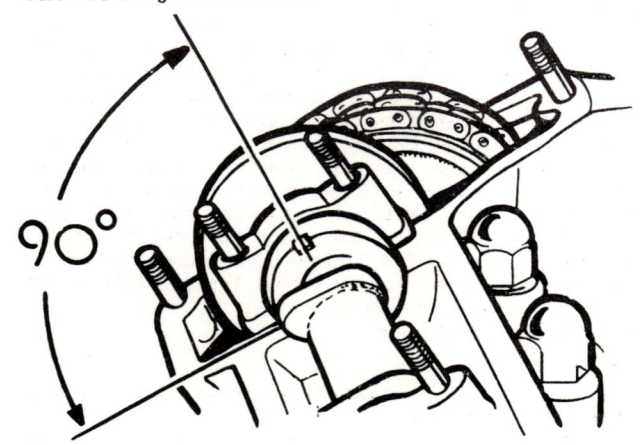

Fig. 1.36. Positioning of the camshaft for assembly

67 Cylinder head - refitting

1 Make sure that the top face of the cylinder block and the mating face of the cylinder head are clean.
2 Check that No.6 piston is at TDC with the distributor rotor arm opposite No.6 segment of the distributor cap.
3 Fit a new cylinder head gasket to the cylinder block and make sure that it is seated right down on the top of the block and that the side marked TOP is uppermost. You will probably find that the new cylinder head gasket with which you have been supplied is of an improved type. The new gasket is thicker than the previous type and is of asbestos compound coated steel-backed construction. The two faces are treated with a varnish

Chapter 1/Engine

finish which eliminates the use of any jointing compound.
4 Check that the slot in the camshafts are at 90° to the camshaft cover face and accurately position them by engaging the valve timing gauge (photo). The valve timing gauge passes over the camshaft and rests on each side of the camshaft cover face and at the same time a projection engages in the camshaft slot to ensure that the slot is indeed perpendicular, Fig. 1.37 illustrates the gauge in position. It may be possible for you to borrow a gauge and apart from this no suitable alternative can be proposed for the accurate positioning of the camshafts required in actual valve timing (see Section 68).
5 Fit the cylinder head, note that the second cylinder head stud from the front on the left hand side is a dowel stud.
6 Fit the sparking plug lead carrier to the third and sixth stud on the right hand side. Fit plain washers to these and to the two front stud positions and D washers to the remaining studs.
7 Fit the fourteen dome nuts and the nuts at the front underside of the head and screw them down finger tight.
8 Using a torque wrench (photo), tighten down the nuts a turn at a time, in the sequence shown in Fig. 1.7, to a torque of 54 lb.f.ft (7.5 kg.f.m) for the old pattern cylinder head gasket or to a torque of 58 lb.f.ft (8.0 kg.f.m) if the later pattern gasket is fitted.

Fig. 1.37. The valve timing gauge in position

67.4 The valve timing gauge

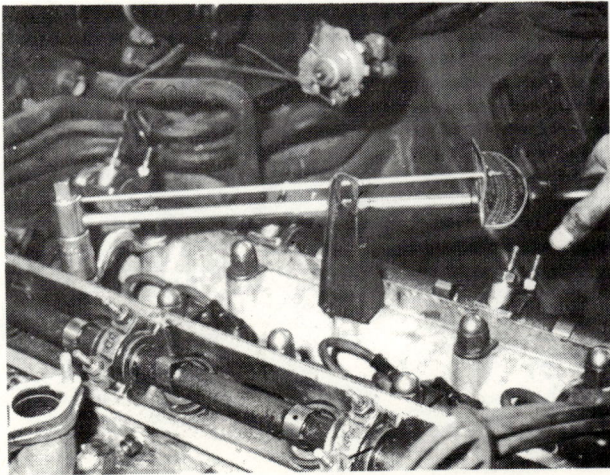

67.8 Using a torque wrench to tighten cylinder head nuts

68 Valve timing (Fig. 1.38)

1 It is important to tension the top timing chain before attempting to check or set the valve timing. If the engine is in course of reassembly, fit the sprockets to their respective camshafts. Tighten down on the securing setscrews but do not lock them at this stage.
2 By access through the breather aperture in the cylinder head, slacken the locknut securing he serrated plate (see Fig. 1.39) press the locking plunger inwards and rotate the serrated plate, by engaging a tool (a pair of round nosed pliers will be found suitable), in an anti-clockwise direction. Turn the engine each way slightly and check the chain tension;. When tensioned correctly there should be slight flexibility on both outer sides below the camshaft sprockets. The cahin should not be dead tight. Release the locking plunger to the serrated plate and securely tighten the plate locking nut.
3 It is in the foregoing operation that you may run into trouble after fitting the new type of cylinder head gasket. The increased thickness of the gasket may result in undue tightness of the top timing chain and if this occurs you will have to fit a replacement idler eccentric shaft (Part No. C27189).
4 Remove the locking wire (if applicable) from the setscrews securing the camshaft sprockets, it may be necessary to rotate the engine to gain access to both the screws. Remove the setscrews.
5 If the engine has been rotated you will have to bring No. 6 piston back to TDC with the distributor rotor arm opposite the No. 6 segment in the distributor head.
6 Look at the left hand side of the clutch housing and you will find a spring loaded cover which, when pushed to one side, uncovers a section of the starter ring. You will find an arrow marked on the crankcase and a similar arrow on the starter ring and when these arrows are in line (Fig. 1.40) No. 6 piston is exactly at TDC. Check also that the rotor arm is opposite No. 6 segment in the distributor cap.
7 Tap the camshaft sprockets off the flanges of the camshafts and make provision for them not to fall into the sump (see Section 10, paragraph 5).
8 Accurately position the camshafts with the valve timing gauge (refer to paragraph 4, of section 66) so that the slots are perpendicular to the face of the camshaft cover.
9 Refer to Fig. 1.41. Withdraw the circlips retaining the adjuster plates to the camshaft sprockets and press the adjusting plates forward until the serrations disengage. Replace the sprockets on the flanges of the camshafts and align the two holes in the adjuster plate with the two tapped holes in each camshaft flange. Engage the serrations of the adjuster plate with the serrations in the sprockets. **It is important** that the holes are in

exact alignment otherwise, when the setscrews are fitted, the camshafts will be moved out of position.

10 If difficulty is experienced in aligning the holes correctly, the adjuster plates should be turned through 180° and this, due to their construction, will facilitate alignment.

11 Refit the circlips to the sprockets and secure the sprockets with the setscrews turning the engine as necessary, after fitting one screw to give access to the other one.

12 Turn the engine and carry out a check on the correctness of the valve timing.

13 If the timing is satisfactory, lock the setscrews with locking wire. It is advisable to put some rag in the front cover aperture if you have to cut off the ends of the locking wire to prevent them falling into the sump.

14 Refit the flanged plug to the cylinder head at the rear of the left hand camshaft (photo) and the generator at the rear of the right hand camshaft (photo).

15 Place new gaskets on the camshaft cover faces. Refit the camshaft covers and tighten down the dome nuts.

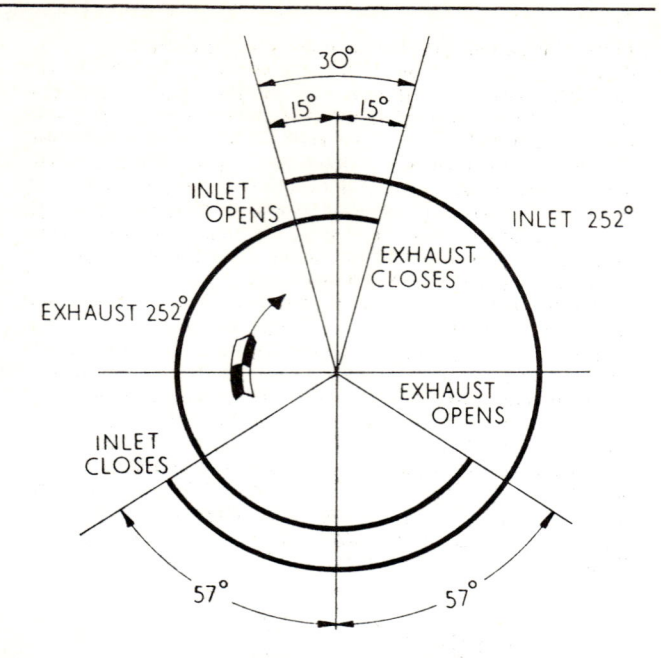

Fig. 1.38. Valve timing diagram

68.14a Plug at rear of left hand camshaft

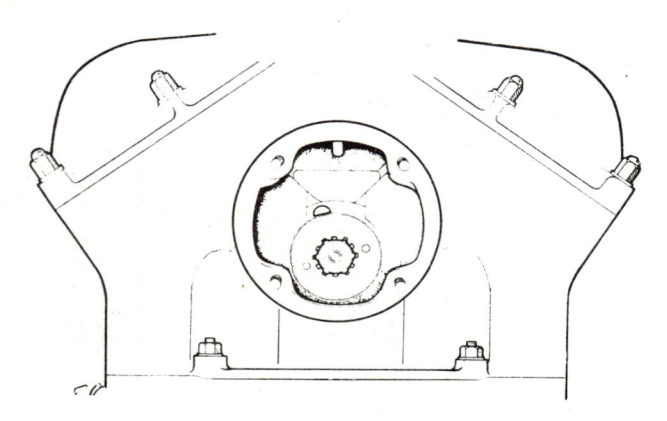

Fig. 1.39. The serrated plate for adjustment of the top timing chain

68.14b Generator at rear of right hand camshaft

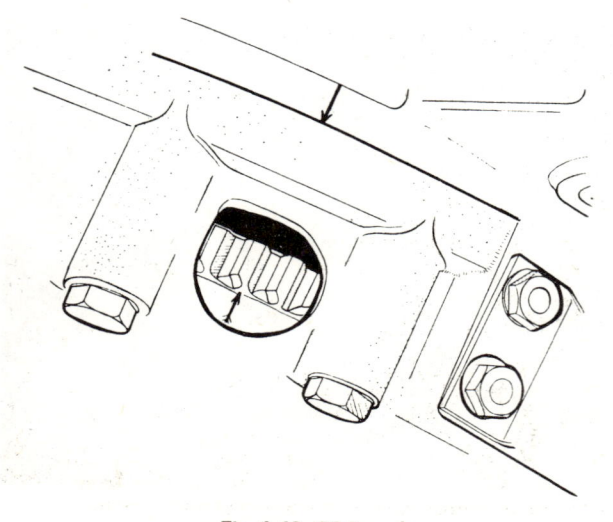

Fig. 1.40. TDC marks

Chapter 1/Engine

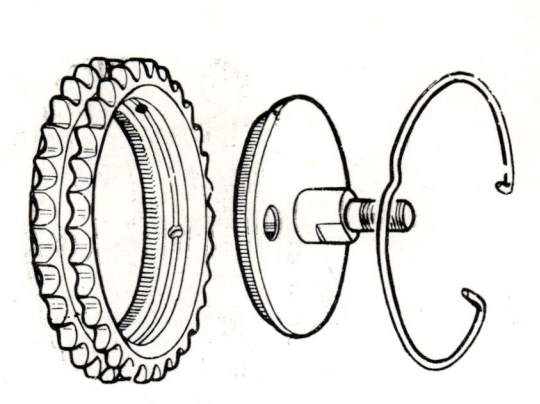

Fig. 1.41. The camshaft sprocket assembly

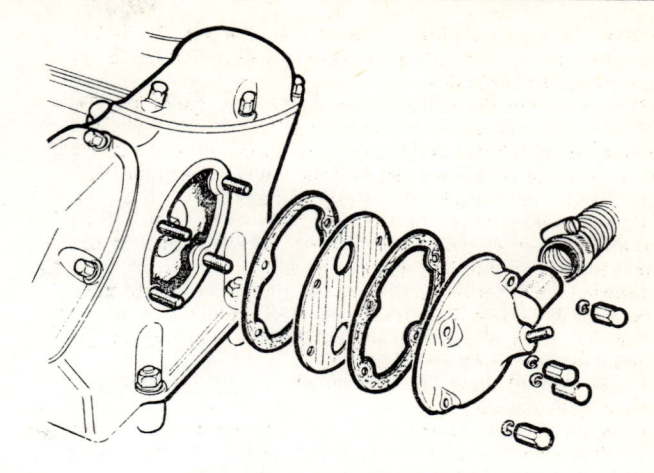

Fig. 1.42. The engine breather assembly

69 Inlet manifold - refitting

1 Clean the mating faces of the inlet manifold and the cylinder block.
2 Place a new gasket over the inlet manifold studs on the cylinder block. Offer up the inlet manifold and tighten down evenly on the nuts.
3 The thermostat can be replaced at this stage or it can be fitted after the engine has been installed. Fit the thermostat and, using a new gasket, refit the water outlet pipe and secure with the three nuts and spring washers.

70 Exhaust manifold - refitting

1 Clean the mating surfaces of the exhaust manifold and the cylinder block.
2 Place a new gasket over each set of exhaust manifold studs on the cylinder block. Tighten down evenly on the nuts to each manifold. Do not forget to fit the clip for the automatic transmisssion dipstick tube (if applicable), and its distance piece, on the rear nuts of the rear manifold.

71 Cylinder head oil feed pipe - refitting

Fit the cylinder head oil feed pipe from the tapped hole in the main oil gallery to the two tapped holes in the cylinder head. Secure the pipe with its three banjo bolts, it is advisable to fit a new copper washer on both sides of each banjo.

72 Water pump - refitting

1 Clean the mating faces of the water pump and the timing cover. They may be treated with sealing compound if desired.
2 Fit a new gasket over the studs in the timing cover.
3 Fit the water pump to the timing cover and secure with the six bolts and the three nuts and spring washers to the studs. Tighten down evenly all round.

73 Engine breather assembly (Fig. 1.42) - refitting

1 Clean the mating faces of the engine breather and of its location at the front of the cylinder block.
2 Fit a new gasket over the studs in the cylinder head, fit the gauze filter followed by the second gasket and then followed by the front cover.
3 Fit, and tighten down evenly, the four dome nuts.

74 Crankshaft damper assembly cone and distance piece - refitting

1 Fit the distance piece over the front end of the crankshaft.
2 Place the Woodruffe key in position and then assemble the crankshaft damper cone to the front end of the crankshaft. Make sure that it does not fall out during subsequent movement of the engine.

75 Gearbox and clutch housing - refitting

1 Offer up the gearbox and clutch housing to the crankcase and make sure that it is fully home. Support the weight to avoid placing strain on the gearbox shaft.
2 Replace the setbolts and nuts securing the clutch housing to the engine starting at the top and working downwards.
3 Offer up the starter and secure in position with the two nuts and bolts.

76 Engine replacement

The engine could be replaced by one man and a suitable hoist but the job will be made much easier, and safer, with the help of an assistant until such time as the engine is secured on its mountings.
Generally replacement is the reverse sequence to removal.
1 Remove the top nut and rubber washer from the engine stabiliser, pass the stem of the stabiliser through its bracket on the front bulkhead and then refit the top washer and nut. Screw down on the nut to the extent of ensuring that the stabilizer will not foul the engine when it is being moved into position.
2 Ensure that all leads, cables etc., are tucked out of the way in the engine bay. It is easy to trap one and so cause additional work when the engine is replaced.
3 Fit the lifting plate to the engine in the manner described in Section 5, paragraph 6.
4 Hoist the engine into position in the engine bay with the car positioned as for removal of the engine (i.e. rear raised or front lowered).
5 Guide the gearbox into its tunnel and then support it on a trolley jack. Start lowering the engine at the front and raising at the rear and at the same time inching the engine rearwards.
6 Before the engine is lowered onto its mountings, you may find it convenient to reconnect the starter motor leads (photo) as these will be easier to get at now than when the engine is on its mountings.

Chapter 1/Engine

7 Raise the engine slightly at the rear to clear the torsion bar mounting and then complete the rearward movement to bring the front mountings in line.
8 Now assemble the components of the rear mounting and offer the mounting up to the gearbox. Place a jack under the mounting plate (photo) and lift upwards to compress the spring and at the same time make sure that the fixing holes in the plate align with the mounting holes in the underbody.
9 Raise the jack under the rear mounting to bring it into its final position and at the same time lower the engine at the front so that the front mountings are also aligned.
10 Replace the setbolts securing the rear mounting and the bolts to the front mounting. Remove the jack and lifting tackle.
11 Replace parts and make all connections generally in the reverse order to that set out in Section 5.
12 Check that all drain taps are closed and that the sump and gearbox drain plugs are tight.
13 Bring the car back to its normal horizontal position.
14 Refill the cooling system and carefully check all water unions for leaks. Tighten up as necessary.
15 Refill the engine and the gearbox with the recommended amount and type of lubricant (photo). Access to the gearbox filler plug is provided by a rubber plug located in the left hand side of the gearbox cover (photo).
16 Reconnect the engine stabiliser to the engine. Now refer to Fig. 1.43. and adjust the engine stabiliser by screwing up the lower flanged washer 'D' until it contacts the bottom of the rubber mounting 'C'. The washer 'D' is slotted in its upper face and can be screwed up the pin by engaging a thin bladed screwdriver in the slot through the centre hole of the rubber mounting. Fit the upper flanged washer 'B' and tighten down on the self locking nut 'A'. Do not overtighten the lower flanged washer as this can cause vibration and/or fouling of the gearbox in its cowl due to the engine having been pulled up on its mountings.
17 Replace the battery (photo) and/or reconnect the terminals.

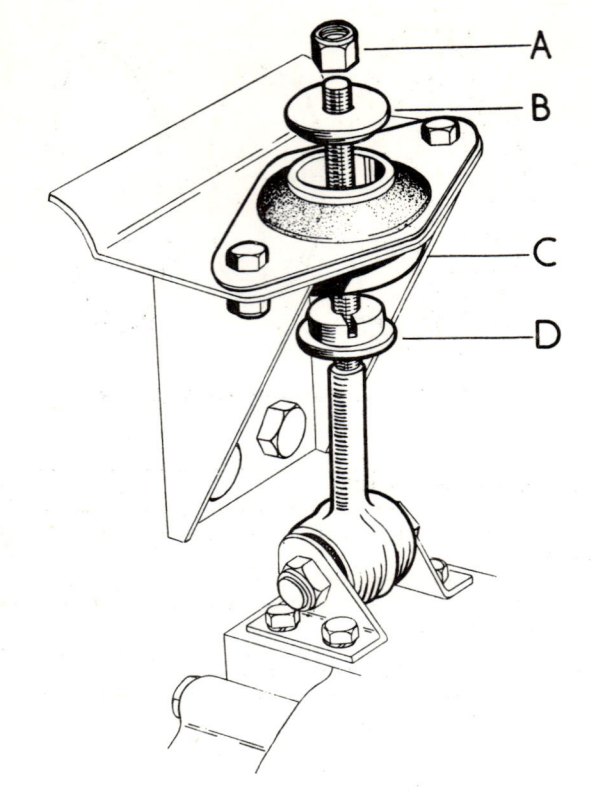

Fig. 1.43. The engine stabiliser

77 Engine initial start-up after overhaul and major repair

Refer to Chapter 3 and follow the instructions in Chapter 4 for timing of the ignition.
Make sure that the battery is fully charged and that all lubricants, fuel and coolant are replenished.
Switch on the ignition and allow the petrol pump time to fill the carburettor float chambers.
Start the engine. As soon as it fires and runs, keep it going at a fast tick-over (no faster). Watch the oil pressure gauge, after a very short wait, whilst the oil filter is being filled, it should register around 40 lbs per sq.in. If, after about 30 seconds running, no oil pressure is registered, switch off the engine and investigate the cause; it may be that you have not fully tightened a union or the filter canister is not correctly seated on the rubber sealing ring in the head of the filter.
Bring the engine up to its normal working temperature, as it warms up there will be odd smells and some smoke from parts getting hot and burning off oil deposits. Look round carefully for water and oil leaks.
When the engine running temperature has been reached, adjust the carburettors as described in Chapter 3.
Stop the engine and wait for a few minutes, check again for leaks. Make sure, before road testing the car, that you are getting full braking power from the servo mechanism. Check your connection if the braking power is weak.
Road test the car to check that the timing is correct and is giving the desired smoothness and power. Do not race the engine. If new bearings or pistons or rings have been fitted, it should be treated as a new engine and run in at reduced revolutions for the first 500 miles.

76.6 The starter motor connections

76.8 Supporting the gearbox on the rear mounting

76.15a Refill engine with lubricant

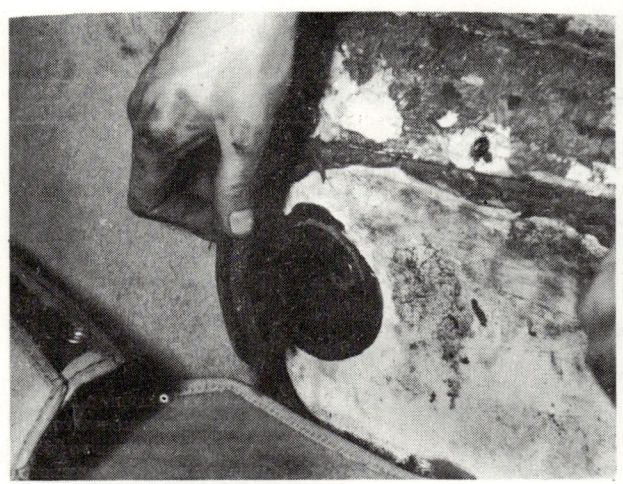

76.15b Access to gearbox filler plug

76.18 Replacing the battery

78 Fault diagnosis - Engine

Symptom	Reason/s	Remedy
Engine fails to turn over when starter switch operated	Flat or defective battery	Charge or replace battery. Push-start car.
	Loose battery leads	Tighten both terminals and earth ends of earth lead.
	Defective starter solenoid or switch or broken wiring	Check and rectify.
	Engine earth strap disconnected	Check and retighten strap.
	Jammed starter motor drive pinion	Place car in gear and rock from side to side.
	Defective starter motor	Remove and recondition.
Engine turns over but will not start	Ignition system damp or wet	Wipe dry the distributor cap and ignition leads.
	Ignition leads to spark plugs loose	Check and tighten at both spark plug and distributor cap ends.
	Shorted or disconnected low tension leads	Check the wiring on the CB and SW terminals of the coil and to the distributor.
	Dirty, incorrectly set, or pitted contact breaker points	Clean, file smooth, and adjust.
	Faulty condenser	Check contact breaker points for arcing, remove and fit new item.
	Defective ignition switch	Replace switch.
	Ignition leads connected wrong way round	Remove and replace leads to spark plugs in correct order.
	Faulty coil	Remove and fit new coil.
	Contact breaker point spring earthed or broken	Check spring is not touching metal part of distributor. Check insulator washers are correctly placed. Renew points if the spring is broken.
	No petrol in petrol tank	Refill tank!
	Vapour lock in fuel line (in hot conditions or at high altitude)	Blow into petrol tank, allow engine to cool, or apply a cold wet rag to the fuel line.
	Blocked float chamber needle valve	Remove, clean, and replace.
	Fuel pump filter blocked	Remove, clean, and replace.
	Choked or blocked carburettor jets	Dismantle and clean.
	Faulty fuel pump	Remove, overhaul, and replace.
	Too much choke allowing too rich a mixture to wet plugs	Remove and dry spark plugs or with wide open throttle, push-start the car.
	Float damaged or leaking or needle not seating	Remove, examine, clean and replace float and needle valve as necessary.
	Float lever incorrectly adjusted	Remove and adjust correctly.
Engine stalls and will not start	Ignition failure - sudden	Check over low and high tension circuits for breaks in wiring.
	Ignition failure - misfiring precludes total stoppage	Check contact breaker points, clean and adjust. Renew condenser if faulty.
	Ignition failure - in severe rain or after traversing water splash	Dry out ignition leads and distributor cap.
	No petrol in petrol tank	Refill tank.
	Petrol tank breather choked	Remove petrol cap and clean out breather hole or pipe.
	Sudden obstruction in carburettor(s)	Check jet, filter, and needle valve in float chamber for blockage.
	Water in fuel system	Drain tank and blow out fuel lines.
	Ignition leads loose	Check and tighten as necessary at spark plug and distributor cap ends.
	Battery leads loose on terminals	Check and tighten terminal leads.
	Battery earth strap loose on body, attachment point	Check and tighten earth lead to body attachment point.
Engine misfires or idles unevenly	Engine earth lead loose	Tighten lead.
	Low tension leads to SW and CB terminals on coil loose	Check and tighten leads if found loose.
	Low tension lead from CB terminal side to distributor loose	Check and tighten if found loose.
	Dirty, or incorrectly gapped plugs	Remove, clean, and regap.
	Dirty, incorrectly set, or pitted contact	Clean, file smooth, and adjust.

Chapter 1/Engine

Symptom	Reason/s	Remedy
	breaker points	
	Tracking across inside of distributor cover	Remove and fit new cover.
	Ignition too retarded	Check and adjust ignition timing.
	Faulty coil	Remove and fit new coil.
	Mixture too weak	Check jets, float chamber needle valve and, filters for obstruction. Clean as necessary. Carburettor incorrectly adjusted.
	Air leak in carburettor	Remove and overhaul carburettor.
	Air leak at inlet manifold to cylinder	Test by pouring oil along joints. Bubbles indicate leak. Renew manifold gasket as appropriate.
	Incorrect valve clearances	Adjust to take up wear.
	Burnt out exhaust valves	Remove cylinder head and renew defective valves.
	Sticking or leaking valves	Remove cylinder head, clean, check and renew valves as necessary.
	Weak or broken valve springs	Check and renew as necessary.
	Worn valve guides or stems	Renew valve guides and valves.
	Worn pistons and piston rings	Dismantle engine, renew pistons and rings.
Lack of power & poor compression	Burnt out exhaust valves	Remove cylinder head, renew defective valves.
	Sticking or leaking valves	Remove cylinder head, clean, check, and renew valves as necessary.
	Worn valve guides and stems	Remove cylinder head and renew valves and valve guides.
	Weak or broken valve springs	Remove cylinder head, renew defective springs.
	Blown cylinder head gasket (accompanied by increase in noise)	Remove cylinder head and fit new gasket.
	Worn pistons and piston rings	Dismantle engine, renew pistons and rings.
	Worn or scored cylinder bores	Dismantle engine, rebore, renew pistons and rings.
	Ignition timing wrongly set. To advanced or retarded	Check and reset ignition timing.
	Contact breaker points incorrectly gapped	Check and reset contact breaker points.
	Incorrect valve clearances	Check and adjust.
	Incorrectly set spark plugs	Remove, clean and regap.
	Carburation too rich or too weak	Tune carburettor for optimum performance.
	Dirty contact breaker points	Remove, clean, and replace.
	Fuel filters blocked causing fuel starvation	Dismantle, inspect, clean, and replace all fuel filters.
	Distributor automatic balance weights or vacuum advance and retard mechanisms not functioning correctly	Overhaul distributor.
	Faulty fuel pump giving top end fuel starvation	Remove, overhaul, or fit exchange reconditioned fuel pump.
Excessive oil consumption	Badly worn, perished or missing valve stem oil seals	Remove, fit new oil seals to valve stems.
	Excessively worn valve stems and valve guides	Remove cylinder head and fit new valves and valve guides.
	Worn piston rings	Fit oil control rings to existing pistons or purchase new pistons.
	Worn pistons and cylinder bores	Fit new pistons and rings, rebore cylinders.
	Excessive piston ring gap allowing blow-up	Fit new piston rings and set gap correctly.
	Piston oil return holes choked	Decarbonise engine and pistons.
	Leaking oil filter gasket	Inspect and fit new gasket as necessary.
	Leaking tap cover gasket	Inspect and fit new gasket as necessary.
	Leaking tappet chest gasket	Inspect and fit new gasket as necessary.
	Leaking timing case gasket	Inspect and fit new gasket as necessary.
	Leaking sump gasket	Inspect and fit new gasket as necessary.
	Loose sump plug	Tighten, fit new gasket if necessary.
Unusual noises from engine	Worn valve gear (noisy tapping from top cover)	Inspect and renew parts as necessary.
	Worn big end bearing (regular heavy knocking)	Drop sump, if bearings broken up clean out oil pump and oilways, fit new bearings. If bearings not broken but worn fit bearing

Symptom	Reason/s	Remedy
	shells.	
Worn chain and gear (rattling from front of engine)	Remove timing cover, fit new timing wheels and timing chain.	
Worn main bearings (rumbling and vibration)	Remove crankshaft, if bearing worn but not broken up, renew. If broken up strip oil pump and clean out oilways.	
Worn crankshaft (knocking, rumbling and vibration)	Regrind crankshaft, fit new main and big end bearings.	

Chapter 2 Cooling system

Contents

Anti-freeze mixture ... 16	Radiator header tank - removal and refitting ... 7
Cooling fan motor - removal and refitting ... 10	Radiator - inspection and cleaning ... 6
Cooling system - draining ... 2	Radiator - removal and refitting ... 5
Cooling system - filling ... 4	Thermostat - removal, testing and replacement ... 11
Cooling system - flushing ... 3	Water pump belt - adjustment, removal and replacement ... 12
Fault diagnosis ... 17	
Fan thermostatic switch - removal and refitting ... 9	Water pump - dismantling and overhaul ... 15
General description ... 1	Water pump - general ... 13
Radiator header tank - inspection and cleaning ... 8	Water pump - removal and refitting ... 14

Specifications

Type	Pressurised system thermostatically controlled
Total capacity of coolant (including heater)	32 Imp. pints, 38½ US pints (18.18 litres)
Coolant pump - type	Centrifugal
- drive	Belt
Coolant pump belt Angle of 'V'	36°
Coolant pump to engine speed ratio	0.9 : 1
Cooling system temperature control	Thermostat in inlet manifold

Thermostat data:

Jaguar Part No.	Opening temperature Degrees C	Remarks
C.20766	70 - 75	
C.20766/1	79 - 84	For use only in cold climates

Auxiliary cooling	Fan driven by electric motor
Auxiliary cooling temperature control	Thermostatic switch located in header tank
Thermostatic switch: Operating temperatures	Cut-in 80° (approx) - cut-out 72° C (approx)
Radiator type:	3.8 and early 4.2 — Cross flow with 10 cooling fins per inch (4 fins/cm)
	Later 4.2 — Vertical flow

Radiator cap:

Make and type		A.C. relief valve
Release pressure:		Early 3.8 models - 4 lbs per sq. in.
		3.8 models commencing at chassis numbers:
		RH drive — LH drive
	Fixed head coupe	861091 — 888241
	Open 2 seater	850657 — 879044
		— 9 lbs per sq. in.
		4.2 models with standard equipment - 7 lbs per sq. in.
		4.2 models with air conditioning system - 13 lbs per sq. in.
Release depression		½ lb

1 General description

Water circulation is assisted by an impeller type pump mounted on the front cover of the engine, the system is pressurised and is thermostatically controlled. Water is circulated from the radiator by the water pump and flows through the cylinder block and the cylinder head water passages to, in the case of earlier models, a separate radiator header tank via the inlet manifold water jacket and is then returned to the radiator. The system is pressurised by means of the header tank filler cap which incorporates a pressure relief valve designed to hold pressures (see Specifications) above atmospheric pressure. As the coolant heats up so it expands and when a pressure equal to, or greater than, the operating pressure of the relief valve in use is reached so the valve opens and allows the excess coolant to escape via the radiator overflow pipe. But, later 4.2 models are fitted with an expansion tank and in this system the overflow pipe from the radiator is connected to the tank so that instead of coolant being lost it is now collected in the expansion tank. As the engine cools so the coolant contracts and because of the pressure differential flows back from the expansion tank into the top of the radiator and thus the need for frequent topping up of the coolant is eliminated. In those systems employing an expansion tank, the combined filler cap/relief valve is fitted to the expansion tank whilst a plain cap is fitted to the radiator, THESE CAPS MUST NOT BE INTERCHANGED.

The thermostat, located in the inlet manifold water jacket, cuts off the coolant in the engine from that in the radiator until such time as the coolant reaches a certain temperature at which, determined by the thermostat setting, the thermostat opens and allow free passage of the coolant around the system. The purpose of the thermostat is to ensure that the engine is brought quickly to its most efficient working temperature.

If the temperature of the coolant rises above 80ºC a fan driven by an electric motor is brought into operation, the motor is controlled by a thermostatic switch mounted in the radiator header tank; the motor automatically cuts out when the temperature of the coolant falls to about 72ºC.

2 Cooling system - draining

1 The engine cooling system is provided with two drain taps, one located in the base of the radiator block (Fig 2.1) and the other positioned on the rear left hand side of the cylinder block below the exhaust manifold (Fig 2.2).
2 Place the car on level ground.
3 If possible, wait until the engine is cold. Unscrew and remove the header tank/expansion chamber filler cap. REMEMBER the system is pressurised so DO NOT remove the cap whilst the engine is hot without taking precautions to prevent injury to yourself by the hot liquid which will be thrown out when the pressure is released. If any liquid containing anti-freeze mixture comes into contact with painted surfaces remove it immediately by washing with clean water otherwise the surface will be damaged.
4 Place the heater control at HOT.
5 Open the radiator drain tap. It some times happens that this tap becomes blocked with sludge accumulated in the bottom of the radiator and it may be possible to clear the tap by poking with a piece of wire but if this is not successful, the tap will have to be removed. Unscrew the tap by use of a spanner on the spanner flats but be careful not to be scalded by the coolant if the engine is at its normal temperature.
6 Open the cylinder block drain tap. This tap may also be blocked by scale and sludge in the cylinder block passages; if poking with wire fails to clear it, remove it by unscrewing with a spanner. Again take precautions if the engine is hot.

3 Cooling system - flushing

1 After prolonged use it is possible that the cooling system will gradually deteriorate in efficiency as the radiator becomes choked with rust scale, deposits from the water and other sediment. The symptom of deterioration is boiling of the engine or high operating temperatures which cannot be accounted for by fuel ignition and other faults. To clean the systems out, remove the filler cap, the bottom hose to the radiator, the cylinder block drain tap and the thermostat (see Section 11). Replace the thermostat housing. Leave a hose running in the radiator filling hole for about fifteen minutes.
2 Reconnect the bottom hose, refit the cylinder block drain tap and the thermostat. Refill the system and at the same time add a proprietary cleaning compound. Beware of splashing the compound on paintwork as this could cause damage. The engine must now be run for the period prescribed by the makers of the compound and this should loosen all sediment and sludge which can now be removed by draining. Thoroughly flush out the system and refill with soft water.
3 In very bad cases it may be necessary to reverse flush the radiator. This can be done with the radiator in position by placing a hose in the bottom hose union of the radiator. Water under pressure is forced through the orifice and out of the filler cap hole.
4 The hose is then removed and is now placed in the filler cap hole and the radiator washed out as described above.

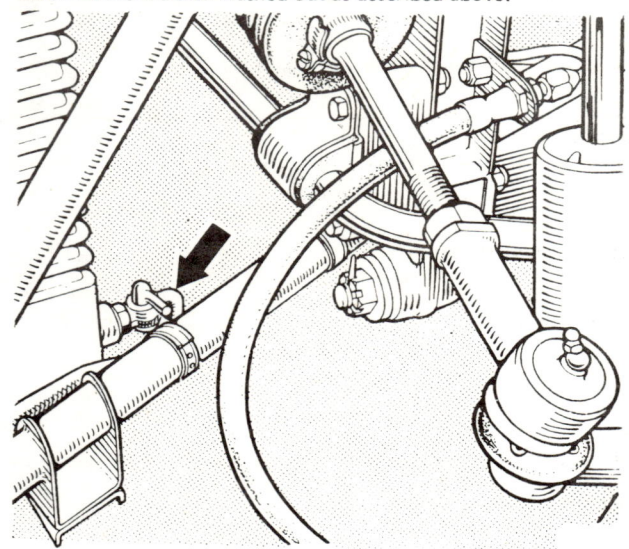

Fig. 2.1. The radiator drain tap

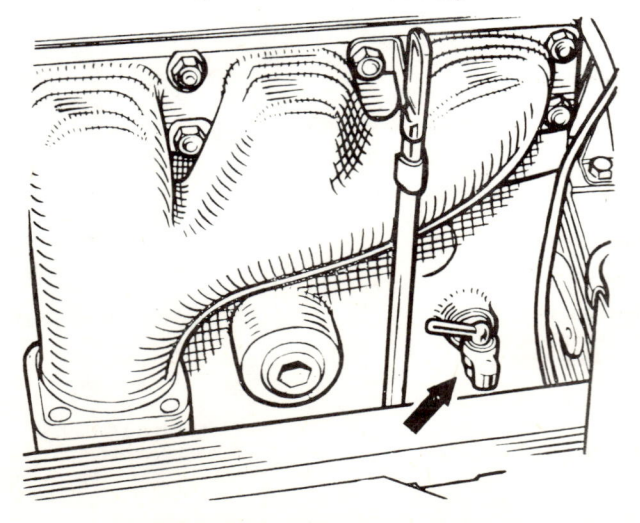

Fig. 2.2. The cylinder block drain tap

Chapter 2/Cooling system

4 Cooling system - filling

1 Place the heater control in the HOT position.
2 Close both drain taps. Fill the system slowly to ensure that no airlocks develop. It is recommended that rain water is used in the system.
3 On those models fitted with an expansion chamber, remove both the radiator and the expansion chamber filler caps. Fill the radiator to the bottom of the filler neck and then replace the cap.
   Top up the expansion tank to the halfway mark, replace the pressure cap.
4 Start the engine and run at a fast idle speed for a few minutes, stop the engine and top up the radiator as necessary. However, for those cars with expansion tanks the coolant level must always be checked at the expansion tank and **not** at the radiator top tank. The level in the tank should be checked when the engine is cold and the tank should be topped up to the halfway mark.

5 Radiator - removal and refitting

1 Drain the radiator and cylinder block in the manner described in Section 2.
2 Refer to Chapter 12 and follow the instruction for removal of the bonnet.
3 Slacken the clips and remove the right and left hand water hoses running from the header tank to the top of the radiator (photo).
4 Slacken the clips securing the rubber hose which connects the bypass pipe to the union at the right hand side of the radiator (photo) and disconnect the hose. However, you may find it easier to also slacken the clip at the inlet manifold water jacket and then remove the pipe complete.
5 Remove the two self locking nuts and bolts securing the radiator steady brackets to the header tank support bracket.
6 On those models equipped with an expansion chamber, the radiator cowl will now have to be disconnected from the radiator and this is done by removing the six setscrews which secure the cowl to the side brackets of the radiator matrix. Now disconnect the fan thermostat switch cables at the cable junction.
7 If air conditioning equipment is fitted to the car, remove the two setscrews which secure the condenser unit to the radiator matrix. Push the condenser unit to one side.
   DO NOT DISTURB THE HOSE CONNECTIONS AT THE CONDENSER UNIT. IT IS MOST UNWISE FOR AN UNQUALIFIED PERSON TO ATTEMPT TO DISCONNECT OR REMOVE ANY PART OF THE AIR CONDITIONING SYSTEM.
8 Now move below the radiator. If your car is an automatic transmission model, the first task is to disconnect the oil cooler pipes at **their** unions in the bottom of the radiator. After disconnecting the pipes, immediately blank them off, and also the unions, to prevent loss of transmission fluid and the ingress of dirt.
9 Slacken the clips of the rubber hose connecting the water pipe from the pump to the bottom of the radiator and then work the hose off the radiator union. As with the bypass pipe, you may find it easier to also disconnect the hose at the pump and remove the pipe completely.
10 Refer to Fig. 2.3 and remove the two bolts which secure the duct shield located between the lower edge of the radiator and the subframe cross tube.
11 Remove the two self locking nuts (Fig. 2.4), the washers and the mounting rubbers which secure the bottom of the radiator to the subframe.
12 Carefully lift out the radiator taking care not to damage the matrix on the cooling fan blades. Take care also not to damage the matrix of the air conditioning condenser unit, if fitted. Collect the remaining mounting rubbers and the spacers.
Important Always store the radiator block in an upright position to guard against sediment which may have collected in the bottom of the tank passing into the small core passages and subsequently causing a blockage.
13 On cars, other than those specified as paragraph 6 above, the radiator cowl can now be removed from the radiator. As illustrated in Fig. 2.5, the cowl is removed by taking off the two self locking nuts which secure it to the bottom of the radiator followed by removal of the two self locking nuts which secure it to the two steady brackets at the top of the radiator. Remove the cowl and the sealing rubber.
14 Refitting the cowl to the radiator and refitting of the radiator to the car is the reverse sequence to the above. Ensure that the correct number of washers are fitted to the bottom bolts securing the radiator to the subframe.

6 Radiator - inspection and cleaning

1 Examine the top and bottom tanks for damage and leaking especially at the seams. Any leaks or possible weakness can be repaired with a compound such as Cataloy, the application of heat to the radiator, soldering for instance, is not recommended for the home enthusiast as this may result in breaking other soldered seams.
2 Examine the core for damage and corrosion. It may be possible to repair leaks from physical damage using Cataloy but if leaks are present due to corrosion it is best to replace the radiator as any repair that is made will only effect a temporary remedy. A replacement radiator can be obtained from your Jaguar agent on an exchange basis.
3 When the radiator is out of the car it is advantageous to reverse flush it in the manner described in Section 3. Clean the outside of the radiator by hosing down the matrix with a strong jet of water to clean away road dirt, dead flies etc.
4 Inspect the radiator hoses for cracks, internal and external perishing and cuts on the exterior from the hose clips. Change the hose if its condition is at all doubtful. Examine the hose clips for rust and damage and replace as necessary.

7 Radiator header tank - removal and refitting

1 Drain the radiator and the cylinder block in the manner described in Section 2.
2 Remove the two water hoses on 4.2 litre cars, or single hose in the case of 3.8 litre models running from the header tank to the top of the radiator (photograph 5.3)
3 Slacken the clip and remove the hose running from the tank to the thermostat outlet on the inlet manifold water jacket (photo).
4 Identify the leads to the fan thermostatic switch and then disconnect them (photo).
5 Identify the leads to the thermostatic switch relay (early models) and then pull them off the lucar tags.
6 Remove the two self locking nuts and bolts securing the tank to the radiator steady brackets.
7 Remove the two bolts securing the header tank support bracket to the cross member (Fig.2.6). Collect the rubber mounting pads, distance collars and washers.
8 Lift off the header tank complete with the mounting bracket.
9 The mounting bracket can be removed from the tank by taking out the four securing setscrews.
10 Refitting is the reverse of the above procedure but when remaking the connection to the fan thermostatic control switch, the black/red cable should be attached to the centre connector with the black wire to the earth connection.

8 Radiator header tank - inspection and cleaning

1 Examine the tank for damage and leaks, especially at the seams. Any leaks or possible weakness can be repaired with a compound such as Cataloy, the application of heat to the tank, soldering for instance, is not recommended for the home enthusiast as this may result in extending the damage or in breaking other seams.

5.3 Remove top water hoses

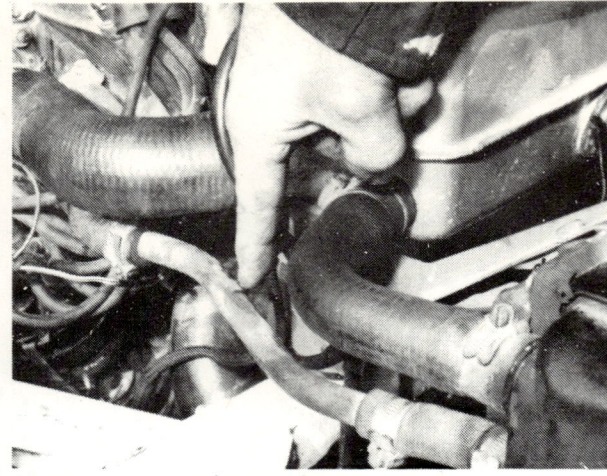

5.4 The by-pass hose (4.2 litres)

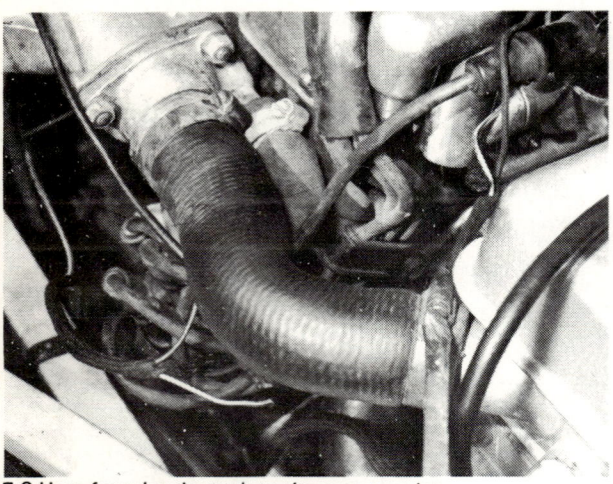

7.3 Hose from header tank to thermostat outlet

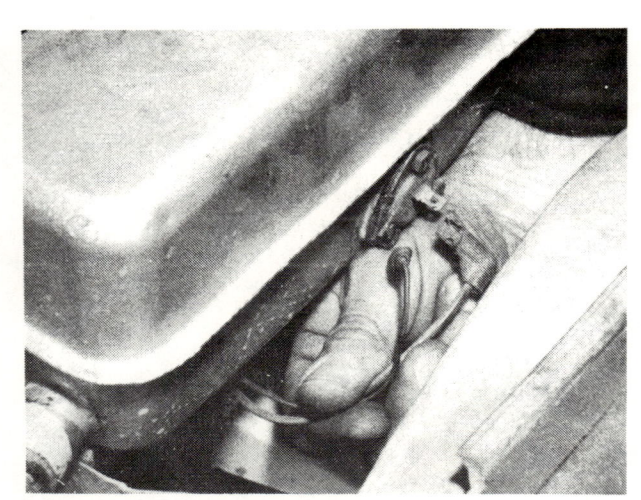

7.4 Leads to fan thermostatic switch

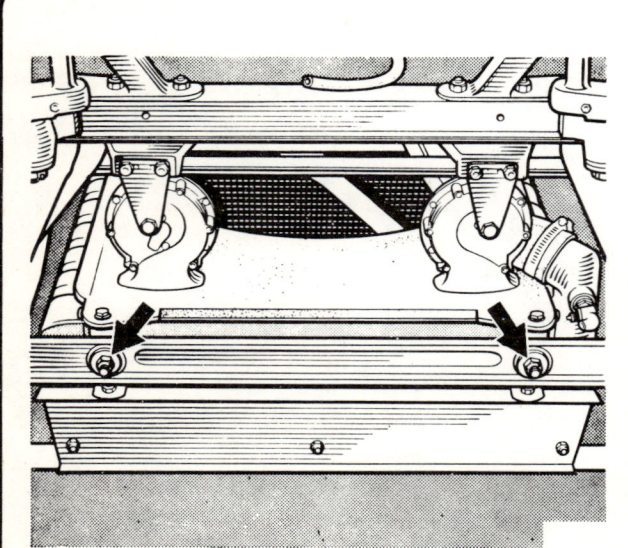

Fig. 2.3. Bolts securing the duct shield

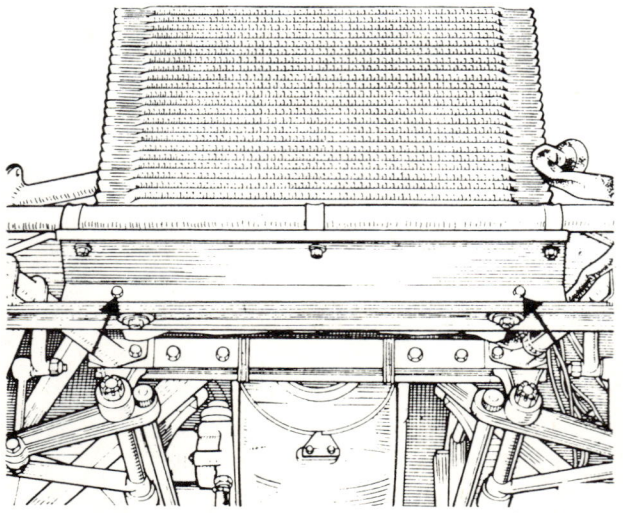

Fig. 2.4. Attachment of radiator to subframe

Chapter 2/Cooling system

2 Examine the inside of the tank for corrosion and acccumulation of sediment. Sediment can be cleaned out by vigorous washing.
3 Inspect the hoses for cracks, internal and external perishing and cuts on the exterior from the hose clips. Change any hose if its condition is at all doubtful. Examine the hose clips for rust and damage and replace as necessary.

9 Fan thermostatic switch - removal and refitting

1 Drain sufficient coolant from the system to empty the header tank.
2 Identify and then disconnect the two electrical connections to the switch by pulling them off the Lucar tags.
3 Remove the three setscrews and washers and then withdraw the switch and the cork gasket (Fig.2.7). The switch is located in the top of the tank in early model cars.
4 Refitting is the reverse of the above procedure but use a new gasket. Connect the black/red cable to the centre tag of the switch and connect the black cable to the earth connection.

10 Cooling fan motor - removal and refitting

Early 3.8 and 4.2 litre models are equipped with a single fan and motor mounted on the front sub assembly. Later 4.2 models have twin fan motors secured to mounting brackets on the radiator cowl.
1 Disconnect the negative lead on the battery.
2 Identify and then disconnect the leads to the thermostatic switch by pulling them off their Lucar tags.
3 The single fan motor is removed by taking off the four self locking nuts which secure the fan motor to the front sub-assembly. Now withdraw the electric motor and the fan blades from the right hand side between the radiator and the frame assembly.
4 The twin fan motors are removed by first taking off the six setscrews which secure the cowl to the radiator.
5 Remove the cowl complete with the fan motors and their mounting brackets.
6 Remove the three nuts and setscrews which secure each fan mounting bracket to the cowl and then detach the bracket assembly.
7 Remove the four nuts and washers securing each motor to its bracket and then detach the motor.
8 Refitting in each case is the reverse of the removal procedure. Reconnect the electric leads to the switch as described in Section 9, paragraph 4.

11 Thermostat - removal, testing and replacement

The thermostat is located in the inlet manifold water rail forward of the carburettors as illustrated in Fig. 2.8 for 3.8 litre and Fig. 2.9 for 4.2 litres.
1 To remove the thermostat first partially drain the system, approx 8 pints will be sufficient, collect the coolant in a suitable container if it is desired to re-use.
2 Slacken the clip securing the hose to the thermostat housing and remove the hose.
3 Remove the two nuts securing the outlet pipe to the housing and remove the pipe and gasket. The thermostat will now be visible in its housing.
4 Remove the thermostat from its housing, it is possible that it will be securely held in place by scale in which case careful levering with a small screwdriver will be necessary but do not lever on the circular valve which will be seen on the top face of the thermostat.
5 Clean the thermostat and ensure that the small hole on the valve is clear. If the valve is open it indicates that the thermostat is unserviceable and should be replaced with a new item of similar operating temperature, this figure will be seen on the top

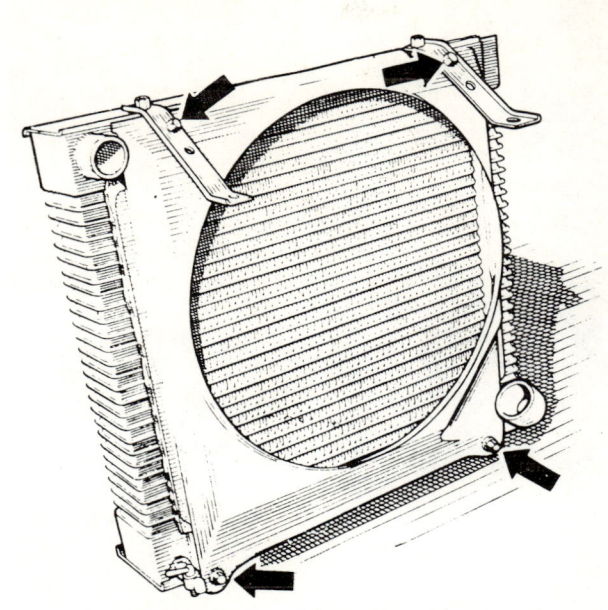

Fig. 2.5. Attachment of the cowl to the radiator

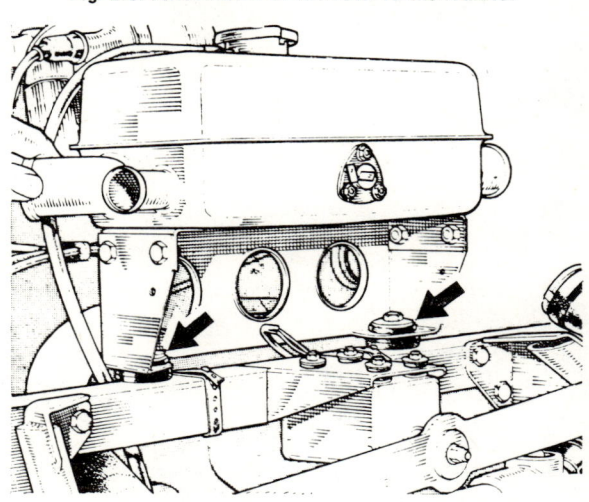

Fig. 2.6. Header tank mounting points

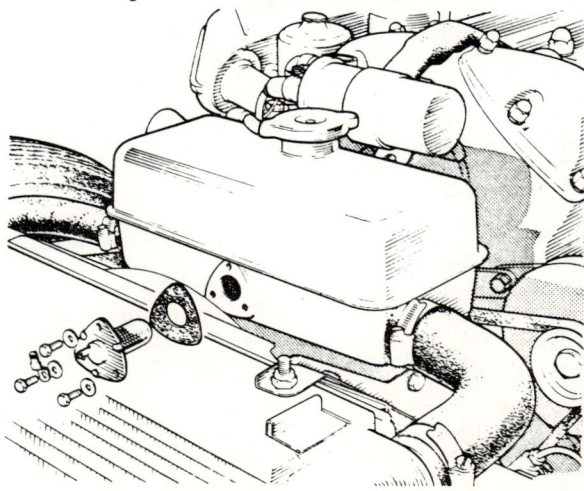

Fig. 2.7. Fan motor thermostatic switch

Chapter 2/Cooling system

side of the thermostat.

6 If correct operation of the thermostat is in doubt test it by immersing it together with a 0 - 100°C thermometer in a container of cold water. Heat the water, keeping it stirred, and observe if the operation of the valve is in close agreement to the temperature marked on the body of the thermostat. Allow the water to cool down and check that the valve closes correctly.

7 If the operation is satisfactory, the thermostat may be refitted in the reverse order to the above. A new gasket should be fitted between the elbow pipe and the thermostat housing.

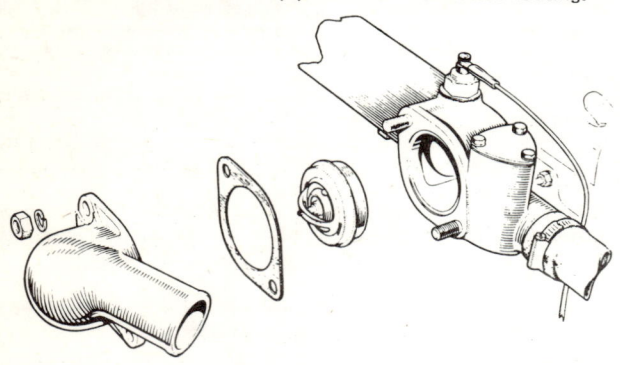

Fig. 2.8. Exploded view of the thermostat and housing — 3.8 litre cars

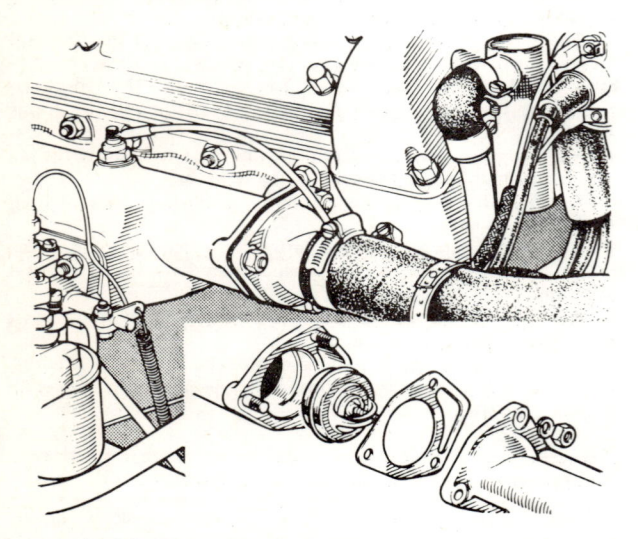

Fig. 2.9. The thermostat and housing — 4.2 litre cars

12 Water pump belt - adjustment, removal and replacement

1 To adjust the belt on early model cars, slacken the two mounting bolts the dynamo/alternator and also the adjusting link bolt as shown in Fig.2.10. Pull the dynamo/alternator outwards until the belt can be flexed about ½" (12.7 mm) either way at a point midway between the pulleys and now tighten the mounting and link bolts. Adjustment is not necessary on later model cars fitted with a jockey pulley.

It is important to keep the belt correctly adjusted as slackness of the belt will cause slip with the possible result of a squealing noise from the belt and/or a reduced rate of charge. On the other hand too much tension on the belt will create undue wear of the belt, the pulleys and the dynamo/alternator bearings.

2 To remove the belt, first slacken the dynamo/alternator mounting and link bolts and then push the dynamo/alternator towards the engine to relieve tension on the belt. If an automatic tensioner is fitted, press it against its spring to relieve belt tension.

3 Work the belt, by hand, off the dynamo/alternator pulley; do not use any form of leverage or the belt will be damaged.

4 Clear the belt from the crankshaft damper pulley and then remove it from the car.

5 Examine the belt. If it appears to be worn, cracked, or is obviously stretched it should be renewed.

6 Refitting of the belt is the reverse of the above procedure followed by adjustment of the tension as described in paragraph 1.

Fig. 2.10. The dynamo/alternator mounting points

A Link bolt (captive nut)
B Top mounting bolt
C Bottom mounting bolt

13 Water pump - general

An exploded view of the water pump fitted to 3.8 and 4.2 litre models is given in Figs.2.11 and 2.12 respectively. It will be seen that except for differences in the pump body (in the 3.8 litre car the bypass hose connects to the pump body and not to the radiator as in the case of the 4.2 litre car) and differences in the pulley carrier, the two pumps are identical, thus the sectioned view of the pump in Fig. 2.13 can be taken to cover both models.

The pump is of centrifugal vane impeller type with the impeller mounted on a steel spindle which in turn runs in a double row of ball bearings. The bearings are sealed at their ends to exclude all dirt and to retain the lubricant. The main seal on the pump spindle is located in the pump housing by a metal cover and the carbon face maintains a constant pressure on the impeller by means of a thrust spring inside the seal. A hole is drilled in the top of the body casting and this acts as an air vent and lead into an annular groove in the casting into which stray water is directed by a rubber thrower on the pump spindle whilst a drain hole, at the bottom of the groove, serves as an outlet and prevents seepage into the bearing.

14 Water pump - removal and refitting

1 Disconnect the battery.
2 Drain the cooling system.
3 Refer to Section 7 and remove the header tank.
4 Remove the water pump drive belt in the manner described in Section 12.
5 Slacken the clips of the water hoses connecting to the pump and then disconnect the hoses from the pump.

Chapter 2/Cooling system

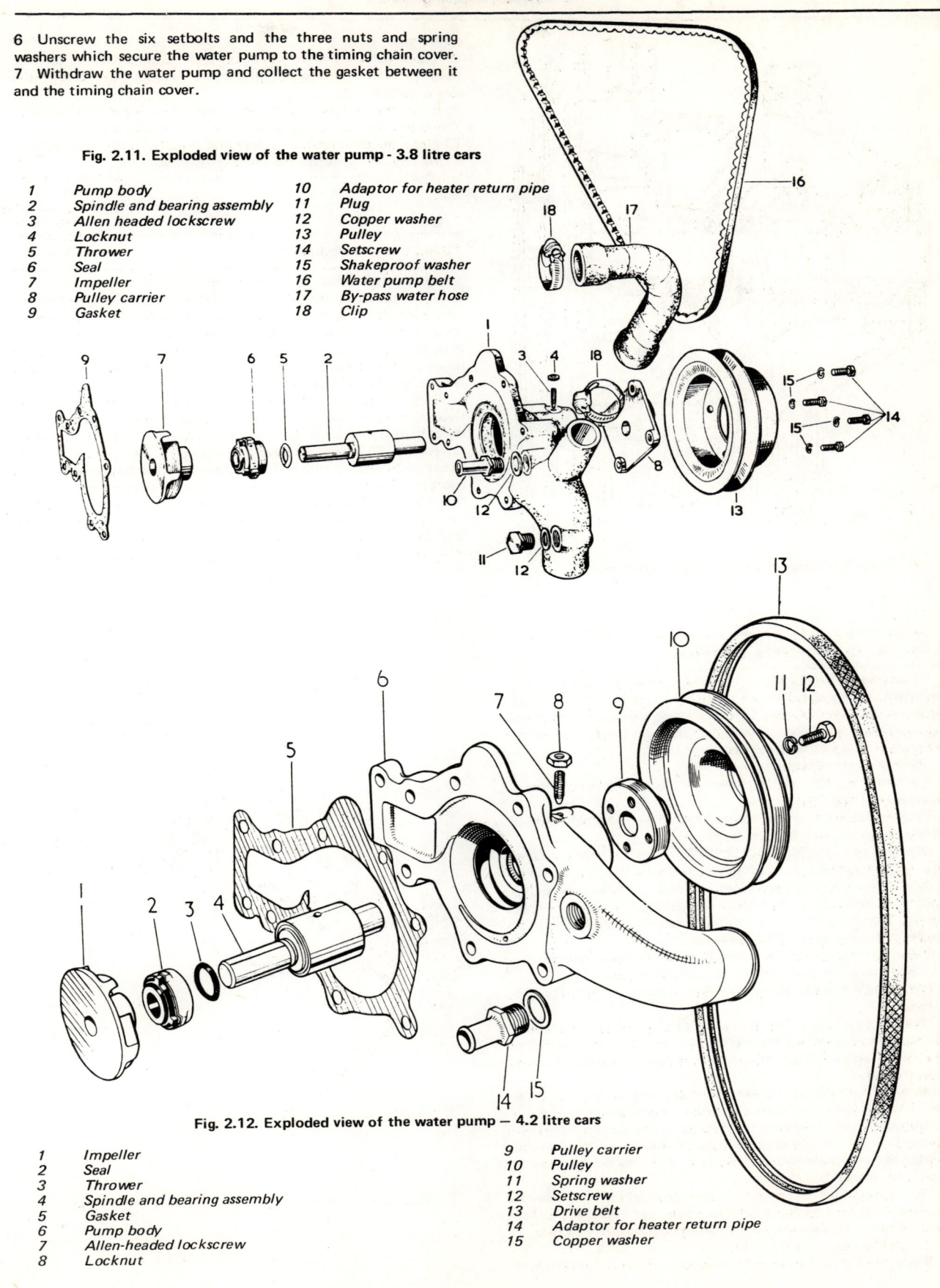

6 Unscrew the six setbolts and the three nuts and spring washers which secure the water pump to the timing chain cover.
7 Withdraw the water pump and collect the gasket between it and the timing chain cover.

Fig. 2.11. Exploded view of the water pump - 3.8 litre cars

1	Pump body	10	Adaptor for heater return pipe
2	Spindle and bearing assembly	11	Plug
3	Allen headed lockscrew	12	Copper washer
4	Locknut	13	Pulley
5	Thrower	14	Setscrew
6	Seal	15	Shakeproof washer
7	Impeller	16	Water pump belt
8	Pulley carrier	17	By-pass water hose
9	Gasket	18	Clip

Fig. 2.12. Exploded view of the water pump — 4.2 litre cars

1	Impeller	9	Pulley carrier
2	Seal	10	Pulley
3	Thrower	11	Spring washer
4	Spindle and bearing assembly	12	Setscrew
5	Gasket	13	Drive belt
6	Pump body	14	Adaptor for heater return pipe
7	Allen-headed lockscrew	15	Copper washer
8	Locknut		

Chapter 2/Cooling system

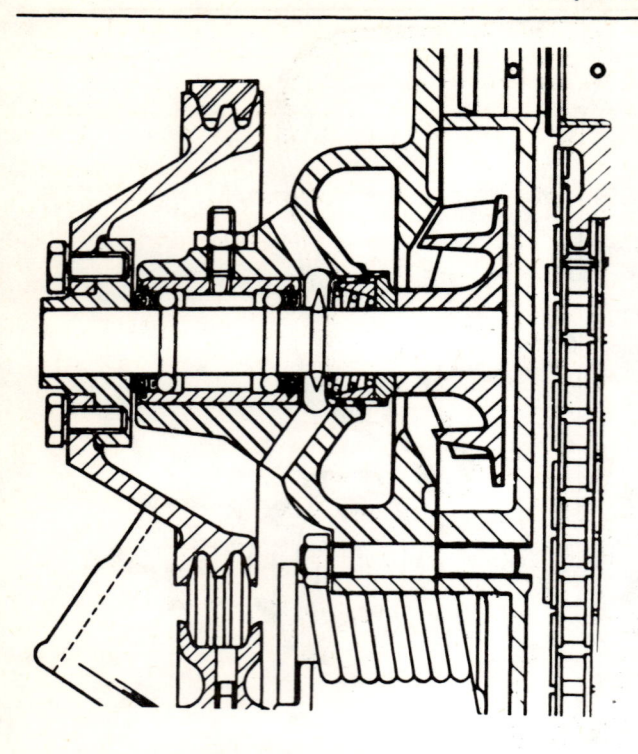

Fig. 2.13. Sectioned view of the water pump

15 Water pump - dismantling and overhaul

If the water pump starts to leak, show signs of excessive movement of the spindle or is noisy during operation it can be dismantled and overhauled. Before starting this task, make sure that individual parts are available but the best plan really is to obtain and fit an exchange assembly.

To dismantle the water pump.
1 Remove the fan hub by means of a suitable extractor as illustrated in Fig.2.14.
2 Slacken the locknut and remove the Allen head screw which retains the pump bearing outer race.
3 Obtain a piece of tube 1 3/32" (27.77 mm) outside diameter and 31/32" inside diameter. Register this with the front face of the outer race of the pump bearing and drift out the pump spindle, the impeller and bearings assembly from the front of the housing. This assembly must not be pushed out by means of the spindles or the bearing will be damaged.
4 Press out the spindle from the impeller as illustrated in Fig. 2.15 and remove the seal and rubber water thrower.
5 The spindle and bearing assembly cannot be dismantled any further.
6 Thoroughly clean all parts of the pump except the spindle and bearing in paraffin. The bearing is a permanently sealed and lubricated assembly and, therefore, must not be washed in any circumstances.
7 Inspect the bearings for excessive play and remove any burrs, rust or scale from the shaft with fine emery paper after taking the precaution of covering the bearing with a cloth to prevent ingress of dirt. If there are any signs of wear or corrosion in the bearing bore or on the face in front of the impeller the housing should be renewed.
8 To reassemble the pump, install the spindle and bearing assembly into the pump body from the rear and line up the location hole in the bearing with the tapped hole in the body.
Fit the locating screw and locknut.
9 Place the rubber thrower in its groove on the spindle in front of the seal.
10 Coat the outside of the brass seal housing with a suitable water resistant jointing compound and fit into the recess in the pump casting.
11 Push the seal into its housing with the carbon face towards the rear of the pump and ensure that it is seated correctly.
12 Press on the impeller as shown in Fig.2.16 until the rear face of the impeller is flush with the end of the spindle.
13 Press the fan hub on to the spindle until it is flush with the end.

16 Anti-freeze - mixture

During the winter months an anti-freeze compound with an inhibited ethylene glycol base should be used in the proportions laid down by the manufacturers of the anti-freeze mixture. It should be remembered, if an anti-freeze mixture is not used, that it is possible for the radiator to freeze-up whilst the car is being driven even though the water in the radiator was not frozen before the car was started.

Before adding anti-freeze solution, check all water unions and the tightness of the cylinder head bolts. Flush out the system described in Section 3 and allow the system to drain. Close all drain taps. To ensure satisfactory mixing of the water and anti-freeze solution, measure the recommended proportions into a container and fill the system from this container rather than add the solution direct to the system. If 'topping up' is necessary during the period that anti-freeze is in use, remember that the addition of straight water will dilute the mixture and so that required degree of protection against frost damage will be lost.

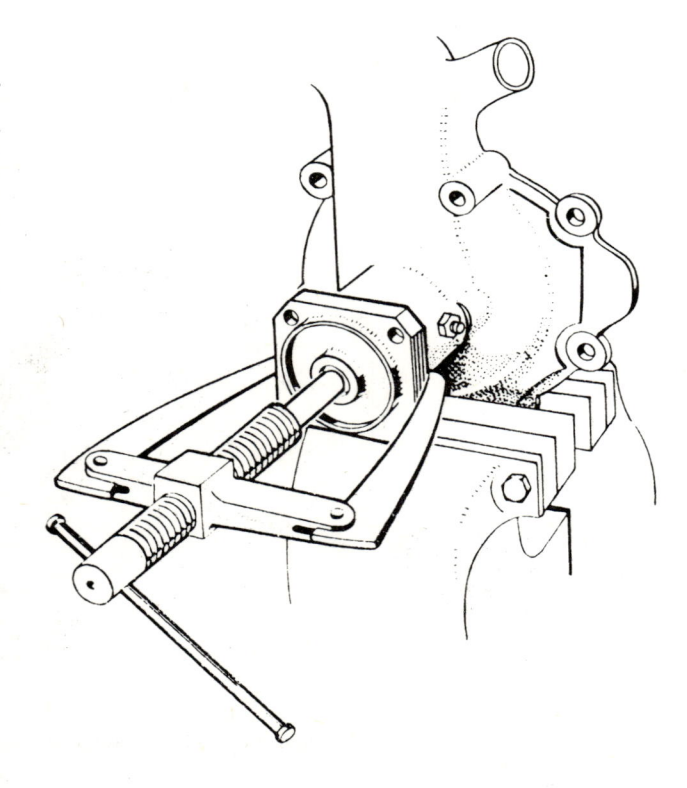

Fig. 2.14. Withdrawing the fan hub from the spindle

Chapter 2/Cooling system

17 Fault diagnosis

Symptom	Cause	Remedy
Overheating	Insufficient water in cooling system	Top up radiator
	Water pump belt slipping	Tighten belt to recommended tension or replace if worn.
	Radiator core blocked	Reverse flush the radiator.
	Thermostat not opening properly	Remove and fit new thermostat.
	Faulty electrical connections to thermostatic switch	Check and rectify
	Faulty fan motor thermostatic switch	Check by substitution.
	Faulty fan motor	Check and replace.
	Ignition advance and retard incorrectly set (accompanied by loss of power and perhaps misfiring)	Check and reset ignition timing.
	Incorrect fuel/air mixture	Tune carburettors.
	Exhaust system partially blocked	Check exhaust pipe for obstruction.
	Oil level in sump too low	Top up to correct level.
	Blown cylinder head gasket (water/steam being forced down the radiator overflow pipe under pressure)	Remove cylinder head and fit new gasket.
	Engine not yet 'run-in'	Run-in slowly and carefully.
	Brakes binding	Check and adjust brakes.
Engine running 'cold'	Thermostat jammed open	Remove and renew thermostat.
	Incorrect grade of thermostat fitted	Remove and replace with correct type of thermostat.
	Thermostat missing	Check and fit correct thermostat.
Leaks in system	Loose clips on water hoses	Check and tighten clips.
	Top or bottom water hoses perished	Check and replace any faulty hoses.
	Radiator leaking	Remove radiator and repair.
	Thermostat gasket leaking	Inspect and renew gasket.
	Pressure cap spring worn or seal ineffective	Renew pressure cap.
	Cylinder wall or head cracked	Dismantle engine and despatch to engineering works for repair.
	Core plug corroded	Remove old plug and fit new item.

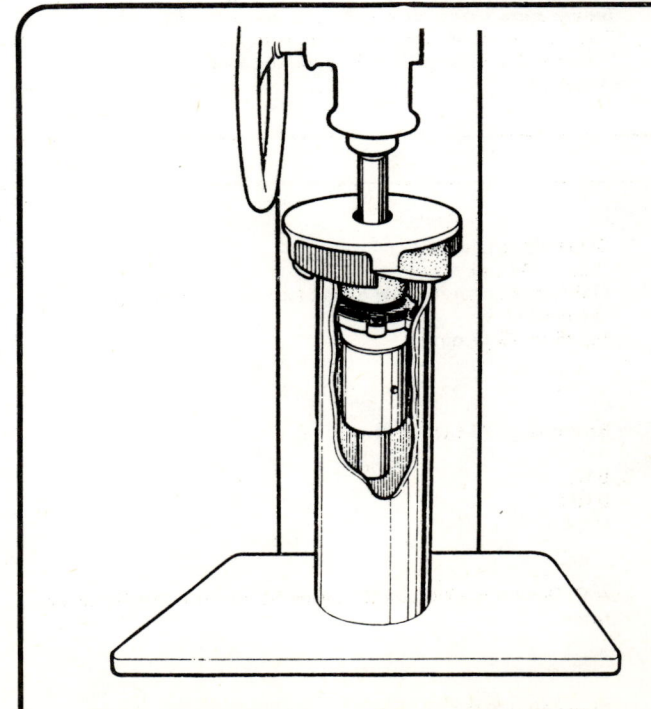

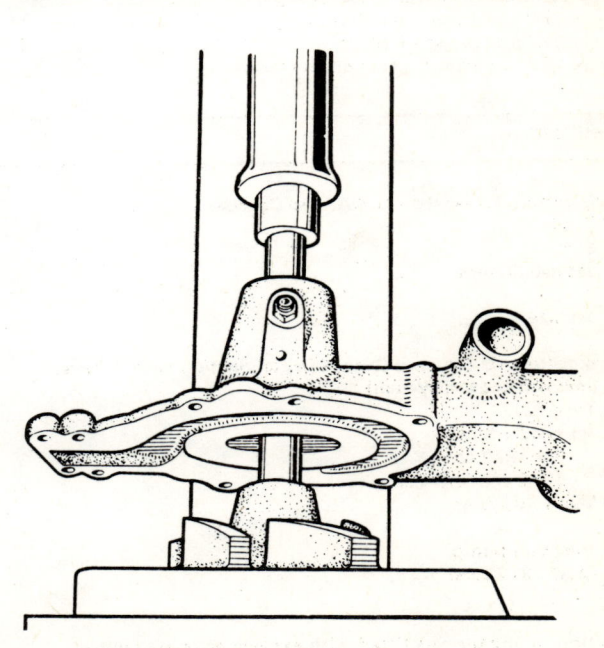

Fig. 2.15. Removing the impeller from the spindle

Fig. 2.16. Fitting the impeller to the spindle

Chapter 3 Fuel system, carburation and exhaust emission control system

Contents

Air cleaner and filter element - removal and refitting ... 38	Lucas 2FP fuel pump - removal and refitting ... 3
Air delivery pump - removal and refitting ... 35	Secondary throttle housing - dismantling and re-assembly ... 43
Air duct - removal and refitting ... 37	Secondary throttle housing - removal and refitting ... 42
Air rail - removal and refitting ... 36	Stromberg 175 CD2SE carburettor - general ... 27
AUF 301 fuel pump - dismantling ... 8	Stromberg carburettor - dismantling and re-assembly ... 31
AUF 301 fuel pump - examination ... 9	Stromberg carburettor - removal and refitting ... 30
AUF 301 fuel pump - fault finding ... 11	Stromberg carburettor - tuning and adjustment ... 32
AUF 301 fuel pump - general information ... 6	SU carburettors - adjustment and tuning ... 25
AUF 301 fuel pump - re-assembly ... 10	SU carburettor - dismantling and re-assembly ... 16
AUF 301 fuel pump - removal and refitting ... 7	SU carburettor - examination and repair ... 17
Carbon canister - removal and refitting ... 34	SU carburettor - float chamber flooding ... 20
Carburettor items that must not be changed ... 28	SU carburettor - float chamber needles ... 21
Carburettor mixture control warning light - setting ... 26	SU carburettor - float needle sticking ... 19
Check valve - removal and refitting ... 41	SU carburettors - general information, adjustment and tuning ... 24
Exhaust emission control - general information ... 33	SU carburettor - jet centering ... 22
Fault diagnosis ... 46	SU carburettor - needle replacement ... 23
Fuel tank gauge unit - removal and refitting ... 13	SU carburettor - piston sticking ... 18
Fuel tank - removal and refitting ... 12	SU carburettor - removal and refitting ... 15
General description ... 1	SU HD.8 carburettors - general information ... 14
Gulp valve - removal and refitting ... 40	Temperature sender unit - removal and refitting ... 39
Inlet manifold - removal and refitting ... 44	Test equipment - emission control systems ... 29
Lucas 2FP fuel pump - fuel flooding ... 5	Thermostatic vacuum switch - removal, testing and refitment ... 45
Lucas 2FP fuel pump - fuel starvation ... 4	
Lucas 2FP fuel pump - general information ... 2	

Specifications

Carburettors (other than U.S.A. and Canada)
- Type ... SU HD.8 (triple)
- Size ... 2 in. (5.08 cm)
- Jet needle type ... UM (stamped on the side or top face of the parallel portion of the needle)
- Jet size ... 0.125 in. (3.17 mm)

Carburettors (U.S.A. and Canada in conjunction with exhaust emission control system)
- Type ... Stromberg 175 CD2SE (twin)
- Jet needle type:
  - Pre 1972 cars ... B1E
  - 1972 and later cars ... B1BT
- Manifold system ... Duplex

Air injection pump (post 1972 cars) ... A.C. Delco type 7803943D 2 vane fitted with centrifugal air filter

Ignition timing for cars fitted with exhaust emission control system:
- Pre 1972 cars ... Static 5° B.T.D.C.
  - 1000 rpm 10° B.T.D.C.
  - 1200 rpm 13 - 17° B.T.D.C.
  - 1600 rpm 22 - 26° B.T.D.C.

Chapter 3/Fuel system, carburation and exhaust emission control system

1972 and later models	2900 rpm 29 - 33° B.T.D.C. 3700 rpm 33 - 37° B.T.D.C. Static 10° B.T.D.C. 1400 rpm 1 - 3½° A.T.D.C. 2100 rpm 7½ - 10° A.T.D.C. 2500 rpm 10½ - 11° A.T.D.C. 5500 rpm 12½ - 14½° A.T.D.C.
Fuel pump	
Early 3.8 litre cars	Lucas 2FP
Later 3.8 litre and all 4.2 litre models	AUF.301
Fuel tank capacity	
Standard	Single tank - 14 Imp. galls, 16¾ U.S. galls (63.64 litres)
4.2 litre export to U.S.A. and Canada	Two tanks, each tank 10 Imp. galls, 12 U.S. galls (45.5 litres)

maintenance:

	RH Drive	LH Drive
Open 2 seater	1E.1905	1E.16057
Fixed Head Coupe	1E.21662	1E.34772
2 + 2	1E.50143	1E.77701

1 General description

A layout of the standard fuel system is shown in Fig.3.1. It consists of a fuel tank mounted at the rear of the car from which fuel is pumped to the carburettors by, in the case of earlier 3.8 litre models, a Lucas 2FP electrical pump located inside the tank. All later models employ an AUF 301 electrical pump mounted on brackets adjacent to the fuel tank. Interposed between the fuel tank and the carburettors is a fuel feed line filter which is mounted on the bulkhead at the right hand side of the engine compartment, the filter is of the glass bowl type with a flat filter gauze which can be removed for cleaning, however, from the chassis numbers quoted below 4.2 litre models have a filter assembly which incorporates a renewable filter element which should be replaced at the intervals given in Routine

All models other than those exported to the USA and Canada are fitted with triple SU HD8 carburettors but to meet USA and Canadian exhaust emission requirements the engines of 4.2 litre cars exported to those countries have been modified by the fitment of twin Stromberg carburettors, a Duplex Manifolding System and a revised ignition timing as indicated in Specifications at the beginning of this Chapter.

The operation of the individual components making up the fuel and carburation systems is described elsewhere in this Chapter.

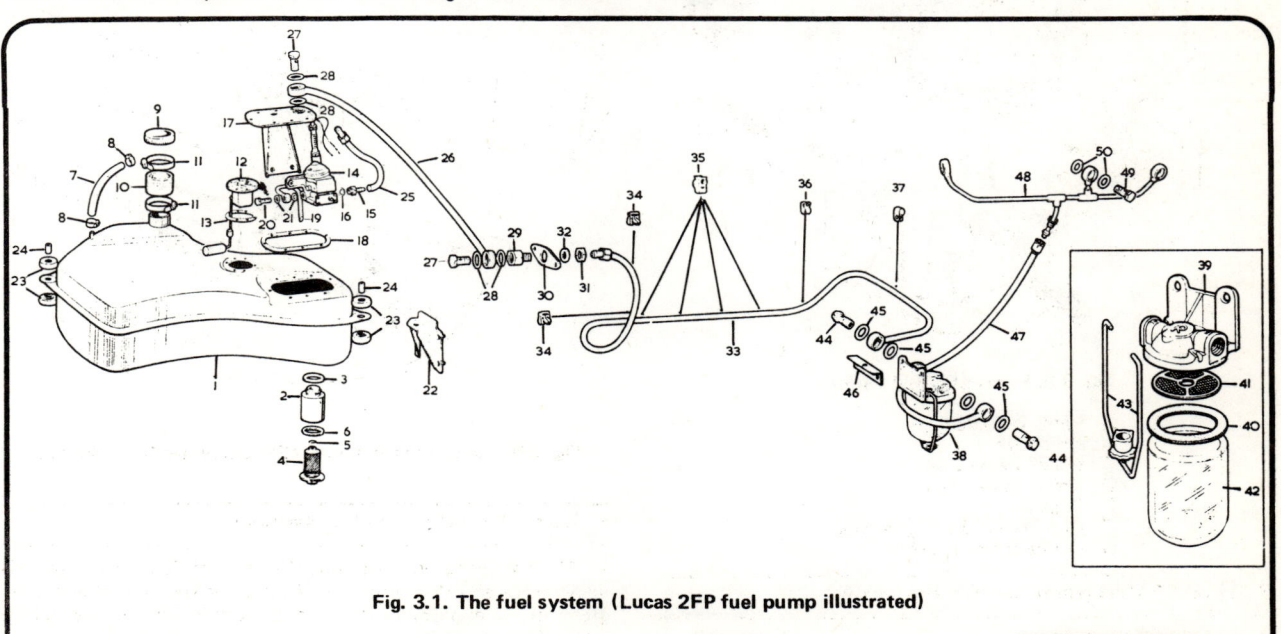

Fig. 3.1. The fuel system (Lucas 2FP fuel pump illustrated)

1	Fuel tank	13	Gasket	26	Pipe
2	Sump assembly	14	Fuel pump	27	Banjo bolt
3	Washer	15	Union	28	Fibre washer
4	Filter and drain plug assembly	16	Fibre washer	29	Connector
		17	Mounting bracket	30	Mounting plate
5	'O' ring	18	Gasket	31	Nut
6	Washer	19	Pipe	32	Brass washer
7	Hose	20	Banjo bolt	33	Pipe
8	Clip	21	Washer	34	Clip
9	Filler cap	22	Mounting bracket	35	Clip
10	Hose	23	Rubber pad	36	Clip
11	Clip	24	Distance piece	37	Clip
12	Tank element	25	Pipe	38	Filter assembly

39 Filter casting	45 Fibre washer
40 Sealing washer	46 Mounting bracket
41 Filter gauze	47 Pipe
42 Bowl	48 Feed pipe
43 Retaining strap	49 Banjo bolt
44 Banjo bolt	50 Fibre washer

Chapter 3/Fuel system, carburation and exhaust emission control system

2 Lucas 2FP fuel pump - general information

The Lucas 2FP fuel pump is fitted to earlier model 3.8 litre cars and is illustrated in Fig.3.2. It is a cumulative type centrifugal pump gdriven by a permanent field electric motor; it is fully sealed and is mounted inside the fuel tank. Electrically the pump is under the control of the ignition switch and will commence to operate as soon as the ignition is switched on and will deliver fuel to the carburettors at a pressure of approximately 2 lbs per sq in (0.14 kg/cm^2) all the time it is running. A 5 amp fuse, located in the fuse pack behind the instrument panel, is incorporated in the pump electrical circuit as a safety measure in the event of a fault developing in the pump or its connections and it is essential, if the fuse blows, to replace it with one of the same value. Under no circumstances should you use a higher rated fuse. The 2FP fuel pump is a sealed unit so servicing is confined to those operations given in Sections 4 and 5.

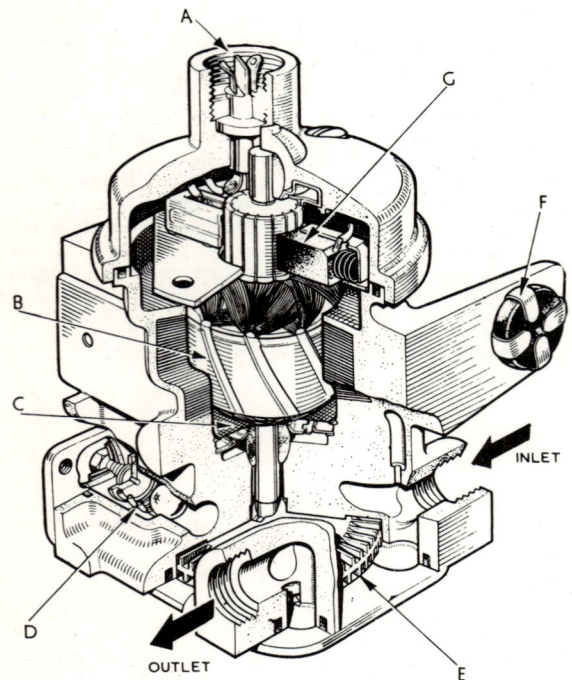

Fig. 3.2. Lucas 2FP fuel pump

- A Cable terminals
- B Armature
- C Gauze flame trap
- D Relief valve
- E Impeller
- F Anti-static earthing washer
- G Commutator brushes

3 Lucas 2FP fuel pump - removal and refitting

1 Disconnect the battery.
2 Raise the boot lid, remove the carpet from the floor of the boot and then remove the two floor panels by unscrewing the setscrews.
3 Remove the cover of the cable connector block which is located in the spare wheel compartment. Withdraw the connectors, identify the cable connections and then disconnect them. If the cable colours can be identified you will find that like colours are connected.
4 From this point onwards make sure that no-one smokes or brings a naked light close to the car.
5 Drain the fuel tank by removing the drain plug and then working the filter off the pump inlet pipe.
6 Remove the eight setscrews which secure the fuel pump carrier plate to the tank.
7 Carefully lift out the pump (Fig.3.3) and the carrier plate taking care not to damage the joint. Make provision to immediately cover the opening to the tank.
8 Disconnect the fuel delivery pipe union from the pump.
9 Remove the nut holding the braided cable conduit to the carrier plate.
10 Remove the two bolts which hold the pump to the carrier plate and separate the two items.
11 Inspect the carrier plate/tank sealing gasket and renew it if its condition is doubtful because leaks at this point will allow escape of petrol fumes into the car and petrol leaks if the tank is full.
12 Refitment of the pump is the reverse of the removal sequence however, if you are fitting a new pump, note that a star washer is provided on one of the petrol proof grommets on the mounting feet. This washer provides an earthing path from the pump to the mounting bracket via the fixing bolt and so prevents a build up of electrostatic charges on the pump unit. Also, when fitting the inlet pipe ensure that it is central in the filter sump before finally fitting the pump to the fuel tank.
13 When you have the fuel pump in position in the tank, refit the filter and drain plug assembly. Lubricate the filter O-ring and then feed the filter onto the fuel inlet pipe. Make sure that the drain plug cork sealing washer is in good condition before refitting the drain plug.

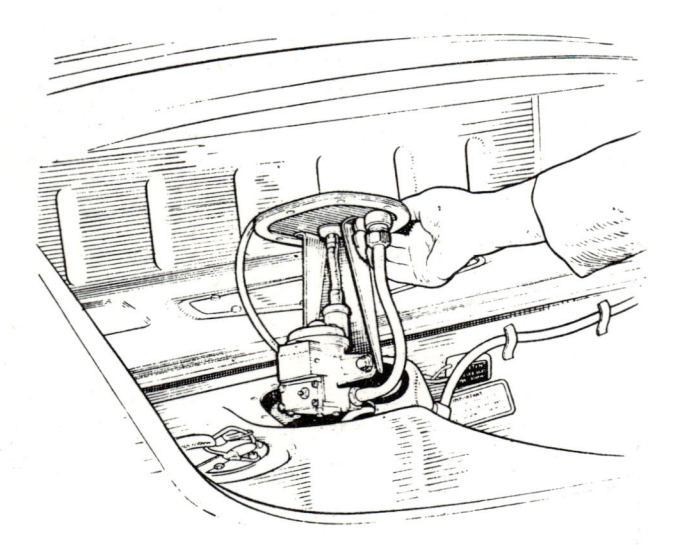

Fig. 3.3. Removing the Lucas 2FP fuel pump from the tank

4 Lucas 2FP fuel pump - fuel starvation

1 The following assumes that the level of fuel in the tank is satisfactory and that you have cleaned the fuel line filter and its gauze or, if applicable, that the renewable filter gauze is in a satisfactory condition.
2 Check the fuse behind the instrument panel and replace if necessary but if the fuse blows again, check for a short circuit in the feed cable or pump unit. Replace the pump unit if faulty or repair the cable if applicable.
3 If the fuse has not blown, go to the cable connectors contained in the rubber anti-flash block located in the spare wheel compartment. Check the voltage and current available at the terminal ends using a first grade voltmeter and an ammeter. With the ignition switched on, the voltage should be 12 volts and the current should not exceed 1.8 amperes.
4 If no voltage is shown, make sure that the fault is not due to a broken or intermittent connection in the switch, feed or earth.

Chapter 3/Fuel system, carburation and exhaust emission control system

5 If no current or excessive current is recorded it is an indication that the pump is faulty. A replacement unit must be fitted as repair is not possible.

5 Lucas 2FP fuel pump - fuel flooding

1 It is assumed that you have checked the needle valves in the carburettor float chambers and have found that these are clean and unworn.
2 Disconnect the fuel line at the carburettor end and then connect a pressure gauge to the fuel line.
3 Switch on the ignition and the pressure recorded should be 2 - 2½ lbs sq in (0.14 to 0.17 kg/cm^2).
4 If a higher pressure than 2½ lbs sq in (0.17 kg/cm^2) is recorded you will have to adjust the pump relief valve (a screw and locknut on the pump cover plate, Fig.3.4, to bring the pressure to the design limits.
5 Remove the pump from the fuel tank in the manner described in Section 3.
6 Connect the pressure gauge to the outlet connection of the pump using a short length of pipe.
7 Submerge the pump in a container of clean paraffin (kerosene) and then connect the pump cables to a fully charged battery. The **black cable** on the pump **must** be connected to the **positive** battery terminal.
8 To reduce pressure, turn the setting screw in an anti-clockwise direction after slackening the locknut.
9 When you have adjusted the pump to deliver at the correct pressure, refit it to the car as described in Section 3.
Warning: When bench testing the fuel pump, extinguish all naked lights or flames in the vicinity and do not allow the cables to spark when making connections, to this end it is advisable to connect a switch into the test cable circuit to enable you to switch off when connecting the pump to the battery.

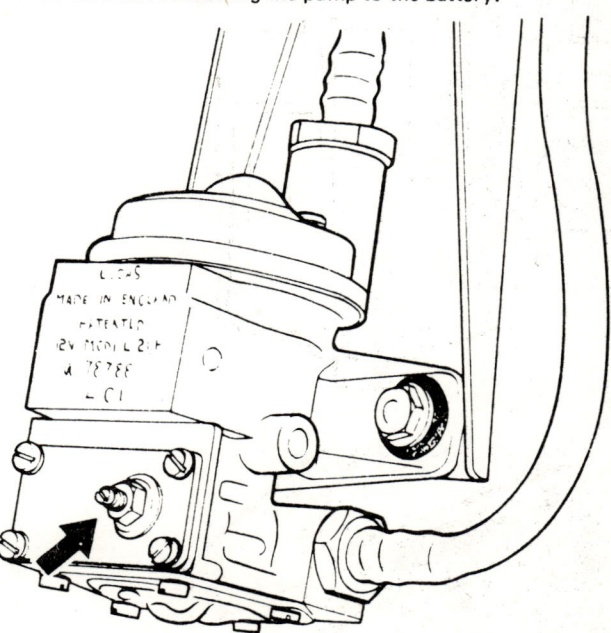

Fig. 3.4. Fuel pump relief valve adjusting screw (arrowed)

6 AUF 301 fuel pump - general information

The AUF 301 fuel pump illustrated at Fig.3.5 consists of three main assemblies: the main body casting "A", the diaphragm armature and magnet assembly "M" contained within the housing and the contact breaker assembly housed within the end cap "T2". A non-return valve assembly "C" is fixed to the end cover moulding to aid the circulation of air through the contact breaker chamber. The main fuel inlet "B" is maintained in communication with an inlet air bottle "I".

Communication with the main pumping chamber "N" is provided by an inlet valve assembly, this assembly comprises a Melinex valve disc "F" permanently assembled within a pressed steel cage, which, in turn is held in place by a valve cover "EI", while the outlet from the pumping chamber is provided with an identical valve assembly reversed in direction. Inlet and outlet valve assemblies and filters are held in position by a clamp plate "H", both valve assemblies may be removed by detachment of the clamp plate after removing the self tapping screws shown on the lower diagram in Fig.3.5. A filter "E" is provided upstream of the inlet valve assembly. The delivery chamber "O" is bounded by a flexible plastic spring loaded diaphragm "L" contained by the vented cover "P". The rubber sealing ring "L2" seals the diaphragm "L".

The magnetic unit consists of an iron coil housing an iron core "Q", an iron armature "AI" which is provided with a central spindle "PI" and is permanently united with the diaphragm assembly "LI", a magnet coil "R" and a contact breaker assembly comprising parts "P2", "UI", "U", "TI", "V" etc. Between the coil housing and the armature are located 11 spherically edged rollers "S" which locate the armature "AI" centrally within the coil housing and so allow absolute freedom of movement in a longitudinal direction.

The contact breaker consists of a bakelite pedestal moulding "T" carrying two rockers "U" and "UI" which are both hinged to the moulding at one end by the rocker spindle "Z" and interconnected at their top ends by two small springs arranged to give a "throw-over" action. A trunnion "P2" is carried by the inner rocker and the armature spindle "PI" is screwed into the trunnion. The outer rocker "U" is fitted with two tungsten points which contact with two tungsten points carried by the spring blade "V" which is connected to one end of the coil whilst the other end of the coil is connected by a short length of wire, "X", to one of the screws which hold the pedestal moulding onto the coil housing and this provides an earth return to the body of the pump which must, in turn, be thoroughly earthed to the body of the vehicle by the earthing terminal provided on the flange of the coil housing.

The action of the pump is that when it is at rest the outer rocker "U" lies in position illustrated with the tungsten points in contact. When the ignition is switched on, current passes from the connector "W" through the coil and back to the blade "V", through the points and so to earth, thus energising the coil and attracting the armature "AI". The armature, together with the diaphragm assembly then retracts and so creates a vacuum to suck fuel from the tank into the pumping chamber "N" through the inlet valve. When the armature is close to the end of its stroke, the throw-over mechanism operates and the outer rocker moves rapidly backwards, thus separating the points and breaking the circuit. The spring "SI" then pushes the aramture and diaphragm away from the coil housing and so forces fuel through the delivery valve at a rate according to engine requirements. As the armature approaches the end of its stroke the throw-over mechanism again operates, the tungsten points again make contact and the cycle of operation is repeated. The spring blade "V" rests against the small projection moulding "T" and it should be set so that when the points are in contact it is deflected away from the moulding. The extent of the gap at the points should be approximately 0.030" (0.75 mm) when the rocker "U" is manually deflected until it contacts the end face of the coil housing.

7 AUF 301 fuel pump - removal and refitting

1 Access to the fuel pump is gained by removing a plate at the right hand side of the luggage compartment. The location of the plate for the Fixed Head Coupe and Open 2 Seater models is shown in Fig.3.6.

Fig. 3.5. AUF 301 fuel pump

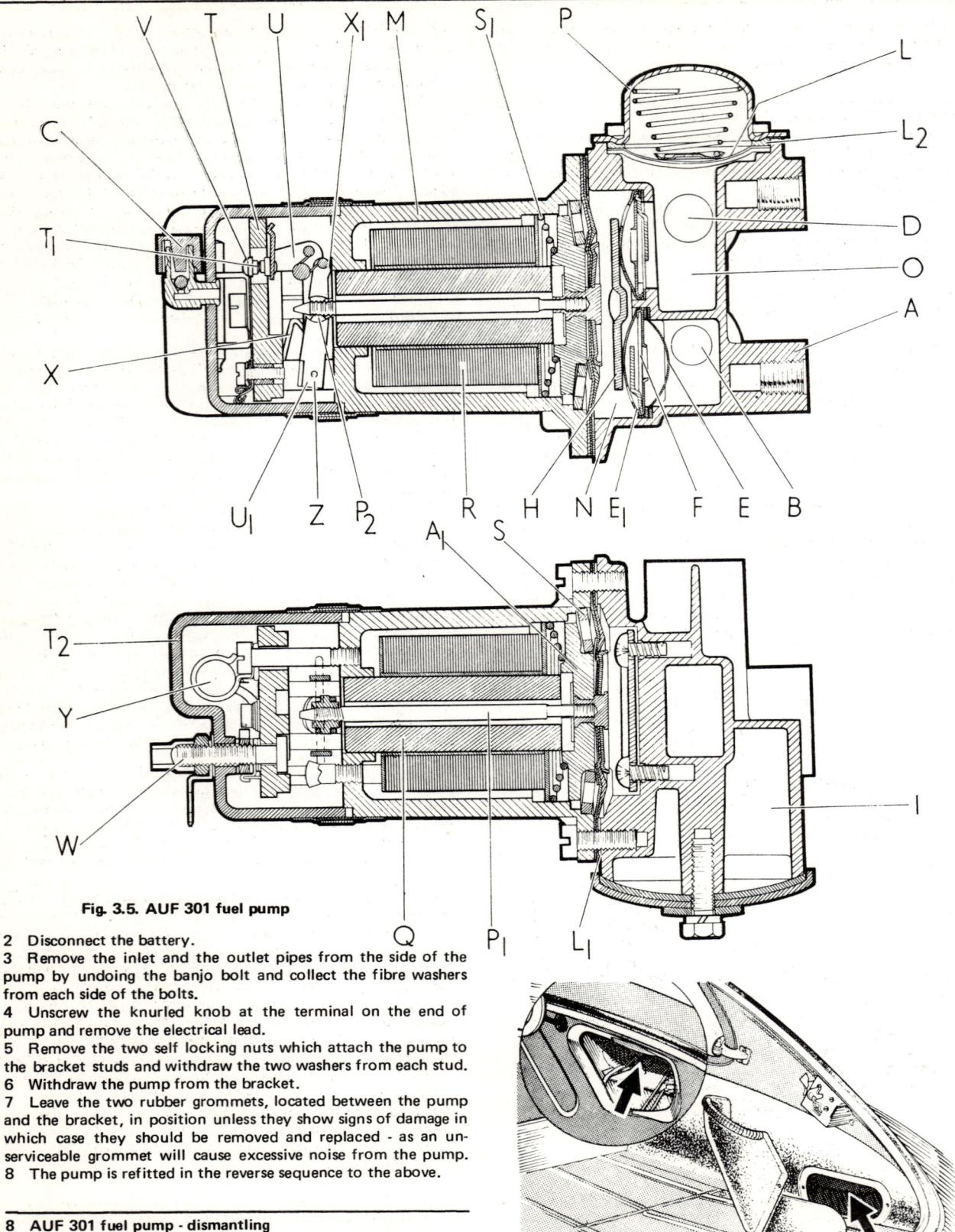

2 Disconnect the battery.
3 Remove the inlet and the outlet pipes from the side of the pump by undoing the banjo bolt and collect the fibre washers from each side of the bolts.
4 Unscrew the knurled knob at the terminal on the end of pump and remove the electrical lead.
5 Remove the two self locking nuts which attach the pump to the bracket studs and withdraw the two washers from each stud.
6 Withdraw the pump from the bracket.
7 Leave the two rubber grommets, located between the pump and the bracket, in position unless they show signs of damage in which case they should be removed and replaced - as an unserviceable grommet will cause excessive noise from the pump.
8 The pump is refitted in the reverse sequence to the above.

8 AUF 301 fuel pump - dismantling

1 Ensure that the exterior of the pump, your hands and the bench are clean.
2 Refer to Fig.3.7.

Fig. 3.6. Location of the fuel pump (fixed head Coupe)
Inset shows Open 2 seater

Chapter 3/Fuel system, carburation and exhaust emission control system

3 Remove the insulated sleeve (33) and then unscrew the terminal nut (32) and connector (31) and take off the shakeproof washer. If a tape seal (43) is fitted to the cover, remove it and then withdraw the cover (29).
4 Take out the screw (24) holding the contact blade (22) to the pedestal (16) and then remove the condenser (25) from its clip.
5 The washer (23), the terminal tag (11) and the contact blade can now be removed.
6 Using a thick bladed screwdriver to avoid damaging the screw heads, unscrew the coil housing securing screws (7).
7 Remove the earthing screw (9) and the coil housing (6) can now be removed from the body (1).
8 The diaphragm and spindle assembly (2) is now removed by unscrewing on the diaphragm in an anti-clockwise direction until the armature spring (5) pushes the diaphragm away from the coil housing. It is advisable to hold the housing over the bench during this operation so that the 11 brass rollers (3) which will be released, do not fall to the floor. The diaphragm and spindle are serviced as a unit and should not be separated.
9 Now remove the end cover seal washer (21) and remove the terminal nut (20) followed by the lead washer. The lead washer may have flattened on the terminal tag and thread, and may have to be cut away with a knife.

10 Remove the two screws (28) which hold the pedestal to the coil housing and take off the earth terminal tag (13) together with the condenser clip (26).
11 Tip the pedestal and so withdraw the terminal stud (17) from the terminal tag (12).
12 The pedestal (16) can now be removed with the rocker mechanism (15) attached.
13 Separate the pedestal and rocker mechanism by pushing out the hardened steel pin (14).
14 Remove the two Phillips screws (35) which secure the valve clamp plate (34) and now remove the valve caps (36), the valves (37 and 38), the sealing washers (39) and the filter (40).
15 Remove the single 2BA screw (48) securing the inlet air bottle cover (45), collect the spring washer, the dished washer (46), the cover and the joint (44).
16 The delivery flow-smoothing device should only be removed if its operation is faulty and if the necessary equipment for pressure testing after assembly is available. The device is removed by taking out the four screws (52) which secure the vented cover (51). Remove the cover and then withdraw the diaphragm spring (59), rubber O-ring (53), spring cap (58), diaphragm (57), barrier (54), diaphragm plate (56) and the sealing washer (55).

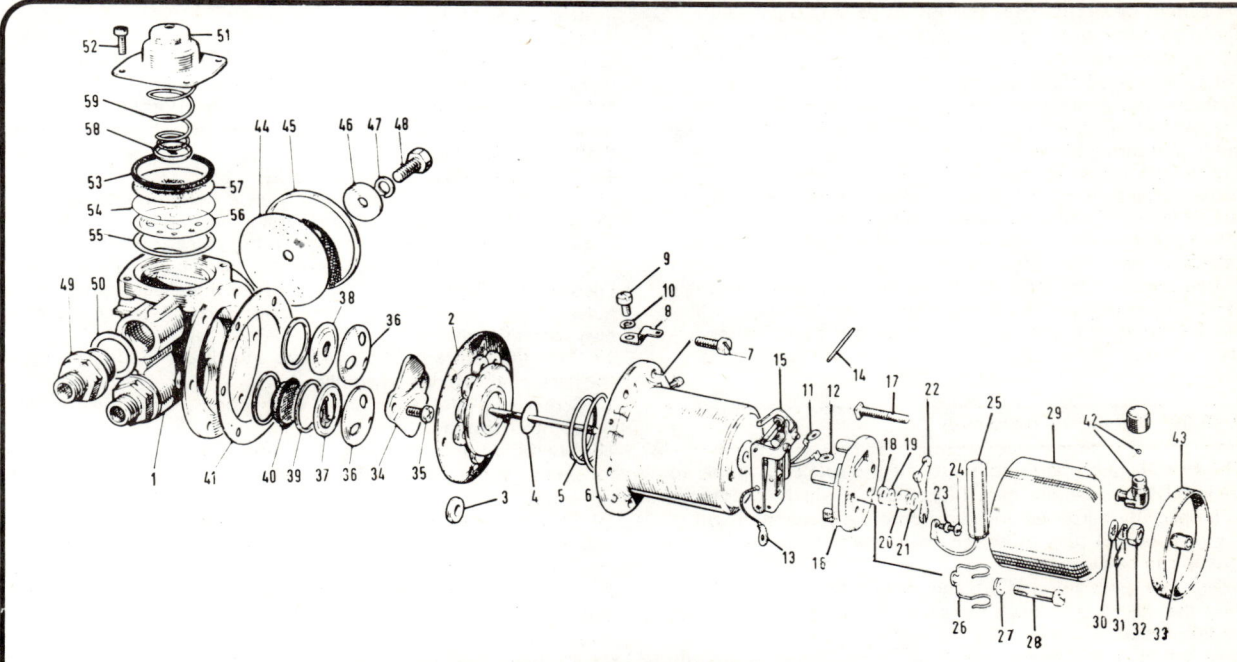

Fig. 3.7. Exploded view of the AUF 301 fuel pump

1	Pump body	21	Washer	41	Gasket
2	Diaphragm and spindle assembly	22	Contact blade	42	Vent valve
3	Roller - armature centralising	23	Washer	43	Sealing band
4	Washer - impact	24	Screw	44	Joint
5	Spring - armature	25	Condenser	45	Inlet air bottle cover
6	Housing - coil	26	Clip	46	Dished washer
7	Screw - securing housing - 2 BA	27	Spring washer	47	Spring washer
8	Connector - earth	28	Screw	48	Screw
9	Screw - 4 BA	29	End cover	49	Outlet connection
10	Spring washer	30	Shakeproof washer	50	Fibre washer
11	Terminal tag	31	Lucas connector	51	Cover
12	Terminal tag	32	Nut	52	Screw
13	Earth tag	33	Insulating sleeve	53	'O' ring
14	Rocker pivot pin	34	Clamp plate	54	Diaphragm barrier
15	Rocker mechanism	35	Screw	55	Sealing washer
16	Pedestal	36	Valve cap	56	Diaphragm plate
17	Terminal stud	37	Inlet valve	57	Diaphragm
18	Spring washer	38	Outlet valve	58	Spring end cap
19	Lead washer	39	Sealing washer	59	Diaphragm spring
20	Terminal nut	40	Filter		

9 AUF 301 fuel pump - examination

1 You may find that the parts in contact with fuel are coated with a substance similar to varnish, this is the result of gum formation in the fuel. The coating will have a strong stale smell and may have attacked the neoprene diaphragm. Brass and steel parts can be cleaned by boiling in a 20% solution of caustic soda followed by dipping in a strong nitric acid solution and then well washed in boiling water. Light alloy parts should be soaked in methylated spirits and then cleaned.
2 All fibre and cork washers, gaskets and "O" sealing rings should be replaced as a matter of course.
3 Examine the diaphragm for signs of deterioration.
4 Examine the plastic valve assemblies for kinks or damage to the valve plates. A good way to check these is to blow and suck with the mouth and any malfunction will soon be apparent.
5 The valve is retained and kept in position by the narrow tongue on the valve cage which is bent over. Make sure that this tongue is not distorted and that it allows the valve to lift about 1/16'' (1.6 mm).
6 Examine the air inlet bottle cover for damage and the valve recesses in the body for corrosion or pitting. If damage of this nature cannot be rectified, the body must be replaced.
7 Clean the filter with a brush and renew it if damaged.
8 Make sure that the coil lead tag is secure and that the insulation is in good condition.
9 Examine the contact breaker points for signs of burning or pitting which, if present, will entail renewal of the rocker assembly and spring blade.
10 Check the pedestal for cracks or other damage paying particular attention to the narrow ridge on the edge of the rectanglar hole on which the contact blade rests.
11 Check the non-return vent valve in the end cover. Make sure that the small ball valve is free to move.
12 If you removed the delivery flow-smoothing device, examine all parts for damage particular attention to the diaphragm which should be replaced if you are at all doubtful as to its serviceability.

10 AUF 301 fuel pump - reassembly

1 The first stage of reassembly is to fit the rocker assembly to the pedestal. The steel pin (item 14 in Fig.3.7) which holds those items is specially hardened and should not be replaced by other than a genuine SU part otherwise early malfunctioning of the fuel pump can be expected.
2 Invert the pedestal and fit the rocker assembly to it by pushing the steel pin through the small hole in the rockers and the pedestal struts.
3 Now position the centre toggle with the spring above the spindle on which the white rollers run with the inner rocker spindle in tension against the rear of the contact point. This positioning is important in order to obtain the correct 'throw over' action, it is also essential that the rockers are perfectly free to swing on the pivot pin and that the arms are not binding on the legs of the pedestal. It is permissible, if necessary, to square up the rockers with a pair of thin nosed pliers.
4 Assemble the square headed terminal stud to the pedestal and see that its head fits snugly into the recess at the back of the pedestal.
5 Refer to Fig.3.8 and fit the spring washer (1) followed by the terminal tag (2). Now fit the lead washer (3) and the coned nut (4) with the coned face towards the lead washer. Tighten down on the nut and then fit the end cover seal washer (5).
6 Assemble the pedestal to the coil housing by fitting the two screws (6) making sure that the spring washer (7) on the left hand screw is between the pedestal and the earthing tag (8). When a condenser is fitted, its wire clip base should be placed under the earthing tag and the spring washer (7) can be omitted.
7 Tighten down on the screws but do not overtighten them or you will crack the pedestal. Make sure that the earthing tag (8) does not turn whilst you are tightening its screw as this may strain and break the earthing flex.
8 Now place the armature spring into the coil housing so that its larger diameter is towards the coil.
9 Make sure that the impact washer is fitted to the armature, this is a small neoprene washer which fits into the recess in the armature. With this washer in position, fit the diaphragm by inserting the spindle in the hole in the coil and screwing it into the threaded trunnion in the centre of the rocker assembly.
10 Screw in on the diaphragm until the rocker will not 'throw over' but do not confuse this with jamming the armature on the coil housing internal steps.
11 When the diaphragm is in position, hold the pump with the rocker end downwards and fit the 11 brass centralising rollers by turning back the edge of the diaphragm and dropping them into the coil recess.
12 Fit the contact blade and check that when the outer rocker is pressed onto the coil housing, the contact blade rests on the narrow rib or ridge which projects slightly above the main face of the pedestal. If it does not, slacken the attachment screw, swing the blade clear of the pedestal and bend it downwards a sufficient amount so that when it is repositioned it will rest against the rib lightly. Do over-tension the blade as this will restrict travel of the rocker mechanism.
13 Refer to Fig.3.9. Correct positioning of the blade will give a gap of 0.035'' $\pm$ 0.005'' (0.88 $\pm$ 0.125 mm) between the pedestal and the tip of the spring blade. Now check the gap between the rocker finger and the coil housing and adjust, if necessary, by bending the stop finger to obtain a gap of 0.70'' $\pm$ 0.005'' (1.75 $\pm$ 0.125 mm).
14 Carefully remove the contact blade.
15 Hold the coil housing assembly in a horizontal position as shown in Fig.3.10 and push firmly and steadily inwards on the diaphragm spindle. Unscrew the diaphragm continually pressing and releasing on the spindle until the rocker just 'thows over', unscrew the diaphragm to the nearest hole and again a further 4 holes (about two thirds of a complete turn). The diaphragm is now correctly set.
17 Press the centre of the armature and fit the retaining fork at the back of the rocker assembly.
17 Screw the inlet and outlet connections, with their sealing rings, into the body.
18 Assemble the outlet valve joint washer into the outlet recess followed by the outlet valve, tongue side downwards, and the valve cap.
19 Place the inlet valve joint washer in position then the filter, dome side downwards, then another joint washer followed by the valve assembly, tongue side upwards and then the valve cap.
20 Check that the valves are settled correctly in their recesses and then fit the clamp plate and secure it to the body with the two screws.
21 Fit the inlet air bottle cover, with its joint washer, and tighten down on the central screw.
22 The first step in reassembly of the delivery flow-smoothing device, if you removed it, is to fit the sealing washer in the bottom of the recess. Now fit the perforated diaphragm plate, dome side downwards and then the plastic barrier followed by the rubber diaphragm. Place the "O" section sealing ring into the recess and ensure that it seats evenly. Place the diaphragm spring with the large end towards the vented cover, into the cover and then place the spring end-cap on the small end of the spring. Fit the spring and cap assembly onto the diaphragm and secure with the four retaining screws. Pressure test the pump and check that there is no leakage.
23 Fit the joint washer to the body so that the screw holes are aligned.
24 Offer up the coil housing to the body, make sure that it seats correctly and that the cast lugs on the coil housing are at the bottom, and then fit the six screws finger tight. Fit the Lucar connector to the earthing screw. During this operation, make sure that the rollers do not become displaced, otherwise the diaphragm will be damaged. Finally tighten down on the screws by diametrical selection.

Chapter 3/Fuel system, carburation and exhaust emission control system

25 Fit the contact blade and the coil lead to the pedestal and secure with the washer and screw but if a condenser is in use the washer may be omitted but the condenser tag should be placed under the coil lead tag.

26 Now adjust the contact blade so that the points are a little above the points on the rocker when closed and so that, when the contact points make and break, one pair of points completely covers the other. This adjustment can be achieved by use of the slot provided for the attachment screw. Tighten the screw when the correct adjustment is obtained.

27 Tuck all spare cable into position so that the rocker mechanism will not be fouled. Fit the end cover sealing washer to the terminal stud and then fit the bakelite end cover and secure it with the brass nut and its lockwasher.

28 Fit the terminal tag or connector and the insulated sleeve.

29 It is advisable to test the pump at this stage and we feel that the easiest way to do this is to loosely replace the pump in the car, connect the inlet and outlet fuel pipes and make the necessary electrical connections. Disconnect the inlet pipe at the carburettors and place the end in a receptacle to catch the fuel when the pump operates. Switch on the ignition and check the pump for correct operation. Testing in this manner will avoid the necessity to make up test leads and to have a quantity of petrol exposed in the workshop.

30 If the pump operates satisfactorily, replace the rubber sealing band over the junction of the end cover with the body and then seal with adhesive tape.

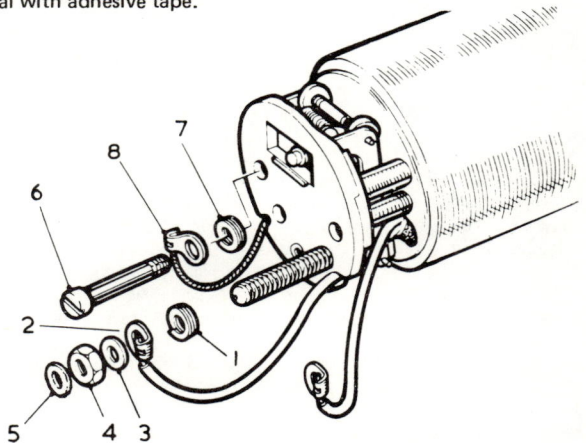

Fig. 3.8. Attaching the pedestal to the coil housing

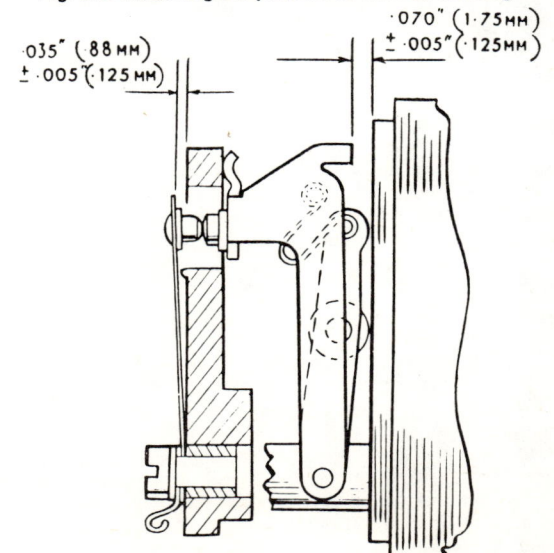

Fig. 3.9. Rocker and contact clearance

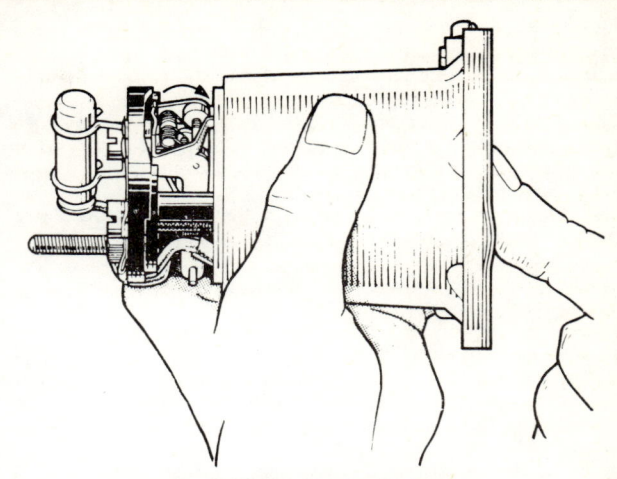

Fig. 3.10. Setting the diaphragm

11 AUF 301 fuel pump - fault finding

1 If fuel is not reaching the carburettors, first check that the fuel tank venting is not blocked.

2 Disconnect the delivery pipe to the carburettor and switch on the ignition. If the pump works, the trouble is probably due to a sticking needle in the float chamber.

3 If the pump does not work, disconnect the lead from the terminal and strike against the body of the pump to see if it sparks and, therefore, if any current is available. If there is current there, remove the cover of the pump and touch the terminal with the lead when the points are in contact. If the pump fails to work it may be due to dirt on the contact faces. This may be cleaned off by inserting a thin piece of card between the contact and working it to and fro.

4 If the pump still fails to work after cleaning the contacts, check that the bottom filter is not blocked.

5 If the filter is clear, slacken off the inlet pipe union and if the pump operates when the ignition is switched on the trouble is probably due to an obstruction in the pipe line to the tank. It may be possible to clear the obstruction by blowing down the pipe with a foot pump.

6 If the pump does not operate after slackening the inlet union, slacken the outlet union. Switch on and if the pump operates the fault will be due to an obstruction in the pipe line between the pump and the carburettors. Check that the glass petrol filter unit in the engine compartment is clear. If it is clear, the fault will lie in the pipe line itself and a blow down with a foot pump will probably clear the trouble.

Note: Under no circumstances must compressed air be applied to the pump as this will damage the valves.

7 If, when either union is slackened off, the pump does not operate or only works slowly or spasmodically, then the trouble is due to a fault in the pump itself, such as a stiffened up diaphragm or undue friction in the throw over mechanism or a combination of both.

8 To check the above, unscrew the six screws and detach the coil housing and rocker unit from the main body. Take care not to lose any of the rollers under the diaphragm.

9 Press on the diaphragm gently and firmly and observe whether the throw over mechanism seems to operate freely. If it does not, lubricate sparingly, with a drop of thin oil, where the steel spindles pass through the brass rockers.

10 Now, to restore the original pliability of the diaphragm, ruckle each of the two fabric layers vigorously between the thumb and fingers. Reassemble and carefully set the throw over as described in Section 10 paragraph 15.

11 If the pump becomes noisy in operation, look for an air leak on the suction side. The simplest way to check for this is to

Chapter 3/Fuel system, carburation and exhaust emission control system

disconnect one of the carburettor feed pipes and allow the pump to discharge petrol into a glass jar until the pipe is immersed in petrol. If bubbles are seen to come through the pipe there must be an air leak which must be traced and cured.

12 If the pump keeps operating but does not deliver any fuel, the fault is probably due to a piece of dirt trapped under one of the valves which will mean dismantling and cleaning the pump as already described.

13 If the pump overheats in operation, the fault will probably be due to an obstruction on the suction side.

12 Fuel tank - removal and refitting

1 Disconnect the battery.
2 For 3.8 litre early models, remove the luggage compartment floor covering and take out the six setscrews to the floor panels. Remove the fuel pump cables from the terminal block noting that like colours are connected.
3 For other models, remove panels as provided (depending on model) to gain access to the top of the fuel tank.
4 Identify and then remove the cables to the fuel gauge in the top of the tank.
5 At this stage you must decide whether or not to drain the tank and this decision will, of course, be influenced by the amount of fuel in the tank and what containers are available to accept the fuel. One gallon of petrol weighs about 10 lbs.
6 If you decide to drain the tank, our advice is to empty it by siphoning because you will find it most difficult to control the flow if the drain plug is removed.
7 Remove the banjo bolt from the fuel feed pipe and collect the fibre washers from each side of the pipe union.
8 Slacken the filler pipe clips and push the pipe up to the filler neck.
9 Slacken the clip securing the breather pipe to the tank union and disconnect the pipe.
10 Remove the four bolts securing the boot lock and remove it.
11 Slacken the clips and remove the boot channel drain tube.
12 Remove the four nuts and the body strengthening plate from the back of the tank mounting bracket inside the rear suspension aperture.
13 The tank is held to the car by three pointed end mounting bolts with spring washers and special flat washers. There are two bolts at the front of the tank and one at the rear. Slacken each mounting bolt, take the weight of the tank, remove the bolts and lower the tank to the ground.
14 Collect the rubber mounting pads and spacers.
15 The method of removing the Lucas 2FP fuel pump from the tank has already been described in Section 3. If it is desired to remove the fuel feed pipe from later model cars, this is done by taking out the eight screws securing the cover plate assembly to the tank. Note that the screws are locked by copper washers. Remove the sump drain plug and then lift out the fuel feed pipe complete with the filter unit. The filter unit is attached to the pipe by a union nut and olive.
16 Refitting is the reverse of the removal procedure but do make sure that all fibre, cork and other sealing washers are in good condition.

13 Fuel tank gauge unit - removal and refitting

1 Proceed as indicated in Section 12 to gain access to the top of the fuel tank.
2 Disconnect the battery.
3 Identify and then remove the cables to the gauge unit.
4 Remove the six setscrews and the twelve copper washers securing the gauge unit to the top of the tank.
5 Remove the gauge unit taking care not to damage the float arm.
6 Before refitting the unit, carefully scrape away all traces of the old gasket from the top of the tank and, if it is being refitted, from the mating face of the unit. Make sure that no pieces of gasket fall into the tank.

7 Apply a suitable sealing compound to both faces of the new gasket and then place it on the fuel tank boss with the holes in line.
8 Insert the gauge unit in the tank so that the float is towards the rear of the car and then secure it in position with the six screws and the twelve copper washers tightened down evenly.
9 From this point onwards the refitting sequence is the reverse of the removal procedure.

14 SU HD 8 carburettors - general information

"E" Type cars, other than those intended for export to the USA and Canada, are equipped with triple SU HD 8 carburettors. These carburettors differ from the earlier types in that the jet glands are replaced by a flexible diaphragm and the idling mixture is conducted along a passage way, in which is located a metering screw, instead of being controlled by a throttle disc.

A sectioned view of the carburettor is shown in Fig.3.11. The jet (18) is fed through its lower end and is attached to a synthetic rubber diaphragm (10) by means of the jet cup (9) and the jet return spring cup (13) between which the centre of the diaphragm is compressed. The diaphragm is held at its outer edge between the diaphragm casing (14) and the float chamber arm. The jet is controlled by the jet return spring (12) and the jet actuating lever (15), the latter has an external adjusting screw which limits the upward travel of the jet and thus controls the mixture adjustment; screwing in on the screw will enrich the mixture and unscrewing will weaken it.

The throttle spindle is sealed by a cork gland (25) used in conjunction with a dished retaining washer (28), a spring (27) and a shroud (26). No servicing should be required on this assembly but, if you ever do require to remove it, this can be done by removing the throttle spindle and disc.

The carburettor idles on the main jet and the mixture is conducted along the passage way (8) which connects the choke space to the other side of the throttle disc. The quantity of mixture passing along the passage way is controlled by the "slow-run" valve (5) but the relative richness of the mixture is determined by the jet adjusting screw. Thus, when the engine is idling, the throttle remains completely closed against the bore of the carburettor.

A manual mixture control is provided which operates on all three carburettors and a reminder that this is in operation, is provided by a red warning light adjacent to the mixture control slide on the facia panel. The manual control mechanism is arranged to operate in two stages. The first stage provides a degree of throttle opening when the mixture control cable moves the jet lever ("A" in Fig.3.12) towards the rear of the engine, the cam (B) is operated and this pulls the fast idle push rod (C) downwards, the fast idle screw (D) then presses down on the fast idle lever (E) and so opens the throttle slightly. The second stage brings the jet lever back further thus opening the throttle further and bringing the lever (F) into contact with the arm of the mixture adjusting screw (G) which is moved upwards and so moves the jet and the diaphragm downwards away from the jet needle. If you ever disconnect the mixture control rods (H), they must be reconnected when all three jet levers are against their stops and the fork ends of the rods must be adjusted to line up with the clevis pin holes in the jet levers.

15 SU carburettors - removal and refitting

1 Disconnect the battery. This is a safety precaution to avoid pumping petrol into the engine bay if the ignition is inadvertently switched on when the fuel lines are disconnected.
2 We found it easier, especially on 4.2 litre models, to remove the inlet manifold and carburettors as a complete assembly.
3 Refer to Chapter 2 and drain the cooling system in the manner described.
4 Slacken the clip securing the hose from tje inlet manifold to the header tank and remove the hose from the manifold.

Chapter 3/Fuel system, carburation and exhaust emission control system

5 Disconnect the two electrical connections from the thermostat fan control in the header tank.
6 Remove the two butterfly nuts at the carburettor trumpets and remove the air cleaner elbow.
7 Remove the carburettor trumpet from the carburettors by taking off the six nuts and spring washers together with the three gaskets.
8 Disconnect the throttle linkage at the rear carburettor.
9 Slacken the clips and disconnect the heater pipes at the water manifold and below the inlet manifold.
10 On those cars fitted with automatic transmission, disconnect the 'kickdown' cable at the rear of the cylinder head.
11 Disconnect the mixture control inner and outer cables.
12 Disconnect the main fuel feed pipe below the centre carburettor.
13 Remove the suction pipe from the front carburettor.
14 Unclip the hose to the breather pipe.
15 Remove the throttle return springs.
16 The inlet manifold complete with the carburettors can now be removed.
17 Remove the banjo union bolt from each carburettor float chamber and collect the fibre washers from each side of the bolt. Take the filters out of the float chambers and store them in a safe place.
18 Remove the four nuts and spring washers securing each carburettor to the inlet manifold and remove the carburettors together complete with the return spring brackets.
19 Separate the carburettors by moving the split pin and then taking out the clevis pin securing the mixture control linkage to each carburettor.
20 Refitting is the reverse of the removal sequence except that new gaskets should be fitted to the inlet manifold, to each side of the heat insulating plate and also to the carburettor trumpet flanges.
21 When refitting the mixture control rods, make sure that they are adjusted so that the clevis pin can be entered with the jet levers against the stops as illustrated in Fig.3.12.
22 Finally, if your car is equipped with automatic transmission, you will have to adjust the 'kickdown' cable, the method of effecting this adjustment is described in Chapter 6.

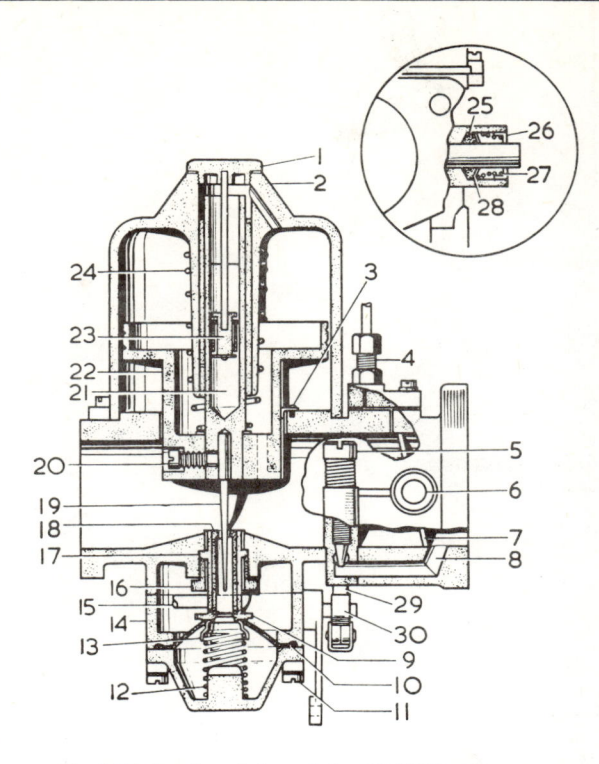

Fig. 3.11. Sectioned view of the SU HD8 carburetter

1	Damper cap	16	Nut - jet bearing
2	Suction chamber	17	Jet bearing
3	Piston guide	18	Jet
4	Union for vacuum advance/retard	19	Needle
5	Slow running volume screw	20	Needle retaining screw
6	Throttle spindle	21	Oil reservoir
7	Throttle butterfly	22	Piston
8	Slow run passage	23	Damper
9	Jet cup	24	Piston return spring
10	Diaphragm	25	Throttle spring gland
11	Float chamber securing screw	26	Shroud for spring
12	Jet return spring	27	Spring
13	Return spring cup	28	Washer
14	Diaphragm casing	29	Push rod
15	Actuating lever	30	Cam

16 SU carburettor - dismantling and reassembly

1 An exploded view of the carburettor is given in Fig.3.13.
2 Thoroughly clean the outside of the carburettor.
3 Clean down on the top of the bench, it is advisable to lay out a sheet of paper on which all parts can be placed as they are removed.
4 Unscrew the damper and remove it together with its washer.
5 Using a small file or scriber, mark the suction chamber so that it can be refitted in its original position.
6 Remove the three suction chamber securing screws and remove the chamber from the body leaving the piston in position. Be careful when lifting off the suction chamber not to apply side loads to the piston otherwise the piston needle may be bent.
7 Lift the piston spring from the piston noting which way round it is fitted.
8 Remove the piston and invert it over a container to allow the oil in the damper bore to drain out. Place the piston in a safe place so that the needle will not be damaged or that the piston will not roll onto the floor. It is suggested that the piston is placed in a suitably sized jar with the needle inside, so acting as a stand.
9 It is recommended that, unless absolutely necessary, the needle is not separated from the piston. However, if the needle must be removed, slacken the retaining screw in the side of the body and remove the needle.
10 Mark the position of the float chamber lid in relation to the body. Undo the cap nut and remove it together with the washer. Lift off the float chamber lid.

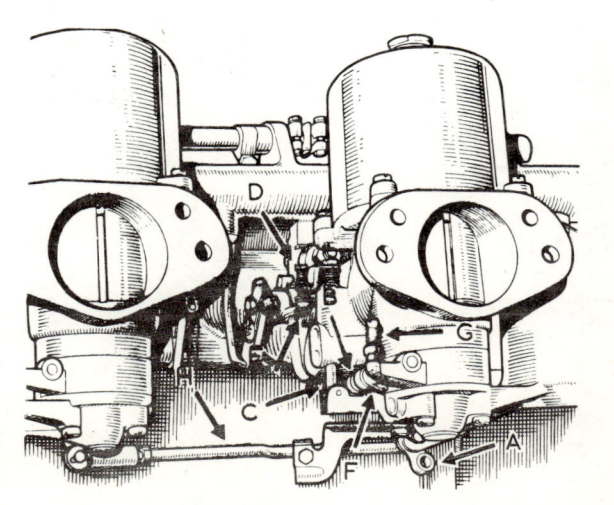

Fig. 3.12. SU carburetter controls

Fig. 3.13 Exploded view of the HD8 carburetters

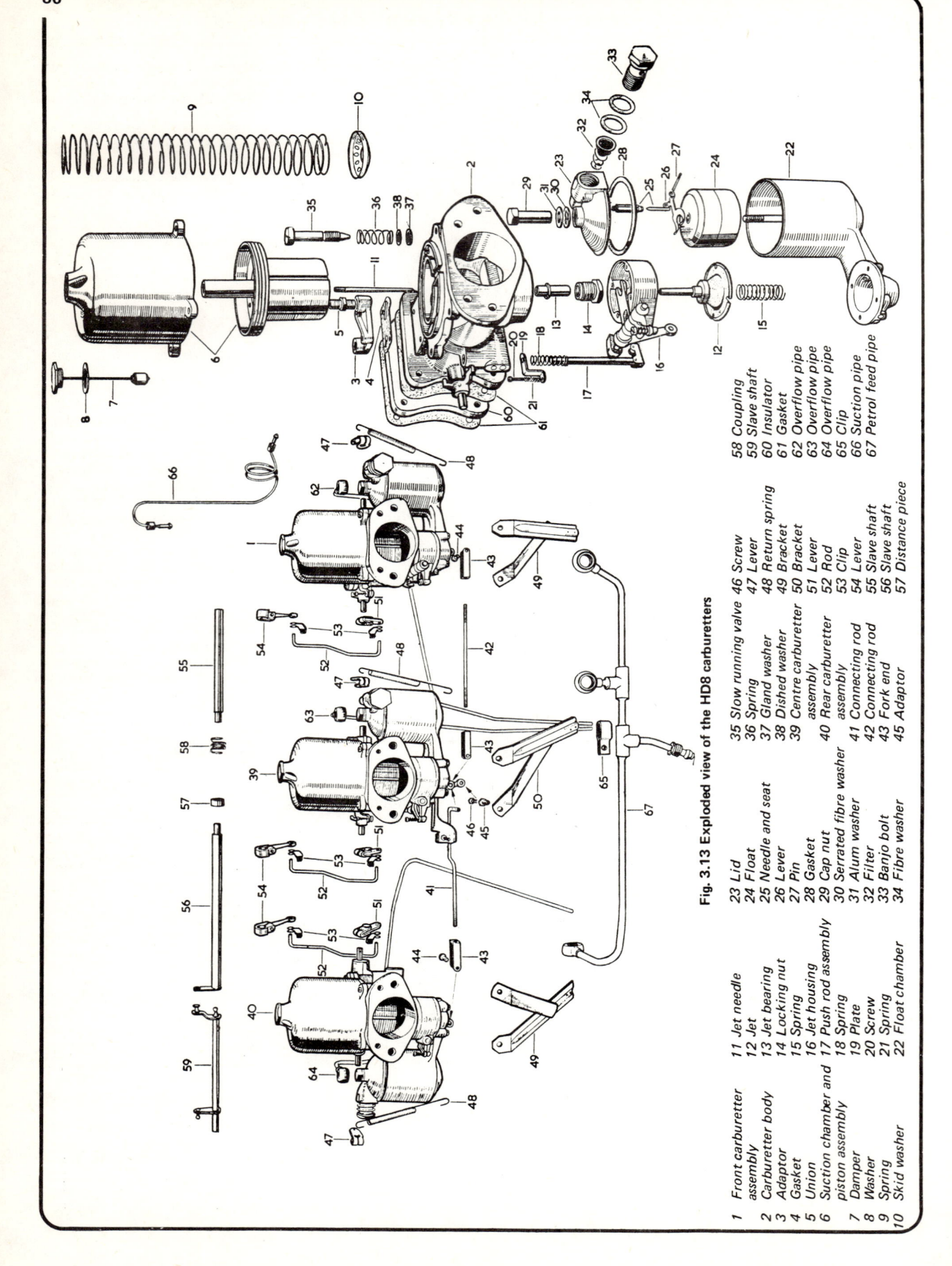

1 Front carburetter assembly
2 Carburetter body
3 Adaptor
4 Gasket
5 Union
6 Suction chamber and piston assembly
7 Damper
8 Washer
9 Spring
10 Skid washer
11 Jet needle
12 Jet
13 Jet bearing
14 Locking nut
15 Spring
16 Jet housing
17 Push rod assembly
18 Spring
19 Plate
20 Screw
21 Spring
22 Float chamber
23 Lid
24 Float
25 Needle and seat
26 Lever
27 Pin
28 Gasket
29 Cap nut
30 Serrated fibre washer
31 Alum washer
32 Filter
33 Banjo bolt
34 Fibre washer
35 Slow running valve
36 Spring
37 Gland washer
38 Dished washer
39 Centre carburetter assembly
40 Rear carburetter assembly
41 Connecting rod
42 Connecting rod
43 Fork end
45 Adaptor
46 Screw
47 Lever
48 Return spring
49 Bracket
50 Bracket
51 Lever
52 Rod
53 Clip
54 Lever
55 Slave shaft
56 Slave shaft
57 Distance piece
58 Coupling
59 Slave shaft
60 Insulator
61 Gasket
62 Overflow pipe
63 Overflow pipe
64 Overflow pipe
65 Clip
66 Suction pipe
67 Petrol feed pipe

Chapter 3/Fuel system, carburation and exhaust emission control system

11 Withdraw the pin from the float chamber lever and remove the lever. You will find that the pin is serrated and can be removed in one direction only.
12 Using the correct sized spanner, unscrew the brass valve body and remove it together with the float needle.
13 Lift the float out of the float chamber and place it in a position where it will not be damaged.
14 Remove the four setscrews securing the float chamber to the body.
15 Separate the float chamber from the body and this will free the jet spring, the jet and diaphragm and the jet housing which may now be lifted out of the body.
16 Unscrew the jet bearing nut and lift out the jet bearing.
17 Unscrew the slow running control valve from the body and collect the spring and the gland washers. Note the position of the gland washers for reassembly.
18 No further dismantling of the carburettor is necessary, indeed, it is rarely that dismantling beyond the stage decided in paragraph 13 will ever be required.
19 Reassembly is the reverse of the above. Fit new washers throughout and, if you removed the jet needle, refer to Section 23 and reassemble it in the manner described. Also, the jet needle must be centred as described in Section 22.
20 Finally, and before fitting the suction chamber, fill the piston damper bore to within ¼" of its top with SAE 20 engine oil, wipe any spillage off the outside of the piston. After fitting the suction chamber, raise the piston by means of the lifting pin and check that it falls back smartly on to the upper face of the body. Any sluggishness, assuming all other factors to be correct, will probably be due to oil on the outside of the piston.

17 SU carburettor - examination and repair

The SU carburettor, generally speaking is most reliable and it is very rarely that you would have to completely dismantle it. However, after a long period of use some deterioration must be expected, therefore, when the time arrives for a major overhaul of the engine, serious consideration should be given to replacing the carburettors with factory reconditioned items. The carburettor may develop one or more of several faults which may not be readily apparent without careful examination. The common faults to which the carburettor is prone are:-
1 Piston sticking.
2 Float needle sticking.
3 Float chamber flooding.
4 Water and dirt in the carburettor.

In addition the following parts are susceptible to wear after high mileage and as they will affect fuel consumption they should be checked and rectified at, say every other 10,000 mile servicing.
a) The carburettor needle: if the carburettor has not been correctly assembled at sometime so that the needel has not been truly central in the jet orifice it will be found that the needle will have a tiny ridge on it. If this is noted, the needle must be replaced with one of a similar type (identification letters are stamped on the flat of the needle). As the needles are made to very fine tolerances, no attempt should be made to clean out the ridge or to rub down the needle with emery cloth. If the needle requires cleaning this can be done by rubbing very lightly with metal polish.
b) The carburettor jet: If the needle is worn it is likely that the rim of the jet will be damaged where the needle has been striking it. It should be renewed as wear in the jet will result in high fuel consumption. The jet may also become worn or ridged on the outside where it has been sliding up and down between the jet bearing every time the choke is pulled out. Renewal is the only remedy.
c) The edges of the throttle and the choke tube may become worn. Renew as necessary.
d) The washers fitted to the base of the jet and under the float chamber lid may deteriorate and leak after long use and result in fuel leakage.
e) After high mileage the float chamber needle and seat may become ridged and if this occurs, flooding of the float chamber becomes a distinct possibility. Renew both the needle and the brass seating.

18 SU carburettor - piston sticking

1 The hardened piston rod which slides in the centre guide tube of the suction chamber is the only part which should make contact with the suction chamber.
2 Corrosion of the piston rod is not uncommon and this will prevent free movement of the piston. The corrosion can be cleared by careful rubbing with metal polish or, in extreme cases, by very light rubbing with 00 crocus paper.
3 Check that the rim of the piston is not burred as the result of a knock or having been dropped. Burrs can be removed by rubbing with fine emery cloth.
4 After high mileage wear in the centre guide tube may allow the piston to touch the wall of the suction chamber and this will cause obstruction to free movement of the piston.
5 Great care should be taken to remove only the minimum amount of metal when freeing the piston as the parts are made to very fine tolerances and too large a gap will cause air leakage and will upset the function of the carburettor. Clean down the walls of the suction chamber and the piston rim and ensure there is no oil on them. A trace of light oil may be applied to the piston rod.
6 If the piston is sticking, under no circumstances try to clear it by stretching the return spring.

19 SU carburettor - float needle sticking

1 If the float needle sticks, the carburettor will soon run dry and the engine will stop.
The easiest way to check for a sticking needle is to disconnect the fuel inlet pipe to the carburettor, check that the gear lever is in "neutral" or, for automatic transmission that it is in "N" or "P", guide the fuel pipe into a wad of rag or into a container, and press the starter solenoid button. If fuel is passed, the fault is almost certainly a sticking needle.
2 Remove the float chamber lid, dismantle the needle valve and clean the housing and float chamber thoroughly.

20 SU carburettor - float chamber flooding

If fuel emerges from the small breather hole in the cover of the float chamber this is known as flooding. It can be caused by the float chamber needle not seating properly in its housing and this is usually due to a piece of dirt or foreign matter which has passed the filters and has become jammed between the needle and its seating in the housing. Alternatively the float may have developed a leak so that it is not rising to operate the float needle lever, this fault can be determined by removing the float and shaking it, any sound of liquid inside the float indicates that it is faulty.

It may be that the setting of the float needle lever in relation to the float chamber cover, is incorrect. Refer to Fig.3.14. The lever clearance should be as shown (use the shank of a 7/16" drill as the test bar) when the lever is lightly pressed on to the needle. If adjustment is required, hold the flat portion with a pair of pliers and bend only at the position shown.

When carrying out the above check, make sure that the spring loaded plunger "A" of the needle is not compressed.

21 SU carburettors - float chamber needles

Commencing at engine No RA 2464, 3.8 litre models, the carburettors are fitted with "Delrin" float chamber needles. These needles can be identified by their white plastic body and incorporation of a spring loaded pin to overcome needle flutter due to engine rock when idling causing flooding with consequent rough slow running or stalling. These needles are used in conjunction with a modified seat, float lever fork and float chamber lid. The modified float chamber lid assembly, which can be identified by having AUD 2283 or 2284 embossed on the inside of the lid, is fully interchangeable with the previous type as a complete assembly. However, the Delrin type needle and seat can be used to replace the previous type provided the original lever fork is retained. These lever forks are not interchangeable and must be kept to their respective lids. The old type needle and seat must not be fitted to the modfied float chamber lid.

At engine number 7E 2226, 4.2 litre models, "Viton" tipped needles are fitted and these can be identified by having a black rubber tip. If used as replacements for the previous Delrin type needle, the needle and seat assembly must be used.

The method of setting the float needle lever when Delrin and Viton type needles are fitted is as described in Section 20.

22 SU carburettor - jet centring

Warning: Take care not to bend the needle when carrying out this operation.
1 Remove the carburettor from the engine.
2 Remove the piston damper.
3 Remove the four setscrews securing the float chamber to the carburettor body, detach the float chamber and remove the jet housing and the jet.
4 Using a ring spanner, slacken the jet locking nut approximately half a turn.
5 Refer to Fig.3.15. Replace the jet and diaphragm assembly. Push the jet and diaphragm assembly as high as possible with hand pressure and at the same time press the piston down onto the jet bridge, using a pencil or a piece of rod for this. Centralisation will be helped by lightly tapping on the side of the carburettor body.
6 Tighten the jet locking nut.
7 The actual centring must be carried out with the setscrew holes in the jet diaphragm and carburettor in alignment. After tightening the jet locking nut the jet dipahragm must be kept in the same position relative to the carburettor body and to do this it is advisable to mark one of the corresponding jet diaphragm and carburettor setscrew holes with a soft pencil. Centring will be upset if the diaphragm is moved radically after tightening the jet nut.
8 The jet is correctly centred when the piston falls freely and hits the jet "bridge" with a metallic click. Check if there is any difference in the sound of the piston hitting the bridge with the jet in its highest and lowest positions. If there is any difference in the sound, the procedure for centralising the jet will have to be repeated.
9 If difficulty in centring the jet is encountered after carrying out above procedure, it is permissible to lower the jet needle slightly in the position to make centralising more positive. The needle must, however, be restored to its normal position when checking the centralisation.
10 Top up the damper with SAE 20 engine oil.

23 SU carburettor - needle replacement

The needle size is determined during engine development and will provide the correct mixture strength unless extremes of temperature, humidity or altitude are encountered. A different needle to that specified may be required if any alteration to the standard specification of the exhaust system, air cleaner, camshaft or compression ratio is made.
1 Remove the suction chamber and piston assembly.
2 Slacken the needle clamping screw in the side of the body of the piston and pull out the needle. If the needle is tight it can probably be loosened by moving it inwards and then pulling out.

3 The needle type letter is stamped on the shank of the needle, check that this corresponds with the item being fitted.
4 Fit the needle to the piston assembly so that it is positioned as shown in Fig.3.16, so that the bottom edge of the groove is level with the bottom edge of the piston rod.
5 Correct positoning of the needle in relation to the piston is essential otherwise the fuel/air mixture to the engine will be upset.

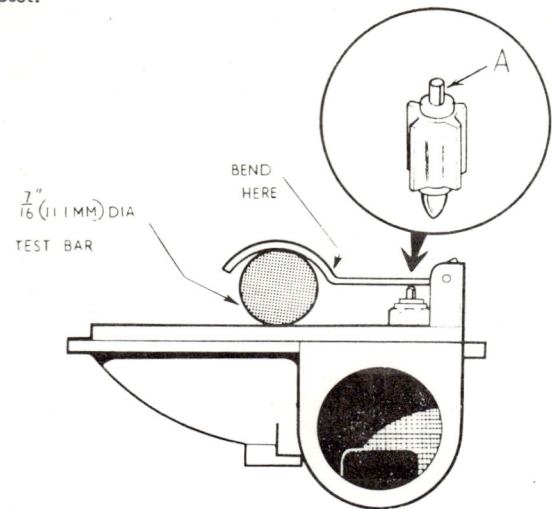

Fig. 3.14. Checking the setting of the float lever

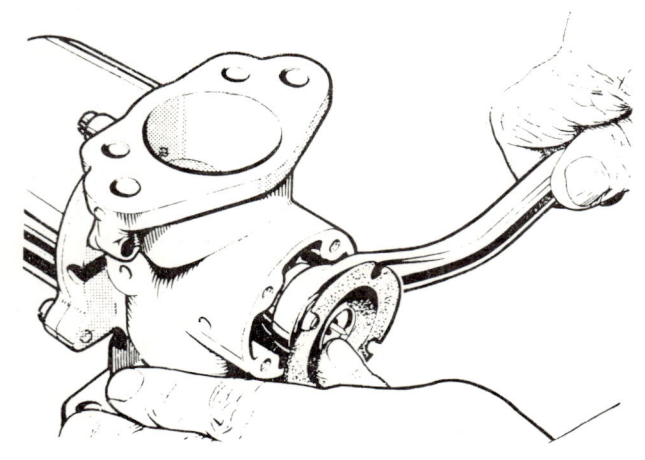

Fig. 3.15. Centring the jet

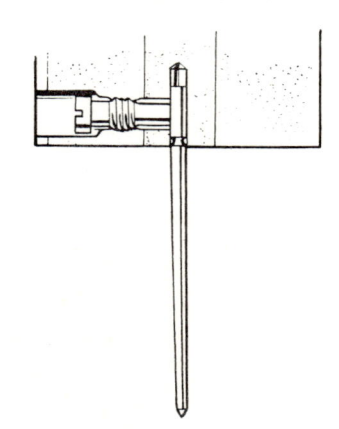

Fig. 3.16. Location of needle in piston

Chapter 3/Fuel system, carburation and exhaust emission control system

24 SU carburettors - general information on adjustment and tuning

It is useless to attempt carburettor tuning until the cylinder compressions, valve clearances, spark plug gaps and contact breaker gaps have been tested, checked and adjusted as necessary. The distributor centrifugal advance mechanism and vacuum advance operation should be checked and ignition timing set to the correct figure. The ignition timing is important since if retarded or advanced too far the setting of the carburettors will be affected. Ensure that the needles are correctly located in the pistons (see Section 23). Check over the carburettors and ensure that the pistons are free in the suction chambers and that the piston dampers are topped up with engine oil SAE 20. Lubricate the throttle controls and check for free operation and travel. Check that petrol filters are clean.

25 SU carburettors - adjustment and tuning

1 Remove the air intake.
2 Release the three pinch bolts securing the two piece throttle levers to the carburettos throttle spindles.
3 Working on one carburettor at a time, full rotate the throttle spindle in a clockwise direction (looking from the front) to close each throttle butterfly valve. With the throttle held closed, tighten the pinch bolt but keep the two piece throttle lever in the midway position.
4 Now operate the accelerator linkage and observe if all throttles are opening simultaneously by noting the movement of the full throttle stops on the left hand side of the throttle spindles.
5 Slacken the manual mixture control cable and check that the jet levers are against their stops, if not, adjust on the mixture control rods to bring about that condition.
6 Screw down the slow running volume screws ("A" in Fig.3.17) onto their seatings and then unscrew them two full turns. These screws only need to touch the seatings, do not overtighten or the seatings will be damaged.
7 Now remove the suction chambers and pistons and unscrew the mixture adjusting screws (C) until each jet is flush with the bridge of its carburettor. Replace the piston and suction chamber, lift each piston on the lifting pin (Fig.3.18) and check that it falls back smartly onto the bridge of its carburettor. Now turn down each mixture adjusting screw 2½ turns.
8 Start the engine and run until it has attained its normal running temperature.
9 The carburettors must now be balanced (synchronised) by adjusting on the slow running volume screws (A) until they are sucking equally. This can best be judged by applying a balance meter to the carburettor air inlet and adjusting on the screws until the readings are the same. Alternatively, listen to the hiss of each carburettor (use a piece of tube as illustrated in Fig.3.19: a piece of old bicycle tube is ideal) and adjust on screws "A" until the hiss from each carburettor is judged to be the same.
10 Keep checking as above and continue adjusting until, with the carburettors balanced, the engine of 3.8 litre cars and those with automatic transmission is idling at 500 rpm. The engine of 4.2 litre cars with standard transmission should be adjusted to idle at 700 rpm.
11 When you are satisfied with the balance of the carburettors, the mixture should be adjusted by screwing each mixture adjustment screw (C) (Fig.3.17) up (weaker) or down (richer) by the same amount, until the fastest idling speed, consistent with even firing is obtained.
12 As you adjust the mixture, the engine will probably run faster so you will have to return to the slow running volume screws (A) and re-adjust (keeping the carburettors in balance) until the desired idling speed is obtained.
13 Now check the mixture strength by lifting the piston of the front carburettor about 1/32" (0.8mm) when, if:
a) The engine speed increases and continues to run faster, this indicates that the mixture is too rich.
b) The engine speed immediately decreases, this indicates that the mixture is too weak.
c) The engine speed momentarily increases very slightly, this indicates that the mixture is correct.
14 Repeat the above for the other two carburettors and after adjustment recheck the front carburettor since the carburettors are interdependant.
15 If the mixture is correct, the exhaust note should be regular and even. If it is irregular, with a splashy type of misfire and colourless exhaust, the mixture is too weak. If there is a rythmical type of misfire in the exhaust beat together with a

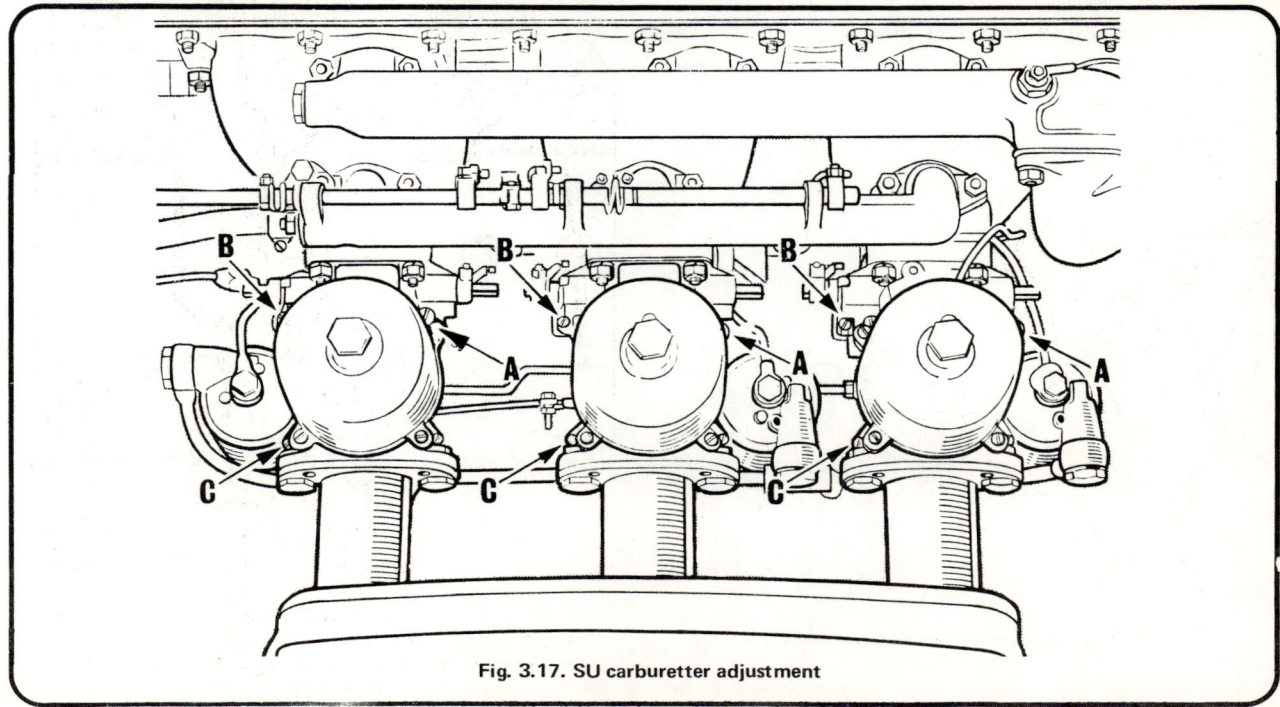

Fig. 3.17. SU carburetter adjustment

blackish exhaust, the mixture is too rich.

16 When all adjustments are satisfactory, reconnect the mixture control cable and allow 1/16" (1.5 mm) free travel at the bottom of the facia control before the jet levers begin to move.

17 The next step is to adjust the fast idle setting. Set the mixture control knob on the facia to the highest position in the slide immediately short of the position where the mixture adjusting screw levers (C) begin to move. This will be about the mid-travel position of the control knob and approximates to 5/8" (16 mm) movement at the bottom of the jet levers. Adjust the fast idle screws (B) (Fig.3.17) on the throttle stops to give an engine speed of about 1,000 rpm when hot. You may find it easier to carry out the foregoing operation by lightly slipping a 0.002" (0.051 mm) feeler gauge under each screw when the mixture control knob is at the bottom of the slide, the advantage of this method is that you can be sure that each carburettor is adjusted by the same amount.

18 Finally replace the air intake.

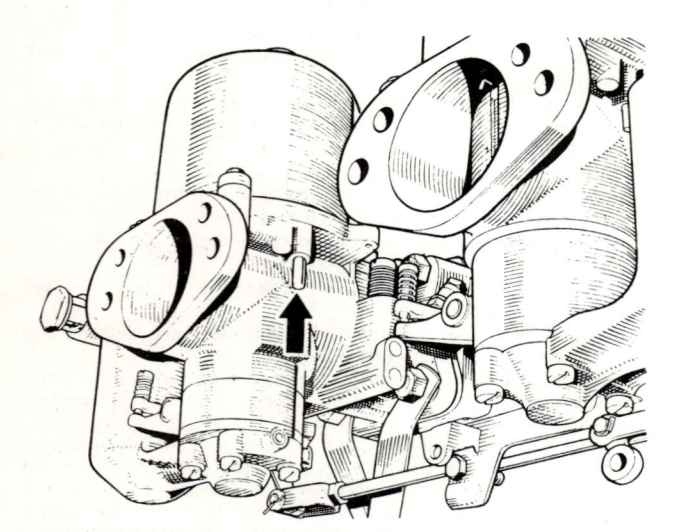

Fig. 3.18. The piston lifting pin

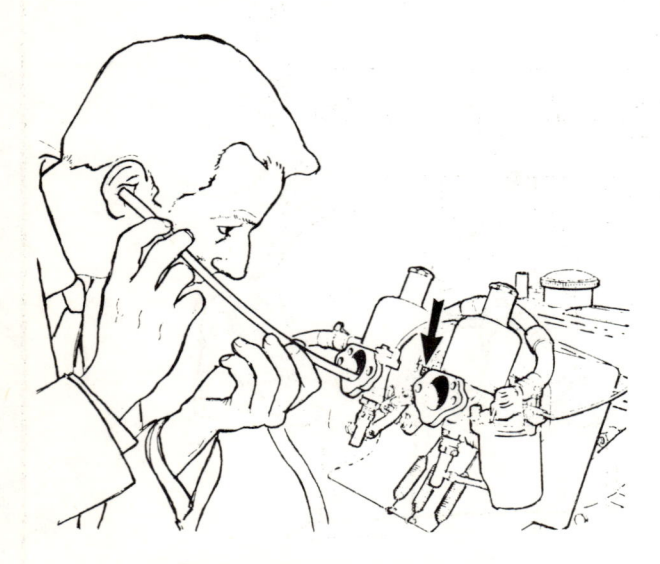

Fig. 3.19. Balancing SU carburetters by ear

26 Carburettor mixture control warning light switch - setting

1 Remove the dash casing below the control slide by taking out the drive screws and, for right hand drive cars, the screwed bezels of the odometer and the clock setting drives.

2 Set the lever on the control slide about ¼" (6.350 mm) from the bottom limit of its travel when a click will be heard.

3 Now adjust the two nuts on the threaded shank of the switch until the warning light ceases to glow when the ignition is switched on.

4 Work the lever up and down once or twice and make any final adjustment as may be necessary.

5 Replace the dash casing and the screwed bezels of the odometer and clock setting drives, if applicable.

27 Stromberg 175 CD2SE carburettor - general

The Stromberg 175 CD2SE carburettor, a sectional view of which is given in Fig.3.20, is fitted to cars for export to the USA and Canada and is used in conjunction with the exhaust emission control system.

It is a development of the constant depression carburettor operating on the principle of varying the effective areas of choke and jet orifice in accordance with the degree of throttle opening, engine speed and engine load. Fuel passes into the float chamber via a **needle** valve which controls flow of the fuel by twin floats on a common arm. Clearance around the piston in its vertical

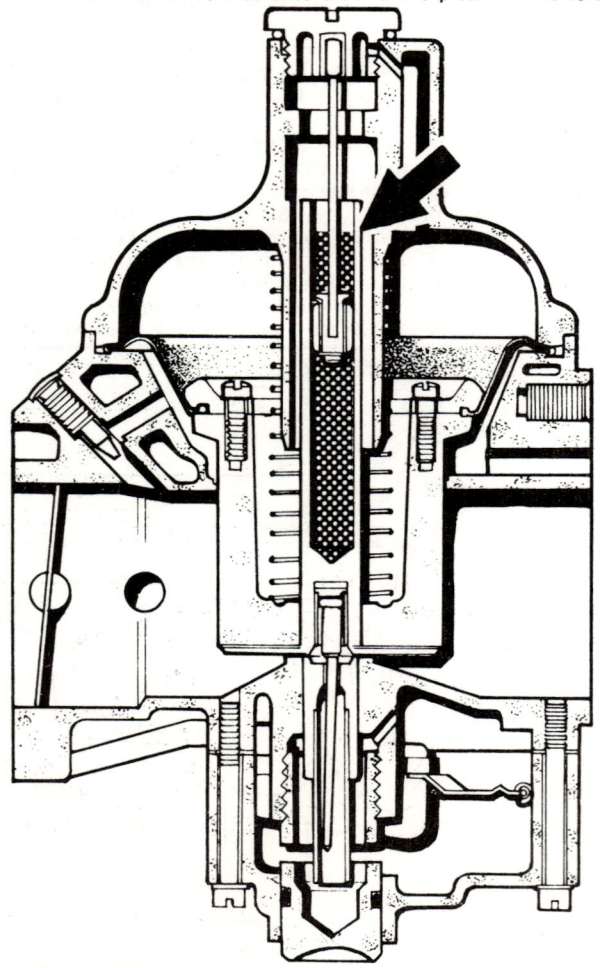

Fig. 3.20. Sectioned view of the Stromberg 175 CD2SE carburetter. Hydraulic damper arrowed. Note the offset jet needle

Chapter 3/Fuel system, carburation and exhaust emission control system

bore permits air to leak into the mixing chamber and thus lower depression. A drilling is taken from the atmospherically vented region beneath the diaphragm to meet a further drilling that breaks into the mixing chamber downstream of the piston, and adjusting screw with a conical tip is inserted into this drilling and is adjusted, and sealed by the manufacturer to bring each carburettor to a common 'leak' datum. **This sealed screw must not be disturbed in any circumstances.**

The mixture is enriched, for cold starting, by a lever on the side of each carburettor which is operated by the choke control on the instrument panel. The lever rotates a disc in the starting device in which a series of holes of different diameters are drilled; in the fully rich position all these holes are in communication with the starting circuit and will thus provide the richest mixture. Fuel is drawn from the float chamber via a vertical drilling adjacent to the central jet, through the starting device and into the throttle body between the piston and the throttle disc. At the same time the cam on the choke lever will open the throttle disc beyond the normal idle position and so give a fast idle speed. As the choke is gradually pushed to the "off" position, fewer and smaller holes will limit the fuel fed from the float chamber and so the mixture is gradually weakened to the point when the choke is fully home at which the mixture strength is governed by the Factory setting of the main jets and the idling speed is determined by the setting of the throttle stop screw. A Control on each carburettor enables the choke to be varied for summer and winter conditions and takes the form of a spring loaded plunger operating against the cam, this device is illustrated in Fig.3.21. To check the setting, note the position of the stop cross-pin; if it is lying in the horizontal slot in the casting the choke is set for winter operation. To adjust it for summer operation, depress the spring loaded pin and turn it through 90° and in this condition it should be at right angles to the slot. There is no separate circuit for idling; fuel is provided by the jet orifice and the amount is controlled by the jet/needle relationship established during manufacture. However, the idling speed can be adjusted by a throttle stop screw which limits the closure of the throttle when the accelerator pedal is released.

The jet/needle relationship not only governs the correct idle mixture but also the correct mixture strength throughout the whole range of engine conditions. Although, during development of the carburettor it was found desirable to have the needle central in the jet, this was not practicable to achieve so the needle has been biased permanently to one side of the jet to rub lightly against the jet orifice. The needle profile has been evolved to compensate for the known air leak, consistency being maintained by factory adjustment of the sealed screw already mentioned, and therefore a constant fuel/air ratio is maintained. To meet emission control requirements, carburettors must be kept within very narrow "flow bands" and so all carburettors fitted to Jaguar cars have the jets pressed into position to a predetermined depth thereby eliminating any possible mal-adjustment in service. As the needle profile has been developed from exhaustive testing, it is vitally important, from engine emission control aspects, that only the recommended needle is used. The needle identification number is stamped on the shank and this is "BIE" for cars manufactured prior to 1972 and "BIBT" for cars manufactured in 1972 and later.

Variations in the mixture strength caused by heat transfer to the carburettor castings are significant in the context of the extraordinary precision demanded by emission requirements so a temperature compensator (Fig.3.22) is incorporated to cater for this condition. An air flow channel permits air passing through the carburettor to bypass the bridge section. A bimetallic blade regulates the movement of a tapered plug which adjusts the quantity of air bypassed to the mixing chamber. The compensator is attached to the body of the carburettor by two screws and the joint is sealed to ensure that no leakage can occur. The assembly is present and unless necessary, due to the tapered plug sticking, should not be readjusted in service. If malfunctioning is suspected and the tapered plug moves freely (when tested carefully by hand) with the engine hot or cold, the compensator assembly must be changed for a new unit.

At any point in the throttle range, a temporary enrichment is required when the throttle is suddenly opened. This is achieved by a hydraulic damper inside the hollow guide rod of the piston. The guide rod is filled with Zenith Lube Pack or SAE 20 engine oil to within ¼" of the end of the rod. The immediate upward movement of the piston when the throttle is suddenly opened is resisted by the damper and for this brief period a temporary increase in the depression over the jet orifice is achieved and so the mixture is enriched.

High emissions will occur during periods of engine over-run if the fuel/air mixture in the combustion chambers is not of sufficient strength (when diluted by exhaust gases) to support combustion. To overcome this, a device is fitted to the carburettors which consists of a bypass formed in the carburettor around the primary throttle under the control of a vacuum operated valve the location of which is shown in Fig.3.23.

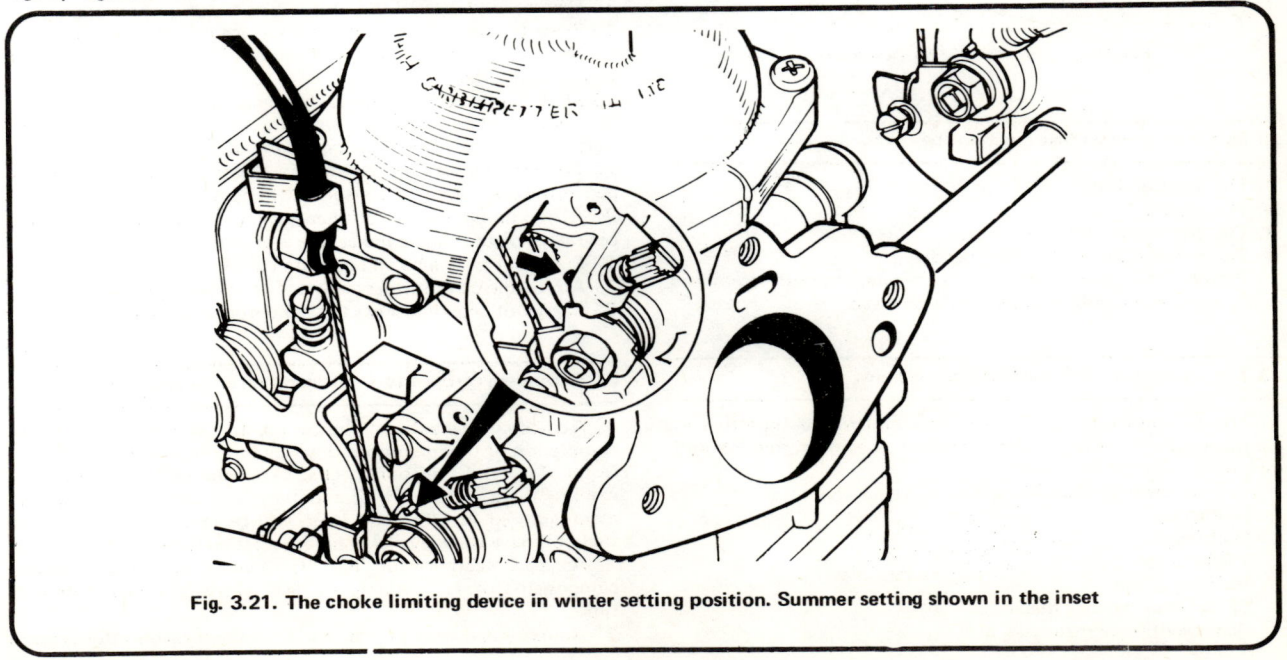

Fig. 3.21. The choke limiting device in winter setting position. Summer setting shown in the inset

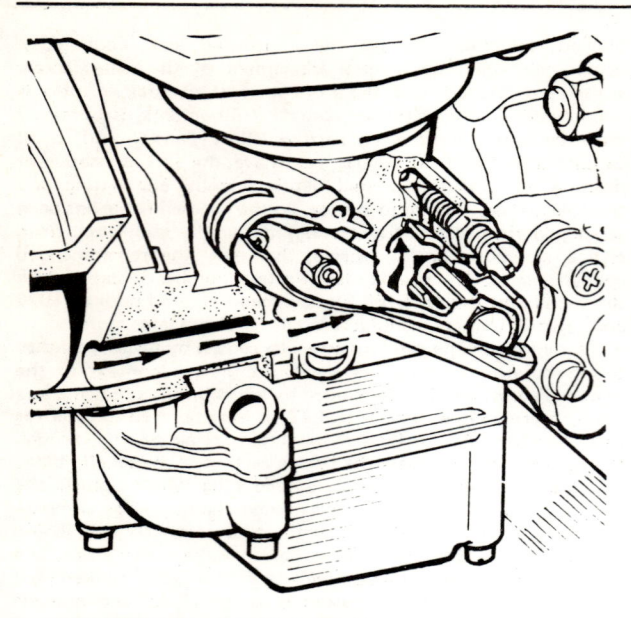

Fig. 3.22. Cut-away of the temperature compensator

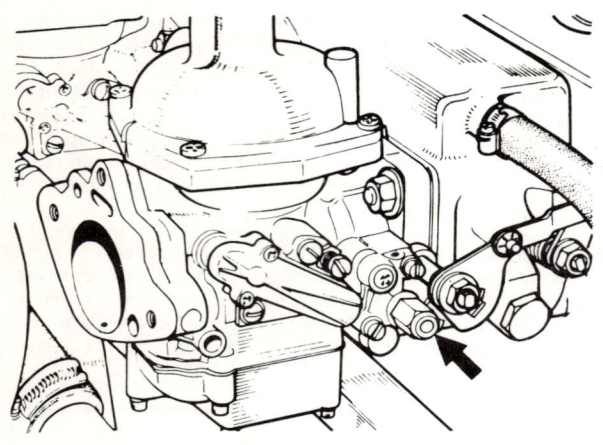

Fig. 3.23. The throttle by-pass valve

28 Carburettor items that must not be changed

a) The jet assembly.
b) The air valve.
c) The depression chamber cover.
d) The position of the metering needle.

If any of the above items require changing, the sub-assemblies or the complete carburettors must be renewed.

29 Test equipment - emission control systems

The recommended equipment for servicing carburettors in conjunction with the exhaust emission control system should include at least the following:-
  Ignition analyser/oscilloscope
  Ohmmeter
  Voltmeter
  Tachometer
  Vacuum gauge
  Carburettor balance meter
  Cam dwell angle meter
  Ignition timing light
  Engine exhaust combustion analyser
  Cylinder leak tester
  Distributor advance tester

The following equipment covers most of the requirements for engine testing and tuning of vehicles fitted with exhaust emission control systems:
  Oscilloscope engine tuning set and exhaust gas analyser, Type 1020 or 720 manufactured by the Sun Electric Corporation.

It is appreciated that you will have some, but not all, of the above equipment, therefore, whenever you carry out any work which disturbs the exhaust emission control system you must take your car to a garage holding the necessary equipment and have it tested to ensure that it conforms to the appropriate regulations. The text of the following Sections will indicate where testing is necessary following the work described.

30 Stromberg carburettor - removal and refitting

1 An exhaust gas CO content analysis check will be required after the work described in this Section is completed.
2 Refer to Section 38 and remove the air cleaner.
3 Refer to Fig. 3.24.
4 Disconnect the crankcase breather pipe (2) from each carburettor.
5 Disconnect the vacuum pipe (3) from each bypass valve.
6 Slacken the clip and disconnect the fuel inlet pipe from the teepiece (4).
7 Slacken the clamp bolts and withdraw the inner choke cables (5).
8 Disconnect the outer choke cables from the clips (6).
9 Prise off the spring clip (7) securing the front throttle lever to the connecting link but do not disconnect the lever at this stage.
10 Prise off the spring clip (8) securing the rear throttle lever to the connecting link but do not disconnect the lever at this stage.
11 On those cars fitted with automatic transmission, release the spring clip (9) and disconnect the 'kick-down' from the throttle lever.
12 Remove the nuts and the spring washers (10) securing the carburettors to the secondary tthrottle housing.
13 Withdraw the carburettors slightly and then disconnect the front and rear throttle levers (11) from the connecting links.
14 Lift off the carburettors taking care not to bend the throttle spindle connecting clamps.
15 Remove and discard the carburettor flange gaskets (13).
16 Place the carburettors on a flat surface in an upright position.
17 Mark the relative position of the throttle spindles (15) to the connecting rod.
18 Slacken one bolt (16) on each connecting clamp.
19 Remove the clips (17) securing the teepiece to the carburettors and they may now be separated.
20 The carburettors are refitted in the reverse sequence to the above but use new carburettor flange gaskets and teepiece clips.
21 After completion of refitment, carry out a check for correct operation of the choke and accelerator controls.
22 Carry out an engine and ignition diagnostic check.
23 Adjust the carburettors as described in Section 32.
24 Carry out an exhaust gas CO content analysis check.

31 Stromberg carburettors - dismantling and reassembly

1 A check of the exhaust emission system will be required after completion of the operations described in this Section.
Note: Two red emission packs, part number 11791 for cars manufactured before 1972 and part number JS 478 for cars manufactured in 1972 and later, will be required so make sure that you have these available before starting work.

As each carburettor has its own individually matched components, it is advisable to deal with one carburettor at a time.
2 An exploded view of the rear Stromberg carburettor is given

Chapter 3/Fuel system, carburation and exhaust emission control system

Fig. 3.24. Stromberg carburetters - removal and refitting

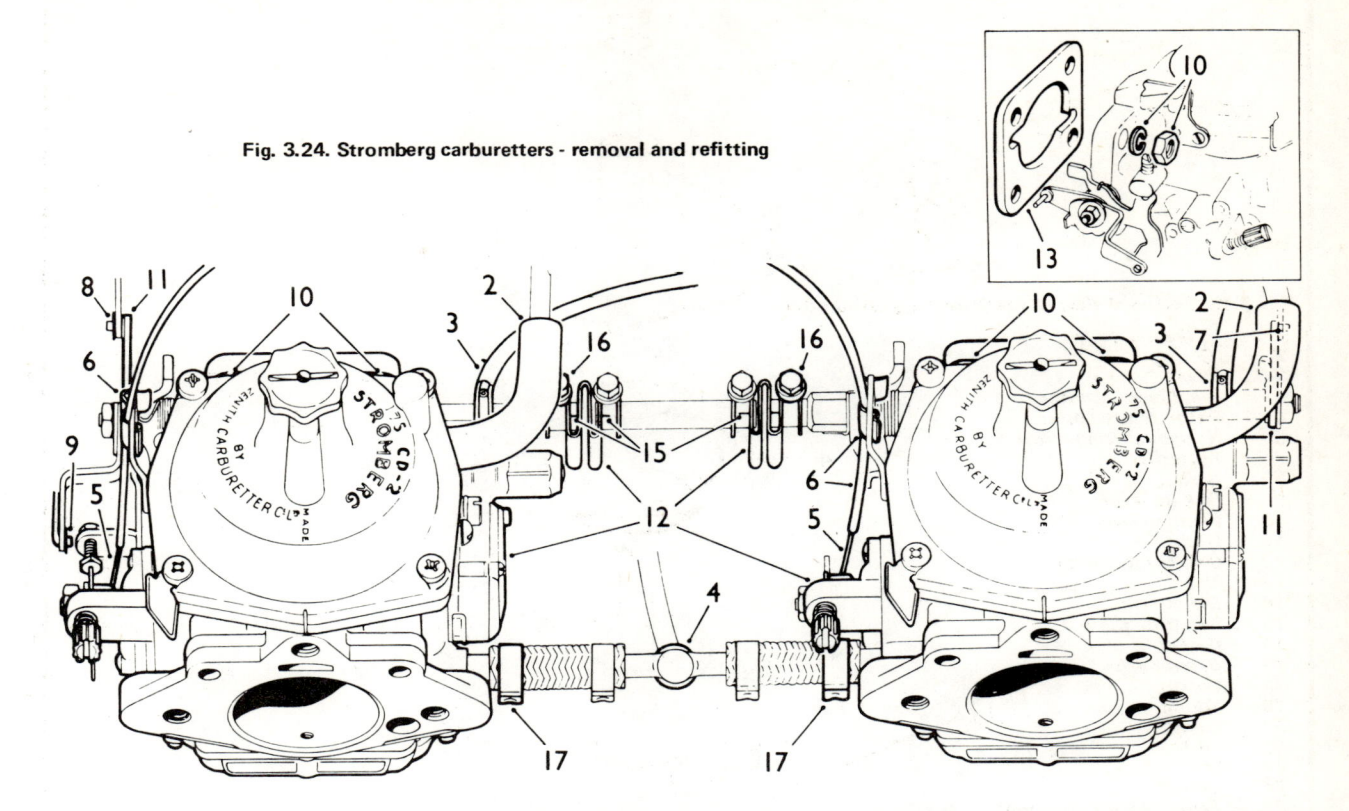

in Fig.3.25. Step by step operations for dismantling the carburettors, referred to by number in the text, are shown in Fig.3.26 to which reference should be made during the following operations.

3 Remove the carburettors from the car as described in Section 30.
4 Remove the lead plug from the cover securing screw (2).
5 Mark the relative positions of the cover to the carburettor body (3).
6 Remove the cover securing screws (4) and lift off the cover.
7 Remove the piston return spring (5).
8 Withdraw the piston and diaphragm assembly (6).
9 Remove the screws securing the diaphragm retaining rim to the piston (7), lift off the ring and remove and discard the diaphragm. **Do not remove** the needle retaining screw and needle. The position of the needle is determined during manufacture and must not be altered.
10 Remove the screws and washers (8) securing the float chamber to the carburettor body and then lift off the float chamber.
11 **Do not** remove the jet cover unless there is evidence of leakage in which case the cover and O-ring assembly must be renewed. The jet cover is removed by bending back the plastic retaining lugs and then removing the jet cover (9) together with the O-ring. Now discard these items.
12 Remove and discard the float chamber gasket (10).
13 Disengage the float hinge pin from the spring clip and then lift out the float (11).
14 Unscrew the needle valve (12). Discard the valve and the alloy seating washer.
15 Remove the nut, serrated washer and plain washer securing the throttle lever to the throttle spindle. Withdraw the throttle lever (13) followed by the bush and actuating link.
16 Mark the relative position of the throttle disc (14) to the spindle.
17 Remove the screws (15) securing the throttle disc and slide the disc out of the throttle spindle.

18 Note the position of the spindle return spring (16) and then withdraw the throttle spindle.
19 Remove the screws and lockwashers securing the bypass valve and cover (17) to the carburettor body, remove the gasket and discard it.
Note: The bypass valve is correctly adjusted during manufacture and this adjustment will be lost if any attempt is made to separate the valve cover from the body. If it is suspected that the diaphragm is damaged, the complete bypass valve assembly must be replaced.
20 Prise the spindle seals (18) out of the body and discard them.
21 Remove the screws securing the temperature compensator (19) to the body and withdraw the compensator. Remove and discard the seals.
22 Some carburettors may have a spring loaded screw (shown as 20 in Fig.3.26) adjacent to the temperature compensator. This is merely a blanking screw to a hole provided for machining purposes in manufacture. Do not take it out and do make sure that it is fully tightened.
23 Reassembly is generally a reversal of the dismantling procedure, however, correct reassembly of the carburettor is so essential that we have decided to give the reassembly sequence in detail.
24 New seals, gaskets etc referred to in the text will be found in the servicing packs.
25 Refer to Fig.3.27.
26 Fit new seals to the temperature compensator assembly (21) and then fit it to the carburettor body.
27 Remove the screws securing the temperature compensator cover (22), lift off the cover and check that the tapered plug is free to move in the housing (23). If movement is satisfactory, replace the cover, but if the plug appears to stick the compensator assembly must be renewed.
28 Press new spindle seals (24) into the carburettor body.
29 Position a new gasket (25) on the bypass valve body, fit the bypass valve and secure with the screws and lock washers.
30 Check the tapped holes (26) in the throttle spindle for burrs

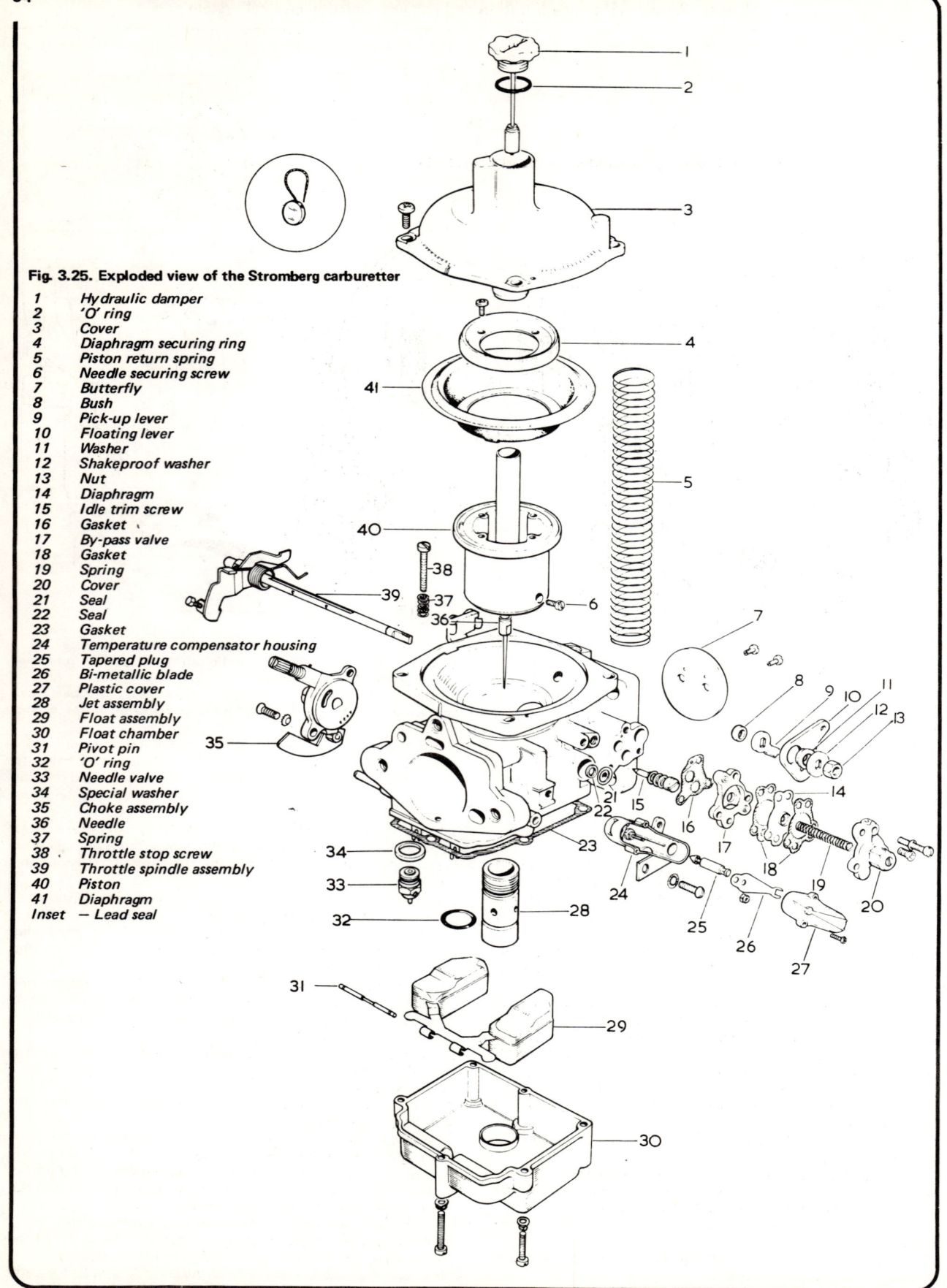

Fig. 3.25. Exploded view of the Stromberg carburetter

1	Hydraulic damper
2	'O' ring
3	Cover
4	Diaphragm securing ring
5	Piston return spring
6	Needle securing screw
7	Butterfly
8	Bush
9	Pick-up lever
10	Floating lever
11	Washer
12	Shakeproof washer
13	Nut
14	Diaphragm
15	Idle trim screw
16	Gasket
17	By-pass valve
18	Gasket
19	Spring
20	Cover
21	Seal
22	Seal
23	Gasket
24	Temperature compensator housing
25	Tapered plug
26	Bi-metallic blade
27	Plastic cover
28	Jet assembly
29	Float assembly
30	Float chamber
31	Pivot pin
32	'O' ring
33	Needle valve
34	Special washer
35	Choke assembly
36	Needle
37	Spring
38	Throttle stop screw
39	Throttle spindle assembly
40	Piston
41	Diaphragm
Inset	— Lead seal

Chapter 3/Fuel system, carburation and exhaust emission control system

which, if present, should be removed by light stoning.
31 Fit the throttle spindle (27) making sure that the return spring is correctly positioned.
32 Fit the actuating link (28), the bush, throttle lever, plain washer, serrated washer and nut, **do not** overtighten the nut. During this operation, check for ovality or wear of the bush which, if found, will mean renewing the bush.
33 Slide the throttle disc (29) into the spindle and fit the securing screws. Adjust the position of the disc until it fully closes and then tighten the screws.
34 Fit new needle valve (31) and using a new alloy seating washer tighten down into position, however, do not overtighten on the valve.
35 Refit the float (32) and now invert the carburettor body, ensure that the needle valve is closed and then measure the distance from the face of the carburettor body to the highest point of the float as shown in Fig.3.28. Dimension "A" should be 0.65" + 0.02" (16.5 + 0.5 mm). It is permissible to bend the float arm slightly to obtain this setting.
36 Fit a new gasket to the float chamber, then fit the chamber (33) and tighten the securing screws by diagonal selection to avoid distortion.
37 Fit a new O-ring to the jet cover (35) after smearing the ring with petroleum jelly. Now press the jet cover into the float chamber and ensure that it is correctly seated.
38 Check by means of a straight edge, as shown in Fig.3.29, that the needle is not bent or distorted in any way. Remember that the needle is permanently biased to one side and this must not be confused with damage to the needle. If the needle is found to be damaged a new piston and needle assembly must be fitted.

The correct needle size will be found in Specification at the beginning of this Chapter.
39 Fit a new diaphragm to the piston (37) and ensure that the lip on the underside of the diaphragm is seated in the recess in the piston.
40 Fit the diaphragm retaining ring (38) and tighten the securing screws evenly.
41 Fit the piston assembly (39) in the carburettor body and ensure that the lip on the outer periphery of the diaphragm is seated in the recess in the body.
42 Position the piston return spring (40) on the piston.
43 Fit the cover (41) in the position marked before dismantling and tighten the screws by diagonal selection to avoid distortion.
44 Drive a new lead plug (42) into the cover securing screws.
45 Refit the carburettors to the car.
46 Carry out an engine and ignition diagnostic check.
47 Tune and adjust the carburettors as described in Section 32.
48 Carry out an exhaust gas CO content check.

Important:

To ensure that the engine exhaust emissions are kept at the required level, the following items must not be changed:
a) The jet assembly.
b) The air valve.
c) The depression chamber cover.
d) The position of the jet needle.

The following items must not be adjusted:
a) The temperature compensator.
b) The piston return spring (the correct spring has no colour coding).
c) The bypass valve spring.

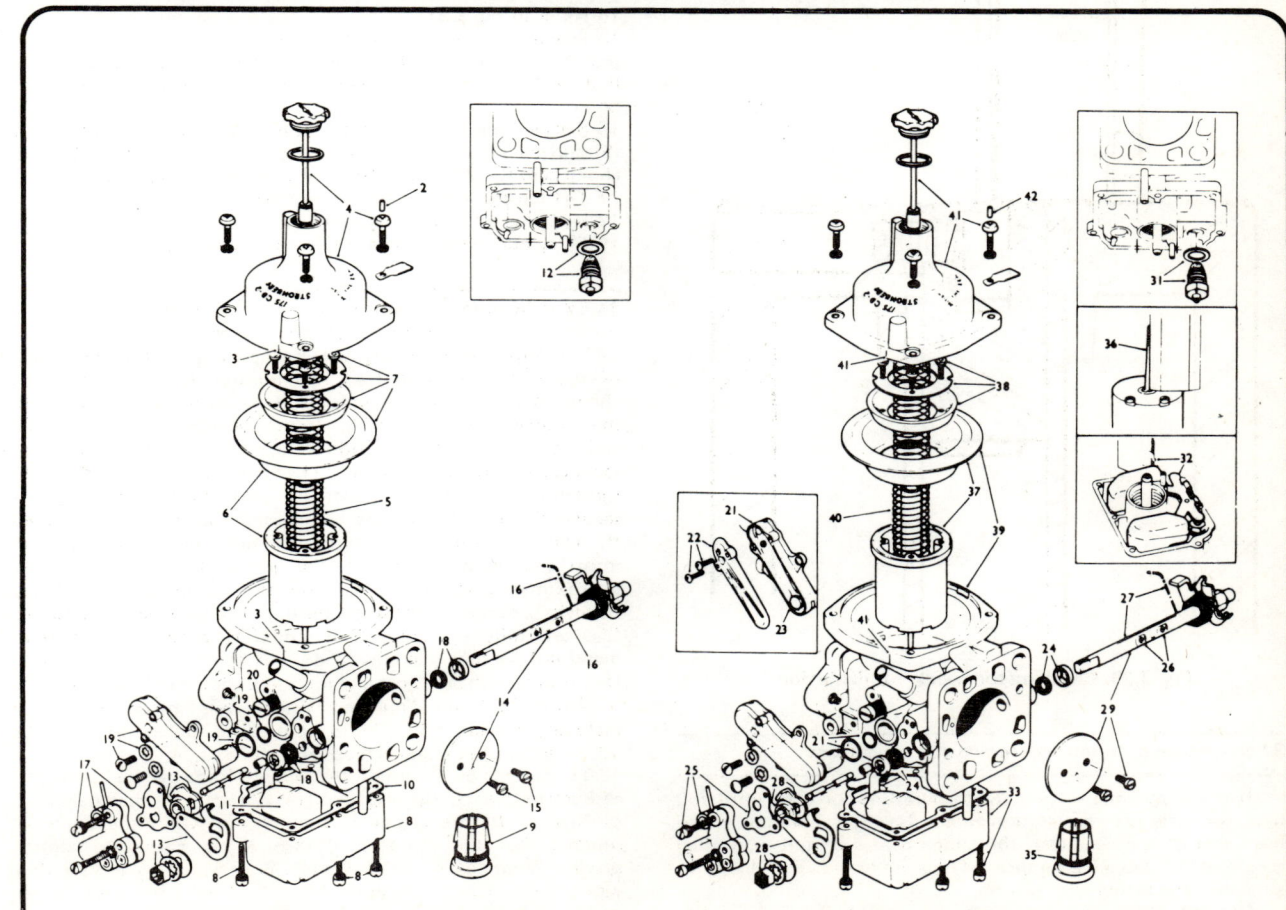

Fig. 3.26. Dismantling the Stromberg carburetter

Fig. 3.27. Assembling the Stromberg carburetter

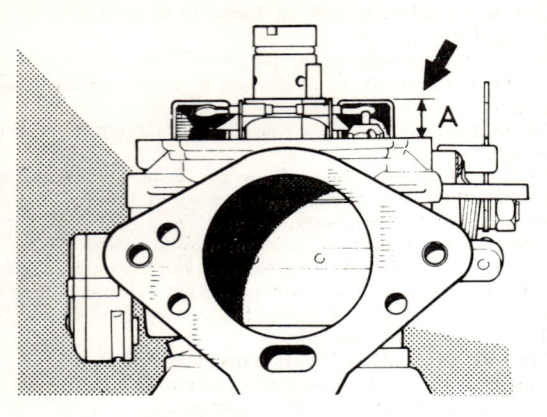

Fig. 3.28. Checking the height of the float

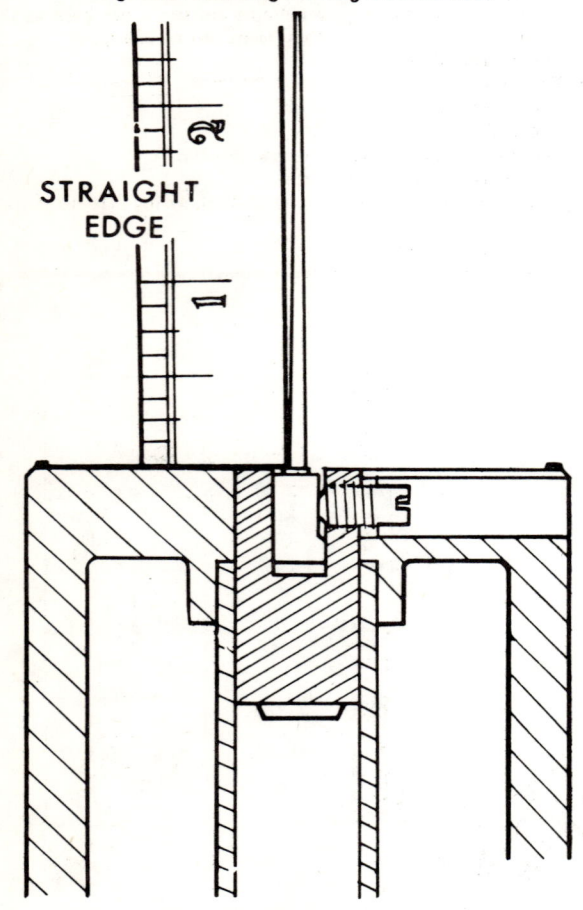

Fig. 3.29. Checking the needle for distortion

32 Stromberg carburettors - tuning and adjustment

It is not possible to tune the carburettors correctly until you have correctly set the ignition timing (see Specifications at the beginning of the Chapter), the contact breaker and spark plug gaps and the tappet clearances. These items are more likely to alter than the carburettor setting.
1 Remove the air cleaner and filter element.
2 Run the engine until it attains its normal running temperature.
3 Refer to Fig.3.30.

4 Make sure that the piston dampers (3) are topped up to within ¼" of the top with SAE 20 engine oil.
5 Slacken off the idle trim screw (4) on each carburettor until it no longer contacts the throttle lever.
6 Slacken off the choke inner cable clamping bolts (5), make sure that each choke is fully closed and then retighten the clamp bolts.
7 Slacken the throttle spindle clamping bolts (6) and ensure that the throttles are fully closed and then retighten the clamp bolts.
8 Screw in on the idle trim screw (7) on the rear carburettor until contact is just made with the throttle lever and then carry on screwing down to give an idle speed of 750 rpm for cars fitted with manual transmission and 650 rpm for cars with automatic transmission.
9 Screw in the idle trim screw on the front carburettor until contact is just made with the throttle lever.
10 Check the synchronisation of the carburettors using a balance meter.
11 If you do not intend to adjust the fast idle speed, refit the air cleaner assembly and carry out an exhaust gas CO content analysis.
12 To adjust the fast idle setting first carry out operations 1, 2 and 4 above.
13 Make sure that the choke is in the correct position for the prevailing conditions. Refer to Fig.3.21 and note the position of the pin (2). If it is lying in a horizontal position, the choke is set for winter running and if vertical it is in the position for summer use. Depress the plunger and rotate the pin to the correct position.
14 Refer to Fig.3.31.
15 Slacken off the choke inner cable clamping bolts (4); push the choke control knob on the facia panel, fully closed. Ensure that each choke is fully closed and then retighten the clamp bolts.
16 Slacken the locknut on the rear carburettor fast idle screw (5) and turn it until the gap between the head of the screw and choke cam (6) is 0.067" (1.6 mm) as shown in Fig.3.32.
17 Tighten the locknut and recheck the gap.
18 Refit the air cleaner assembly.
19 Carry out an exhaust gas CO content analysis.

33 Exhaust emission control - general information

The Duplex manifolding system employs the metering system of the carburettors to feed fuel mixture to the combustion chambers through two manifolds. Two throttles are employed, the primary throttle being in its normal position in the carburettor and the second is located in a water jacketed secondary throttle housing. The linkage between the two throttles is so arranged that on part throttle opening the secondary throttle remains closed and the mixture is routed through the primary mixture pipe and thence to the inlet manifold downstream of the secondary throttles. After approximately 25° of primary throttle opening, the secondary throttle comes into operation until, at full throttle, both butterflies are fully opened and the mixture passes through both manifolds so that maximum power is maintained. A schematic layout of the carburettors and linkage is shown in Fig.3.33.

From 1972 an improved system was introduced which includes, amongst other things, control for fuel evaporation loss - this part of the system is illustrated in Fig.3.34.

To prevent air pollution by vapours from the fuel tank and carburettor vents, the vapours are stored in a charcoal filled canister whilst the engine is stopped but whilst the engine is running they are disposed of through the crankcase control system. Venting of the fuel tank is designed to ensure that vapours are passed through the control system even when the car is parked on an inclined surface. A capacity limiting device is built into the fuel tank to ensure that sufficient free capacity is available to accommodate fuel which might otherwise be displaced as a result of high ambient temperatures.

Chapter 3/Fuel system, carburation and exhaust emission control system

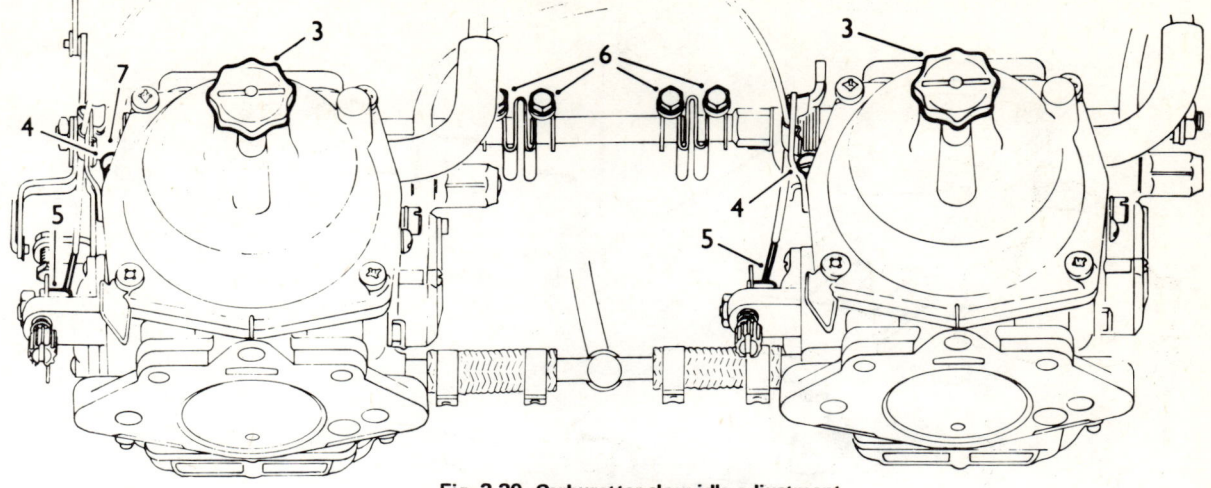

Fig. 3.30. Carburetter slow idle adjustment

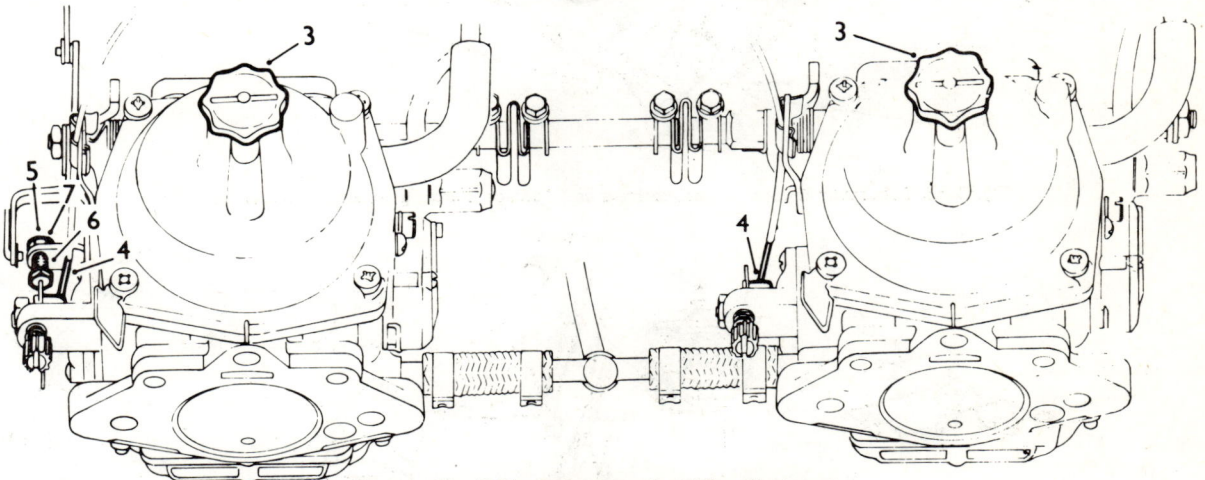

Fig. 3.31. Carburetter fast idle adjustment

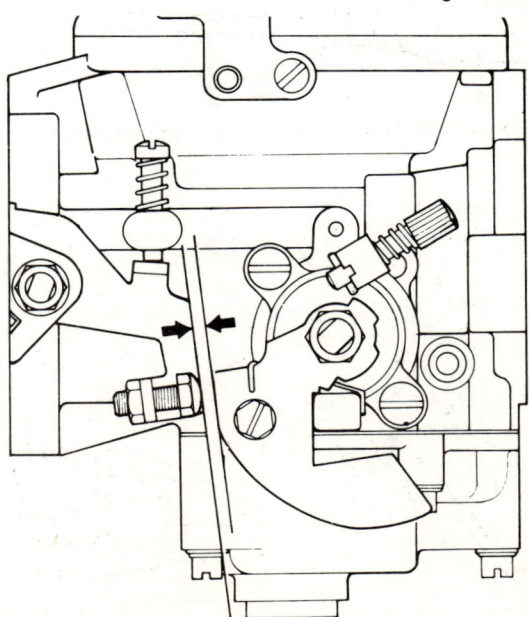

Fig. 3.32. Adjustment of the fast idle screw

34 Carbon canister - removal and refitting

1 Refer to Fig.3.35.
2 Slacken the clip and disconnect the outlet hoses (1).
3 Disconnect the inlet hose (2).
4 Remove the nut (3), the lockwasher and plain washer which hold the canister to the wing valance. Access to the bolt head is gained through the wheel arch.
5 Withdraw the canister (4).
6 Refitting is the reverse of the above but do not overtighten the securing nut and do make sure that the hoses are in good condition and that the clips are tight.

35 Air delivery pump - removal and refitting

1 An exhaust emission test must be carried out after refitting the pump to the car.
2 The layout of the pump is given in Fig.3.36.
3 If your car is fitted with air conditioning equipment, **on no account** must you disconnect the outlet pipe from the air conditioning compressor pump as the gas contained in the system can cause blindness. So, before, starting work, study Fig.3.36 carefully and ensure that you do not confuse the two pumps and the pipe layout.
4 Slacken the clip on the air pump delivery pump (1).
5 Remove the nut, washer, spacer and bolt securing the

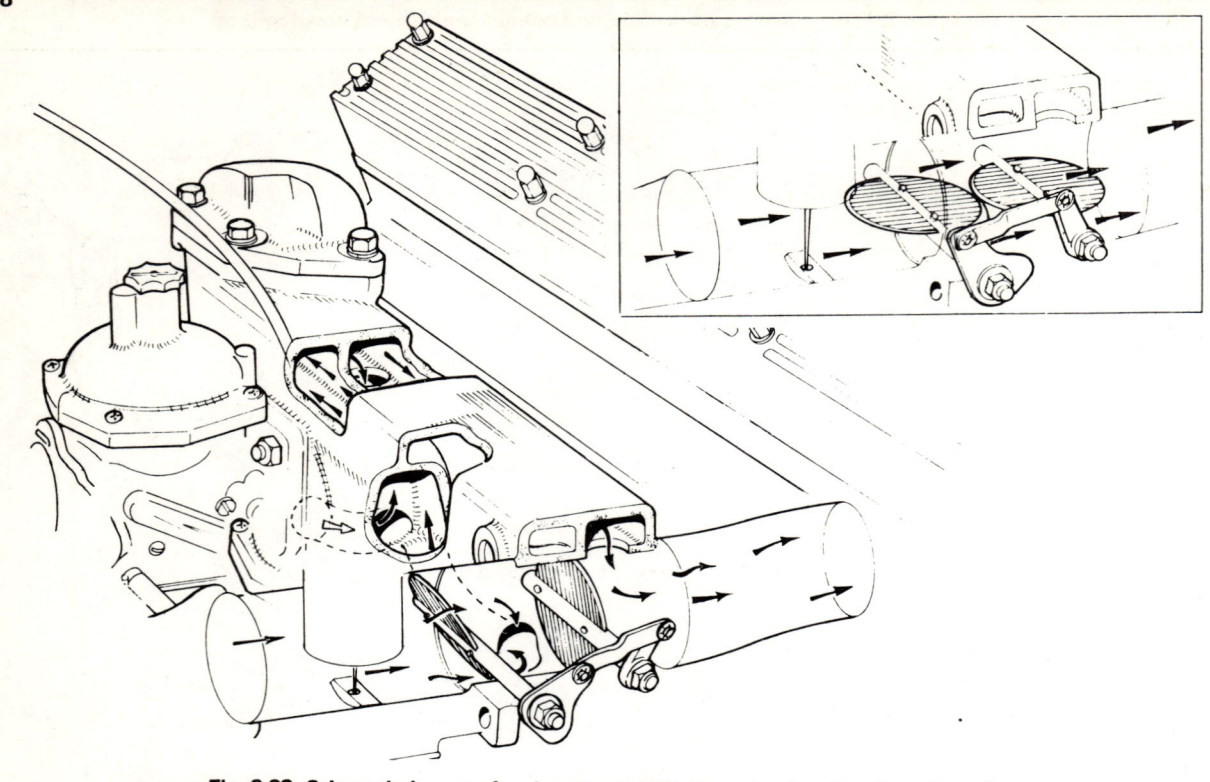

Fig. 3.33. Schematic layout of carburetters and linkage showing direction of gas flow

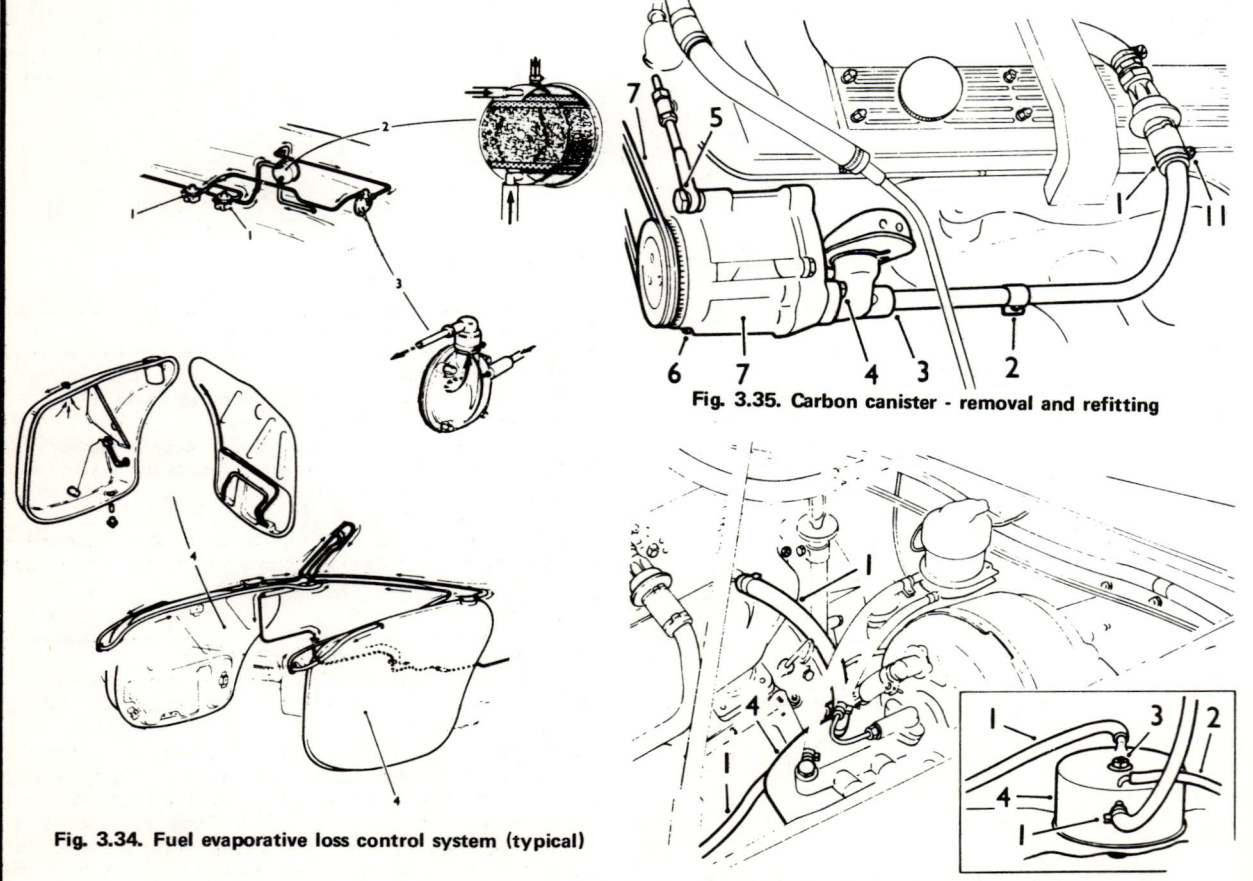

Fig. 3.35. Carbon canister - removal and refitting

Fig. 3.34. Fuel evaporative loss control system (typical)

Fig. 3.36. Air delivery pump - removal and refitting

Chapter 3/Fuel system, carburation and exhaust emission control system

delivery pipe (2) to the exhaust manifold heat shield.
6 Pull out the air delivery pipe from the elbow on the pump (3).
7 Remove the bolts, and collect the washers, which hold the outlet elbow (4) to the pump. Remove the elbow.
8 Remove the bolts, washers and spacers from the trunnion (5).
9 Remove the nut and washer on the pump mounting bolt (6).
10 Swing the pump inwards towards the engine and disengage the drive belt.
11 Withdraw the pump mounting bolt and collect the washers and spacers and then lift out the pump (7).
12 Refit the pump in the reverse sequence to the above but use a new gasket on the pump outlet elbow and O-ring inside the outlet elbow.
13 The belt will have to be adjusted before final tightening of the mounting bolts. Refer to Fig.3.37.
14 Slacken the adjuster locknut (2) and now tighten, or slacken, on the adjuster nut to give a belt tension so that when a load of 2.2 lbs (1.0 kg) is applied on the upper and midway portion of the belt (arrowed at 3) a deflection of 0.15" (4.0 mm) is obtained.
15 When you are satisfied with the belt tension, tighten down on the adjuster locknut and the mounting bolts (4 and 1).
16 Before finally tightening the clip (11) securing the air delivery pipe to the check valve inlet, start the engine and check that the pump does deliver air. If satisfactory, tighten the clip.
17 No servicing or overhaul of the air delivery pump is possible, so, in the event of your check at paragraph 16 being negative and if you are sure that the belt is correctly adjusted, there is no alternative but to obtain, and fit, a service exchange unit.
18 Finally carry out an exhaust emission check.

36 Air rail - removal and refitting

1 The location of this item is shown in Fig.3.38.
2 Follow the instructions given in Section 37 for removal of the air duct.
3 Slacken the clip securing the check valve hose (2) to the air rail.
4 Remove the nuts and spring washers (3) holding the air delivery pipe to the steady brackets.
5 Unscrew the union nut (4) securing the air pipe to the cylinder head; disengage the air rail from the steady brackets and withdraw it.
6 Refitting is the reverse of removal, but make sure that the clips and hoses are in good condition and be careful not to cross thread the union nut (4) when reconnecting.

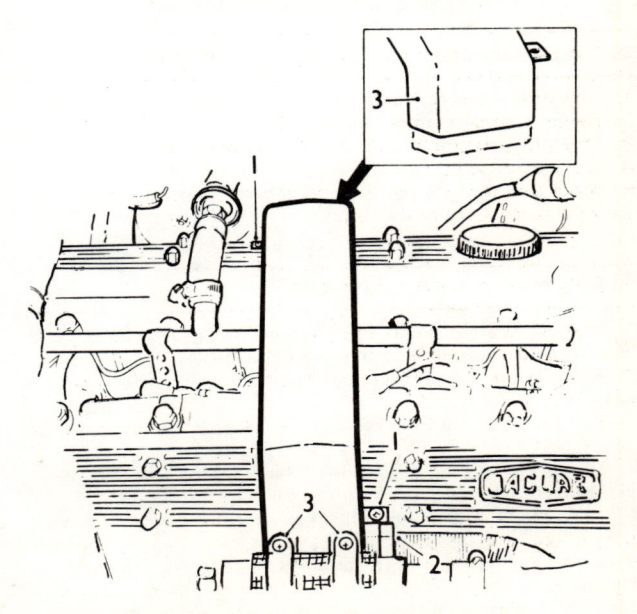

37 Air duct - removal and refitting

1 Refer to Fig.3.39.
2 Take out the screws (1) which secure the air duct to the camshaft covers. Collect the spring amd plain washers.
3 Slacken the clip (2) securing the air delivery pipe to the air duct.
4 Remove the screws and spire nuts (3) which secure the heat exchanger (if applicable) to the air duct and then withdraw the duct.
5 The air duct is refitted in the reverse manner to the above.

38 Air cleaner and filter element - removal and refitting

1 The layout of the air cleaner assembly is shown in Fig.3.40.
2 Slacken the nut (1) which secures the inner end of the stay to the mounting bracket.
3 Unscrew the nut and then remove the bolt and washer (2) which holds the outer end of the stay to the mounting bracket and now swing the stay clear of the top of the air cleaner.
4 Slacken the clip (3) and disconnect the flexible inlet pipe.
5 Slacken the clip (4) to disconnect the air duct flexible pipe.

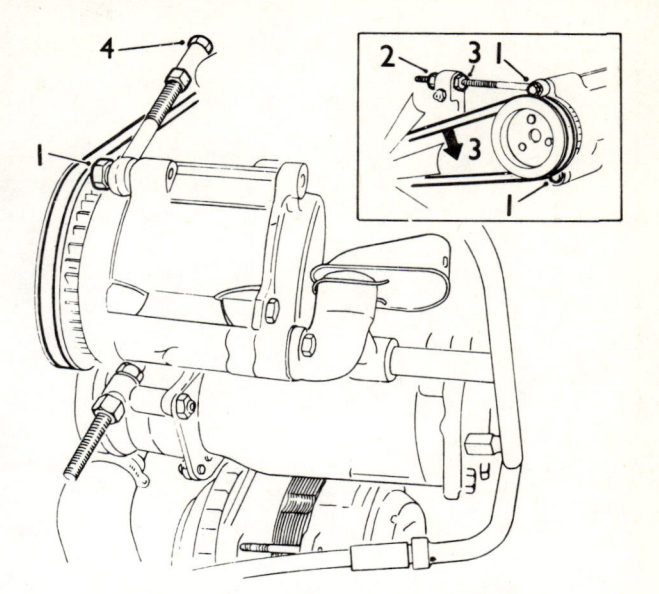

Fig. 3.37. Air delivery pump belt adjustment

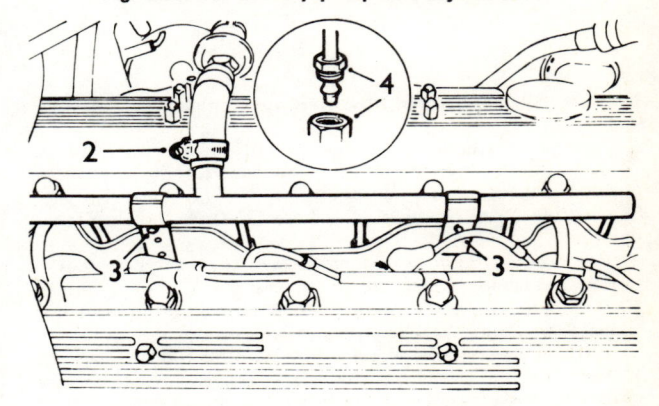

Fig. 3.38. Air rail - removal and refitting

Fig. 3.39. Air duct - removal and refitting

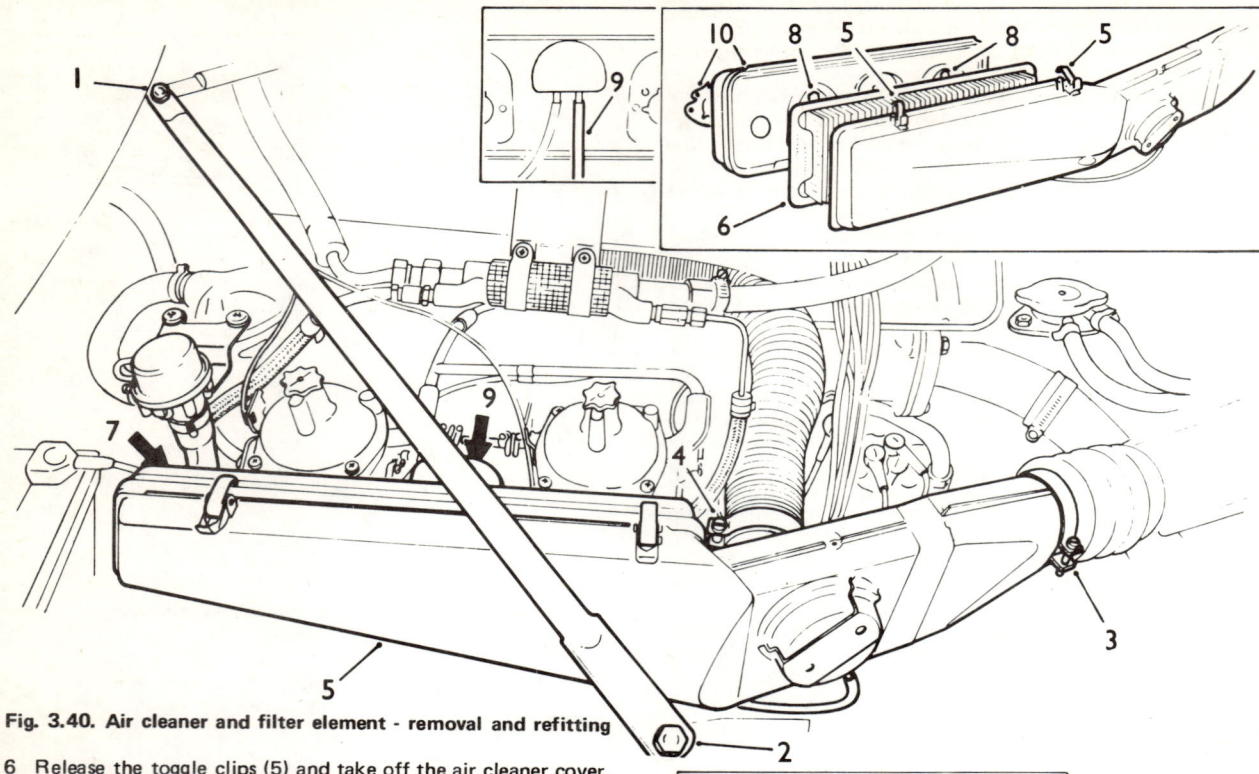

Chapter 3/Fuel system, carburation and exhaust emission control system

Fig. 3.40. Air cleaner and filter element - removal and refitting

6   Release the toggle clips (5) and take off the air cleaner cover.
7   Lift out the filter element (6). If applicable, refit the element in the reverse sequence to the above.
8   Carefully prise the gulp valve hose (7) from the air cleaner backplate.
9   Remove the bolts (8) which secures the backplate to the carburettor flanges, collect the spring washers.
10 Move the backplate away from the carburettors and disconnect the vacuum pipe from the thermal sensing switch (9).
11 Lift out the backplate and collect, and discard, the gaskets.
12 Refit the air cleaner in the reverse sequence to the above using new gaskets between the backplate and the carburettors and ensuring that the cover is correctly positioned.

39 Temperature sender unit - removal and refitting

1   This item is shown at Fig.3.41.
2   Remove the air cleaner (Section 38).
3   Carefully prise the clip (2) off the retaining pipes.
4   Lift out the sender (3).
5   Make sure that the clip is in good condition and then refit in the reverse sequence to the above.

40 Gulp valve - removal and refitting

1   Remove the clip (1 in Fig.3.42) which secures the throttle housing air delivery hose to the gulp valve outlet.
2   Remove the clip (2) securing the air delivery hose to the gulp valve inlet.
3   Pull the inlet manifold sensor tube (3) from the union on the gulp valve.
4   Remove the four screws (4) which secure the gulp valve mounting bracket and manifold top cover to the manifold.
5   Lift off the gulp valve and top cover (5). Collect and discard the gasket.
6   Refit in the reverse manner to the above but use red Hermetite sealing compound on the new top cover gasket and the hose clips. The hose clips may be reuseable but we recommend that you replace them.

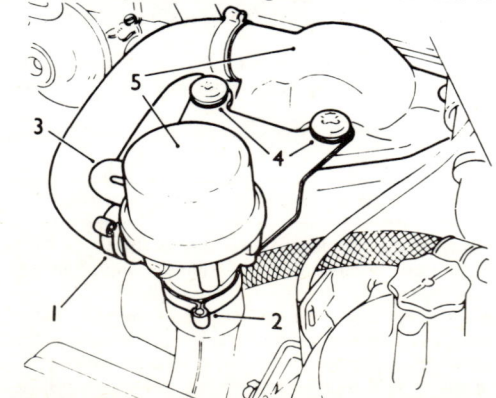

Fig. 3.41. Temperature sensing unit - removal and refitting

Fig. 3.42. Gulp valve - removal and refitting

41 Check valve - removal and refitting

1 This item is shown in Fig. 3.43.
2 Remove the clip (1) securing the air delivery pipe hose to the check valve inlet.
3 Slacken the jubilee clip (2) holding the check valve outlet to the air delivery pipe rail and lift out the check valve.
4 Refit in the reverse manner to the above but use a new clip to secure the air delivery pipe hose.

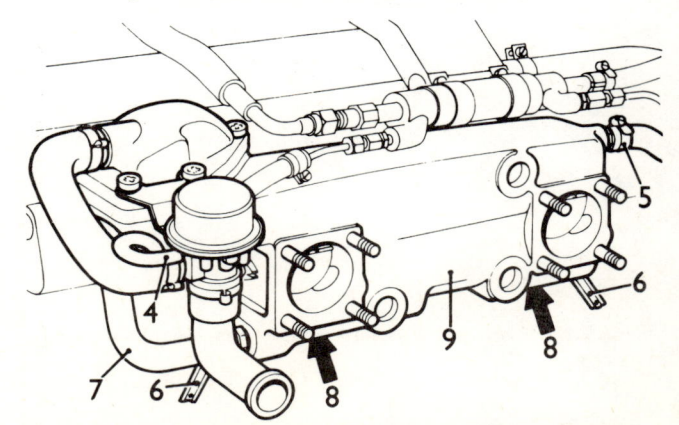

42 Secondary throttle housing - removal and refitting

1 Remove the radiator header tank cap. Open the radiator drain tap and drain off about one gallon of coolant which should be conserved if antifreeze is in use.
2 Remove the carburettors as described in Section 30.
3 Refer to Fig.3.44.
4 Disconnect the inlet manifold hose (4) from the gulp valve.
5 Slacken the clip and disconnect the water inlet hose (5).
6 Take note of which hole in the spring anchor brackets the secondary throttle return springs (6) are secured to and then detach the springs.
7 Slacken the clips and disconnect the water outlet hose (7).
8 Take off the nuts and spring washers (8) securing the secondary throttle housing and the spring anchor brackets to the inlet manifold.
9 Withdraw the secondary throttle housing (9) complete with the gulp valve and anchor brackets. Collect and discard the gaskets.
10 Refit the secondary throttle housing in the reverse sequence to the above but use new gaskets between the housing and the inlet manifold.

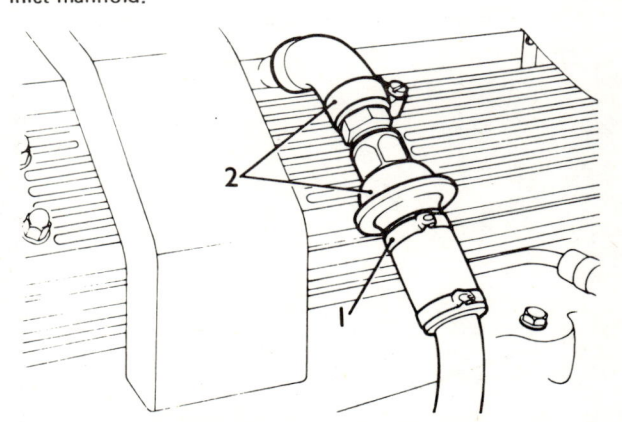

Fig. 3.43. Check valve - removal and refitting

Fig. 3.44. Secondary throttle housing - removal and refitting

43 Secondary throttle housing - dismantling and reassembly

1 Follow the instructions given in Section 42 for removal of the housing from the car.
2 Refer to Fig.3.45 and remove the nut and washer (2) which secure the bellcrank to the front secondary throttle spindle. Remove the bell crank.
3 Take off the nut and washer (3) securing the bell crank to the rear secondary throttle spindle. Remove the bell crank.
4 Mark the relative positions of each throttle disc (4) to the spindle and of each spindle to the secondary throttle housing.
5 Take out the screws (5), rotate the spindles and withdraw the discs.
6 Withdraw the spindle (6) from the secondary throttle housing.
7 Prise the throttle spindle seals (7) out of the spindle bush and discard them.
8 The first step in reassembly is to press new spindle seals into position in the spindle bush (8 in Fig.3.46).
9 Check the throttle spindles for distortion using a straight edge (9). Any spindle found to be distorted or damaged must be replaced.
10 Fit the spindles into the housing (10).
11 Fit the throttle discs (11) to the spindles and adjust them until they close fully, hold them in this position and then tighten the two securing screws.
12 Fit the bell cranks (13) to the throttle spindles but do not overtighten the securing nuts, otherwise the thread will be stripped.
13 Refit the housing to the car as described in Section 42.

44 Inlet manifold - removal and refitting

1 Follow the instructions given in Sections 37 and 42 for removal of the air duct and the secondary throttle housing.
2 Slacken the clip and remove the water outlet hose from the union (5 in Fig.3.47).
3 Lift the crankcase breather pipe (6) out of the retaining clips and withdraw the pipe from the rubber elbow.
4 Disengage the HT lead harness from the retaining clip (7).
5 Slacken the clip and disconnect the radiator top hose (8).
6 Slacken their clips and remove the hoses (9) from the filler neck of the header tank.
7 Slacken the clip securing the water pump hose (10) to the header tank.
8 Take off the three nuts and spring washers (11) securing the throttle linkage. Withdraw the linkage and support in frame.
9 Disconnect the pipes (12) and the lead to the water temperature transmitter. These are located on the underside of the manifold.
10 Disconnect the pipes from the thermostatic switch (13).
11 Remove the nuts and spring washers securing the manifold to the cylinder head, note the position of the two pipe clips (14) and remove them.
12 Withdraw the inlet manifold from the studs. Disconnect the water pump hose (15) from the underside of the header tank and the manifold can now be removed.
13 Remove the manifold gasket and discard.
14 The inlet manifold is refitted in the reverse sequence to the above - using a new gasket between the manifold and the cylinder head.

45 Thermostatic vacuum switch - removal, testing and refitment

1 Remove the header tank filler cap and drain off about one gallon of coolant. Make provision to collect the coolant if antifreeze is in use.
2 Refer to Fig.3.48, slacken the clip and disconnect the flexible hose from the air cleaner (2).
3 Disconnect the vacuum pipes (3) from the switch.

Chapter 3/Fuel system, carburation and exhaust emission control system

4 Slacken the nut (4) and unscrew the switch from the inlet manifold.
5 Clean the exterior of the switch and then connect suitable lengths of tubing to the vacuum pipe unions.
6 Blow down pipe "D" and air should be ejected from pipe "C". If no air is ejected, the switch is unserviceable and must be renewed.
7 If the above test is satisfactory, immerse the switch in a bath of oil but keep the unions clear by retaining the tubing on unions "C" and "D" and by fitting another piece of tubing to union "F".

8 Heat the oil to a temperature of 104.4°C (220°F) and blow down pipe "D" when air should be ejected from "E" which will indicate that the switch is serviceable. However, if no air is ejected from "E" or if it ejected from "C" the switch is unserviceable and must be replaced.
9 If tests prove the switch to be satisfactory clean off all traces of oil and refit it in the reverse order to the removal sequence but the switch must be fitted in the attitude shown in Fig.3.48 and to enable this to be done the switch is allowed to turn independently of the union nut. Tighten the union nut when the switch is fully home and properly positioned.

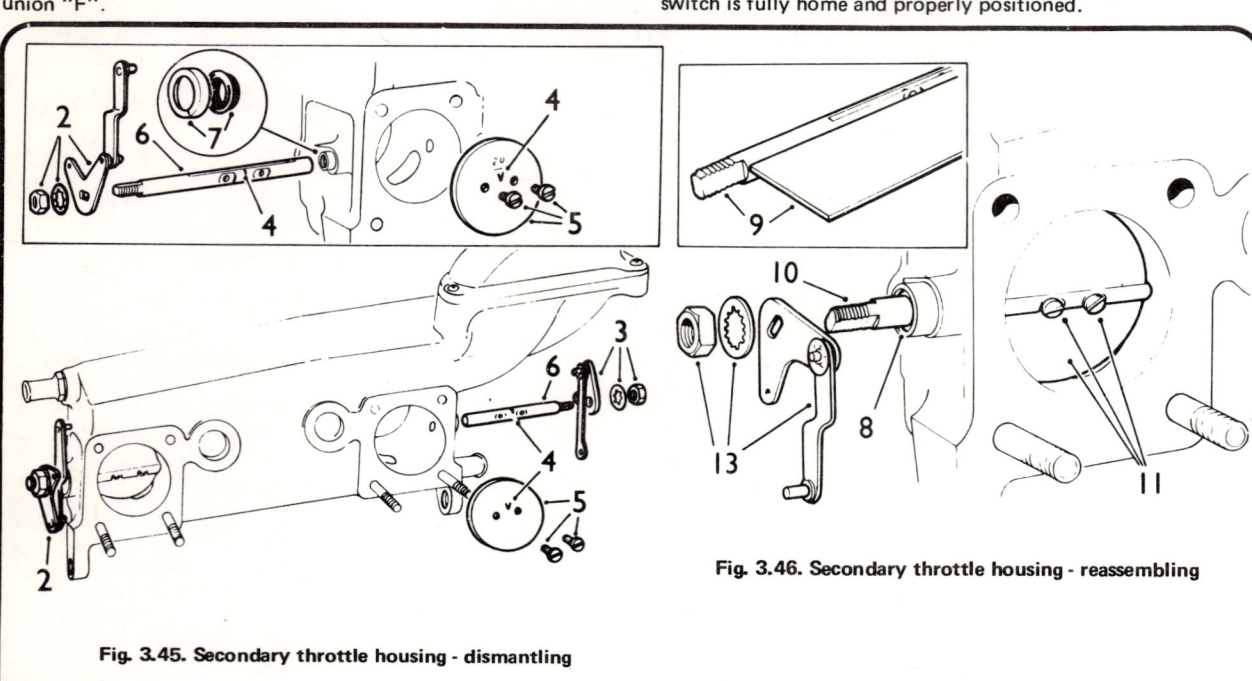

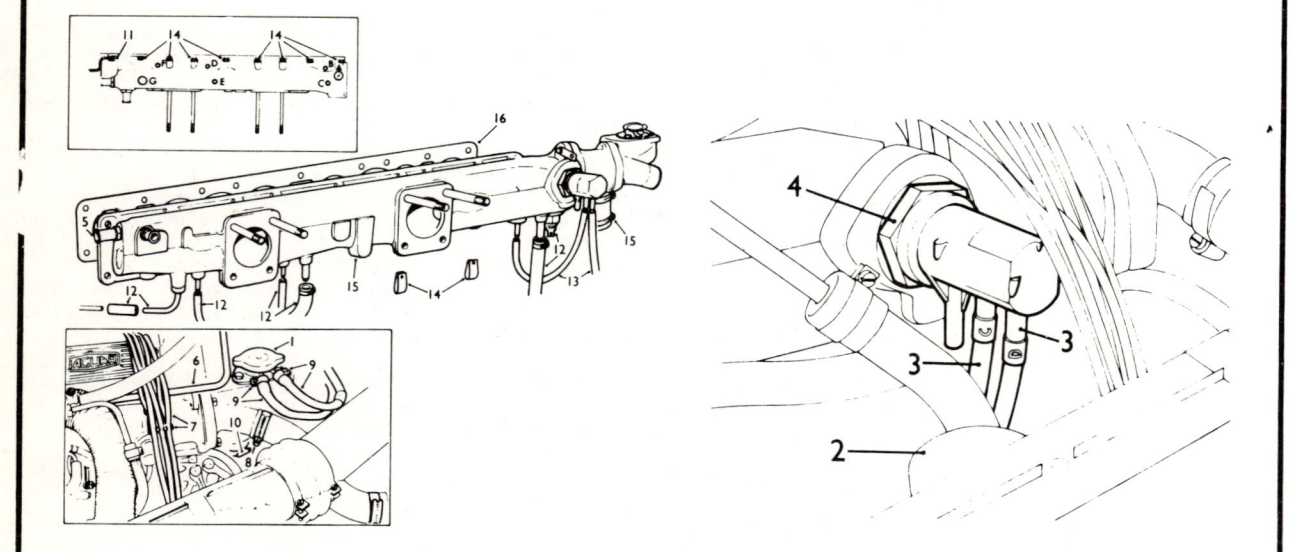

Fig. 3.45. Secondary throttle housing - dismantling

Fig. 3.46. Secondary throttle housing - reassembling

Fig. 3.47. Inlet manifold - removal and refitting

Fig. 3.48. Thermostatic vacuum switch - removal and refitting

Chapter 3/Fuel system, carburation and exhaust emission control system

46 Fault diagnosis

The checks and procedures enumerated below must be carried out on both carburettors.

Symptom	Reason/s	Remedy
Erratic or poor idling	1. Float height incorrect	1. Check float height and rectify.
	2. Dirty or worn needle valve	2. Wash valve in petrol and check that the filter gauze is clean. Replace the valve if sticking or worn.
	3. Piston sticking	3. Clean piston and rod and lubricate with clean engine oil. Top up damper to correct level.
	4. Carburettor inlet obstructed	4. Check that air cleaner and cover are correctly fitted and that gaskets are not causing trouble.
	5. Diaphragm damaged	5. Check and renew if necessary.
	6. Temperature compensator not operating correctly	6. Remove compensator cover and check for freedom of tapered plug. Renew the assembly if plug is sticking.
	7. Leakage at induction manifold joints	7. Check and renew gaskets if necessary.
	8. Leakage from vacuum pipe connections	8. Check connections for security and rectify.
	9. Carburettors not tuned correctly	9. Follow instructions in Section 32.
Hesitation or flat spot	Check items 1 - 9 above	If trouble still persists, check that piston return spring is not broken.
Air delivery pump inoperative or delivering low amount of air	1. Driving belt slipping	1. Check and adjust belt.
	2. Poor hose connections	2. Check all connections for tightness and rectify as necessary.
	3. Symptoms persist after checking 1 and 2	3. Pump unserviceable.

Chapter 4 Ignition system

Contents

Condenser - removal, testing and replacement 4	Distributor - removal and replacement 6
Contact breaker points - adjustment 3	Fault diagnosis - engine fails to start 13
Contact breaker points - removal and replacement ... 2	Fault diagnosis - engine misfires 14
Distributor - dismantling 7	General description 1
Distributor - inspection and repair 8	Ignition system - fault finding 12
Distributor - lubrication 5	Ignition - timing 10
Distributor - reassembly 9	Spark plugs and leads 11

Specifications

Spark plugs:

	3.8 litre	4.2 litre
Type	Champion UN12Y	Champion N.11Y
Gap	0.025 in. (0.64 mm)	

Firing order   1 5 3 6 2 4
(No. 1 cylinder being at rear of engine unit)

Distributor:

Make		Lucas
Type	DMBZ.6A	22D6
Cam dwell angle	35° ± 2°	34° ± 3°
Contact breaker gap	0.014 in. – 0.016 in. (0.36 – 0.41 mm)	
Contact breaker spring tension	18 – 24 ozs (512 – 682 gms)	

Ignition timing:

8 to 1 compression ratio	9 deg. BTDC
9 to 1 compression ratio	10 deg. BTDC
U.S.A. only	5 deg. BTDC (static)
	10 deg. BTDC at 1,000 rpm.

1 General description

The ignition system is based on the supply of low tension voltage from the battery to the ignition coil where it is converted into high tension voltage. The high voltage is powerful enough to jump the spark plug gap in the cylinders under high compression pressures - provided that the ignition system is in good working order and that all adjustments are correct.

The ignition system comprises two individual circuits known as the low tension (LT) and the high tension (HT) circuits. The LT circuit which is sometimes referred to as the primary circuit, comprises the battery, the lead to the control box, the lead to the ignition switch and from there to the low tension or primary coil windings of the coil (terminal SW), and the lead from the low tension coil windings (terminal CB) to the contact breaker points and condenser in the distributor.

The HT circuit consists of the high tension or secondary coil windings, the heavy ignition lead from the centre of the coil to the centre position in the distributor cap and thence via a carbon brush to the rotor arm and then through the spark plug leads to the spark plugs.

The system functions as follows. Low tension voltage is changed in the coil into high tension voltage by the opening and closing of the contact breaker points in the low tension circuit. HT voltage is then fed via the carbon brush in the centre of the distributor cap to the rotor arm of the distributor. The rotor arm revolves inside the distributor cap and each time it comes into line with one of the six metal segments in the cap, which are connected to the spark plug leads, the opening and closing of the contact breaker points causes the HT voltage to build up, jump the gap from the rotor arm to the appropriate metal segment and so via the lead to the spark plug, where it finally jumps the spark plug gap before going to earth. The contact breaker points consist of one fixed and one free point. The free point bears on the shaft which carries the rotor arm and movement of this point is governed by the shape of the shaft which is hexagonal at the position where the point bears. As the shaft revolves the free

Chapter 4/Ignition system

contact breaker point moves over one of the humps of the hexagon and is so brought out of contact with the fixed point.

The ignition is advanced and retarded automatically to ensure that the spark occurs at the right moment for the particular load at the prevailing engine speed.

The ignition advance is controlled both mechanically and by a vacuum operated system. The mechanical system comprises two lead weights, which act in the same manner as a governor, and which due to centrifugal force, move out from the distributor shaft as the engine speed rises. As they move outwards they rotate a cam relative to the distributor shaft and so advance the spark. The weights are held in position by two light springs and it is the tension of the springs which is largely responsible for correct spark advancement.

The vacuum control consists of a diaphragm, one side of which is connected via a small bore tube to the inlet manifold and the other side to the contact breaker plate. Depression in the inlet manifold, which varies with engine speed and throttle opening, causes the diaphragm to move carrying with it the contact breaker plate and thus advancing or retarding the spark. A fine degree of control is achieved by a spring in the vacuum assembly. It will be seen from the specification at the beginning of the Chapter that the type of distributor used varies with the model of the car. There is little or no outward difference in either of them and the above description and the following sections apply to each.

2 Contact breaker points - removal and replacement

1 If the contact breaker points are burned, pitted or badly worn, they must be replaced.
2 To remove the points first spring back the two clips (one at each side) holding the distributor cap to the body of the distributor. Lift off the cap and place it so that it is held clear of the distributor. Remove the rotor arm.
3 Unscrew the terminal nut securing the spring loaded free contact breaker arm. Remove both leads from the stud and the top insulating bush.
4 Lift off the contact breaker arm and remove the large fibre washer from the terminal pin.
5 Remove the two screws in the case of the DMBZ6A distributor, or the single screw in the case of the 22D6 distributor, which hold the adjustable breaker arm and remove it.
6 If the contact points are dirty or are pitted they may be polished by use of a fine carborundum stone but it is essential that the faces are kept square, and flat. Wipe away all dust after cleaning using a non-fluffly cloth moistened in petrol.
7 If a new set of points is being fitted it is essential that the faces of the points are thoroughly cleaned with a non-fluffy cloth moistened in petrol in order to remove the preservative which will have been applied to them.
8 To replace the points, first position the adjustable contact breaker plate and secure it with its screw(s) and spring and flat washer.
9 Fit the large fibre washer to the terminal pin and then, bending the spring of the free contact breaker arm between the thumb and two fingers, refit the fibre of the arm to its pin and the eye of the spring to the terminal pin.
10 Insert the flanged nylon bush over the terminal pin and into the eye of the spring with the condenser lead immediately under its head and the low tension lead under that. Fit the steel washer and screw on the securing nut.
11 It is important to use the correct sized spanner for the nut and the correct sized screwdriver for the screws otherwise there is a possibility of these becoming 'chewed-up' and making subsequent removal or replacement, difficult.
12 The gap of the contact breaker points must now be adjusted as described in the following Section.

3 Contact breaker points - adjustment

1 Remove the distributor cap and the rotor arm. If the rotor arm is difficult to move, it is permissible to lever on it, gently and evenly, with a screwdriver.
2 Remove the spark plugs to facilitate turning the engine to bring it to the required position.
3 Rotate the engine (by using a spanner on the crankshaft damper securing nut) until the neck of the fibre body of the spring loaded contact breaker arm is on the peak of one of the hexagonal lobes of the distributor shaft.
4 Measure the gap between the contact breaker points. It should be 0.14" to 0.16" (0.36 to 0.41 mm.) The gap must be adjusted if outside of these limits.
5 Refer to Fig.4.1 which illustrates the DMBZ6A type of distributor. The fixed plate is held by two screws (A) one of which passes through an elongated hole in the plate. Slacken these screws and rotate screw (B) which is an eccentric headed adjusting screw, until the correct contact breaker gap is obtained. Now tighten screws (A).
6 Fig.4.2 illustrates the 22D6 type of distributor. Here the means of securing and adjusting the fixed plate is slightly different. The fixed plate is anchored at one end on a pin and is secured by a single screw (A). Slacken screw (A) and enter a screwdriver blade into one of the notches (B), turn the screwdriver to move the plate to obtain the correct gap. Tighten screw (A).
7 It is an elementary point, but do make sure that the blade of your feeler gauge is clean and free of oil because if the contact points are dirtied the result will be no spark at the plug.
8 Replace the rotor arm, distributor cap and spark plugs.

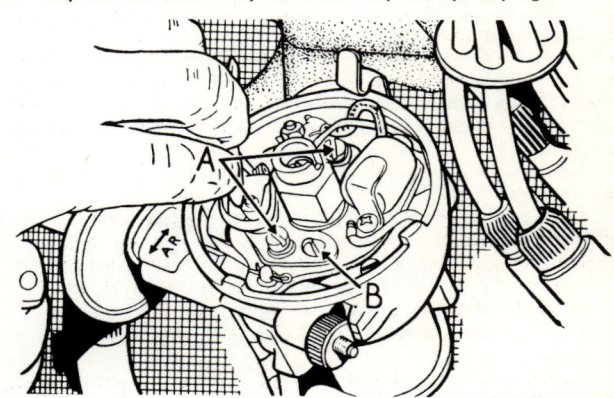

Fig. 4.1. Checking the contact breaker gap - DMBZ6A distributor

A   Screws securing fixed contact
B   Eccentric headed adjusting screw

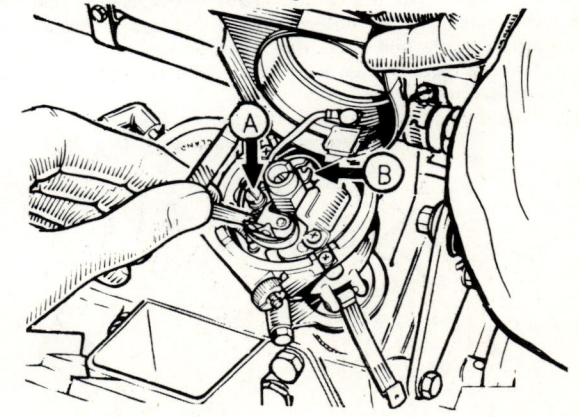

Fig. 4.2. Checking the contact breaker gap - 22D6 distributor

A   Contact plate securing screw
B   Adjusting notches

4 Condenser - removal, testing and replacement

1 The purpose of the condenser (capacitor) is to ensure that when the contact breaker points are open there is no sparking across them which would waste voltage and cause wear.
2 The condenser is fitted in parallel with the contact breaker points. If it develops a short circuit, it will cause ignition failure as the points will be prevented from interrupting the LT circuit.
3 If the engine becomes difficult to start, or begins to miss after several miles running, and the contact breaker points show signs of excessive, then the condition of the condenser must be suspect. A check can be made by separating the points when the ignition is switched 'ON'; if this is accompanied by a flash, it is an indication that the condenser has failed.
4 Without special test equipment, the only sure way to diagnose condenser trouble is to replace the suspect item with a new one and see if there is any improvement. Condensers are not expensive.
5 To remove the condenser from the distributor, remove the distributor cap and the rotor arm. Unscrew the contact breaker arm terminal nut, remove the nut and the flanged nylon bush. Remove the condenser lead.
6 Undo the condenser securing screw and lift away the condenser.
7 Replacement of the condenser is a reversal of the above procedure. Take particular care that the condenser lead does not short circuit against any part of the breaker point.

5 Distributor - lubrication

1 The distributor should be cleaned and lubricated periodically (every 2500 miles is recommended). But do not be over lavish with oil and under no circumstances allow it anywhere near the contacts.
2 Remove the distributor cap and the rotor arm.
3 Refer to Fig.4.3.
4 Lubricate the cam bearing by injecting a few drops of thin machine oil into the rotor arm spindle (A). Do not remove or slacken the screw inside the spindle - a space is provided beneath the head of the screw to allow the passage of thin lubricant.
5 Lubricate the post (B) with one drop of oil.
6 Lightly smear the faces of the cam (C) with clean engine oil or vaseline.
7 Lubricate the centrifugal timing control by injecting a few drops of thin oil through a convenient aperture in the contact breaker base plate.
8 Clean the distributor cap inside and out, with a soft dry cloth. Pay particular attention to the spaces between the terminals. Check that the carbon brush in the head of the cap can move freely in its holder. Rough, burned or blackened contacts can be cleaned with a fine carborundum stone or emery cloth. Remove metallic dust after cleaning the contacts by use of a cloth moistened in petrol.
9 Replace the rotor arm and the distributor cap.

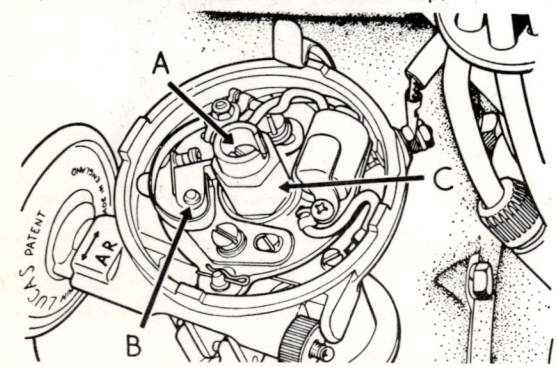

Fig. 4.3. Distributor lubrication points

6 Distributor - removal and replacement

1 For safety reasons, disconnect the battery.
2 Release the clips securing the distributor cap to the body and lift off the cap.
3 Remove the spark plugs.
4 Slowly turn the engine until the static timing marks coincide (see Section 10) and the rotor arm is pointing to the distributor cap segment which is connected to No. 6 spark plug (front plugs).
5 Disconnect the low tension lead from the terminal on the side of the distributor.
6 Detach the vacuum pipe from the distributor advance unit.
7 Undo the screw securing the distributor clamp plate to the cylinder block. Remove the screw and the spring washer. The distributor may now be lifted up with the clamp plate still attached.
8 If it is not wished to disturb the ignition timing, then under no circumstances must the distributor clamp pinch bolt be loosened. Provided the distributor clamp is not moved and the engine is not turned then the distributor can be replaced without losing ignition timing.
9 Replacement is the reverse of the above sequence. If the engine has been turned or the distributor clamp has been disturbed it will be necessary to retime the ignition as described in Section 10.

7 Distributor - dismantling

1 Remove the distributor from the car as described in Section 6.
2 Refer to Fig.4.4 which shows an exploded view of the distributor.
3 Remove the contact breaker points as described in Section 2 and remove the condenser as described in Section 4.
4 Remove the two screws securing the base plate and earth lead. Disconnect the link to the vaccuum control unit and lift off the base plate.
5 Before proceeding any further, take careful note of the relative positions of the rotor arm slot located above the cam and of the offset driving dog. It is possible to assemble these items 180° out on reassembly which means that the distributor would have to be rotated 180° in order to obtain correct timing of the engine and connections cannot be made with the distributor so located.
6 Remove the cam retaining screw ('A' in Fig.4.3) and remove the cam.
7 Lift out the automatic timing control weights and their springs. Note how these are fitted.
8 Remove the circlip securing the knurled advance and retard adjustment nut. Remove the adjusting nut and spring. The vacuum unit can now be withdrawn.
9 Remove the clamp plate (but see Section 6 paragraph 8) by undoing the pinch bolt and sliding the plate off the base of the distributor.
10 To remove the driving dog, knock out the taper pin and lift off the dog and thrust washer. The shaft may now be lifted upwards.

8 Distributor - inspection and repair

1 Thoroughly wash all mechanical parts in petrol and wipe dry using a clean non-fluffy rag.
2 Check the contact breaker points as described in Section 2.
3 check the distributor cap for signs of 'tracking' which will be indicated by a thin black line between the segments. Replace the cap if this defect is noted.
4 Examine the contacts in the cap. If they are rough, burned or blackened, clean them with a fine carborundum stone or fine emery cloth. Thoroughly clean the cap after rectification.

Chapter 4/Ignition system

5 Ensure that the carbon brush in the cap is free to move in its holder and is not worn down. Do not remove the brush needlessly as the spring is usually a tight fit in the holder and will be badly stretched as you pull the brush out.
6 If the metal portion of the rotor arm is badly burned or is loose, renew the arm. Slight burning can be rectified with a fine file but maintain the face square.
7 Examine the fit of the contact breaker plate on the base plate and check the breaker arm pivot for looseness or wear. Renew the plate if necessary.
8 Examine the centrifugal weights and pivot pins for wear and renew the weights or cam assembly if a degree of wear is found.
9 Examine the shaft and the fit of the cam assembly on the shaft. If the clearance appears to be excessive, compare with new items and renew either or both if they show excessive wear.
10 If the shaft is a loose fit in a distributor bush and can be 'rocked', we suggest that a reconditioned distributor is obtained. However, a new bush can be obtained and fitted to the DMBZ6A type of distributor but not to the types 22D6 for which there is no alternative but replacement of the complete distributor. The bearing bush is replaced as follows.
11 Drive out the old bush with a suitable punch.
12 Prepare the new bush for fitting by allowing it to stand completely immersed in a medium viscosity engine oil (SAE30-40) for at least 24 hours. The period of time can be shortened by soaking oil heated to 100°C for 2 hours.
13 Press the new bush into the distributor body using a shouldered mandrel the shank of which should be approximately 0.0005" greater in diameter than the distributor shaft.
14 The bush **must not** be opened out by reaming or by any other means as this will impair the porosity and thus the self lubricating quality of the bush.
15 The ball bearing at the upper end of the shank can be removed with a shouldered mandrel locating on the inner journal of the bearing.
16 When replacing the bearing, the shouldered mandrel must locate on both the inner and the outer journals of the bearing.

9 Distributor - reassembly

1 Reassembly is a straightforward reversal of the dismantling process. Note in addition:
2 Lubricate the centrifugal weights and other parts of the mechanical advance mechanism with thin machine oil. Lubricate the distributor shaft with clean engine oil and smear the cam face with engine oil or vaseline. Do not be too lavish with the oil.
3 Check the action of the weights in the fully advanced and retarded positions, make sure they are not binding.
4 Adjust the micrometer advance and retard adjusting nut to bring the mechanism to the mid position of the timing scale.
5 Finally set the contact breaker points as described in Section 3.

10 Ignition - timing

1 Timing marks are engraved on the front face of the crankshaft damper and are used in conjunction with a pointer as illustrated in Fig.4.5. The engine is timed on No.6 (front) cylinder and when the zero mark on the crankshaft damper is aligned with the pointer No. 6 piston is at TDC. The graduations to the left of the zero mark are in degrees before TDC.
2 The first step in timing the engine is to be sure of the direction of rotation of the rotor arm. It is anticlockwise but to be clear in your own mind we suggest that you remove the distributor cap and, with the ignition switched off, press the button on the starter solenoid and observe the movement of the arm.
3 The engine has to be rotated and brought to rest in a precise position, this cannot be done with the spark plugs in place, so remove them.

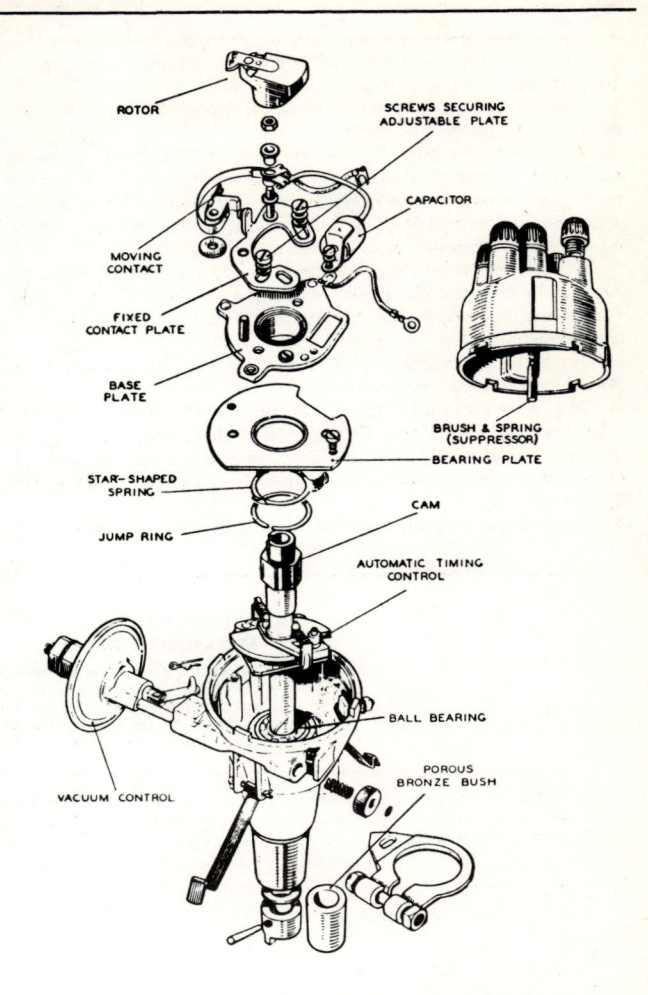

Fig. 4.4. Exploded view of the distributor (Type DMBZ6A illustrated)

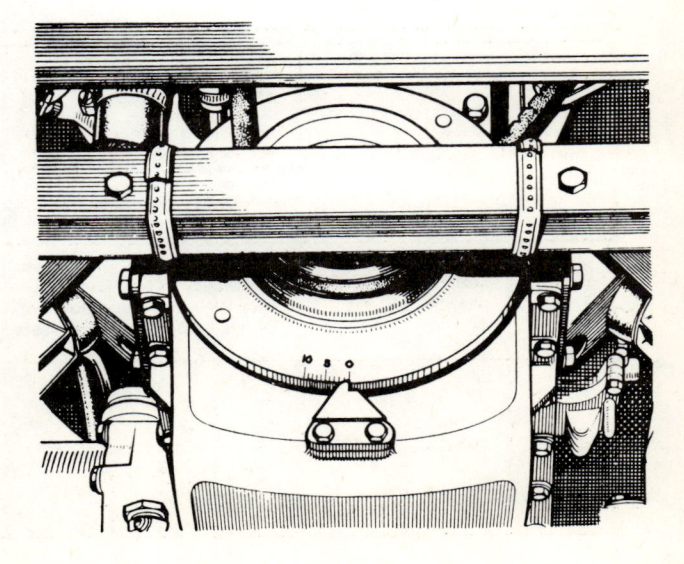

Fig. 4.5. Ignition timing scale on the crankshaft damper

4 Check that the micrometer advance/retard adjustment at the distributor is in the centre of the scale and that the contact breaker points are correctly set.
5 Place the car over a pit or raise the front of the car to give access to the crankshaft damper.
6 We have found that the best method of turning the engine is to engage a socket spanner on the crankshaft damper centre bolt: besides giving close control on the movement of the engine this method means that you are close to the damper, as is necessary, to observe the timing marks.
7 Have an assistant place a thumb over the spark plug hole of No.6 cylinder, turn the engine and when suction is felt and then dies away it means that No. 6 piston is coming up on the compression stroke.
8 Turn the engine slowly until the correct timing mark (see Specifications) is aligned with the pointer.
9 The rotor arm should now be pointing in the direction of No. 6 cylinder segment of the distributor cap. Check that this is so, if the arm is 180° out and the distributor has been stripped, it means that your assembly is incorrect (see Section 7 paragraph 5).
10 Slacken the distributor pinch bolt and turn the distributor, bearing in mind the direction of rotation of the rotor arm, until the contact breaker points are just beginning to open. Tighten the pinch bolt.
11 The only accurate way of judging when the points just open is to connect a 12 volt test lamp with one lead to the distributor terminal (or to the CB terminal of the ignition coil), and the other lead to a good earth. Switch on the ignition and the lamp will light as soon as the points open.
12 The static timing of the engine is now completed but it must be appreciated that this adjustment is nominal and final adjustment should be made under running conditions.
13 With the engine at normal running temperature, accelerate in top gear from about 30 mph and listen for heavy pinking of the engine. If this occurs, the ignition needs to retarded slightly until the faintest trace of pinking can be heard when accelerating very hard. Make your final adjustments by the vernier adjustment at the distributor but a maximum of six clicks to either advance or retard is allowed so if more is required it will have to be done by moving the distributor slightly. Movement of the distributor in the direction of rotation of the rotor arm will retard the ignition whilst movement counter to the direction of rotation of the arm will advance it.

Distributor Type	Lucas Service Number	Lucas Vacuum Unit Number	VACUUM TIMING ADVANCE TESTS The distributor must be run immediately below the speed at which the centrifugal advance begins to function to obviate the possibility of an incorrect reading being registered:		No advance in timing below ins. of mercury	Lucas Advance Springs Number	CENTRIFUGAL TIMING ADVANCE TESTS Mount distributor in centrifugal advance test rig and set to spark at zero degrees at 100 rpm.				No advance in timing below RPM
			Vacuum in inches of mercury and advance in degrees				Accelerate to RPM and note advance in degrees		Decelerate to RPM and note advance in degrees		
			Inches	Degrees			RPM	Degrees	RPM	Degrees	
DMBZ6A	40617A	54410415	20 13 9 7½ 6	7–9 6–8½ 2½–5½ 0–3 0–½	4½	54410416	2,000	12	1,500 1,300 850 650 450	10–12 9–11 7–9 3½–6½ 0–2½	325

Auto advance weights Lucas number 410033/S. One inch of mercury = 0.0345 kg/cm².

Distributor Type	Lucas Service Number	Lucas Vacuum Unit Number	VACUUM TIMING ADVANCE TESTS The distributor must be run immediately below the speed at which the centrifugal advance begins to function to obviate the possibility of an incorrect reading being registered:		No advance in timing below-ins. of mercury	Lucas Advance Springs Number	CENTRIFUGAL TIMING ADVANCE TESTS Mount distributor in centrifugal advance test rig and set to spark at zero degrees at 100 rpm.				No advance in timing below RPM
			Vacuum in inches of mercury and advance in degrees				Accelerate to RPM and note advance in degrees		Decelerate to RPM and note advance in degrees		
			Inches	Degrees			RPM	Degrees	RPM	Degrees	
22 D6	41060A	54415894	20 13 9 7½ 6	7–9 6–8½ 2½–5½ 0–3 0–½	4½	55415562	2,300	8½–10½	1800 1250 800 650 525	8½–10½ 6½–8½ 5–7 2–4 0–1½	300

Auto advance weights Lucas number 54413073. One inch of mercury = 0.0345 kg/cm².

Chapter 4/Ignition system

11 Spark plugs and leads

1 The correct functioning of the the spark plug is vital for the proper running and efficient operation of the engine.
2 The plugs should be removed and thoroughly cleaned and the gap reset at intervals of not more than 2500 miles but more frequent cleaning will be required if the engine is in poor condition and giving rise to heavy fouling of the plugs. The most efficient method of cleaning plugs is by abrasive blasting in the Champion Service Unit but this is not always possible so use of a length of file card fastened to a block of wood is usually an acceptable substitute. Rub the plug vigorously on the card to remove all fouling and make sure that the sparking surfaces of the electrodes are clean and bright, if necessary open the gaps slightly and file the points with a point file keeping the surfaces parallel. After cleaning, blow out the interior of the plug to remove all residue.
3 Use the wire brush to clean the threads.
4 Visually inspect the plug for cracked or chipped insulators, discard any suspect plug.
5 Reset the gap, to the dimension quoted in Specification at the beginning of this Chapter, using the special setting tool as illustrated in Fig.4.6. Do not apply pressure on the centre electrode as insulator fractures may result. Use the tool to obtain parallel sparking surfaces for maximum gap life.
6 Examine the gaskets. If the gaskets were excessively compressed, installed on dirty seats or distorted, leakage has probably occurred during service which would tend to cause overheating of the plug. The gasket should have a clean and flat surface, those which are approximately one half of their original thickness will be satisfactory but thinner ones should be renewed.
7 Finally clean the gasket seats in the cylinder head before installing the plugs to ensure proper seating of the spark plug gasket. Screw in the plug finger tight on to its gasket, if it cannot be seated on its gasket by hand, clean out the cylinder head threads with an old spark plug having two or three vertical flutes filed in the threads. Remember that you are screwing the plug into soft material so every care must be taken against cross threading.
8 Tighten the spark plug to a torque of 27 lb f ft (3.73 kg f m).
9 Examination of the firing end of the plug, noting the type of the deposit and the degree of electrode erosion will give a good indication of faults in the engine or the carburation and ignition system.
10 Refer to Fig. 9.6 for plug maintenance and electrode conditions.
11 A plug in normal condition will be obviously dry and will have light powdery deposits ranging from brown to greyish tan in colour. The electrodes may be worn slightly. All that is required for plugs in this condition is cleaning and regapping.
12 Oil fouling of a plug is usually indicated by wet, sludgy deposits due to excessive oil entering the combustion chamber past worn cylinders, rings and pistons or due to wear in the inlet valve stems or guides. Hotter spark plugs may alleviate oil fouling temporarily but engine overhaul is the only sure remedy.
13 Petrol fouling is indicated by dry, fluffy black deposits which result from incomplete combustion of the air/fuel mixture, the mixture being too rich or excessive use of the mixture control is indicated. In addition, a defective coil, contact breaker points or plug cable can reduce the voltage supplied to the spark plug which will result in incomplete ignition. If the fouling is evident in only a few cylinder it may be that sticking valves is the the cause but evidence of this will be given on 'tick-over' and during normal running.
14 Burned or overheated plugs can be identified by a white burned or blistered insulator nose and badly eroded electrodes. Poor engine cooling or improper ignition timing may be the cause of the fault which can also arise from severe use such as sustained high speed or heavy loads.
15 The plug leads require no routine attention other than being kept clean and wiped over regularly. It is a good plan to remove them from the distributor, at the 10,000 mile servicing, by undoing the knurled terminal knobs or undoing the securing screws, as water can seep into these joints giving rise to a white corrosive deposit which, if present, must be carefully removed.
16 Finally, to get the best results from your engine, renew all plugs at 10,000 mile intervals. If the engine is in good condition the plugs will appear to be quite serviceable, and they probably are, but they have already given a useful life and in time some breakdown in insulation is inevitable even if it has not already occured.

12 Ignition system - fault finding

By far the majority of breakdown and running faults are caused by faults in the ignition system, either in the low tension or in the high tension circuits. There are two main symptons: either the engine will not start or fire or it is difficult to start and misfires. If it is a regular misfire i.e. one or more cylinders are not firing, the fault is almost certainly in the HT circuit. If misfiring is intermittent, the fault could be either in the HT or LT circuits. If the engine stops suddenly, or will not start at all, it is likely that the fault is in the LT circuit. Loss of power and overheating, apart from faulty carburation settings are normally due to faults in the distributor or incorrect ignition timing.

13 Fault diagnosis - engine fails to start

1 If the engine fails to start and it was running normally when last used, first check that there is fuel in the tank. If the engine turns over normally on the starter and the battery is evidently well charged, then the fault may be in either the HT or LT circuit.
2 One of the commenest reasons for bad starting is wet or damp plugs, leads and distributor. Remove the distributor cap, if condensation is visible internally, dry the cap with a rag and wipe over the leads. Replace the cap.
3 If the engine still fails to start, check that current is reaching the plugs by disconnecting each plug lead in turn and holding the end of the lead about 3/16" away from the cylinder block. Switch on the ignition and spin the engine from the starter solenoid (hold the lead with the rubber to avoid shock).
4 Sparking between the lead and the block should be fairly strong with a regular blue spark. If sparking, it is obvious that current is reaching the plugs so remove them, clean and regap. The engine should now start.
5 Spin the engine as before, when a rapid successsion of blue sparks between the end of the lead and the block indicates that the coil is in order and that either the distributor cap is cracked, the carbon brush in the cap is stuck or worn, the rotor arm is faulty, or the contact breaker points are burnt, pitted or dirty. If the points are in bad shape, clean and reset them as described in Section 3.
6 If there are no sparks from the end of the lead, then check out the connections of the HT lead from the coil to the distributor. If that is in order, check the LT lead from the coil to the distributor and then go on to check the leads in the distributor especially between the condenser and the breaker terminal. Make sure that the earth lead is satisfactory.
7 Cases occur of the neck of the fibre body of the spring loaded contact breaker fracturing where it bears on the distributor shaft, check this.
8 If everything is visually in order and the engine still refuses to start, a physical check of the circuit using a 20 voltmeter or a test lamp will have to be made.
9 Turn the engine so that the contact breaker points are fully open, switch on the ignition. Check that current is reaching the starter solenoid switch from the battery. No reading indicates a fault in the cable to the cable to the switch, or in the connections at the switch or at the battery terminals. Alternatively the battery earth lead may not be properly earthed to the body.
10 If in order, check that current is reaching the fuse unit A1

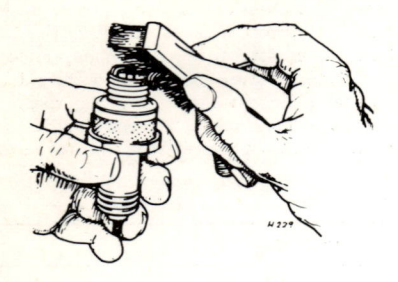

Cleaning deposits from electrodes and surrounding area using a fine wire brush.

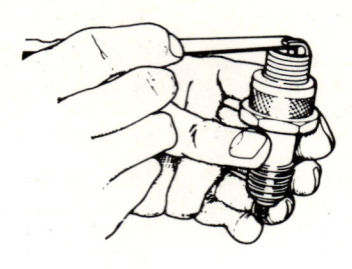

Checking plug gap with feeler gauges

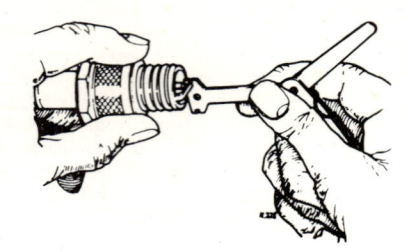

Altering the plug gap. Note use of correct tool.

Spark plug maintenance

White deposits and damaged porcelain insulation indicating overheating

Broken porcelain insulation due to bent central electrode

Electrodes burnt away due to wrong heat value or chronic pre-ignition (pinking)

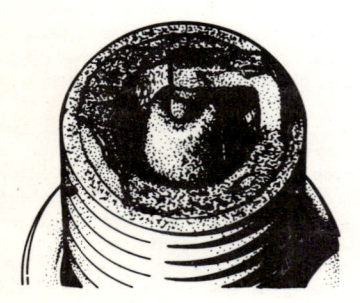

Excessive black deposits caused by over-rich mixture or wrong heat value

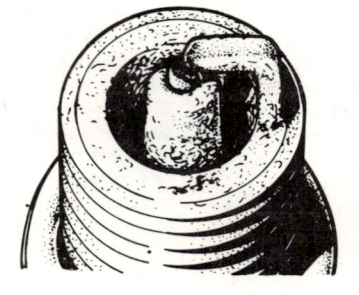

Mild white deposits and electrode burnt indicating too weak a fuel mixture

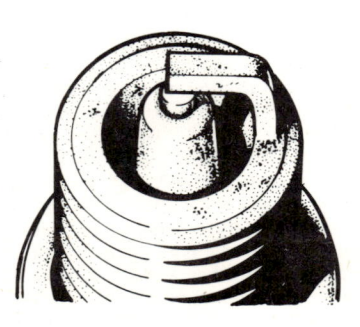

Plug in sound condition with light greyish brown deposits

Spark plug electrode conditions

terminal. Connect the lamp between the fuse unit and earth. If there is no reading, this indicates a loose cable or faulty connection between the solenoid switch and the fuse unit.

11 If in order, check between the control box terminal A1 and earth. No reading indicates a fault in the control box. The control box will have to be replaced.

12 Next check that current is reaching the switch by connecting the lamp to the switch input terminal A and earth. A faulty cable or loose connection is indicated if there is no reading.

13 The next check is between the fuse unit A3 terminal and earth. Again, no reading shows that there is a faulty connection or a broken cable.

14 Now check between the ignition coil terminal SW and earth. A faulty connection or broken cable is indicated if there is no reading.

15 Connect the lamp between the ignition coil terminal CB and earth. If there is no reading, the ignition coil is faulty.

16 Now connect the lamp between the distributor low tension terminal on the cable especially at the point where it joins the tag.

17 The final check of the LT circuit is to connect the lamp across the contact breaker points. No reading means an unserviceable condenser and when this is replaced, the car should start.

14 Fault diagnosis - engine misfires

1 If the engine misfires regularly, run it at a fast idling speed, and short out each plug in turn using a screwdriver with a **wooden or plastic insulated handle.**

2 No difference in the speed of the engine will be noticed when the defective cylinder is short circuited but short circuiting of those cylinders working properly will accentuate the misfire.

3 Remove the plug lead from the suspect cylinder and hold, by the insulation, about 3/16" away from the cylinder block. If the sparking is regular and fairly strong, the fault must lie in the plug.

4 The plug may be loose, the insulation may be cracked, the points may be badly set or the plug very badly fouled. Either renew the plug, or clean it and reset the gap.

5 If there is no spark at the end of the lead or if it is weak and intermittent, check the condition of the lead from the plug to the distributor. Renew the lead if the insulation is cracked or perished. If the lead is in good order, disconnect it at the distributor and see it it is wet. If it is wet, dry it and the housing in the distributor, it will be a good plan to remove all the other leads and make sure that moisture is not present.

6 If there is no spark at the lead, examine the distributor cap carefully for tracking. This can be recognised as a thin black line running between two or more electrodes or between an electrode and some other part of the distributor. These lines are paths which conduct electricity across the cap and let it run to earth. If faulty in this respect, the cap must be replaced.

7 Apart from the ignition timing being incorrect, other causes of misfiring allied to the ignition circuit have already been dealt with in the previous Section.

Chapter 5 Clutch

Contents

Clutch - dismantling ... 7	Clutch - refitting ... 10
Clutch - examination ... 8	Clutch - removal ... 6
Clutch fork and release bearing - removal, examination and refitting ... 11	Clutch slave cylinder - dismantling ... 20
Clutch hydraulic system - bleeding ... 2	Clutch slave cylinder - general description ... 18
Clutch hydraulic system - flushing ... 3	Clutch slave cylinder - reassembly ... 21
Clutch judder ... 27	Clutch slave cylinder - refitting ... 22
Clutch master cylinder - dismantling ... 15	Clutch slave cylinder - removal ... 19
Clutch master cylinder - general description ... 13	Clutch slip ... 25
Clutch master cylinder - reassembly ... 16	Clutch spin ... 26
Clutch master cylinder - refitting ... 17	Clutch squeal ... 24
Clutch master cylinder - removal ... 14	Fault diagnosis and remedy ... 23
Clutch pedal - free travel ... 4	General description ... 1
Clutch - reassembly ... 9	Hydraulic system - general ... 12
	Removing and refitting a flexible hose ... 5

Specifications

	3.8 early models	3.8 later models and early 4.2	4.2 later models
Make	Borg and Beck	Laycock	Borg and Beck
Outside diameter	9.84 in. - 9.87 in. (231-232 mm)	10 in. (254 mm)	10 in. (254 mm)
Inside diameter	6.12 in. - 6.12 in. (153-154 mm)		
Type	Single dry plate	Diaphragm spring	
Clutch release bearing		Graphite	
Operation		Hydraulic	

Clutch thrust springs:
  Number ... 12
  Colour ... Violet
  Free length ... 2.68 in. (68 mm)

Driven plate:
  Type - touring ... Borglite
       - racing and competition ... Arcuate
  Facings - touring ... Wound yarn
          - racing and competition ... Wound yarn cemented
  Damper springs number ... 6
  Colour - touring ... Brown/cream
         - racing and competition ... Buff

Note: In addition to the differences given above, you will find that a different clutch cover assembly and driven plate assembly are used for touring and competition purposes. These items are fully interchangeable and are subject to the same servicing procedures. If you wish to change from one type to the other, we suggest you first consult with your Jaguar agent.

Chapter 5/Clutch

1 General description

The clutch unit fitted to earlier cars is of the Borg and Beck single plate dry type which is hydraulically operated. An exploded view of the assembly is given in Fig.5.2. The clutch assembly comprises a steel cover which is bolted and dowelled to the rear face of the flywheel and contains the pressure plate, pressure plate springs, release levers and the driven plate.

The pressure plate, pressure springs and release levers are all attached to the clutch assembly cover. The driven plate is free to slide along, and is splined to, the first motion shaft of the gearbox and is held in position between the flywheel and the pressure plate by the pressure plate springs. The driven plate is faced on both sides with friction material and has a spring cushioned hub to absorb transmission shocks.

The clutch is actuated hydraulically by a pendant clutch pedal which is connected to the clutch master cylinder by a short pushrod. The master cylinder is mounted on the engine side of the bulkhead. A layout of the clutch operating system is given in Fig.5.3. Depression of the clutch pedal moves the piston in the master cylinder forwards forcing hydraulic fluid through the pipe to the slave cylinder. The piston in the slave cylinder is now moved forward and actuates the clutch release arm by means of a short pushrod, the opposite end of the release arm is forked and carries the release bearing which is a graphite faced disc. As pressure on the clutch pedal continues, the release bearing bears hard on the release lever plate and pushes it forward, this movement rotates the release levers and they in turn pull back the pressure plate away from the driven plate and at the same time compress the thrust springs. The driven plate is now free of the flywheel and consequently there is no drive to the gearbox.

When the clutch pedal is released the thrust springs force the pressure plate into contact with the high friction linings of the driven plate to hold it firmly against the flywheel and so taking up the drive.

As the friction linings on the driven plate wear, the pressure plate automatically moves closer to the driven plate to compensate. This makes the inner ends of the release levers travel further towards the gearbox which decreases the release bearing clearance.

The diaphragm spring clutch fitted to later cars, which is illustrated in Figs.5.4 and 5.5 comprises a steel cover which is dowelled and bolted to the rear face of the flywheel and contains the diaphragm spring, the fulcrum rings and the pressure plate and driven plate, which, as with the other type of clutch is splined to the first motion shaft of the gearbox.

The driven plate is held in position between the pressure plate and the flywheel by the pressure of the diaphragm spring. It has high friction material on both faces and has a spring cushioned hub to absorb transmission shocks.

The action on depressing the clutch pedal is similar to that which occurs with the other clutch except that in this instance the release bearing contacts the release plate which is a fixture on the diaphragm spring. Forward movement of the release plate causes a deflection of the diaphragm spring thus pulling the pressure plate away from the driven plate and freeing the clutch. When pressure on the clutch pedal is released, the diaphragm spring asserts itself to push the pressure plate hard against the driven plate to hold it in tight contact with the flywheel to transmit the drive to the gearbox.

The 4.2 litre car was equipped, on introduction of the model, with the Laycock diaphragm spring clutch as fitted to late production 3.8 litre models. It is known that some enthusiasts have removed this clutch assembly and have fitted the original Borg and Beck model 10A6-G single plate dry type in lieu. If your car has this type of clutch fitted, and if it has given good service, we suggest that you retain it in use. If, however, you decide to replace it with the latest type Borg and Beck diaphragm spring clutch assembly, this should only be done after consultation with your local Jaguar agent as it is quite possible that you will also have to replace the flywheel.

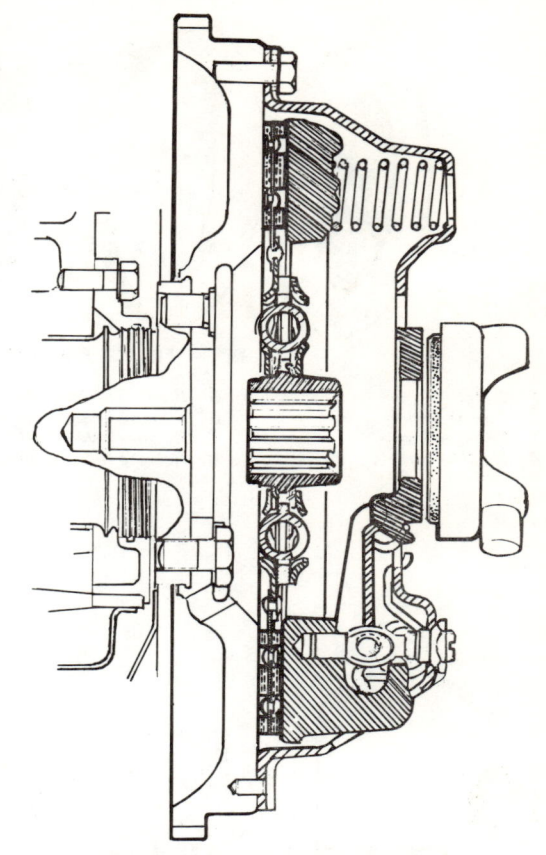

Fig. 5.1. Sectioned view of the 10A6-G type clutch

2 Clutch hydraulic system - bleeding

1 Bleeding the clutch hydraulic system (expelling air) is not a routine maintenance operation and should only be necessary when some portion of the hydraulic system has been disconnected or where, due to a leak, the level of fluid in the hydraulic reservoir has been allowed to drop too low. The presence of air in the system will result in poor clutch operation as, unlike brake fluid, the bubbles of air can be compressed.

2 Thoroughly clean the top of the clutch fluid reservoir (this is shown in Fig. 5.6) and fill the reservoir with hydraulic fluid.
  Note: Castrol Girling Universal Brake and Clutch Fluid is recommended. Where this is not available, only fluid guaranteed to conform to Specification SAE 70 R3 should be used as an alternative.

3 The bleed nipple for the system is located on the slave cylinder on the right hand side of the clutch housing. (Fig. 5.7). Thoroughly clean the exterior of the nipple.

4 Attach a length of rubber tube to the nipple and allow it to hang in a clean glass jar, partly filled with hydraulic fluid.

5 Unscrew the screw nipple one complete turn.

6 Have an assistant in the car to depress the clutch pedal slowly to the full extent of its travel. Tighten the screw nipple whilst the clutch pedal is held depressed.

7 Release pressure on the clutch pedal and repeat operation 5 and 6 until the fluid issuing from the tube is entirely free of air. Take care to replenish the reservoir frequently during these operations because if the fluid level is allowed to drop more than halfway, air will enter the system.

8 When you are satisfied that the system is clear of air, top up the master cylinder reservoir to the 'fluid level' mark.

9 Do not use the fluid which has been bled through the system as this will be aerated. Always use fresh fluid straight from the container.

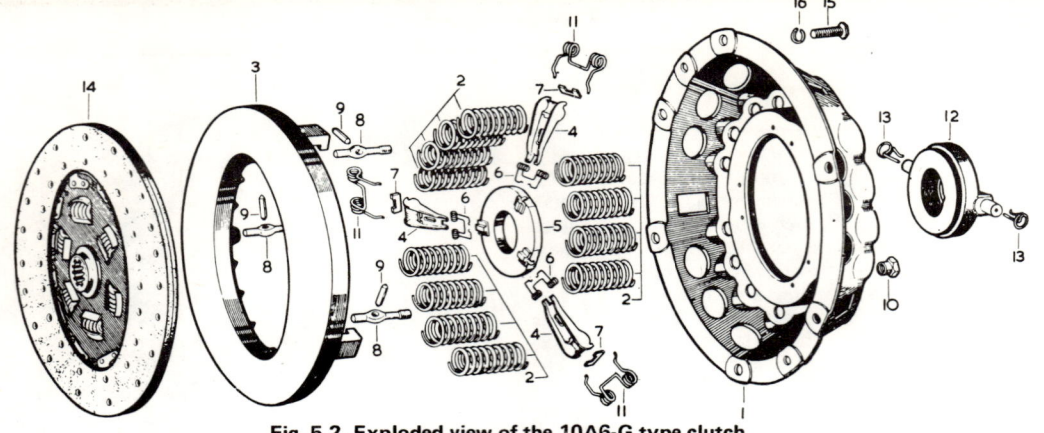

Fig. 5.2. Exploded view of the 10A6-G type clutch

1	Cover	5	Release lever plate	9	Eyebolt pin	
2	Thrust spring	6	Release lever retainer	10	Adjustment nut	
3	Pressure plate	7	Release lever strut	11	Anti-rattle spring	
4	Release lever	8	Release lever eyebolt	12	Release bearing and cup assembly	
13	Release bearing retainer					
14	Driven plate assembly					
15	Securing bolt					
16	Spring washer					

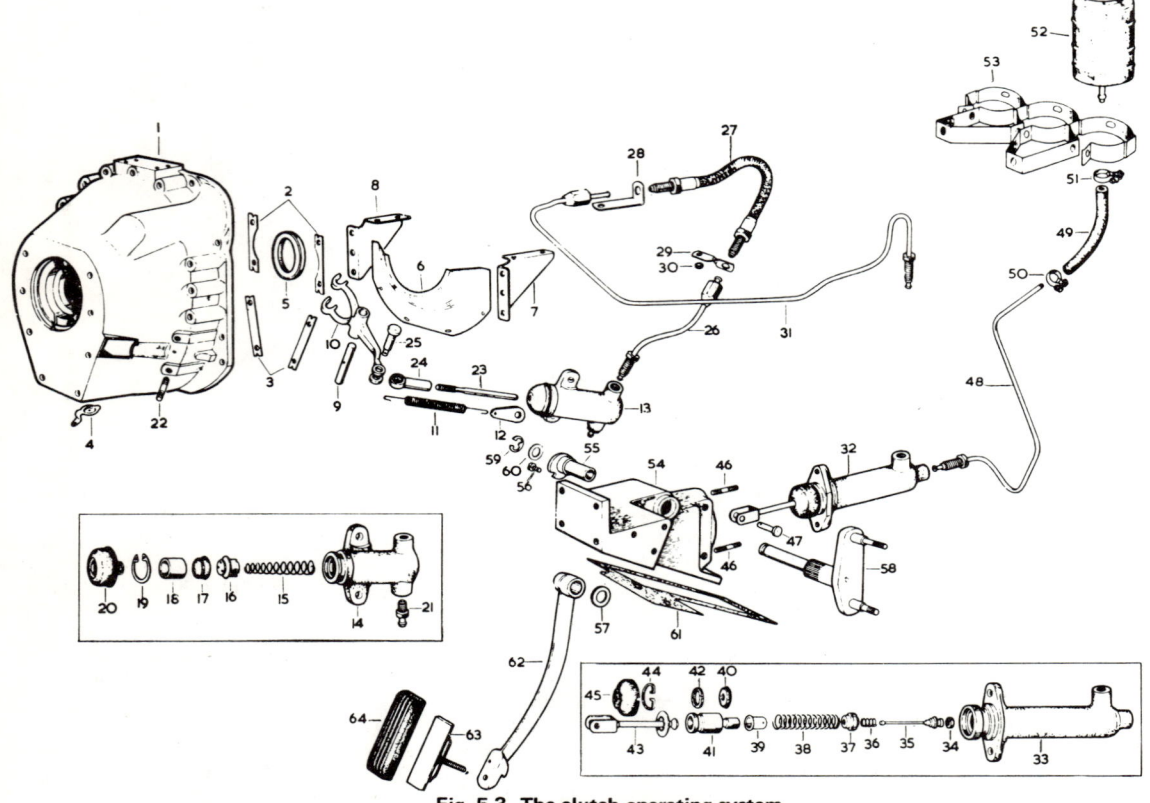

Fig. 5.3. The clutch operating system

1	Clutch housing	17	Seal	33	Master cylinder body	
2	Locking plate	18	Piston	34	Seal	
3	Locking plate	19	Circlip	35	Valve	
4	Timing aperture cover	20	Rubber dust cover	36	Spring	
5	Oil seal	21	Bleeder screw	37	Spring support	
6	Cover plate	22	Stud	38	Main spring	
7	Support bracket	23	Operating rod	39	Spring support	
8	Support bracket	24	Adjuster assembly	40	Cup seal	
9	Shaft	25	Pivot pin	41	Piston	
10	Operating fork	26	Hydraulic pipe	42	Static seal	
11	Return spring	27	Flexible hydraulic pipe	43	Push rod	
12	Anchor plate	28	Bracket	44	Circlip	
13	Slave cylinder	29	Bracket	45	Dust cover	
14	Slave cylinder body	30	Distance piece	46	Stud	
15	Spring	31	Hydraulic pipe	47	Clevis pin	
16	Cup filler	32	Master cylinder	48	Hydraulic pipe	
49	Flexible pipe					
50	Hose clip					
51	Hose clip					
52	Reservoir					
53	Mounting bracket					
54	Clutch pedal housing					
55	Bush					
56	Setscrew					
57	Fibre washer					
58	Pedal shaft					
59	Circlip					
60	Washer					
61	Gasket					
62	Pedal					
63	Pedal pad					
64	Pedal pad cover					

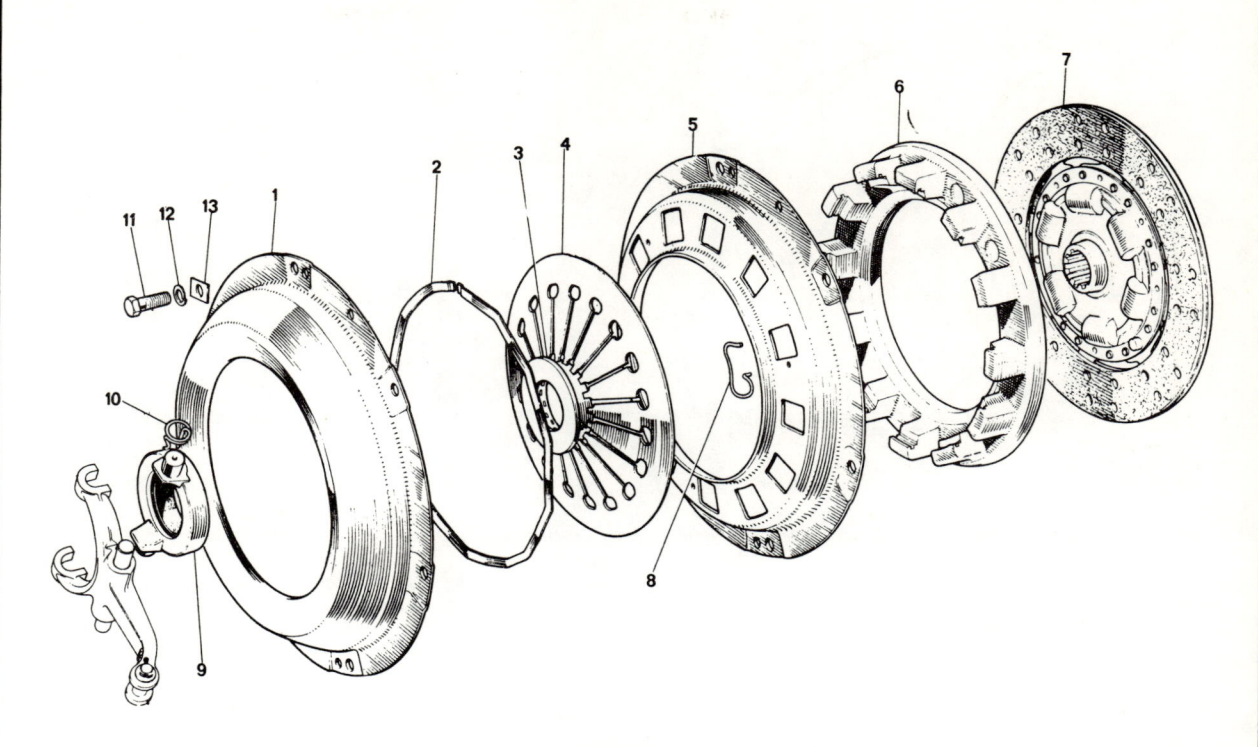

Fig.5.4. Exploded view of the 10" Laycock diaphragm spring clutch

1	Cover	5	Driving plate	8	Clip
2	Spring retaining ring	6	Pressure plate	9	Release bearing
3	Release ring	7	Driven plate assembly	10	Clip
4	Diaphragm spring				

11	Bolt
12	Spring washer
13	Balance weight

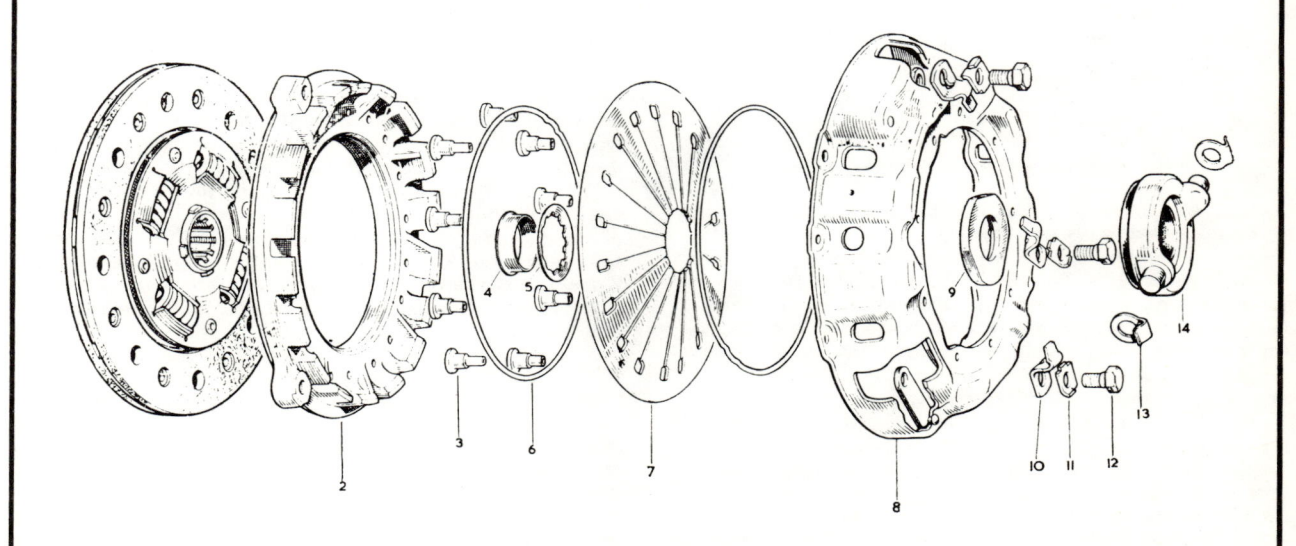

Fig.5.5. Exploded view of the Borg and Beck 10" diaphragm spring clutch

1	Driven plate	5	Belleville washer	8	Cover pressing	11	Tab washer
2	Pressure plate	6	Fulcrum ring	9	Release plate	12	Setscrew
3	Rivet	7	Diaphragm spring	10	Retainer	13	Retainer
4	Centre sleeve					14	Release bearing

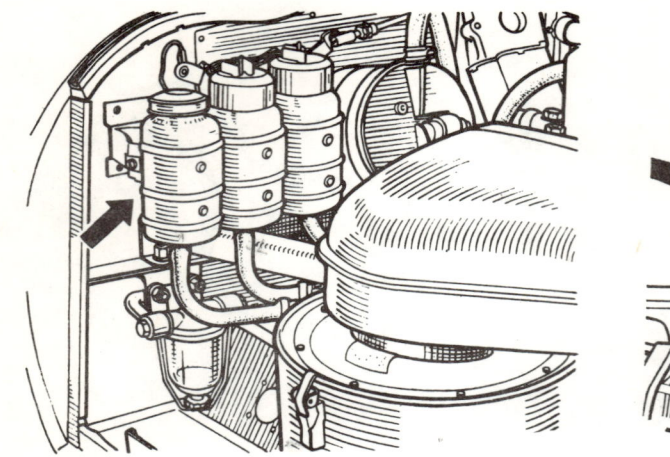

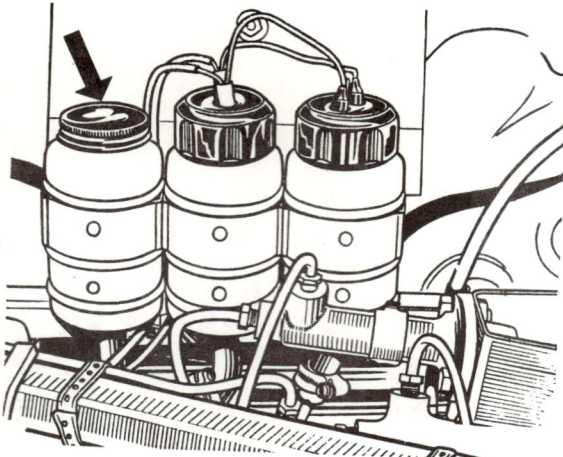

Fig.5.6. The clutch fluid reservoir (identify the left hand view by caption 'RH drive' and the right hand view by 'LH drive')

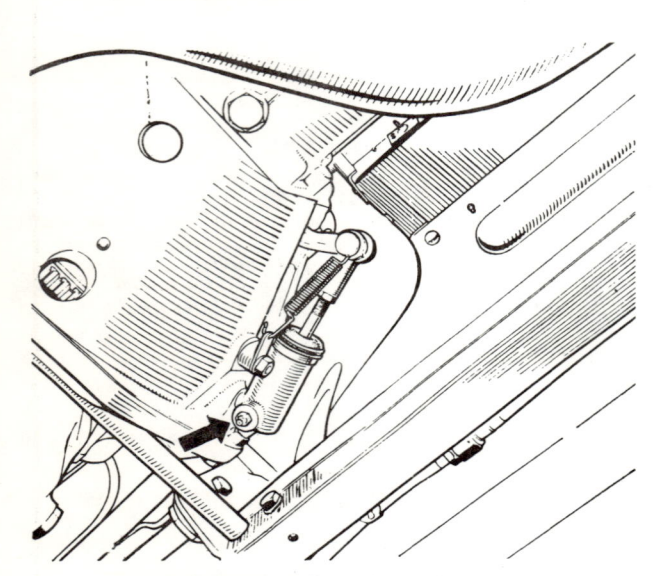

Fig.5.7. Location of the clutch bleed screw

3 Clutch hydraulic system - flushing

After many years of service or if the vehicle has been laid up for some time, it is possible that the fluid in the system will become thick or gummy and this will have an adverse affect on the operation of the clutch. If this fault occurs, the system should be flushed out and refilled - in any event, it is recommended that you do this every five years.

1 Pump all fluid out of the system through the bleed nipple of the clutch slave cylinder. Although the fluid will be discarded we suggest, to avoid making a mess, thay you connect a piece of hose to the bleed nipple and collect the fluid in a container. Slacken the nipple one complete turn and then pump the clutch pedal by depressing it quickly and allowing it to return without assistance.

2 Continue pumping on the clutch pedal until all fluid has been expelled.

3 Fill the fluid reservoir with methylated spirit (industrial quality) and pump it through the system in the manner described above. Keep the reservoir replenished and carry on pumping until at least a quart of spirit has passed through the system.

4 Now slacken the clip securing the low pressure hose to the bottom of the reservoir, remove the hose and allow the remaining spirit to drain into a container.

5 Remove the master cylinder in the manner described in Section 14 and pour off any accumulated spirit. Refit the master cylinder.

6 Fill the system with clean hydraulic fluid (see Section 2, paragraph 2 for Specification) and carry out the bleeding procedure as described in Section 2.

7 The above procedure will not be effective if you have inadvertently topped up the system with mineral oil. In this case rapid deterioration of rubber components can be expected and there is no alternative but to remove and thoroughly clean all non-rubber hydraulic components, including metal pipe line. All rubber components will have to be renewed, as described later in this Chapter.

4 Clutch pedal - free travel

1 Commencing at engine number 7E.4607, the 4.2 litre car was fitted with a hydrostatic clutch operating slave cylinder. Normal clutch wear is automatically compensated for by this type of slave cylinder and providing the operating rod is set as described in Section 22, no clearance adjustment is necessary.

This type of slave cylinder can be readily identified by the fact that it is not provided with a return spring.

2 However, from the introduction of the 4.2 litre Open 2 seater and Fixed Head Coupe at engine number 7E.18356 and the 2 + 2 model at engine number 7E.55558 reversion was made to the same type of non-hydrostatic slave cylinder as is fitted to the 3.8 litre models.

3 For models covered in paragraph 2, refer to Fig. 5.8. For normal road use there should be 1/16" (1.6mm) free travel measured on the operating rod between the slave cylinder and the clutch withdrawal lever. This clearance can be more easily judged if you remove the return spring, then move the operating rod toward the slave cylinder until restriction is felt and then return it to the withdrawal lever held at its fullest extent.

4 Adjustment is effected by slackening the locknut which can be seen against the head of the pushrod. Screwing the rod into the head will increase the free pedal travel whilst screwing it out will decrease the travel.

5 Tighten the locknut when adjustment is correct.

6 For competition and racing, there should be as much free travel of the operating rod as is possible to obtain between the slave cylinder and the clutch withdrawal lever without grating of the gears being experienced when engaging first gear.

7 Correct adjustment of the clutch pedal movement is

Chapter 5/Clutch

important as:-
a) Insufficient free travel may cause a partly slipping clutch leading to burning out of the clutch facings if not corrected. Overtravel of the pedal will also result; this will cause undue internal strain of the clutch assembly and hydraulic cylinders, plus, excessive bearing wear.
b) Too much free pedal movement results in inadequate release movement of the bearing and may produce a dragging clutch condition which make clean gear changes impossible.

5 Removing and refitting a flexible hose

1 Carefully remove all dirt from each union of the flexible hose.
2 Have a clean jar handy to catch any fluid which may drain when the pipe is disconnected.
3 Unscrew the tube nut from the hose union and allow hose to drain.
4 Unscrew the locknut and withdraw the hose from the bracket.
5 Undo the hose at the other connection to the rigid pipe.
6 When refitting a hose it is essential to check that it is not twisted or kinked.
7 Pass the hose union through the bracket. Hold the union with a spanner to prevent the hose from twisting. Fit a shakeproof washer and tighten down on the locknut.
8 Connect the pipe by screwing on the tube nut.
9 Repeat the above for the other end of the pipe.

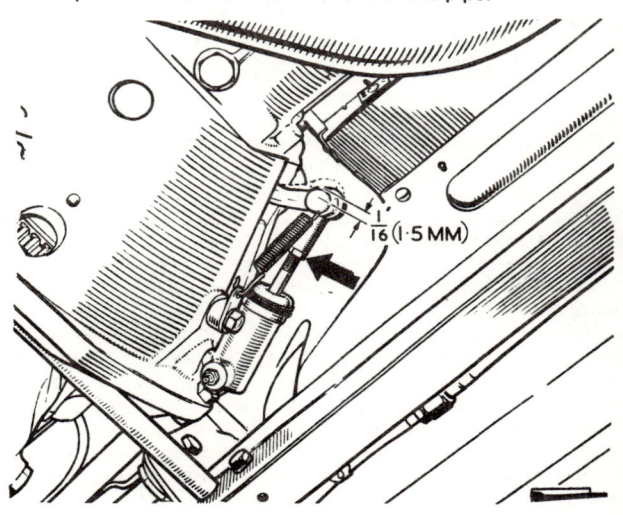

Fig.5.8. Adjustment for free travel of the clutch pedal

6 Clutch - removal

1 Remove the engine and the gearbox from the car and then separate the gearbox from the engine in the manner described in Chapter 1.
2 Look for the balance marks stamped on the clutch and the clutch and the flywheel as depicted in Fig. 5.9 If there are no marks, make your own with a centre punch so that the clutch can be replaced on the flywheel in its original position.
3 Slacken the clutch mounting screws a turn at a time by diagonal selection until the thrust spring pressure is released. Remove the setscrews.
4 Withdraw the clutch assembly from the dowels on the flywheel at the same time taking care that the driven plate, which will not now be supported, does not fall away and get damaged.
5 Do not handle the driven plate with oily hands. If it is to be re-used, place it where it will not be dirtied or damaged.

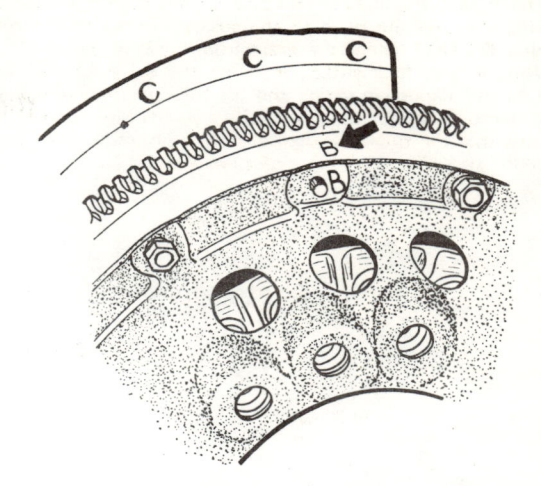

Fig.5.9. Balance marks on the clutch and flywheel

7 Clutch - dismantling

1 The following paragraphs 2 - 8 inclusive, refer to the 10A6G type clutch using thrust springs for the operation of the pressure plate.
2 Before dismantling, mark all the major components for reassembly in their original positions.
3 It is now necessary to evenly compress the thrust springs and to take their weight whilst removing the adjusting nuts from the eyebolts. One way of doing this without the use of special equipment is to bolt the clutch to a flywheel in order to put pressure on the pressure plate (Fig. 5.10) but a press is really the answer. Place the clutch on the bed of the press with wood blocks under the pressure plate in such a manner that the cover can move downwards when pressure is applied, this set-up is shown in Fig.5.11.
4 Having compressed the clutch, unscrew the adjusting nuts. These are locked by staking and considerable torque may be necessary in order to break the lock.
5 Slowly release the clamping pressure when all the nuts are removed.
6 Lift the cover and the thrust spring off the pressure plate and remove the release lever mechanism.
7 Note the positions of the various coloured thrust springs.
8 Fig.5.12 shows how the strut is disengaged from the lever after which the threaded end of the eyebolt and the inner end of the lever are held close together as possible so that the shank of the eyebolt clears the hole in the pressure plate.
9 The following paragraphs refer to the diaphragm spring type of clutch.
10 The diaphragm spring type of clutch is serviced in this country by fitting an exchange unit and as these are readily available from your Jaguar agency it is strongly recommended that you do not attempt to dismantle the assembly.
11 However, individual parts can be obtained for the repair of the clutch and the following instruction for dismantling are given for the benefit of Overseas customers in cases where complete exchange units may not be readily available.
12 It is essential to rigidly observe the following instructions and in particular, attention is drawn to the necessary special tools required.
13 Refer to Fig.5.4.
14 The centrally mounted release plate is held in position by a centre sleeve which passes through the diaphragm spring and the belleville washer into the release plate. To free the plate, collapse centre sleeve with a hammer and chisel as shown in Fig. 5.13 Support the release plate in the locating boss of the special tool, shown at Fig. 5.20, which should be held firmly in a vice.

15 Knock back the locking tabs and remove the three setscrews securing the pressure plate to the straps rivetted to the cover pressing. DO NOT detach the straps from the cover pressing.

16 Using a spot face cutter, machine the shank of the rivets securing the diaphragm spring and the fulcrum rings to separate those items. Drive out the rivets with a standard pin punch. It is essential that the thickness of the cover is not reduced in excess of 0.005" (0.127 mm.) at any point (see Fig. 5.14.).

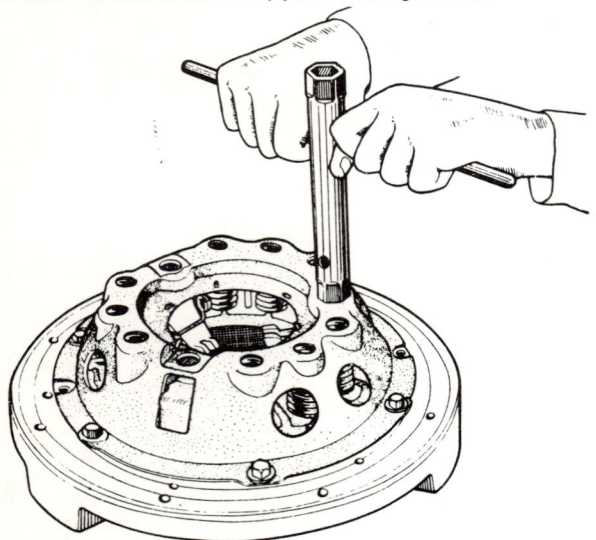

Fig.5.10. Removing the adjusting nuts with the clutch bolted to the flywheel or to a fixture

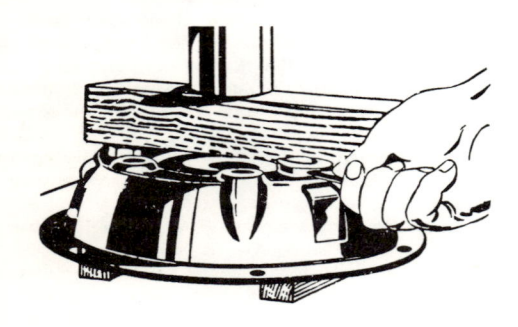

Fig.5.11. Using a press for removal of the adjusting nuts

Fig.5.12. Disengaging the strut from the lever

Fig.5.13. Collapsing the centre sleeve

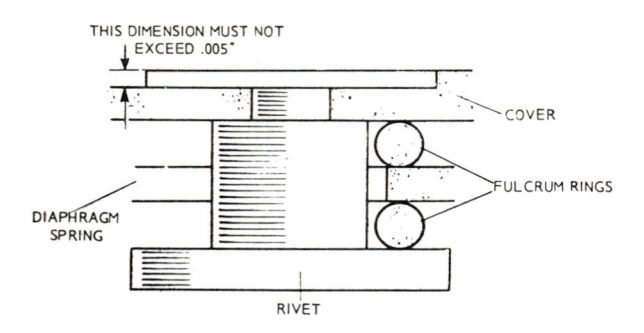

Fig.5.14. Showing maximum permissible reduction in cover thickness

8 Clutch - examination

We advise that you fit a new driven plate, but if you decide against this:-

1 Examine the driven plate friction facings; they will probably have a glazed surface, through which the grain of the material can be clearly seen, and mid-brown in colour. The facings are satisfactory if they are in this condition but if there are dark, highly glazed, patches which hide the grain or if there is a resinous deposit on the facings or if they have a black soaked appearance the indication is that they are contaminated with oil in this event the driven plate assembly should be renewed.

2 Examine the rivets of the driven plate; they should be well below the surface of the friction material and should be secure. Renew the driven plate if the facings are worn or if any rivets are loose.

3 Check the driven plate springs for fracture and security. Check the condition of the splines in the centre hub, excessive wear, which results from faulty alignment, will mean renewing the plate.

4 Examine the thrust springs and check their length. Do not immediately discard any that are under length but check them against one of the correct length by placing end to end in a vice with a metal plate interposed between them, screw up the vice to compress the springs but do not compress them to their fullest extent. Measure their lengths and if the short spring is now under length compared to the other, discard it.

5 The face of the pressure plate should not be ridged or pitted and this applies to the flywheel in the contact area of the driven plate. These parts should also be free from contamination by burnt oil. Photograph 8.5 illustrates a pressure plate exhibiting these faults and which, in addition, is cracked.

6 Check that the flange of the cover is not distroted.

7 In addition to the above points as applicable to diaphragm

spring type clutch, carefully examine the diaphragm spring and the circlips for fracture and distortion.

3.5 Pressure plate ribbed, cracked and with oil contamination

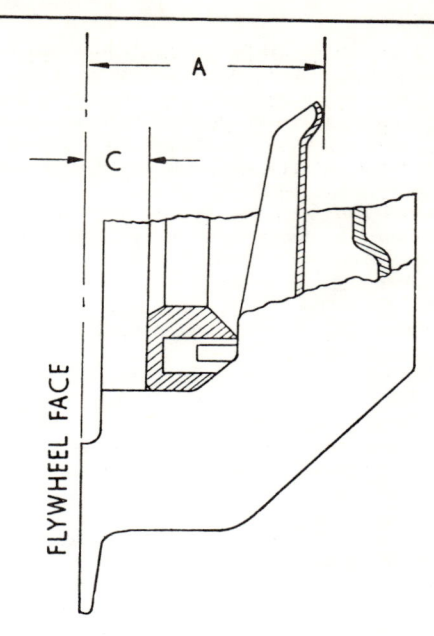

Fig.5.15. Setting the release levers

9 Clutch - reassembly

1 It is essential that all major components are returned to their original positions if balance of the assembly is to be maintained.
2 The instructions contained in paragraphs 3 - 19 inclusive refer to the 10A6G type clutch utilising thrust springs.
3 Fit a pin into an eyebolt and locate the parts within a release lever.
4 Hold the threaded end of the eyebolt and the inner end of the lever as close together as possible, and, with the other hand, engage a strut within the slots in a lug on the pressure plate, push outwards on the other end of the strut towards the rim of the plate.
5 Offer up the lever assembly, first engaging the eyebolt shank within the hole in the plate. Now locate the strut within the groove in the lever.
6 Fit the remaining release levers in the same manner and lightly lubricate all bearing surfaces.
7 The cover now has to be assembled to the pressure using same method, as during dismantling, to compress the thrust springs. Assuming that a press is being used, support the pressure plate on two blocks of wood on the bed of the press.
8 Assemble the thrust springs on the bosses of the pressure plate.
9 Assemble the anti-rattle springs to the cover.
10 Rest the cover on the thrust springs so that the pressure plate lugs are aligned with the slots in the cover.
11 Place a wooden block across the top face of the cover and apply pressure with the press to compress the assembly.
12 Screw the adjusting nuts into an approximate correct position.
13 The release levers must now be set to their correct height and the following procedure assumes that a special setting fixture or gauge plate is not available.
14 Mount the clutch on the flywheel with the driven plate in its normal position or, alternatively, clamp the assembly to any truly flat surface having clearance for the boss of the driven plate.
15 Refer to Fig.5.15.
16 Adjust on the adjusting nuts until the tips of the release levers, dimension 'A', are 1.955" (49.65 mm.) from the flywheel face. Dimension 'C' should be 0.330" (8.381 mm.). These dimensions apply to clutches for normal touring and for racing/competition purposes.

17 Having set the levers, slacken the clamping pressure and turn the driven plate through 90°, reclamp the cover and check the levers again as an insurance against any distortion in the driven plate.
18 When satisfied with the setting of the release levers, lock all adjusting nuts by staking.
19 Fit the release lever retainers and the release lever plate.
20 The following instructions apply to the diaphragm spring type clutch.
21 First check the cover pressing for distortion by bolting the cover firmly to a truly flat surface and then measuring the distance from the cover flange to the machined land inside the pressing. This, as indicated in Fig. 5.16 should not be more than 0.007" (0.2 mm). If this dimension is exceeded the cover must be replaced.
22 Make up a tool to the dimension given in Fig.5.17. Except for the spring all parts can be made out of mild steel.
23 Place the fulcrum ring inside the cover pressing so that the location notches in the fulcrum ring engage a depression between two of the larger diameter holes in the cover pressing. See Fig. 5.18.
24 Place the diaphragm spring on the fulcrum ring inside the cover and align the long slots in the spring whith the small holes in the cover pressing.
25 Locate the other fulcrum ring on the diaphragm spring so that the location notches are diametrically opposed to the location notches in the first ring.
26 Fit new rivets and ensure that the shouldered portion of each seats on the machined land inside the cover.
27 Place the base plate of the tool on to the rivet heads and invert the clutch and base plate (Fig.5.19).
28 Fit the collar of the tool over the large bolt and fit the large bolt complete with spring, spider and collar into the tapped hole in the base. Position the three setscrews on the spider of the tool so that they contact the cover pressing.
29 Tighten down the centre bolt, as depicted in Fig.5.20, until the diaphragm spring is flat and the cover pressing is held firmly by the setscrews.
30 Peen over the rivets with a hand punch (Fig.5.21).
31 Before assembling the pressure plate examine it for wear or damage. If damaged or excessively scored it should be replaced but if this is not possible it is permissible to rectify it by grinding but this must be expertly done as incorrect grinding may affect

operation of the clutch. The pressure plate must not be worked to a thickness of less than 1.070" (27.178 mm).

32 Position the pressure plate inside the cover assembly so that the lugs on the plate engage the slots in the cover pressing.

33 Insert the three setscrews through the straps which are riveted to the cover pressing, tighten down and lock with the tab washers.

34 The pressure plate must now be fitted and for this a special tool is required. The tool number is SSC 805 and it can be obtained from Automotic Products Ltd., Service and Spares Division, Banbury, England. The tool is shown in Fig.5.22 for information.

35 Grip the base plate of the tool in a vice and place the locating boss into the counterbore.

36 Place the release plate, face down, into the counterbore of the locating boss.

37 Apply a little high melting point grease to the tips of the diaphragm spring finger and position the clutch, with the pressure plate friction face upwards, on to the release plate. Ensure that the diaphragm spring fingers locate between the small raised pips on the release plate.

38 Place the belleville washer, concave surface towards the spring, on to the centre of the diaphragm spring and then push the centre sleeve through the spring into the release plate.

39 Drop the special washer of the tool into the sleeve and insert the staking guide into the centre of the assembly. Fit the knurled nut to the thread on the staking guide and tighten down until the whole assembly is solid.

40 Using the special punch, inserted in the slots in the staking guide, stake the centre sleeve in six places into the groove in the release plate. (Fig.5.23).

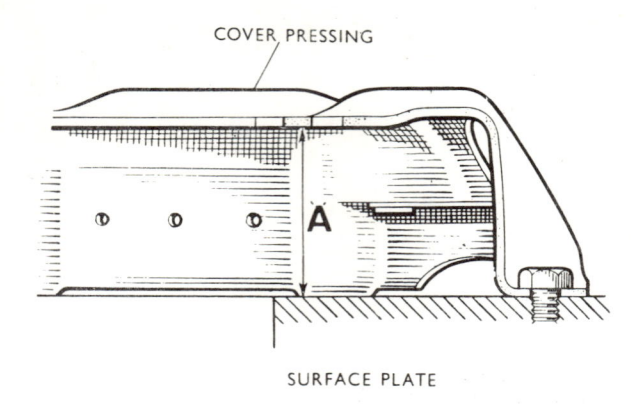

Fig.5.16. Check for distortion of cover pressing. Maximum variation of dimension "A" must not exceed 0.007" (0.2 mm)

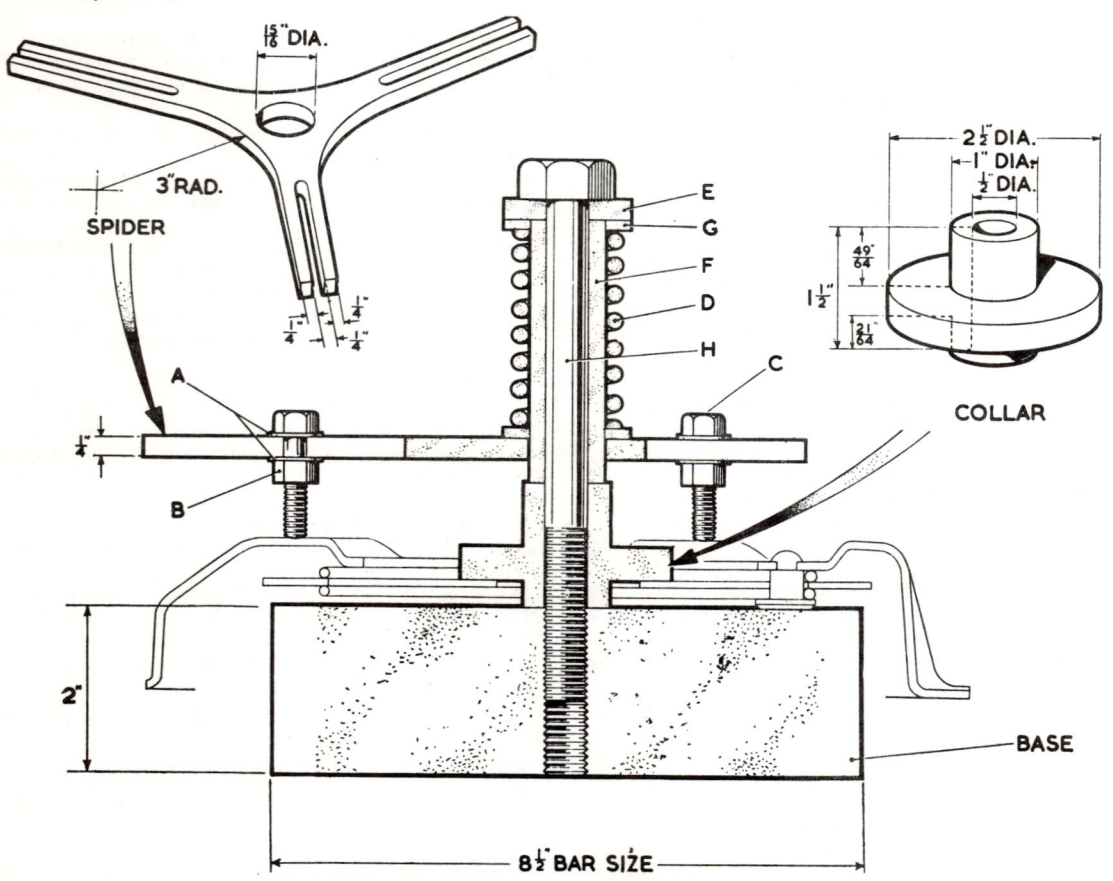

Fig.5.17. Special tool for compressing diaphragm spring

Ref.	Qty.	Description
A	6	¼" flat washer
B	3	¼" nut
C	3	1¼" diameter setscrew
D	1	Spring (minimum load of 100 lbs fitted length)
E	1	Washer ½" I.D. x 1½" O.D. x ¼" thick
F	1	Tube ½" I.D. x 3¼" long
G	2	Washer 7/8" I.D. x 1½" O.D. x 1/8" thick
H	1	Bolt ½" Whit. x 6" long

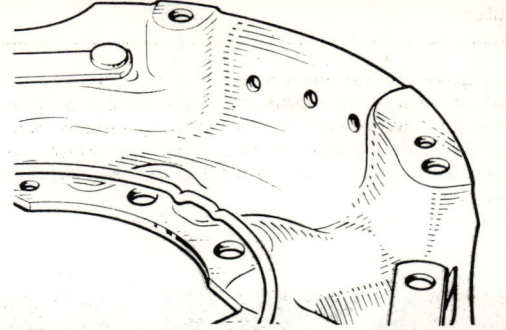

Fig.5.18. Assembly of the fulcrum ring to the cover pressing

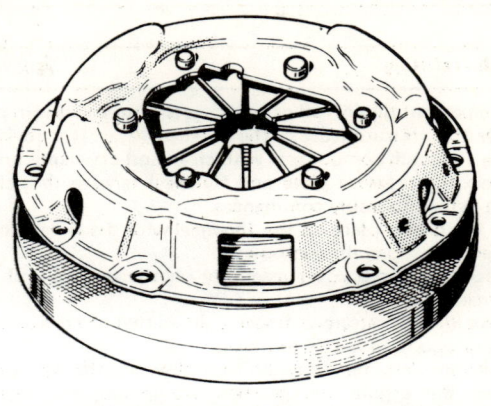

Fig.5.19. Clutch and base plate of tool inverted

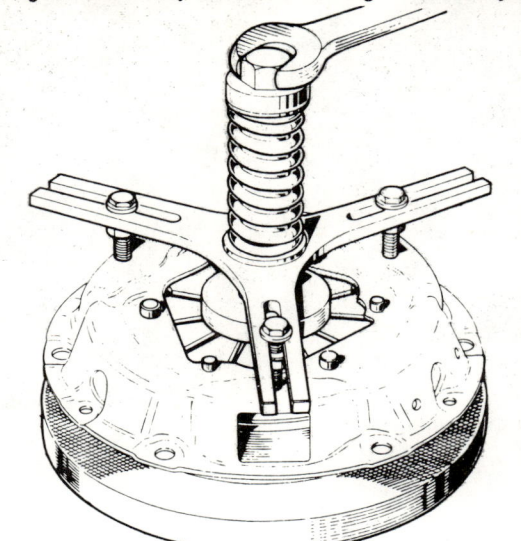

Fig.5.20. Positioning the tool for rivetting

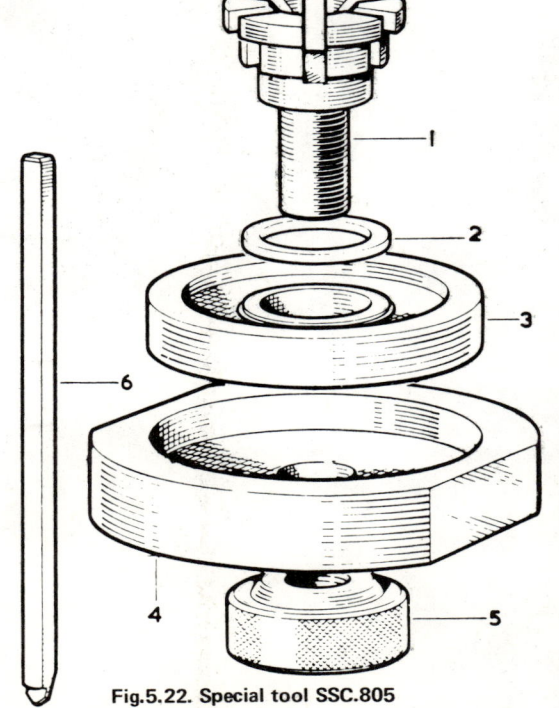

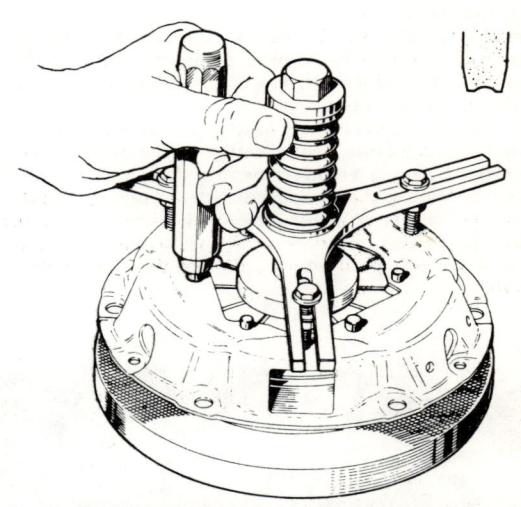

Fig.5.21. Rivet securely with a hand punch

Fig.5.22. Special tool SSC.805

1 Staking guide
2 Washer
3 Locating boss
4 Base plate
5 Knurled nut
6 Punch

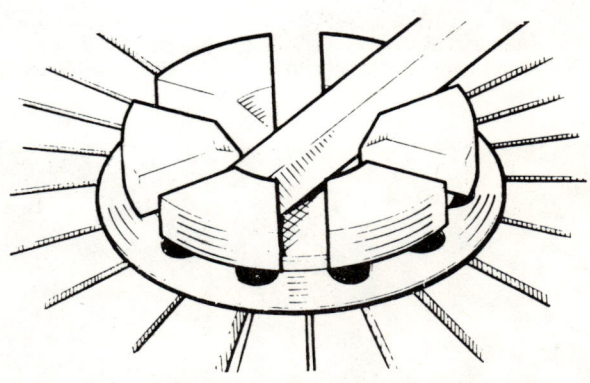

Fig.5.23. Staking the centre sleeve to the release plate

10 Clutch - refitting

1 It is important that no oil or grease gets onto the driven plate linings, or the pressure plate or the flywheel faces. It is advisable to handle all clutch components with clean and dry hands and to wipe down the pressure plate and flywheel faces with a clean, dry cloth before assembly commences.

2 Place the driven plate on the flywheel with the larger part of the splined hub facing the gearbox.

3 Replace the clutch cover assembly on the dowels with the balance marks aligned.

4 Replace the six setscrews finger tight so that the driven plate is loosely gripped and is able to move.

5 The driven plate must now be centralised on the flywheel so that when the engine and gearbox are mated, the gearbox constant pinion shaft splines will pass through the splines in the centre of the driven plate hub.

6 If you have the facilities, turn up a piece of bar to the inside diameter of the splines in the driven plate with a reduced diameter at one end to just enter the constant pinion shaft bearing at the rear of the crankshaft.

7 Insert the bar through the hole in the centre of the clutch and move it until the small diameter enters the constant pinion shaft bearing in the crankshaft. The driven plate is now correctly aligned. Leave the bar in position.

8 An old constant pinion shaft (photo) can be used instead of the bar but failing all else, centralisation of the driven plate can be carried out by using a long screwdriver inserted from the rear of the clutch. Moving the screwdriver sideways or up and down will move the plate in whatever direction is necessary to achieve centralisation. Correct positioning of the driven plate can be judged by viewing its position in relation to the holes at the rear of the clutch.

9 Tighten the setscrews a turn at a time by diagonal selection whilst the bar is still holding the driven plate in position. Tighten down fully and remove the bar. If using a screwdriver or similar tool for centralisation, re-check that the driven plate is still central as there is a tendency for it to move during the tightening operation.

10.8. Centralising the driven plate

11.2A Easing out a spring retainer clip

11.2B Removing the release bearing

11.3 The lever shaft retaining screw

11.4 Removing the release lever shaft

11.5 Withdrawing the release lever

11.6 Release bearing: new left; worn right

Chapter 5/Clutch

11 Clutch fork and release bearing - removal, examination and refitting

1 With the gearbox and engine separated to provide access to the clutch, attention can be given to the release bearing and fork, and lever located in the gearbox bellhousing.
2 To remove the clutch release bearing, ease back the two spring clips located at the ends of the release bearing carrier and lift away the release bearing.
3 Slacken the locknut and using an Allen key, remove the lever shaft retaining screw (photo).
4 Press out the release lever shaft (photo).
5 Remove the release lever (photo).
6 If the graphite release bearing ring is badly worn (photo) it should be replaced by a complete new bearing assembly. Our advice is that a new release bearing is fitted irrespective of the condition of the one removed as a lot of clutch troubles start with this component.
7 Check the fork ends and the lever shaft for wear, renew as necessary.
8 Reassembling the clutch fork and lever assembly and refitting the clutch release bearing is the reverse of the dismantling procedure.

12 Hydraulic system - general

If you have to dismantle any part of the hydraulic system, the operation must be carried out under conditions of scrupulous cleanliness.

Before breaking a hydraulic union, first clean around the union, first clean around the union to remove all traces of road dirt, oil, grease etc. Allow the hydraulic fluid to drain into a clean container, but if you have any doubts as to the cleanliness of the fluid, discard it. Having drained the fluid, blank off the unions.

When a component has to be dismantled, first thoroughly clean the exterior but do not swill it in cleaning agents such as paraffin or petrol, these can be harmful to the rubber parts. Having cleaned the components, wash your hands before starting to dismantle it. Clean off the top of the bench and then cover it with a sheet of paper on which the parts can be placed as they are removed.

Place all metal parts in a tray of clean brake fluid and allow them to soak, after which they can be dried with a clean non-fluffy cloth. Carefully examine all rubber parts and if there is any sign of swelling or perishing they should be renewed - we advocate renewal of rubber items as a matter of course. All internal parts should be dipped in clean brake fluid and assembled wet as the fluid will act as a lubricant. Always assemble the rubber components using the fingers only, if a tool is used (except where special tools are called for) the chances of damaging the item are high.

13 Clutch master cylinder - general description

The master cylinder, Fig.5.24, is mounted on the bulkhead in the engine compartment and is mechanically linked, by means of a clevis pin, to a pendant clutch pedal via the pushrod. It is the master cylinder which, on pressing the clutch pedal, supplies hydraulic pressure to operate the clutch.

The components of the master cylinder are enclosed within the bore of the body at one end of which are two integral pipe connections one taking in low pressure fluid from the reservoir and the other connected to the high pressure side of the system. At the other end of the body is an integral flange provided with two holes to accept the mounting studs.

In the unloaded conditions, a spring holds the piston against a dished washer which is retained by a circlip. One end of the push rod seats in a recess in the head of the piston and the other end of the rod, which is forked, attaches to the clutch pedal. The pushrod is sealed at its point of entry to the cylinder by a rubber dust excluder through which the rod passes; the excluder is retained in a groove in the body of the cylinder.

A cylindrical spring support is located at the inner end of the piston and a small drilling in the end of the support is engaged by the stem of a valve, the larger diameter head of which locates in a central blind bore in the piston. The stem of the valve passes through a vented spring support and between the support and a flange formed on the valve is a small coil spring. A lipped rubber seal fits in a groove around the end of the valve and this assembly forms a recouperation valve to control the flow of fluid to and from the reservoir.

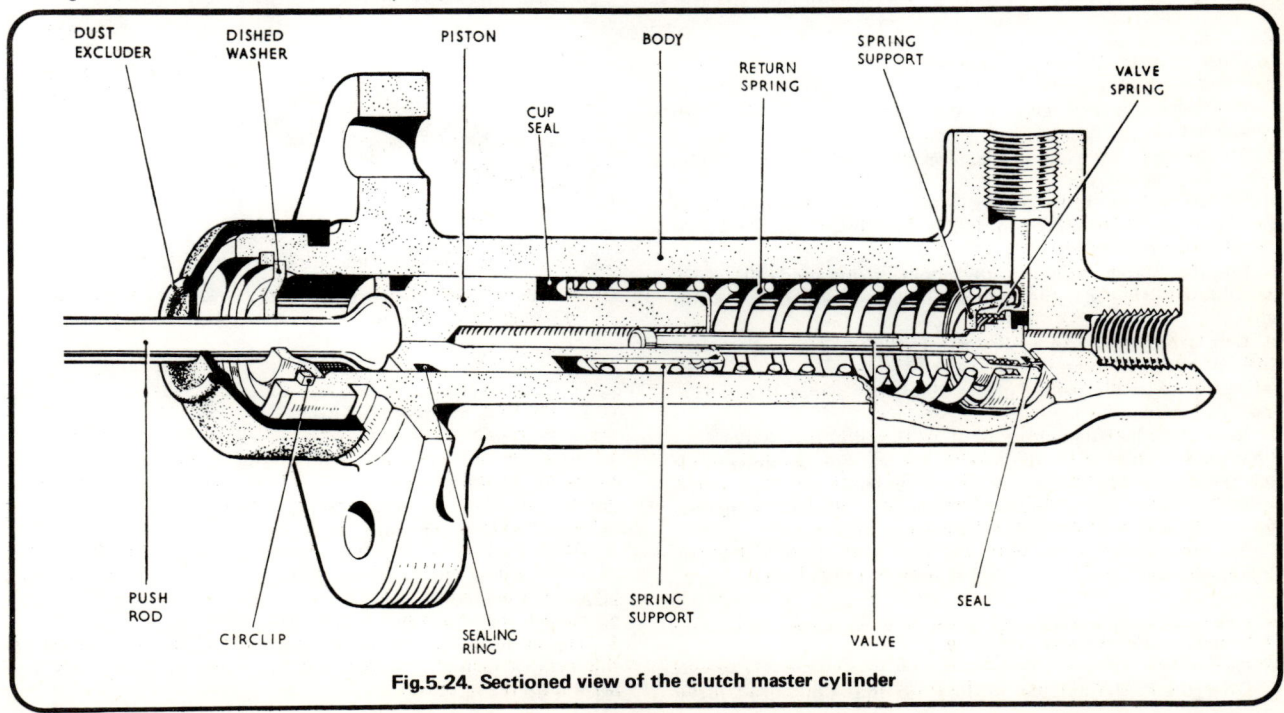

Fig.5.24. Sectioned view of the clutch master cylinder

With no load applied to the clutch pedal, the valve is held clear of the base of the cylinder and in this condition the cylinder is in fluid communication with the reservoir thus permitting recouperation of any fluid loss. When the clutch pedal is pressed, the piston is moved down the cylinder and compresses the main spring; as soon as the piston movement is in excess of the valve clearance the valve closes and isolates the cylinder from the reservoir. Further movement of the pedal puts pressure on the fluid between the master cylinder and the clutch slave cylinder. Removal of load from the pedal reverses the sequence.

14 Clutch master cylinder - removal

1 Drain the clutch fluid reservoir and then detach the inlet and outlet pipes from the master cylinder by undoing the union nuts. (see Section 12).
2 Blank off the pipes and the master cylinder unions.
3 Move inside the car. Take out the split pin which locks the clevis pin securing the master cylinder push rod to the clutch pedal. Remove the clevis pin.
4 In the engine compartment, take off the two nuts with their spring washers which secure the flange of the cylinder body to the mounting studs. Withdraw the master cylinder towards the front of the car.

15 Clutch master cylinder - dismantling

1 Refer to Section 12.
2 Ease the dust excluder out of the groove in the body and then push it up the rod out of the way.
3 Depress the push rod to help relieve the pressure on the circlip and then, using a pair of circlip pliers, remove the circlip.
4 Remove the push rod complete with the dished washer.
5 Withdraw the piston, vigorous shaking may be needed to do this.
6 Take out the spring support, the return spring and the valve assembly.
7 If you intend to renew the rubber seals, they may now be eased out of the locating grooves. If the seals are not to be renewed, we suggest that you do not disturb them.

16 Clutch master cylinder - reassembly

1 Lubricate the new seals and the bore of the cylinder with hydraulic fluid.
2 Fit the seal to the end of the valve and make sure that the lip registers in the groove.
3 Fit the cup seal in its groove in the shank of the piston and the static seal in the groove in the front of the piston. Make sure that they are seating correctly.
4 Fit the spring support, the main spring and the valve assembly ensuring that the head of the valve engages the piston bore.
5 Lubricate the piston with Castrol Rubber Grease and then slide the complete assembly into the cylinder taking care not to damage or twist the seals. The use of a fitting sleeve, if you have one, is advisable.
6 Now assemble the rubber boot to the pushrod followed by the dished washer. Offer the pushrod to the cylinder and ensure that the dished washer seats on the shoulder at the head of the cylinder. Depress the piston against the main spring and then fit the circlip making sure that it fully engages the groove.
7 Fill the rubber boot dust excluder with Castrol Rubber Grease and then reseat it around the head of the cylinder.

17 Clutch master cylinder - refitting

1 Offer the master cylinder to the studs on the bulkhead, make sure that it is the correct way up, and then secure it in position with the nuts and spring washers.
2 Remove the blanking plugs and then reconnect the high and low pressure pipes.
3 Fit the forked end of the push rod in position on the clutch pedal. Insert the clevis pin followed by the plain washer and secure with a new split pin.
4 Fill the reservoir with clean hydraulic fluid and bleed the system as described in Section 2.
5 Check for leaks at the unions and remedy as necessary - further bleeding may be necessary if leaks are present.

18 Clutch slave cylinder - general description

The clutch slave cylinder is the link between the master cylinder and the clutch operating lever. It is essentially a casting with an integral mounting flange and two screwed connecting points one of which (that parallel to the mounting flange) receives the pipe from the master cylinder and the other accepts a bleed nipple.

The clutch slave cylinder is illustrated in Fig.5.25. The body is bored and honed to fine limits and accommodates a piston against the inner face of which is a rubber cup loaded by a cup filler and spring. The travel of the piston is limited by a circlip which fits into a groove at the rear of the bore. The end of the bore is protected against the intrusion of dirt by a rubber boot through which a pushrod, connected to the release bearing operating lever, passes.

Hydraulic pressure from the master cylinder moves the rubber cup, and the piston, rearwards; the piston is bearing on the push rod and this in turn, is moved rearwards and pivots the release bearing operating lever to which it is connected. When the clutch pedal is released, spring pressure on the operating lever moves the components back to the 'at rest' position.

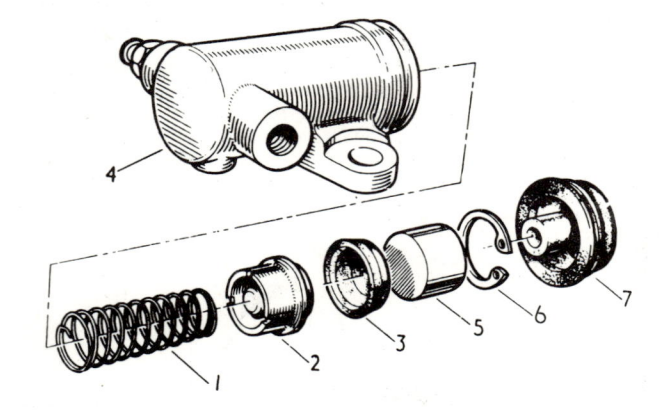

Fig.5.25. The clutch slave cylinder

1 Spring
2 Cup filler
3 Rubber cup
4 Body
5 Piston
6 Circlip
7 Rubber boot

19 Clutch slave cylinder - removal

1 Place the car over a pit or on a ramp or raise the car to give access to the base of the gearbox. If the car is raised on a jack, use an axle stand or otherwise make sure it is safely supported before crawling underneath.
2 Refer to Section 12. Disconnect the pipe from the master cylinder. Allow the fluid to drain into a container and then blank off the unions.
3 Detach the return spring (if fitted).
4 The pushrod may be left in position on the car if desired. In this case, detach the rubber boot from the cylinder body and leave it on the push rod.

Chapter 5/Clutch

5 If you wish to remove the pushrod with the cylinder, take out the split pin locking the clevis pin which secures the push rod to the release bearing operating lever. Remove the flat washer and the clevis pin. In some cases the clevis pin is secured by a spring clip instead of a split pin, compress the spring clip and slide it off the pin (see Chapter 1, photograph 5.54).
6 Remove the nuts securing the cylinder to the clutch housing, lift the cylinder away from the studs and remove from the car.

20 Clutch slave cylinder - dismantling

1 Refer to Section 12.
2 Take out the push rod, if still in position, by easing the rubber boot out of the groove in the body of the cylinder then withdrawing the rod complete with the boot.
3 Remove the circlip from the end of the bore.
4 Apply low air pressure from, say, a foot pump to the open connection to expel the piston and the other parts. It is advisable to hold a piece of rag over the end of the bore to catch the parts as they are pushed out.
5 Remove the bleed nipple.

21 Clutch slave cylinder - reassembly

1 Smear all internal parts and the bore of the body with Castrol Rubber Grease.
2 Fit the spring into the cup filler and slide, spring uppermost, into the bore of the body.
3 Insert the cup washer, lip leading, into the bore and take care not to turn back or damage the lip.
4 Fit the piston with the flat face innermost. Using the pushrod, put pressure on the piston to compress the spring slightly and then fit the circlip.
5 The pushrod can be refitted to the body at this stage if desired. Fit the rubber boot to the pushrod, enter the pushrod in the bore until it contacts the piston and then ease the boot into the groove in the body.

22 Clutch slave cylinder - refitting

1 Offer the slave cylinder up to the studs on the clutch housing at the same time entering the pushrod into the bore if the rod is already attached to the vehicle.
2 Secure the cylinder in position with the two nuts and spring washers.
3 Refit the return spring (where fitted).
4 Fit the rubber boot into the groove in the body (where fitted).
5 Remove the blanking plugs, and refit the pipe to the master cylinder then refit the bleed nipple, do not overtighten the screw.
6 It is important at this stage to check the free travel of the clutch pedal as described in Section 4 and also to adjust the pushrod of the hydrostatic type of slave cylinder (i.e. those cylinders not fitted with a return spring). Proceed as follows:
7 Release the locknut at the fork end of the pushrod.
8 Push the clutch operating lever away from the slave cylinder until resistance is felt and then hold it in this position.
9 Push the operating rod to the limit of its travel in the slave cylinder and adjust the frok end to a dimension of 0.75" (19 mm.) between the centre of the fork end and the operating rod attachment holes as illustrated in Fig.5.26.
10 Tighten the locknut and refit the clevis pin and secure.
11 Now bring the car to a horizontal position and, for both types of cylinder, refill and bleed the system as described in Section 2.

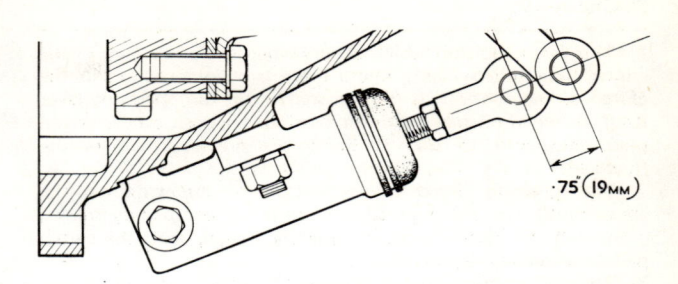

Fig.5.26. Setting dimension for hydrostatic clutch slave cylinder

23 Fault diagnosis and remedy

There are four main faults to which the clutch and the release mechanism are prone. They may occur by themselves or in conjunction with each other. They are clutch squeal, slip, spin and judder.

24 Clutch squeal

1 If, on taking up the drive or when changing gear, the clutch squeals, it is an indication of a badly worn clutch release bearing.
2 As well as regular wear due to normal use, wear of the clutch release bearing is accentuated if the clutch is ridden or held down for long periods in gear with the engine running. To minimise wear of this nature the car should always be taken out of gear at traffic lights or at similar hold-ups.
3 The clutch release bearing is not an expensive item but it is difficult to get at as its replacement requires the removal of the engine and gearbox assembly and work as detailed in Section 11.

25 Clutch slip

1 Clutch slip is a self evident condition which occurs when the clutch driven plate is badly worn or oil or grease have got onto the flywheel or pressure plate faces. It may also be that the pressure plate is faulty.
2 The reason for clutch slip is that due to one or more of the faults above, there is either insufficient pressure from the pressure plate, or insufficient friction in the driven plate to ensure a solid drive.
3 If small amounts of oil get onto the clutch, they will be burnt off under the heat of the clutch engagement and in the process will gradually darken the linings. Excessive oil on the clutch will burn off leaving a carbon deposit which can cause slip, fierceness, spin and judder.
4 If clutch slip is suspected and confirmation of this condition is required, there are several tests which can be made.
5 With the engine in second or third gear and pulling lightly, sudden depression of the accelerator pedal may cause the engine to increase speed without any noticeable increase in road speed. Easing off on the accelerator will cause a drop in engine speed but not in road speed.
6 In extreme cases of clutch slip the engine will race under normal accelerating conditions.
7 If slip is due to oil or grease on the linings a temporary cure can sometimes be effected by squirting carbon tetrachloride into the clutch. The permanent cure is, of course, to renew the clutch driven plate and to investigate and to cure the cause of the oil leak. WARNING: Concentrated Carbon Tetrachloride fumes are extremely dangerous if inhaled.

26 Clutch spin

1 This is a condition which occurs when there is a leak in the clutch hydraulic system, where there is an obstruction in the clutch either in the first motion shaft or in the operating lever itself, or when oil may have left a resinous deposit on the driven plate causing it to stick to either the pressure plate or the flywheel.
2 The reason for clutch spin is that due to one or more of the above faults, the clutch pressure plate is not completely freeing even with the clutch pedal completely freeing with the clutch pedal completely depressed.
3 The symptoms of clutch spin are difficulty in engaging a gear from rest, difficulty in changing gear, and a very sudden take up of the drive at the fully depressed end of the clutch pedal travel as the clutch is released.
4 Check the clutch master cylinder, the slave cylinder and hydraulic connections for leaks. Fluid in one of the rubber boots is a sure sign of a leaking piston seal.
5 If these points are checked and are found to be in order then the fault lies internally in the clutch which will have to be removed for examination.

27 Clutch judder

1 Clutch judder is a self-evident condition which occurs when the gearbox or engine mountings are loose or are too flexible, or when there is oil on the face of the driven plate or when the pressure plate has been incorrectly adjusted.
2 The reason for clutch judder is that due to one of the above faults, the pressure plate is not freeing smoothly from the driven plate and is snatching.
3 Clutch judder is usually most noticeable when the clutch pedal is released in first or in reverse gear and the whole car judder as it moves backwards or forwards.

Chapter 6 Gearbox and automatic transmission

Contents

MANUAL GEARBOX
Gearbox - dismantling ... 3
Gearbox - examination ... 4
Gearbox - reassembly of type EB/JS ... 5
Gearbox - reassembly of EJ, KE and KJS numbered series ... 6
Gearbox - removal and replacement ... 2
General description ... 1

AUTOMATIC GEARBOX
Automatic shift speeds ... 17
Automatic transmission - fault diagnosis ... 18
Automatic transmission - fluid level ... 8
Automatic transmission - general ... 7
Automatic transmission unit - removal and refitting ... 9
Front band - adjustment ... 14
Governor - examination, removal and refitting ... 16
Manual linkage - adjustment ... 12
Oil pan - removal and refitting ... 13
Rear band - adjustment ... 15
Throttle/kickdown cable - adjustment ... 10
Throttle/kickdown cable - removal and refitting ... 11

Specifications

MANUAL GEARBOX	3.8 models	4.2 models
Number of forward speeds ...	4	
Gearbox prefix ...	EB	—
Gearbox suffix ...	JS	—
Gearbox identification number ...	—	EJ 001 onwards
Synchromesh ...	Second, third and top	EJ 001 onwards All four forward gears
Gearbox ratios:		
1st gear ...	3.377 : 1	2.68 : 1
2nd gear ...	1.86 : 1	1.74 : 1
3rd gear ...	1.283 : 1	1.27 : 1
4th (top) gear ...	1 : 1	1 : 1
Reverse ...	3.377 : 1	3.08 : 1
With effect from gearbox identification numbers:		
Open 2 seater and F.H. Coupe ...	KE 101 onwards	
2 + 2 ...	KJS 101 onwards	
Gearbox ratios were changed to:		
1st gear ...		2.933 : 1
2nd gear ...		1.905 : 1
3rd gear ...		1.389 : 1
4th (top) gear ...		1 : 1
Reverse ...		3.378 : 1
Endfloat on mainshaft:		
1st gear ...	—	0.005 to 0.007 inch (0.13 to 0.18 mm)
2nd gear ...	0.002 to 0.004 inch (0.05 to 0.10 mm)	0.005 to 0.008 inch (0.13 to 0.20 mm)
3rd gear ...	0.002 to 0.004 inch (0.05 to 0.10 mm)	0.005 to 0.008 inch (0.13 to 0.20 mm)
Countershaft gear unit endfloat ...	0.002 to 0.004 inch (0.05 to 0.10 mm)	0.004 to 0.006 inch (0.10 to 0.15 mm)

Chapter 6/Gearbox

Automatic transmission:
Type ... Model 8
Max ratio of torque converter ... 2.00 : 1
1st gear reduction ... 2.40 : 1
2nd gear reduction ... 1.46 : 1
3rd gear reduction ... 1.00 : 1
Reverse gear reduction ... 2.00 : 1

Capacities:

MANUAL GEARBOX:
3.8 litre EB/JS gearbox ... 2½ Imp pints (3 US pints, 1.42 litres)
4.2 litre ... 3 Imp pints (3¼ US pints, 2.4 litres)

AUTOMATIC GEARBOX
From dry (2 + 2 only) ... 16 Imp pints (19 US pints, 9 litres)

1 General description

The manual gearbox fitted to both the 3.8 litre and the 4.2 litre models is of the four speed type with - in the case of the EB/JS gearbox - synchromesh on second, third and top gears, the first and reverse gears of this box have spur teeth which slide into mesh.

The later type gearbox (identification number EJ 001 onwards) has baulk ring synchromesh on all gears, with the exception of reverse. The detents for the gears are incorporated in the synchro assemblies - the three synchro balls engaging with grooves in the operating sleeve. A spring loaded ball forming the detent for the reverse gear engages on a groove in the selector rod. Engagement of two gears at the same time is prevented by two interlock balls and a pin which are located at the front of the selector rods. The gears on this box are pressure fed at 5 lbs. per sq. inch by a pump which is driven from the rear of the mainshaft.

The EJ 001 gearbox is superseded for the Open 2 Seater and FH Coupe by boxes with identification numbers KE 101 onwards whilst the 2 + 2 models have boxes with identification number KJS 101 onwards. The helix angle of the gear teeth of these later boxes have been altered for quietness and the taper dog gear lock has been modified to prevent possible jumping out of gear. Apart from these differences the boxes are identical with the EJ 001 series and have the same servicing, dismantling and assembling procedures.

An exploded view of the EB/JS gearbox casing and top cover is given in Fig.6.1 and of the gears in Fig. 6.2. An exploded view of the gears for the EJ 001 (and KE 101 and KJS 101) type of gearbox is shown in Fig.6.3.

Automatic transmission is supplied as an optional extra for 4.2 litre 2 + 2 models. The transmission unit incorporates a hydraulic torque converter in place of the flywheel and clutch assembly as used with the manual gearbox. The converter is coupled to a hydraulically operated planetary gearbox providing three forward speeds and reverse. An exploded view of the automatic transmission unit is given at Fig.6.4.

The automatic transmission control (Fig.6.5) selector lever is mounted centrally in a quadrant on the console. The quadrant is marked from front to rear P,R,N,D2,D1,L, the selector lever can be moved freely between the N and the D2 positions but to move it to the other positions it must be pressed to the right against spring pressure and moved through the gate. With D2 selected from N (neutral) the car will start from rest in second gear and will operate automatically between second and third gears, first gear is not obtainable with this selection. With D1 selected the car starts from rest in first gear and will operate automatically through all three forward gears. Downshifts in either D1 or D2 can be obtained by fully depressing the accelerator ('kickdown' position).

L (lockup) position provides over-riding control with either first or second gear with engine braking in either ratio. When starting from rest in the lockup position, the transmission will start in first gear and will remain in that gear irrespective of road speed or throttle position.

When P (park) is selected, the gearbox is mechanically locked by means of a parking pawl which engages with external teeth formed in the ring gear integral with the driven shaft. The Park position can be used whenever the car is parked with or without the engine running.

In order to ensure that the engine cannot be started when one of the drive positions is selected, a starter inhibitor switch is provided and this switch - when correctly set - ensures that the starter will operate only when the selector is in either the N or P position.

2 Gearbox - removal and replacement

The gearbox and engine must be removed from the car as a complete unit and full instruction for this, and for separating and refitting the box to the engine, will be found in Chapter 1.

3 Gearbox - dismantling

It is assumed that the clutch housing is still fitted to the gearbox but that the clutch release mechanism has been removed from the housing in the manner described in Chapter 5. It is also assumed that you drained the oil from the box prior to removing the engine and the gearbox from the car.

1 Before commencing work, thoroughly clean the exterior of the gearbox using a solvent such as paraffin or 'Gunk'. Finish off gearbox by wiping down the exterior of the unit with a dry non-fluffy rag.
2 The first task is to remove the clutch housing: this is held to the gearbox by eight bolts two of which are secured by locking wire and the remainder by tabbed locking plates.
3 Break the locking wire, knock up the tabs and remove the bolts (photo). The clutch housing can now be lifted off (photo).
4 Remove the six setscrews and the four bolts securing the top cover to the gearbox case and then lift off the top cover (photo). The mainshaft gears are now exposed (Fig 6.6).
5 Engage first and reverse gears to lock the unit. Rest the unit on the bench so that the gearbox flange is uppermost. Remove the split pin locking the flange securing nut (photo).
6 Have an assistant to hold the gearbox from moving and then unscrew and remove the flange securing nut. By using two screwdrivers, or something similar, apply pressure to the flange to lever it off the mainshaft (photo). Lift off the flange and collect the four propeller shaft securing bolts (photo).
7 Remove the housing for the rear oil seal (photo). Remove the locking screw which retains the speedometer driven gear bush in the extension and take out the driven gear and the bearing. Remove the speedometer driving gear (photo).
8 Remove the seven setscrews which secure the rear end cover and remove the cover (photo) taking care nto to pull the layshaft and the reverse gear shaft out of the gearbox. Remove the oil pump drive where fitted (photo).
9 Remove the reverse gear pinion shaft by pulling it to the rear

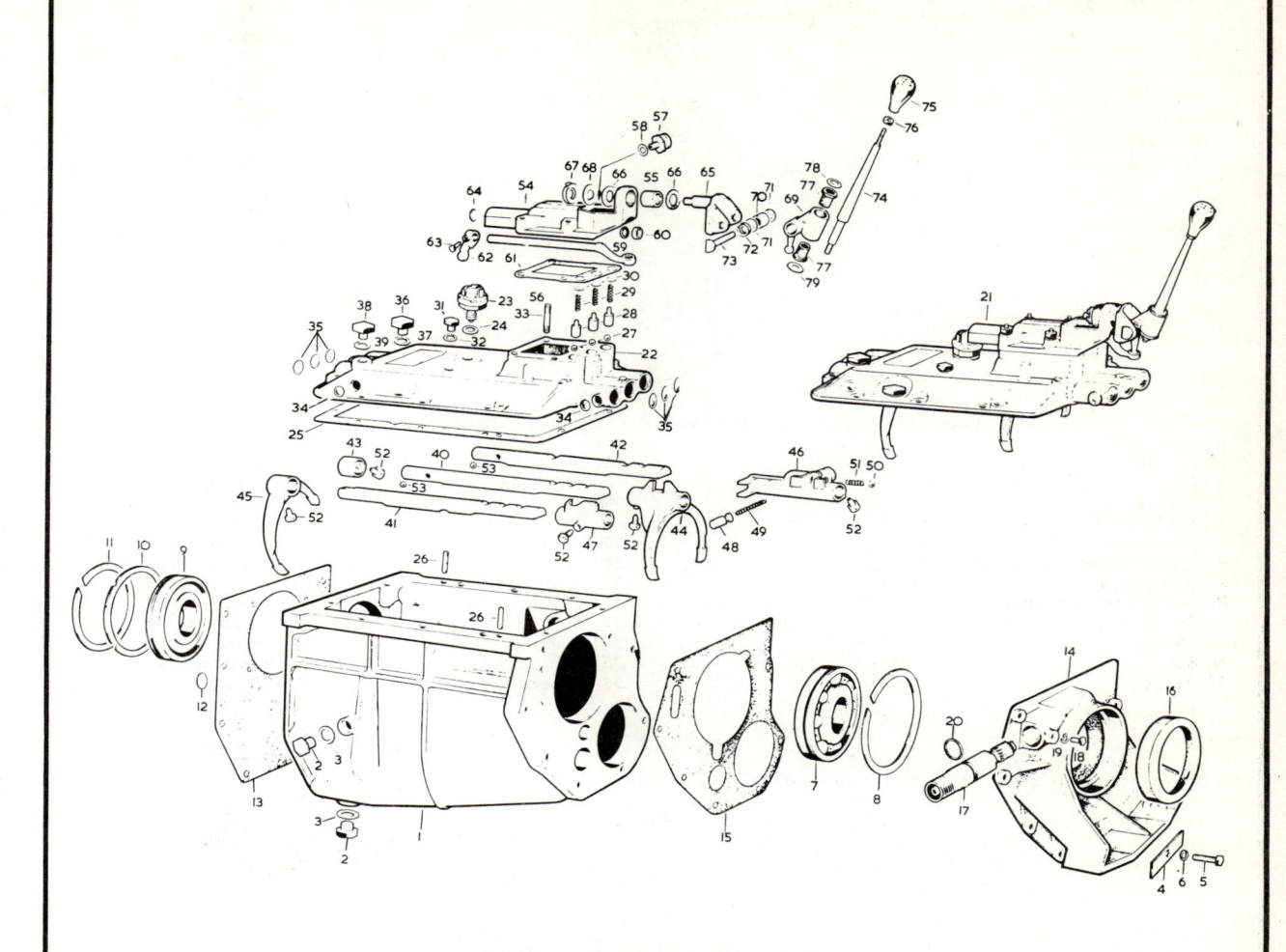

**Fig.6.1. Exploded view of the EB/JS gearbox casing and top cover
(EJ, KE and KJS series similar)**

1	Gearbox case	21	Remote control assembly	40	Striking rod assembly-1st & 2nd gears	60	Retaining clip
2	Drain plug & oil filler plug	22	Top cover	41	Striking rod assembly-3rd & top gears	61	Selector shaft
3	Fibre washer	23	Switch	42	Striking rod - reverse gear	62	Selector finger
4	Locking plate	24	Gasket	43	Stop	63	Screw
5	Setscrew	25	Gasket	44	Change speed fork-1st & 2nd gears	64	Welch washer
6	Spring washer	26	Dowel	45	Change speed fork-3rd & top gears	65	Pivot jaw
7	Ball bearing	27	Ball	46	Change speed fork - reverse gear	66	Washer
8	Circlip	28	Plunger	47	Selector - 3rd and top gears	67	Spring washer
9	Ball bearing	29	Spring	48	Plunger	68	'D' washer
10	Collar	30	Shims	49	Spring	69	Selector lever
11	Circlip	31	Plug	50	Ball	70	Bush
12	Fibre washer	32	Washer	51	Spring	71	Washer
13	Gasket	33	Stud	52	Dowel screw	72	Spring washer
14	Rear end cover	34	Welch washer	53	Ball	73	Pivot pin
15	Gasket	35	Welch washer	54	Housing	74	Gear lever
16	Oil seal	36	Plug	55	Bush	75	Knob
17	Speedometer drive gear	37	Fibre washer	56	Gasket	76	Nut
18	Locking screw	38	Plug	57	Breather	77	Bush
19	Washer	39	Copper washer	58	Fibre washer	78	Washer
20	'O' ring			59	'O' ring	79	Washer

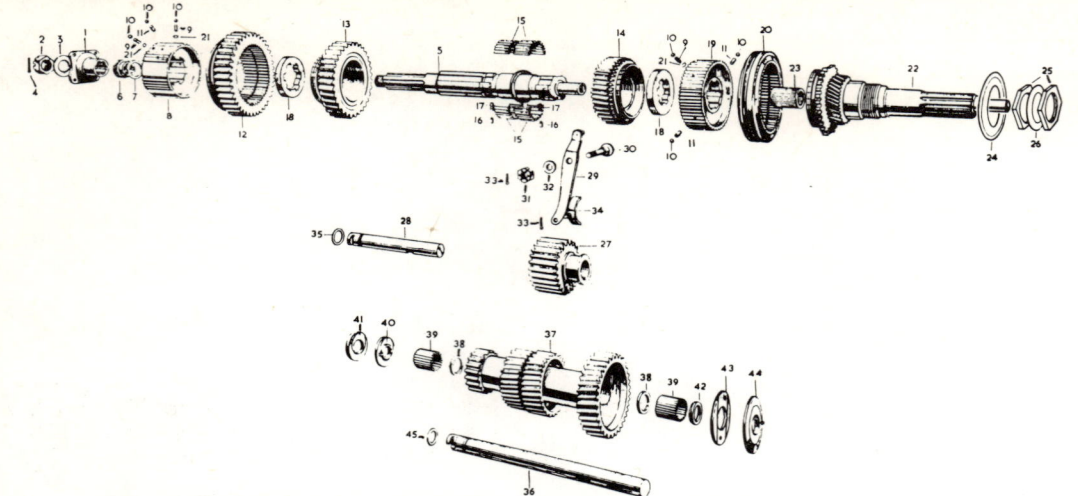

Fig.6.2. Exploded view of the internal components - EB/JS gearbox

1	Flange	12	1st speed gear	23	Roller bearing	35 Sealing ring
2	Nut	13	2nd speed gear	24	Oil thrower	36 Countershaft
3	Washer	14	3rd speed gear	25	Locknut	37 Gear unit on counter-shaft
4	Split pin	15	Needle roller	26	Tab washer	38 Retaining ring
5	Mainshaft	16	Plunger	27	Reverse gear	39 Needle roller
6	Speedometer driving gear	17	Spring	28	Reverse spindle	40 Thrust washer
7	Distance piece	18	Thrust washer	29	Lever	41 Thrust washer
8	Synchronising sleeve - 2nd gear	19	Synchronising sleeve	30	Fulcrum pin	42 Retaining ring
9	Spring	20	Operating sleeve	31	Slotted nut	43 Thrust washer
10	Ball	21	Shim	32	Plain washer	44 Thrust washer
11	Plunger	22	Constant pinion shaft	33	Split pin	45 Sealing ring
				34	Reverse slipper	

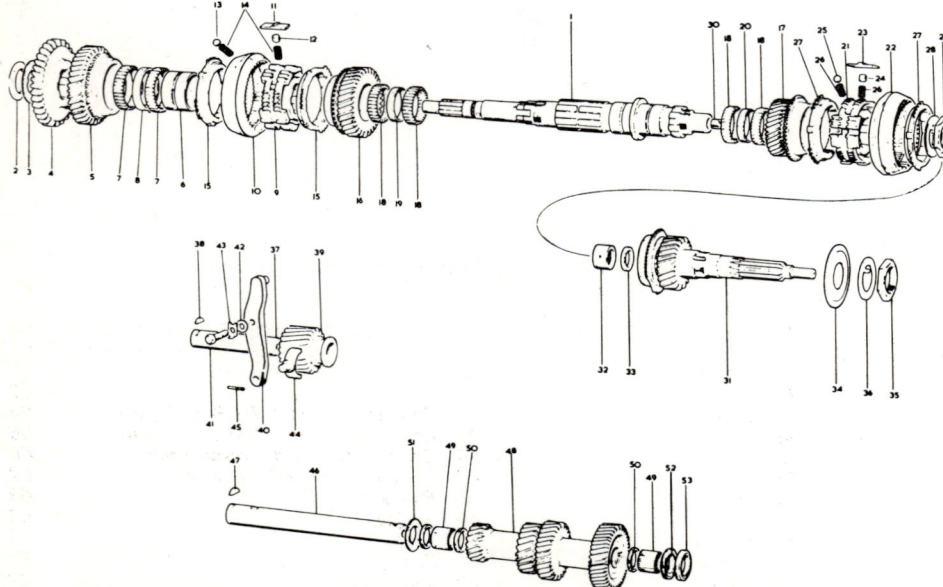

Fig.6.3. Exploded view of the internal components - EJ, KE and KJS series gearboxes

1	Mainshaft	14	Spring	27	Synchro ring	40	Lever assembly
2	Nut	15	Synchro ring	28	Nut	41	Setscrew
3	Tab washer	16	2nd speed gear	29	Tab washer	42	Fibre washer
4	Reverse gear	17	3rd speed gear	30	Plug	43	Tab washer
5	1st speed gear	18	Needle roller	31	Constant pinion shaft	44	Reverse slipper
6	Bearing sleeve	19	Spacer	32	Roller bearing	45	Split pin
7	Needle roller	20	Spacer	33	Spacing	46	Countershaft
8	Spacer	21	Synchro hub	34	Oil thrower	47	Key
9	Synchro hub	22	Operating sleeve	35	Nut	48	Gear unit (cluster)
10	Operating sleeve	23	Thrust member	36	Tab washer	49	Needle roller
11	Thrust member	24	Plunger	37	Reverse spindle	50	Retaining ring
12	Plunger	25	Detent ball	38	Key	51	Thrust washer (rear)
13	Detent ball	26	Spring	39	Reverse idler gear	52	Thrust washer (front)
						53	Thrust washer (outer)

Chapter 6/Gearbox

of the gearbox.

10 Now make up a dummy countershaft to the dimensions given in Fig.6.7.

11 Insert the dummy shaft into the countershaft bore at the front of the gearbox casing and push the layshaft out of the rear of the gearbox. Allow the dummy shaft, which is now inside the layshaft cluster, to drop into the bottom of the box and thus retain the needle roller bearings.

12 Remove the constant pinion shaft followed by the spigot roller bearings.

13 Remove the circlip which retains the mainshaft bearing and then take off the washer followed by the shims.

14 Tap the mainshaft, using a hide faced hammer, towards the front of the gearbox to remove the rear bearing (Fig. 6.8).

15 Push the reverse gear forward to clear the first speed gear on the mainshaft. Now lift the front end of the mainshaft upwards and forwards to remove it complete with all the mainshaft gears (Fig.6.9).

16 The layshaft cluster is now visible in the bottom of the casing.

17 Push the reverse gear rearwards as far as it will go in order to clear the first speed gear on the layshaft. The layshaft gear can now be lifted out. Take note of the inner and outer thrust washers fitted at each end of the gears and take care not to lose any of the needle rollers at each end of the gear unit.

18 Push the reverse gear back into position and then lift it out through the top of the case.

19 **The following paragraphs refer to procedure for dismantling the mainshaft of EB/JS gearboxes.**

20 Withdraw the top/third gear operating and synchronising sleeves by sliding them forward off the shaft.

21 Press the operating sleeve off the synchronising sleeve and collect the six synchronising balls and springs. Remove the interlock plungers and balls from the synchro sleeve.

22 Withdraw the second synchronising sleeve and the first speed gear rearwards off the shaft.

23 Press the first speed gear off the sleeve and collect the six balls and springs and now remove the interlock ball and plunger from the synchronising sleeve.

24 Refer to Fig.6.10. Press in the plunger locking the third speed gear thrust washer and then rotate the washer until the splines lin up. It can now be withdrawn.

25 Remove the third speed gear by sliding it forward off the shaft but be very careful not to lose any of the needle rollers which will emerge as the gear is removed.

26 Remove the locking plunger and spring.

27 Move to the opposite end of the shaft and remove the second speed gear in the same manner as for removal of the third speed gear (paragraphs 24 and 25).

3.3a Removing the clutch housing securing bolts

3.3b Clutch housing separated from the gearbox

3.4 Removing the top cover

3.5 The flange securing nut and split pin

3.6a Levering the flange off the mainshaft

3.6b Removal of the flange and propeller shaft bolts

3.7a The speedometer driving gear

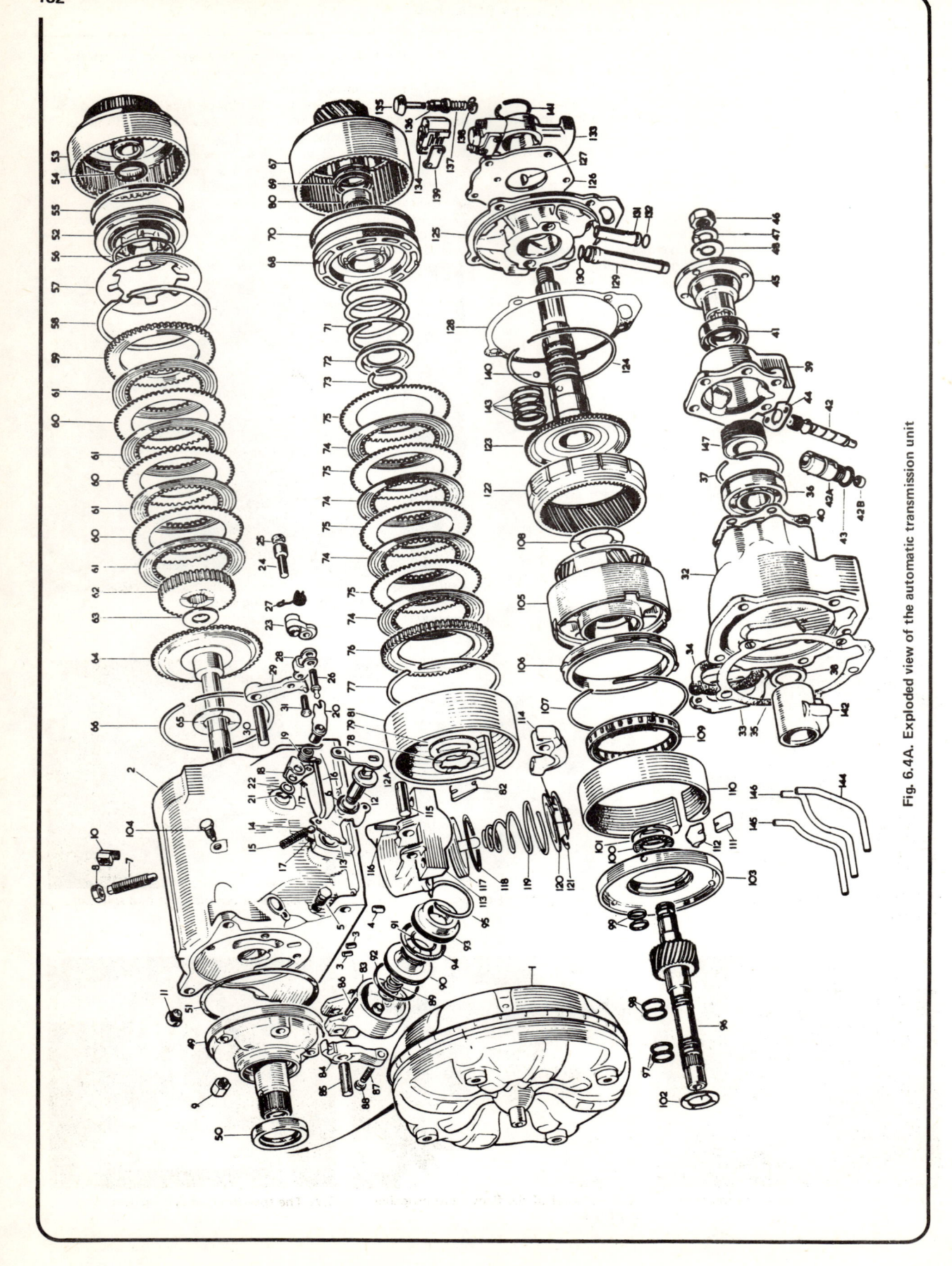

Fig. 6.4A. Exploded view of the automatic transmission unit

Fig. 6.4A. Exploded view of the automatic transmission unit

1 Converter assembly	48 Washer	98 Sealing ring
2 Transmission case assembly	49 Front pump assembly	99 Sealing ring
3 Plug	50 Oil seal assembly	100 Thrust bearing
4 Dowel	51 Sealing ring	101 Race
5 Plug	52 Piston assembly	102 Thrust washer (bronze)
6 Oil seal	53 Cylinder	103 Centre support assembly
7 Screw	54 Sealing ring (inner)	104 Screw
8 Nut	55 Sealing ring (outer)	105 Planetary gears and rear drum assembly
9 Union	56 Split ring	106 Outer race
10 Union	57 Spring	107 Snap ring
11 Breather assembly	58 Snap ring	108 Thrust washer
12 Manual control shaft assembly	59 Pressure plate	109 One way clutch assembly
12A Selector lever	60 Clutch plate (drive)	110 Brake band for rear drum
13 Lever assembly	61 Clutch plate (friction)	111 Strut (servo)
14 3/8" ball	62 Hub	112 Anchor strut
15 Spring	63 Thrust washer (fibre)	113 Body assembly
16 Link	64 Input shaft assembly	114 Lever
17 Clip	65 Thrust washer	115 Shaft
18 Torsion lever	66 Snap ring	116 Roll pin
19 Spring	67 Front drum assembly	117 Piston
20 Forked lever	68 Piston assembly	118 "O" ring
21 Clip	69 Sealing ring (inner)	119 Return spring
22 Washer	70 Sealing ring (outer)	120 Plate
23 Toggle lever	71 Spring	121 Snap ring
24 Toggle pin	72 Seat	122 Ring gear
25 Plug	73 Snap ring	123 Mainshaft assembly
26 Ball pin	74 Clutch plate (friction)	124 Snap ring
27 Spring	75 Clutch plate (drive)	125 Rear pump assembly
28 Link	76 Pressure plate	126 Plate
29 Pawl	77 Snap ring	127 Key
30 Pivot pin	78 Thrust washer (bronze)	128 Gasket
31 Pin	79 Thrust washer (steel)	129 Oil inlet tube
32 Extension case assembly	80 Needle bearing	130 "O" ring
33 Cover plate	81 Brake band	131 Oil outlet tube
34 Gasket	82 Strut (servo)	132 "O" ring
35 Gasket	83 Body	133 Governor assembly
36 Bearing	84 Lever	134 Governor body
37 Anap ring	85 Pivot pin	135 Governor weight
38 Spacing washer	86 Roll pin	136 Governor valve
39 Speedometer housing	87 Screw	137 Spring
40 Gasket	88 Nut	138 Retainer
41 Oil seal assembly	89 Return spring	139 Cover plate
42 Speedometer driven gear	90 Piston assembly	140 ¼" ball
42A Bearing	91 "O" ring (small)	141 Snap ring
42B Oil seal	92 "O" ring (large)	142 Oil collector sleeve
43 "O" ring	93 Piston sleeve	143 Piston ring
44 Plate	94 Sealing ring	144 Oil collector tube (front)
45 Flange	95 Snap ring	145 Oil collector tube (intermediate)
46 Nut	96 Forward sun gear assembly	146 Oil collector tube (rear)
47 Lockwasher	97 Sealing ring	147 Speedometer drive gear

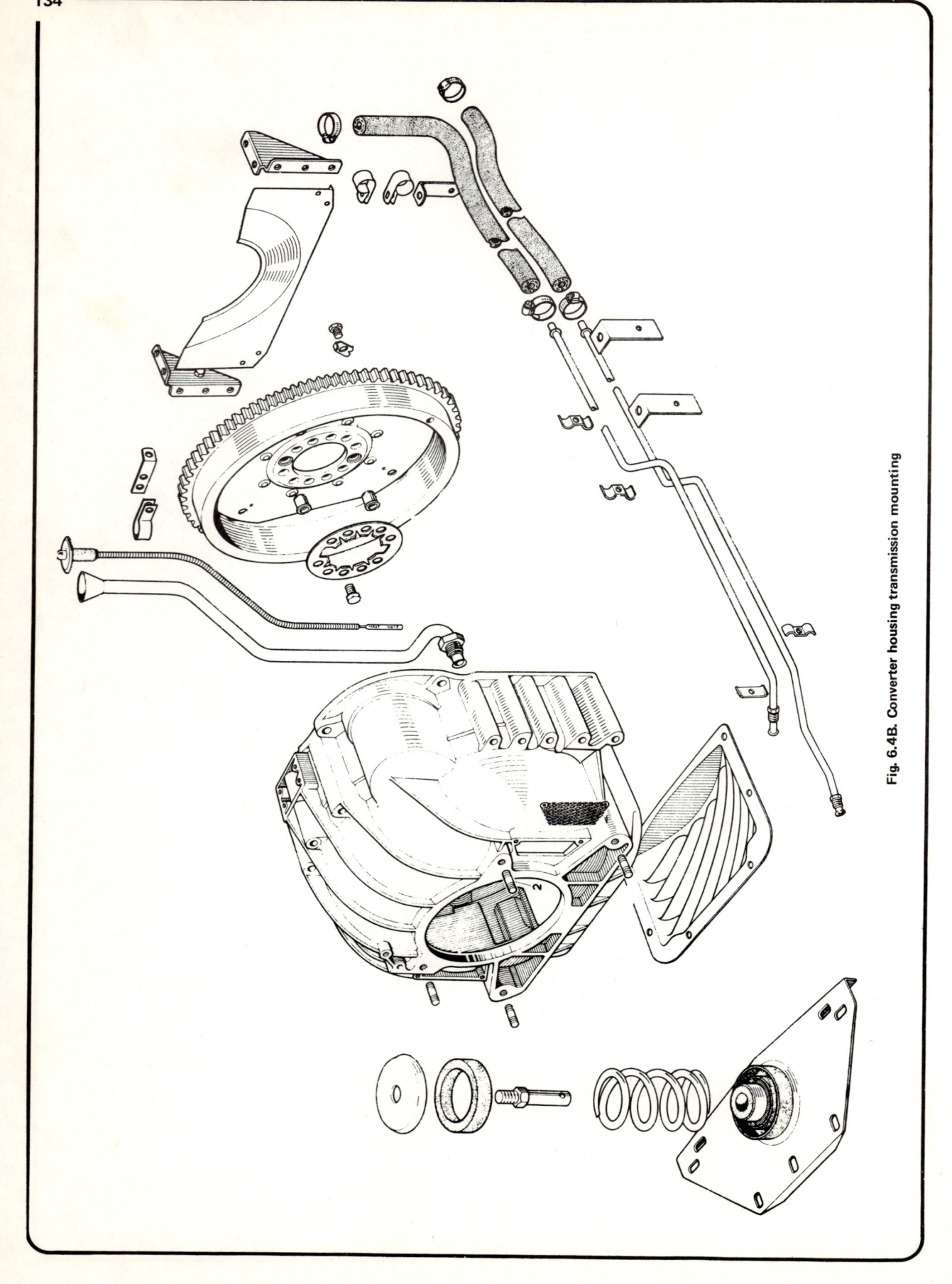

Fig. 6.4B. Converter housing transmission mounting

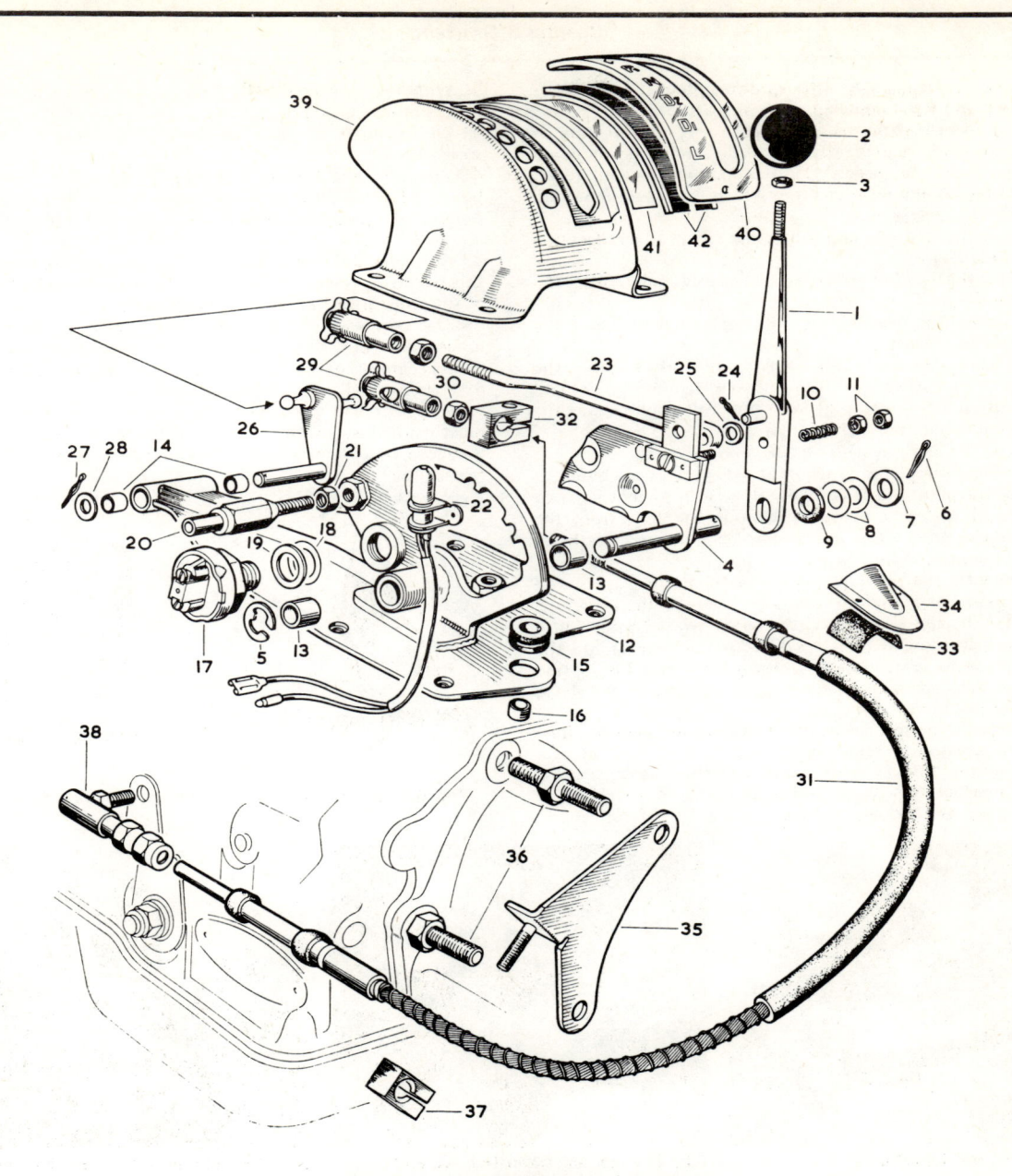

Fig. 6.5. The automatic transmission controls

1	Selector lever assembly	22	Lamp assembly
2	Knob	23	Operating rod assembly
3	Nut	24	Split pin
4	Cam plate assembly	25	Washer
5	Circlip	26	Transfer lever assembly
6	Split pin	27	Split pin
7	Washer	28	Washer
8	Shim	29	Ball joint
9	Washer (rubber)	30	Nut
10	Spring	31	Gear control cable assembly
11	Nut	32	Clamp
12	Mounting plate and selector gate assembly	33	Pad
13	Bush	34	Plate
14	Bush	35	Abutment bracket
15	Grommet	36	Stud
16	Distance tube	37	Clamp
17	Reverse lamp switch	38	Adjustable ball joint
18	Shim	39	Cover assembly
19	Shim	40	Indicator plate
20	Starter cut-out switch	41	Light filter
21	Nut	42	Seal

28 The following paragraphs refer to dismantling the mainshaft of the EJ, KE and KJS numbered series of gearboxes.
29 When dismantling the mainshaft, take very careful note of the positions from which **sets of needle roller bearings** are removed because they are graded in size and **must** be kept in sets for reassembly to the position from which they were removed.
30 Slide off the reverse gear.
31 Withdraw the first gear and collect the 120 needle rollers, the spacer and the sleeve.
32 Withdraw the 1st/2nd gear synchro assembly and collect the two synchro rings.
33 Remove the 2nd gear with its 106 needle rollers. Leave the spacer on the mainshaft.
34 Knock back the tab washer locking the nut securing the 3rd/top synchro assembly to the mainshaft. Remove the nut and withdraw the synchro assembly from the shaft at the same time collecting the two loose synchro rings.
35 Withdraw the 3rd gear with its 106 needle rollers.
36 When dismantling the cynchro assembly it is advisable to completely surround the assembly with a cloth so that none of the balls and springs are lost. Push out the hub from the operating sleeve and collect the synchro balls and springs, thrust members plungers and springs.
37 To dismantle the constant pinion shaft, of the EJ, KE and KJS gearboxes, open the tab washer and remove the large nuts followed by the tab washer and the oil thrower. Tap the shaft smartly against a metal plate to dislodge the bearing and the spacer. The same procedure is followed for the EB/JS box except that in this case double locking is provided in the shape of a locknut in addition to the tab washer.
38 To dismantle the top cover, break the locking wire and then unscrew the selector rod retaining screws.
39 Withdraw the 3rd/top selector rods with the selector, spacing tube and interlock ball. Take note of the loose interlock pin at the front of the 1st/2nd selector rod.
40 Withdraw the reverse selector rod and collect the reverse fork, stop spring and detent plunger.
41 Withdraw the 1st/2nd selector rod with its fork and short spacer tube.
42 The oil pump (Fig.6.11) of the EJ, KE and KJS series gearboxes is housed on the inside face of the rear end cover and the body of the pump is secured to the cover by three countersunk screws (photo) which are locked by staking.
43 Normally there is no necessity to remove the pump but, if this has to be done, the first task is to clear the staked metal out of the screwdriver slot as far as is possible.
44 The best way of removing the screws is to use an impact screwdriver but, failing this, use a screwdriver with some auxiliary means of applying increased torque (photo).
45 Remove the pump body (photo) complete with inner and outer gears. Before the gears are removed from the body, you must mark them in some way, marking ink is suggested, so that they can be correctly mated and fitted the original way up on reassembly.

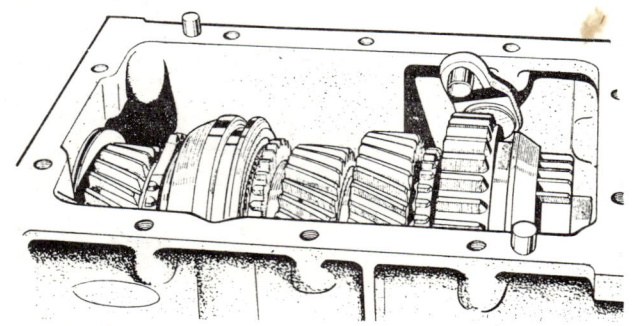

Fig. 6.6. Top cover removed — showing layout of mainshaft gears

3.7b Rear oil seal housing

3.8a The rear end cover (EJ, KE and KJS series)

3.8b The oil pump drive (EJ, KE and KJS series)

3.42 The oil pump body

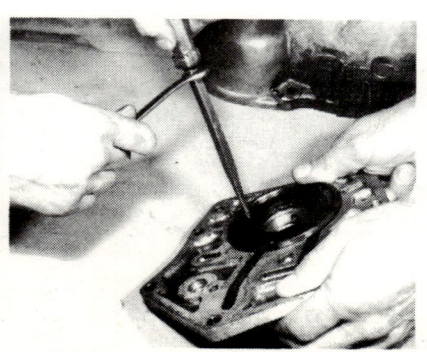

3.44 Removing the oil pump body securing screws

3.45 The oil pump removed

Chapter 6/Gearbox

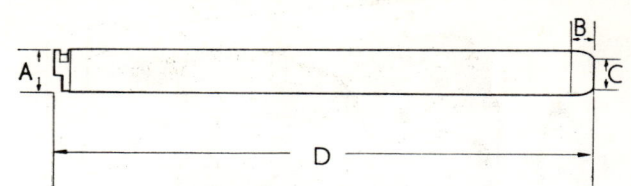

Fig. 6.7. The dummy countershaft

A .979 in. (24.86 mm)
B .5 in. (12.7 mm)
C .75 in. (19.05 mm)
D 11.125 in. (28.25 cm)

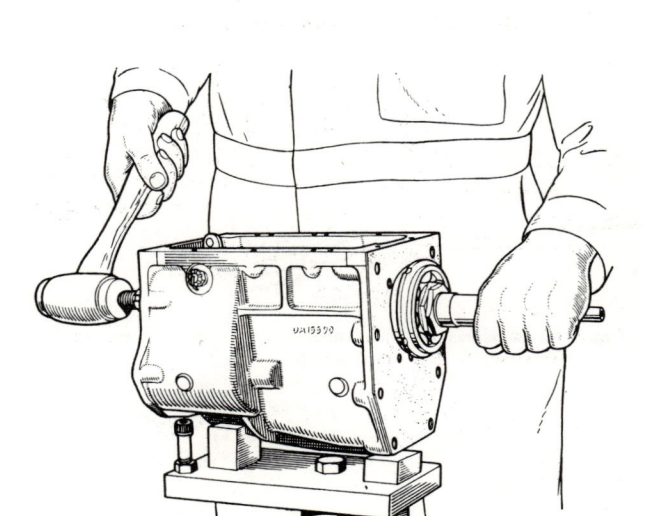

Fig. 6.8. Removing the constant pinion shaft and rear bearing

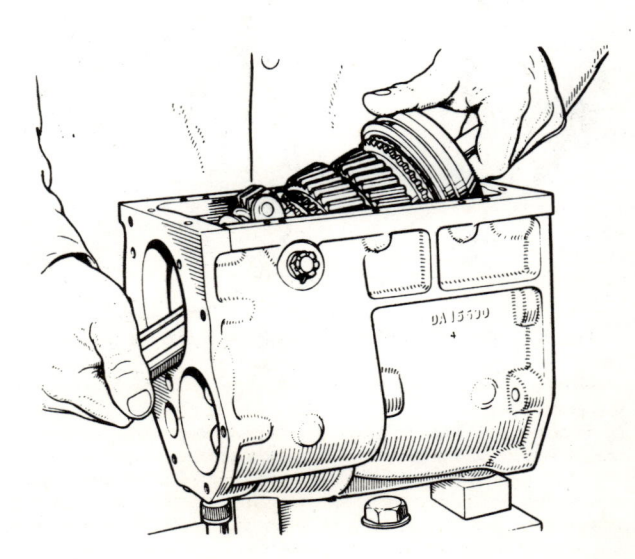

Fig. 6.9. Removing the mainshaft

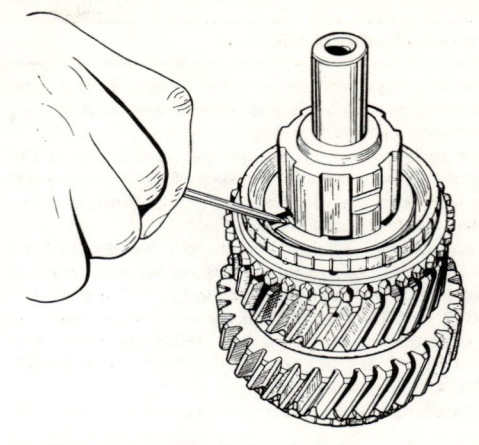

Fig. 6.10. Depressing the 3rd speed thrust washer locking plunger

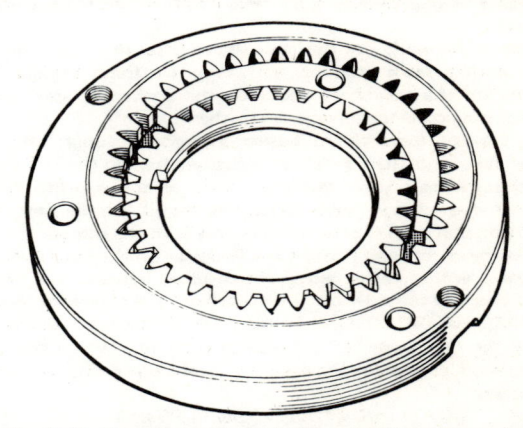

Fig. 6.11. The oil pump — EJ, KE and KJS series gearboxes

4 Gearbox - examination

1 It is assumed that the gearbox has been dismantled because of some malfunction, possibly excessive noise, ineffective synchromesh or failure to stay in a selected gear. The cause of most gearbox faults is failure of the needle rollers on the input or the mainshaft and wear on the synchro rings. These tiems can be replaced but there is always the possibility of some obscure fault remaining even after a visually unserviceable component has been renewed which means that all your work has gone for nothing as the fault, if any, will not be discovered until the gearbox has been re-installed in the car. It is worthwhile, therefore, if faults are found, to enquire about the availability of parts and their cost and it may still be worth considering, even at this stage, fitting an exchange gearbox.
2 Examine the teeth of all gears for signs of uneven or excessive wear and, of course, chipping. If a gear on the mainshaft is in doubtful condition, check that the corresponding gear on the layshaft is not equally damaged.
3 All gears should be a good running fit in the shaft with no signs of rock and the hubs should not be a sloppy fit on the splines.
4 Examine the selector forks for signs of wear or ridging on the faces which contact the operating sleeve.
5 Look for wear on the selector rods.
6 It is difficult to decide on the degree of wear on roller bearings but we advise taking no chances. Considering the work entailed in removing and dismantling the gearbox, it would be very short sighted not to replace all bearings as a matter of course and the same applies to all oil seals and synchroniser rings.

5 Gearbox - reassembly of type EB/JS

1 Liberally coat the layshaft needle roller retaining rings with vaseline and then fit one to each end of the layshaft gear unit followed, at each end, by the 29 rollers.
2 Fit the outer roller retaining ring at the front end and the inner and outer thrust washers at either end of the gear unit and then lower the gears into the case and insert a dummy countershaft to locate the layshaft gears in place.
3 The end float of the layshaft must now be checked by measuring the clearance between the thrust washer and the casing at the rear of the shaft as shown in Fig.6.12. The end float should be 0.002" - 0.004" (0.05 - 0.10 mm). Adjustment is effected by an exchange of thrust washers which are available in thicknesses of 0.152", 0.156", 0.159", 0.162" and 0.164" (3.86, 3.96, 4.04, 4.11 and 4.17 mm).
4 Remove the dummy countershaft and insert a thin rod in its place.
5 Place the reverse gear in position and draw it rearwards as far as possible to give clearance for final positioning of the layshaft gear unit.
6 Now start work on the mainshaft. Liberally coat the 41 needle rollers with vaseline and then fit them behind the shoulder on the mainshaft and slide the 2nd speed gear, with synchronising cone to the rear, on to them.
7 Fit the 2nd speed thrust washer spring and plunger into the plunger hole and then slide the thrust washer up the shaft and over the splines. Align the large hole in the synchro cone, compress the plunger and rotate the thrust washer into the locked position with the cut-away in line with the plunger.
8 Now check the end float of the 2nd gear on the mainshaft by measuring, with a feeler gauge, the distance between the thrust washer and the shoulder on the mainshaft. The clearance should be 0.002" - 0.004" (0.05 -0.10 mm) and if this is not achieved, remove the thrust washer and replace it by one which will give the required clearance. Washers are available in the following thicknesses:

    0.471 in/0.472 in (11.96/11.99 mm)
    0.473 in/0.474 in (12.01/12.03 mm)
    0.475 in/0.476 in (12.06/12.09 mm)

9 The above work in respect of the 2nd speed gear is now repeated to fit the 3rd gear on the opposite side of the shoulder on the mainshaft. End float is checked in the same manner as for the 2nd speed gear and the same range of adjusting shims is available, the holes through which the thrust washer locking plungers are depressed are shown in Fig.6.13.
10 To assemble the 2nd gear synchro assembly, first fit the springs and balls and shims, if fitted to the six blind holes in the synchro sleeve.
11 Refer to Fig.6.14 and fit the 1st speed gear to the 2nd speed synchronising sleeve so that the relieved tooth of the internal splines in the gear are in line with the stop pin in the sleeve.
12 It may be helpful to compress the springs using a jubilee clip and then slide the operating sleeve over the synchronising sleeve until the balls can be heard and felt to engage the neutral position groove.
13 It should require 62 to 68 lbs (28 to 31 kg) pressure to disengage the synchronising sleeve from the neutral position in the operating sleeve. This can be judged, if special equipment is not available, by gripping the operating sleeve in the palms of the hands and then pressing the synchronising sleeve with the fingers until it disengages from the neutral position. It should require firm finger pressure before disengaging and, if necessary, shims can be fitted, or removed from, underneath the springs to adjust the pressure of the balls against the operating sleeve.
14 Now fit the 1st speed gear/2nd speed synchro assembly to the mainshaft using any spline position and check that the synchro sleeve moves freely on the mainshaft when the ball and plunger is not fitted. If there is any restriction, try the sleeve on different splines on the shaft and if there is no improvement, check for burrs on the splines and rectify as required.
15 Take the synchro assembly off the mainshaft, fit the ball and plunger and then refit it to the shaft in the position from which it was removed.
16 Support the shaft in a vice (with protected jaws) and check the interlock plunger by sliding the outer operating sleeve into the 1st gear position as shown in Fig.6.15. Apply slight downward pressure on the synchro assembly and at the same time rotate the 2nd speed gear. It should rotate freely without any tendency for the synchro cones to rub, but if restriction is felt, a longer plunger should be fitted to the synchro sleeve. Plungers are available in the following lengths:-

    0.490 inch, 0.495 inch and 0.500 inch (12.4,12.52 and 12.65 mm).

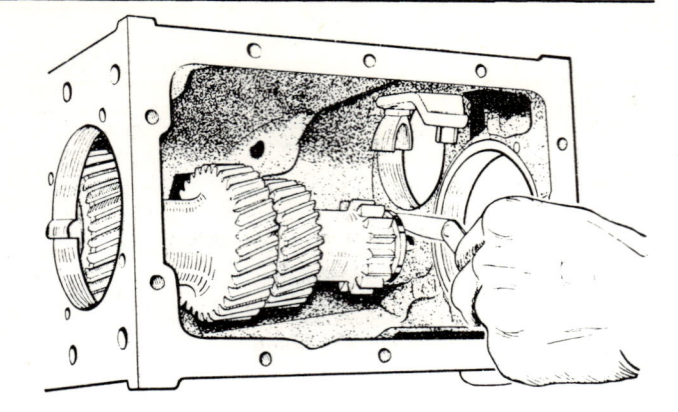

Fig. 6.12. Checking layshaft end float

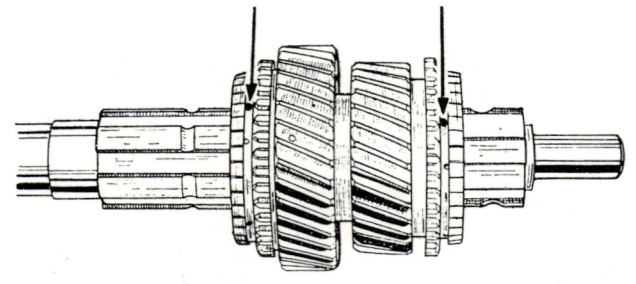

Fig. 6.13. Showing the holes through which the thrust washer locking plungers are depressed

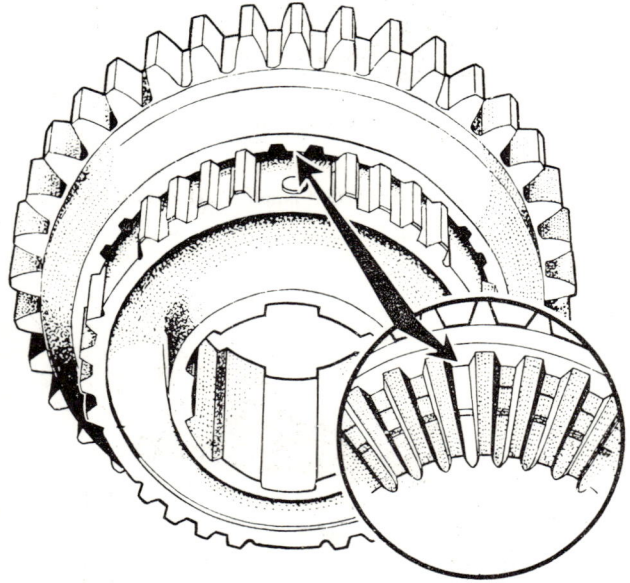

Fig. 6.14. Alignment of the relieved tooth and the stop pin

Chapter 6/Gearbox

17 The 3rd/top synchro assembly is put together, and is tested for operation, in the same manner as the 2nd gear synchro assembly but make sure that the wide chamfer end of the operating sleeve faces the large boss end of the inner synchronising sleeve as shown in Fig.6.16 and that the two relieved teeth in the operating sleeve are in line with the ball and plunger holes as illustrated in Fig.6.17.

18 The 3rd/top synchro assembly is now ready for fitting to the mainshaft.

19 Note the following points when fitting the assembly to the mainshaft:-

a) There are two tranverse grooves on the mainshaft splines and the relieved tooth on the wide chamfer end of the outer operating sleeve **must** be in line with the foremost groove in the mainshaft as shown in Fig.6.18. Incorrect alignment will result in the locking plungers engaging the wrong grooves and so prevent full engagement of top and 3rd gear.

b) The wide chamfer end of the outer operating sleeve must face forwards, that is , towards the constant pinion shaft end of the gearbox.

c) The inner sleeve must slide freely on the mainshaft when the balls and plungers are not fitted. If there is any restriction, check the splines for burns and rectify as neccessary.

20 Fit the two balls and plungers to the holes in the inner synchro sleeve and then fit the assembly to the mainshaft in the manner indicated in the preceding paragraph.

21 Support the mainshaft in a vice (with protected jaws) and check the operation of the interlock plungers by sliding the 3rd/top operating sleeve over the 3rd speed gear dogs as shown in Fig.6.19. With 3rd gear engaged, lift and lower the synchro assembly; it should be possible to move it about 3/32" (2.5 mm) without any drag being felt. If it does not move freely, a shorter 3rd speed plunger should be fitted, this is the plunger which, when looking at the wide chamfer end of the outer operating sleeve, is not opposite the relieved tooth in the operating sleeve. Plungers are available in the following lengths:

    0.490 inch, 0.495 inch and 0.500 (12.4, 12.52, and 12.65 mm).

22 Now slide the operating sleeve into the top gear position as shown in Fig. 6.20 and again lift and lower the synchro assembly; it should be possible to move the assembly about 3/16" (4.5 mm) without any drag being felt and also, with slight downward pressure on the assembly, the 3rd speed gear should be free to rotate without any tendency for the cones to rub.

23 Fit a shorter top gear plunger if the synchro assembly does not move freely when lifted and lowered. A longer top gear plunger should be fitted if the 3rd gear synchro cones are felt to rub, this plunger is the one in line with the relieved tooth in the operating sleeve looking from the wide chamfer end of the outer operating sleeve. Plungers are available in the lengths quoted in paragraph 21.

24 Now for assembly of the constant pinion shaft. Fit the oil thrower followed by the ballrace on to the shaft with the circlip and collar fitted to the outer track of the bearing. Screw on the nut and fit the tab washer and locknut. Finally fit the roller race into the shaft spigot bore.

25 The gears are now ready for assembly to the casing in which the layshaft cluster and the reverse gear have already been positioned. Enter the mainshaft through the top of the casing and move it rearwards through the bearing hole in the case.

26 Fit a new gasket to the front face of the casing and insert the constant pinion shaft at the front of the case with the cutaway portions of the toothed driving member facing the top and bottom of the casing. Tap the shaft to the rear until the collar and circlip on the bearing butt against the casing.

27 Hold the constant pinion shaft in position and tap in the rear bearing complete with its circlip.

28 Lift the layshaft cluster into mesh using the thin rod which should still be in position and then insert the dummy countershaft through the bore in the front face of the casing as illustrated in Fig.6.21.

29 Engage top and first gears. Fit the Woodruff key and the speedometer drive gear to the mainshaft followed by the tab

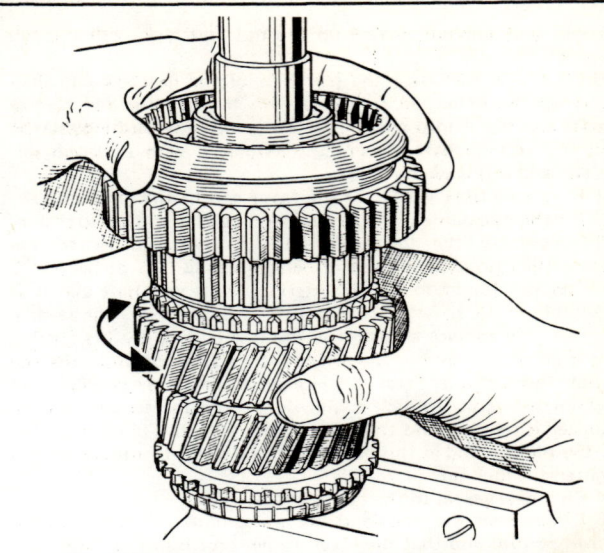

Fig. 6.15. Checking the 2nd speed gear for freedom

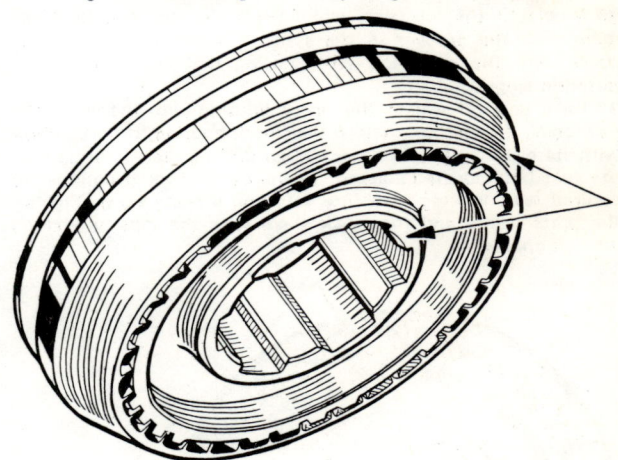

Fig. 6.16. Assembly of the operating sleeve to the inner synchronising sleeve

Fig. 6.17. The relieved tooth must be in line with the ball and plunger holes

washer and locknut. Screw up the nut and lock with the tab washer.

30 Fit a new gasket to the rear face of the gearbox and then offer up the extension complete with the counter and reverse shafts and tap it into position at the same time pushing out the dummy countershaft. Secure the extension with its seven setscrews and spring washers.

31 Fit a new fibre washer at the front end of the countershaft.

32 Fit the speedometer driven gear and bearing to the extension.

33 Reassemble the top cover in the reverse order to the dismantling procedure given in paragraphs 38 - 41 of Section 3 but do not forget to fit the interlock balls and pins and it is advisable to fit new 'O' rings on the selector rods. The reverse plunger will require adjustment and this is done by first fitting the plunger and spring. Now fit the abll and spring and enter the screw and locknut. Press the plunger in as far as possible and tighten the screw to lock it. Slowly slacken the screw until the plunger is released and the ball engages with the circular groove in the plunger and at this point, hold the screw from turning and tighten the locknut.

34 Fit a new gasket to the top of the gearbox.

35 Ensure that the gears and the gear selctors of the top cover are in neutral and that the reverse idler gear is out of mesh with the reverse gear.

36 Now refit the top cover making sure that the selector forks mate with the grooves in the synchro assemblies. Secure the cover with the nuts and bolts noting that the bolts are of different lengths.

37 Fit a new oil seal to the clutch housing with the lip of the seal facing the gearbox. Attach the clutch housing to the gearbox with its eight bolts and three tabbed locking plates noting that the two bolts located adjacent to the clutch fork trunnions are secured with locking wire (use soft iron locking wire). Tighten the bolts evenly and then lock them with the tabs or locking wire, as applicable.

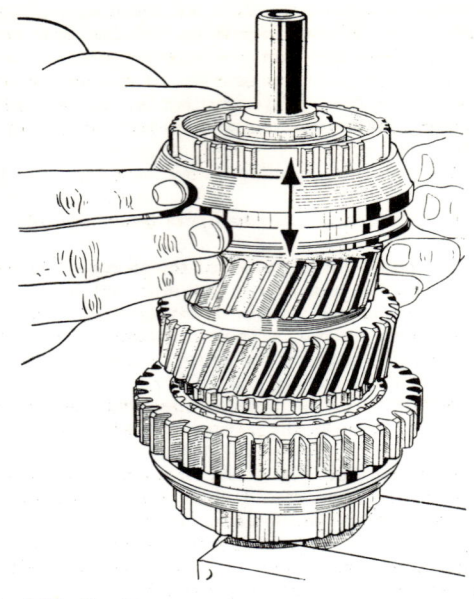

Fig. 6.19. Checking operation of the interlock plungers

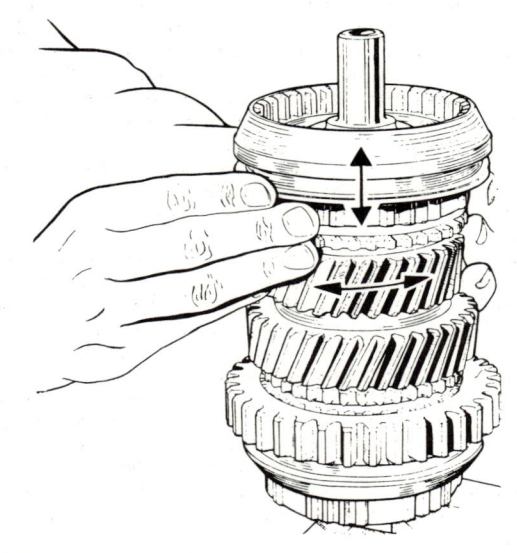

Fig. 6.20. Checking 4th (top gear) interlock plunger and assembly

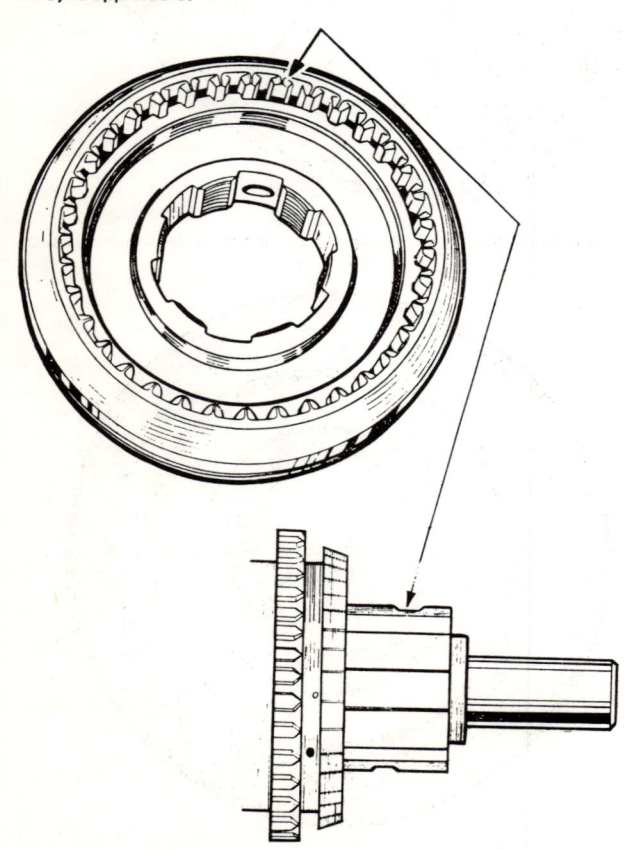

Fig. 6.18. Location of the operating sleeve on the mainshaft

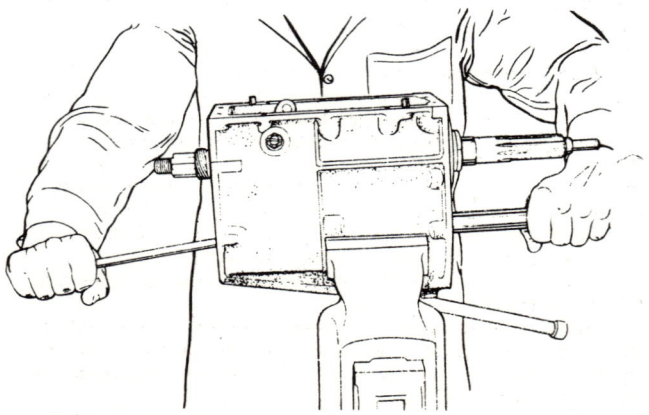

Fig. 6.21. Fitting the dummy countershaft

6 Gearbox - reassembly of EJ, KE and KJS numbered series

1 Start with the synchro assemblies for which the assembly procedure for 1st/2nd and 3rd/top is the same but note that although the 3rd/Top and 1st/2nd synchro hubs are similar in appearance they are not identical. To distinguish them, a groove is cut on the edge of the 3rd/Top hub as illustrated in Fig.6.22.
2 Assemble the synchro hub to the operating sleeve so that the wide boss of the hub is on the opposite side to the wide chamfer end of the sleeve as depicted in Fig. 6.23 and so that the three balls and springs will be in line with the teeth having three detent grooves (Figs.6.24 and 6.25).
3 Pack up the synchro hub so that the holes for the ball and springs are exactly level with the top of the operating sleeve as shown in Fig.6.26.
4 Fit the three springs, plungers and thrust members to their correct positions and press down the thrust members as far as possible. Fit the three springs and balls to the remaining holes. It may help to keep the plungers and balls in position if the springs are liberally coated with vaseline.
5 Compress the springs with a large jubilee clip, a piston ring clamp is ideal, as shown in Fig.6.27. Depress the hub slightly and then push down the thrust members with a screwdriver until they engage the neutral groove in the operating sleeve.
6 Tap the hub down evenly and carefully until the balls can be heard, and felt, to engage the neutral groove.
7 Now start assembling the layshaft cluster gear by fitting one retaining ring in the front end of the cluster.
8 Liberally coat the 29 needle rollers with vaseline and place them in the front position of the cluster followed by the front inner thrust washer but make sure that the peg on the washer locates in the groove machined in the front face of the cluster gear.
9 Fit the retaining ring, the 29 needle rollers and the second retaining ring to the rear of the cluster.
10 Fit the slipper to the reverse idler lever and secure it with a new split pin. Assemble the lever to the casing and secure it in position with its setscrew and lock with the tab washer.
11 Liberally coat the rear thrust washer with vaseline and then place it on its boss in the casing making sure that the peg locates correctly.
12 Coat the front outer thrust washer with vaseline and place it in position on the cluster and now lower the layshaft cluster carefully into position.
13 Insert a dummy countershaft into the bore in the casing and through the cluster gear. The next task is to check the end float of the cluster by measuring the clearance between the rear thrust washer and the cluster. (Fig.6.28).
14 The end float should be 0.004" - 0.006" (0.10 - 0.15 mm). Adjustment is made by replacement of the outer front thrust washer which is available in the following thicknesses:-
    0.152 in., 0.156 in., 0.162 in., and 0.164 in., (3.86, 3.96, 4.04, 4.11 and 4.17 mm.).
15 The constant pinion shaft assembly should now be put together, fitting a new oil thrower, tab washer and locknut. Tighten down on the nut and secure with the tab. Fit the spacer to the other end of the assembly followed by the roller bearing.
16 The mainshaft is reassembled in the reverse order to the dismantling procedure given in Section 3. You may find it helpful to fit a jubilee clip to the shaft in order to prevent the reverse gear sliding off when assembling the shaft to the casing. Do make sure that the correct set of needle rollers are fitted to their individual gears; they are graded on diameter and rollers of one grade only must be used for an individual gear.
17 The end float of the gears must be checked and details of the permissible clearance will be found under 'Specifications' at the beginning of this Chapter. If the end float is found to be excessive it can only be rectified by the fitment of new parts.
18 From this point onwards follow the instructions given in Section 5 paragraphs 25 - 37 to complete assembly of the gearbox. However, in this case the oil pump (Fig.6.11) must be refitted to the rear extension. Refit the gears to the pump in accordance with your marks made when dismantling. Coat the gears and the pump body with oil and then secure the assembly to the extension with the three setscrews and lock them by staking using a centre punch. (photo).
19 After refitting the engine and the gearbox to the car, run the car in top gear as soon as possible in order to attain the necessary mainshaft speed to prime the oil pump.

6.18 Locking the setscrews securing the oil pump body

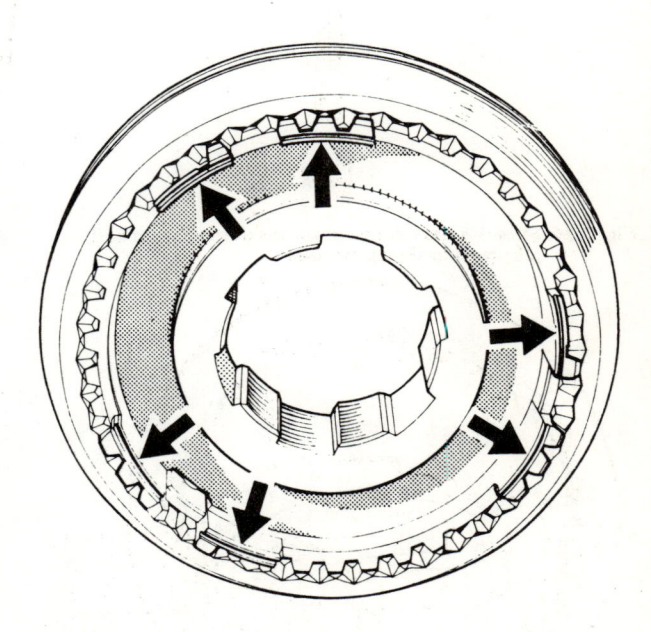

Fig. 6.22. Identification grooves, 3rd/top synchro assembly

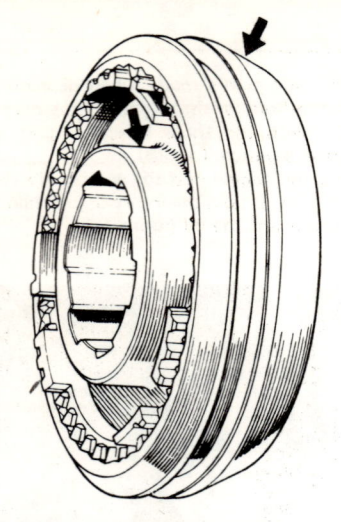

Fig. 6.23. Assembly of the synchro hub and operating sleeve

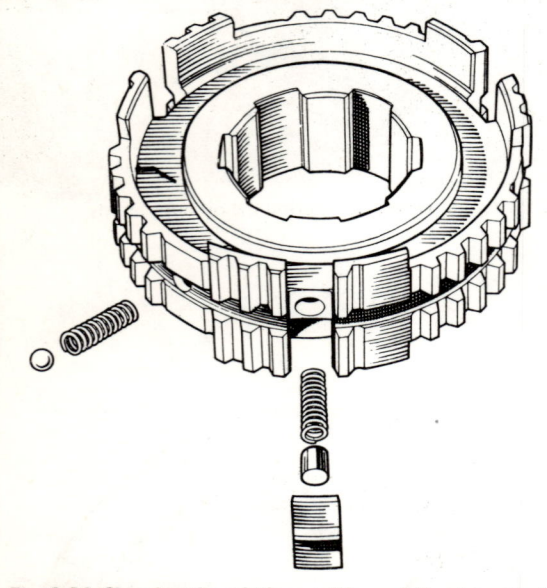

Fig. 6.24. Showing the relative positions of the detent ball, plunger and thrust member

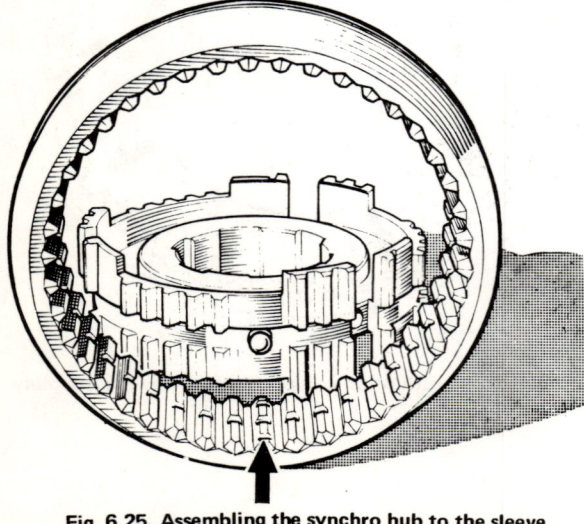

Fig. 6.25. Assembling the synchro hub to the sleeve

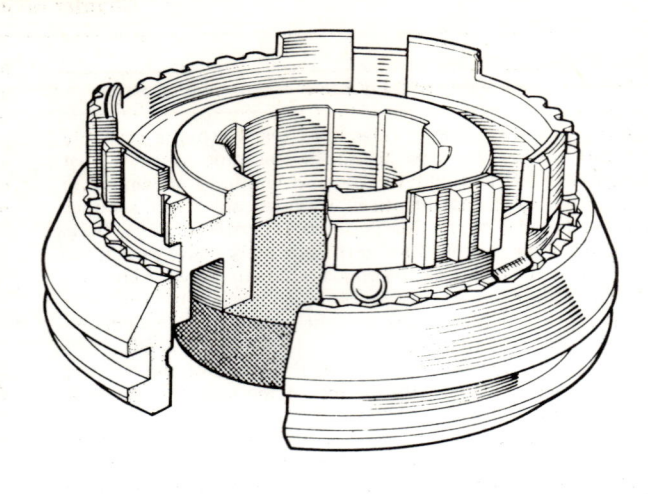

Fig. 6.26. Fitting the springs, plungers and thrust members

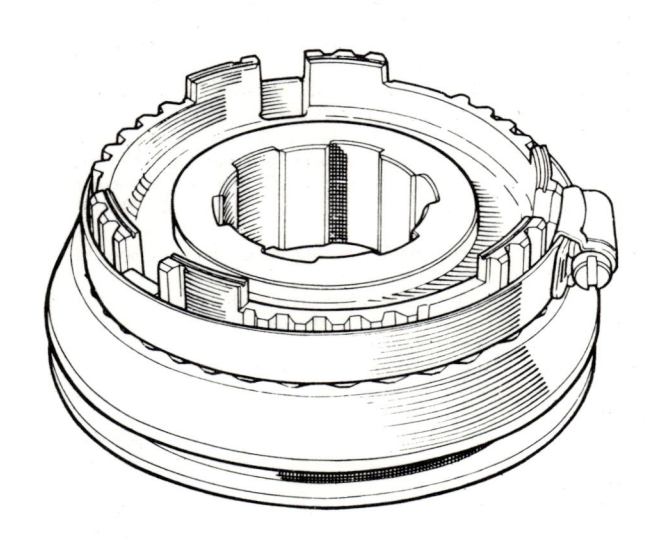

Fig. 6.27. Compressing the springs

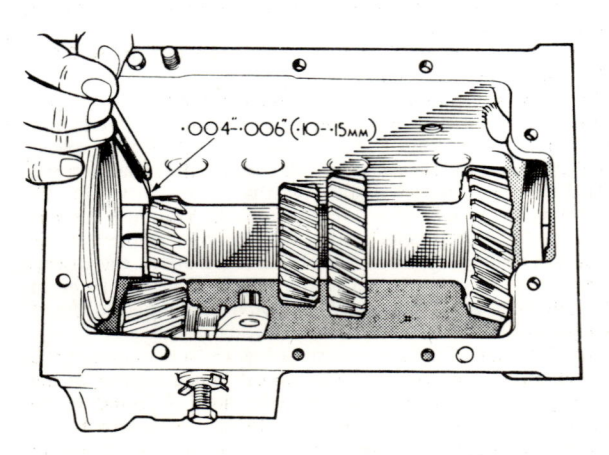

Fig. 6.28. Checking layshaft end float at the rear thrust washer

Chapter 6/Gearbox

Fault diagnosis - manual gearbox

Symptom	Reason/s	Remedy
Weak or ineffective synchromesh	Synchronising cones worn, split or damaged	Dismantle and overhaul gearbox. Fit new gear wheels and synchronising cones.
	Synchromesh dogs worn, or damaged	Dismantle and overhaul gearbox. Fit new synchromesh unit.
Jumps out of gear	Broken gearchange fork rod spring	Dismantle and replace spring.
	Gearbox coupling dogs badly worn	Dismantle gearbox. Fit new coupling dogs.
	Selector fork rod groove badly worn	Fit new selector fork rod.
Excessive noise	Incorrect grade of oil in gearbox or oil level too low	Drain, refill, or top up gearbox with correct grade of oil.
	Bush or needle roller bearings worn or damaged	Dismantle and overhaul gearbox. Renew bearings.
	Gearteeth excessively worn or damaged	Dismantle and overhaul gearbox. Renew gear wheels.
	Laygear thrust washers worn allowing excessive end play	Dismantle and overhaul gearbox. Renew thrust washers.
Excessive difficulty in engaging gear	Clutch pedal adjustment incorrect	Adjust clutch pedal correctly.

7 Automatic transmission - general

Due to the complexity of the automatic transmission unit, if the performance is not up to standard or if a fault develops, it is imperative that rectification work is undertaken by a Jaguar dealer who will have the necessary special equipment and 'know-how' for fault diagnosis and rectification.

The contents of the following Section is, therefore, confined solely to general and servicing information.

8 Automatic transmission - fluid level

It is important that only the correct and recommended type of transmission fluid is used. The specification required is Type A; Castrol TQF being a typical example.

The fluid level should be checked at regular intervals, and in the manner, as described in Routine maintenance at the beginning of this Manual. A combined filler/dipstick tube will be found under the bonnet attached to the rearmost exhaust manifold studs. The capacity of the transmission unit is 16 Imp. pints (19 US pints, 9 litres) and care should be taken not to overfill the unit. If it is found necessary to add fluid frequently, it is an indication that there is leakage in the transmission and an immediate check should be made to prevent damage to the transmission.

9 Automatic transmission unit - removal and refitting

Any suspected fault should be referred to a Jaguar dealer before any attempt is made to remove the engine and transmission unit as it is necessary to diagnose and confirm the fault before the unit is disturbed (See Section 18).

1 To remove the transmission unit, it is necessary to withdraw the engine and transmission as a complete unit from the car. Instructions for this work will be found in Chapter 1.
2 Drain the oil from the transmission unit.
3 Disconnect the kickdown linkage at the operating shaft.
4 Remove the bolts securing the transmission to the converter housing and withdraw the unit.
5 Remove the starter if it is still in position.
6 Withdraw the cover from the front of the converter hosuing and take out the setscrews securing the converter housing to the engine.
7 Remove the four setscrews which secure the torque converter to the flywheel, these are accessible through the starter motor aperture. You will have to rotate the engine to gain access to each setscrew in turn.
8 The flywheel can now be removed by removing the setscrews and locking plate and withdrawing the flywheel from the crankshaft.
9 Refitting is the reverse of the above sequence but it will be necessary to adjust the 'kickdown' cable as described in Section 10.

10 Throttle/kickdown cable adjustment

The shift quality and the correct shift position are controlled by precise movement of the throttle/kickdown cable inrelation to the carburettor throttle shaft movement. The importance, therefore, of correct cable adjustment cannot be over-emphasised.

1 Test the car on a flat stretch of road.
2 Select D1 or D2 and with minimum throttle opening; the 2-3 upshift should occur at 1100 - 1200 rpm. A 'run-up' of 200 - 400 rpm at the change point indicates a **low** operating pressure.
3 Check at full throttle opening: a jerky 2 - 3 upshift or a sharp 2 - 1 downshift when stopping the car in D1 is an indication of **high** operating pressure.
4 Now, in the workshop, install a 0 - 200 lb/sq. in (0 - 14 kg/sq.cm) pressure gauge in the line pressure point at the left hand rear face of the transmission unit. (Fig.6.29).
5 Apply the handbrake firmly and chock the wheels. Start the engine and allow it to reach the normal operating temperature. Now select D1 or D2 and increase the idling speed to **exactly** 1250 rpm when the pressure gauge reading should be 72.5 $\pm$ 2.5 lb/sq in (5.097 $\pm$ 0.175 kg/sq cm.)
6 If the tests show that the setting of the cable is incorrect adjustment is made at the fork end as illustrated in Fig. 6.30.
7 Slacken the fork end locknut and then remove the split pin and the clevis pin.
8 To **lower** the pressure, turn the fork end anti-clockwise and to **raise** the pressure, turn it clockwise. You will find that one full turn will alter the setting by 9 ln/sq.in (0.63 kg/sq.cm).
9 Slight adjustment only should be necessary; excessive adjustment will result in loss of 'kickdown' or an increase in shift speeds.
10 When you are satisfied with the adjustment, reconnect the fork end and tighten the locknut.
11 Restart the engine and check the pressure as described in paragraph 5.
12 Finally, make sure that the butterfly valves of the carburettors are closed at idling speed.
13 If, after repeated attempts to stabilise the change points, the pressure still fluctuates, the throttle/kickdown inner cable may be binding in which case there is no alternative other than to replace it.

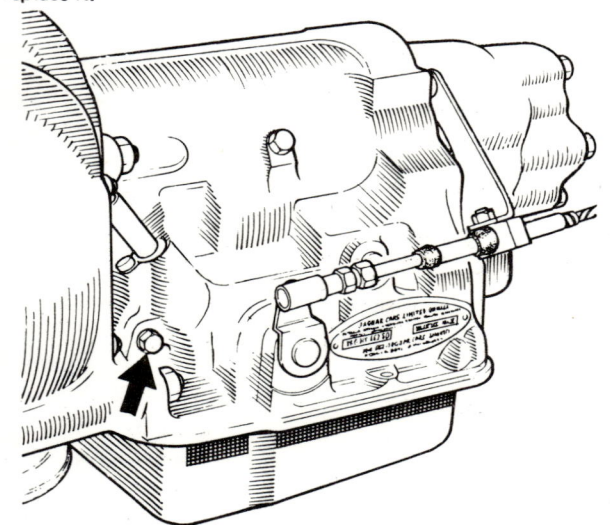

Fig. 6.29. The transmission pressure take-off point

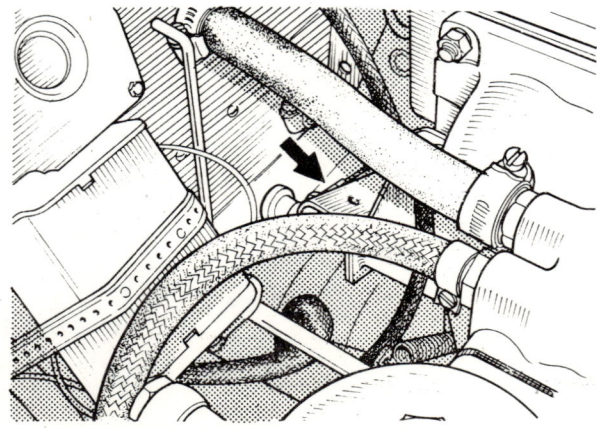

Fig. 6.30. "Kickdown" cable adjustment

Chapter 6/Gearbox

11 Throttle/kickdown cable - removal and refitting

1 Disconnect the cable at the fork end.
2 Undo the setscrew and remove the cable retaining clip.
3 Lift the carpet and underfelt to expose the left hand side of the gearbox tunnel, remove the six screws and the aperture cover plate which will now be exposed.
4 Remove the Allen headed screw, and washer, which retains the outer cable.
5 Withdraw the outer cable and, using a small screwdriver, spring open the clip which secures the inner cable to the control rod operating the kickdown cam in the transmission unit. Withdraw the inner cable.
6 Refitting is the reverse of the above.
7 Now adjust the length of the operating cable to 3.5/16" (84.1mm) between the centre line of the clevis pin and the end of the outer cable.
8 Check that the carburettor butterfly valves are closed and now carry out the adjustments described in Section 10.

12 Manual linkage - adjustment

1 Refer to Chapter 1 and remove the transmission tunnel finisher assembly.
2 Remove the carpet and prize up the sound proofing at the side of the transmission cover.
3 Remove the setscrews and take off the cover plate on the left-hand side of the transmission cover.
4 Loosen the linkage cable locknut and remove the cable from the transmission lever.
5 Refer to Fig. 6.31. Push the transmission lever fully forward to the Lockup detent and then place the selector lever in the Lockup position.
6 Adjust the cable end to fit freely on the transmission lever and then temporarily re-attach it to the lever.
7 Move the selector lever through the various positions and make sure that the gating at L, D1, R and P does not interfere with the transmission lever setting at the detent positions. The transmisssion lever **must** engage the transmission detents positively.
8 When satisfied with the adjustment, secure the cable to the transmission lever and tighten the locknut.
9 Replace the cover plate on the transmission cover, the sound proofing and carpets and finally replace the transmission tunnel finisher assembly.

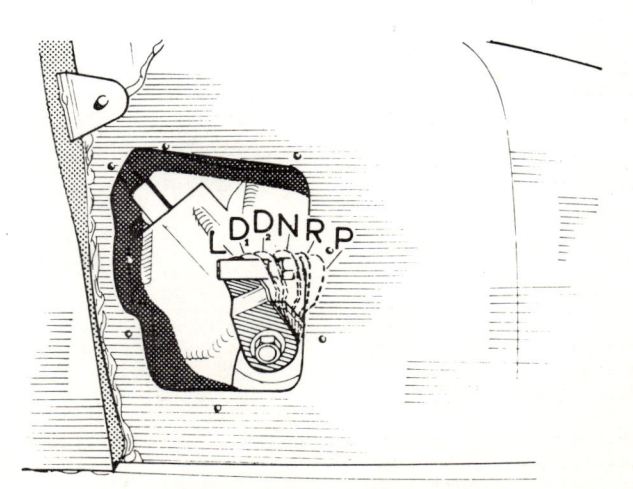

Fig. 6.31. Manual selector linkage adjustment

13 Oil pan - removal and refitting

1 Place the car over a pit or on a ramp or raise the car to give access to the drain plug. If the car has to be raised, make sure that it is safely supported before working underneath it.
2 Remove the drain plug, drain the oil into a container and discard it.
3 Remove the 14 screws securing the pan and withdraw it.
4 A few wear particles in the bottom of the pan are normal but if these, whether ferrous or non-ferrous, appear to be excessive or if pieces of band lining material are found the indication is that further checking is called for.
5 Use a new gasket when refitting the pan and tighten the screws to a torque of 10 - 15 lb.f.ft (1.38 - 2.07 kg.f.m).
6 Finally, refill the transmission unit. Instructions for this work, and of the type of fluid to use, will be found in Section 8.

14 Front band - adjustment

1 The front band should be adjusted after the first 1000 miles (1600 km) of use and at 21,000 miles (35,000 km) intervals thereafter.
2 Refer to Section 13 and follow the instruction for removal of the oil pan.
3 Slacken the locknut of the adjusting screw on the servo (Fig.6.32) apply the lever and ensure that the screw turns freely in the lever.
4 Make up a ¼" (6.44 mm) thick gauge block and then place it between the servo piston pin and the servo adjusting screw.
5 Tighten the adjusting screw with a suitable torque wrench or adjusting tool to a torque of 10 lbs.f. ins (0.12 kg.f.m) and then tighten the locknut to a torque of 20 - 25 lbs. f.ft (2.76 - 3.46 kg.f.m).
6 Remove the gauge block, replace the oil pan (Section 13) and refill the transmission unit (Section 8).

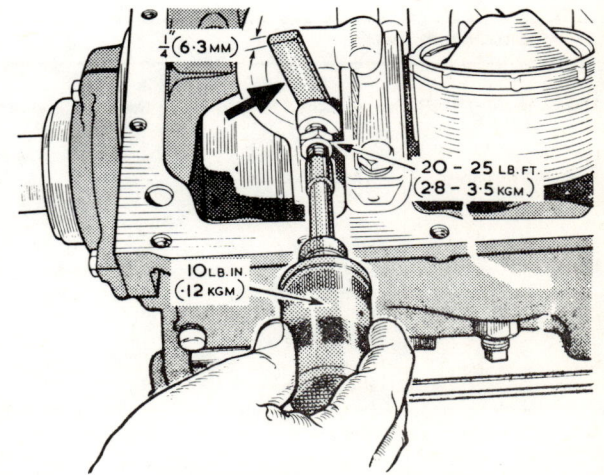

Fig. 6.32. Adjusting the front band

15 Rear band - adjustment

1 The rear band should be adjusted after the first 1000 miles (1600 km) of use and at 21,000 mile (35,000 km) intervals thereafter.
2 The rear band adjusting screw will be found on the right hand side of the casing and an access hole is provided in the transmission cowl. (Fig.6.33)

Chapter 6/Gearbox

3 Slacken the adjusting screw locknut and then back it off three or four turns. Now check that the screw moves freely in the case.
4 Using a torque wrench or special tool for the purpose, screw in on the adjusting screw to a torque of 10 lb.f.ft (1.382 kg.f.m). Now back off the screw exactly 1½ turns and re-tighten the locknut to a torque of 35 6 40 lb.f.ft (4.84 6 5.53 kg.f.m).

16 Governor - examination, removal and refitting

1 The governor can be examined, without removing the oil pan, by taking off the inspection cover but you may have to turn the output shaft to bring the governor opposite the opening.
2 Check for freedom of the valve by pushing and pulling on the governor weight.
3 If the valve appears to be sticking, the governor can be removed for dismantling and cleaning by taking out the two retaining screws but be careful not to drop the screws inside the extension housing.
4 Dismantle the governor and clean all parts using a non-fluffy rag.
5 When reassembling, torque the body plate screws to a torque of 20 - 30 lb.f.ins (0.24 - 0.36 kg.f.m.).
6 When refitting the governor to the transmission, torque the retaining screws to 50 - 60 lb.f.ins (0.60 - 0.72 kg.f.m).
7 Use a new gasket on the inspection cover plate and torque the plate retaining screws to 50 - 60 lb.f.ins (0.60 - 0.72 kg.f.m).

Fig. 6.33. Access for adjustment of the rear band

17 Automatic shift speeds

185 X 15 SP 41 HR Tyres—2.88 : 1 Final Drive Ratio (All countries except USA and Canada)

Selector position	Throttle position	Upshifts 1 – 2	2 – 3	Downshifts 3 – 2	3 – 1	2 – 1
				M.P.H.		
D1	Minimum	7 – 9	12 – 15	8 – 14	—	4 – 8
	Full	38 – 44	66 – 71	23 – 37	—	—
	Kickdown	52 – 56	81 – 89	73 – 81	20 – 24	20 – 24
D2	Minimum	—	12 – 15	8 – 14	—	—
	Full	—	66 – 71	23 – 37	—	—
	Kickdown	—	81 – 89	73 – 81	—	—
L	Zero	—	—	60	—	12 – 20
				K.P.H.		
D1	Minimum	11 – 14	19 – 24	13 – 23	—	6 – 13
	Full	61 – 71	106 – 114	37 – 60	—	—
	Kickdown	83 – 90	130 – 143	118 – 130	32 – 39	32 – 39
D2	Minimum	—	19 – 24	13 – 23	—	—
	Full	—	106 – 114	37 – 60	—	—
	Kickdown	—	130 – 143	118 – 130	—	—
L	Zero	—	—	96	—	19 – 32

Chapter 6/Gearbox

185 X 15 SP 41 HR Tyres – 3.31 : 1 Final Drive Ratio (USA and Canada only)

Selector position	Throttle position	Upshifts 1–2	2–3	Downshifts 3–2	3–1	2–1
				M.P.H.		
D1	Minimum	6 – 8	11 – 13	7 – 13	–	3 – 7
	Full	33 – 40	58 – 62	19 – 33	–	–
	Kickdown	45 – 49	70 – 78	63 – 71	17 – 21	17 – 21
D2	Minimum	–	11 – 13	7 – 13	–	–
	Full	–	58 – 62	19 – 33	–	–
	Kickdown	–	70 – 78	63 – 71	–	–
L	Zero	–	–	60	–	10 – 18
				K.P.H.		
D1	Minimum	9 – 13	18 – 21	11 – 21	–	5 – 11
	Full	53 – 64	93 – 100	31 – 53	–	–
	Kickdown	73 – 80	113 – 126	101 – 114	28 – 34	28 – 34
D2	Minimum	–	18 – 21	11 – 21	–	–
	Full	–	93 – 100	31 – 53	–	–
	Kickdown	–	113 – 126	101 – 114	–	–
L	Zero	–	–	96	–	16 – 29

Note: Shift points are approximate and not obsolute values. Reasonable deviations from the above values are permissible.

18 Automatic transmission - fault diagnosis

It is important to gain as much information as is possible on the precise nature of any fault before any attempt is made to remove and dismantle the transmission unit. In all cases the following road test procedure should be completely carried out as there may be more than one fault but, before carrying out any of the tests, it is important that the car is run long enough for the transmission fluid to attain its normal running temperature; a check must also be made that the fluid in the transmission is at the correct level. If the checks show up a fault, its possible cause will be found in the Section dealing with road test fault diagnosis, it is appreciated that we have not given information on methods of rectification for a large number of the faults enumerated as we feel that this work, as previously stated, should be left to an authorised dealer.

1 Check that the starter will operate only with the selector in 'P' and 'N' and that the reverse light operates only in 'R'.
2 Apply the brakes and with the engine at normal idling speed, select N–D, N–L, N–R. Engagement should be felt in each position selected.
3 Check the engine stall speed as described under the heading 'Converter diagnosis'.
4 Select D1, release the brakes and accelerate with minimum throttle opening. Check for 1-2 and 2-3 shifts. These may be difficult to detect under these conditions so confirmation can be obtained that the transmission is, indeed, in 3rd gear by selecting 'L' when a 3-2 downshift will be felt.
5 At just over 30 mph (48 kph), select 'N', switch off the ignition and allow the car to coast. At 30 mph, switch on the ignition and select 'L' when the engine should start through the rear wheels indicating that the rear oil pump is operating.
6 Stop and restart using full throttle opening, i.e. kickdown position, check for 1 - 2 and 2 - 3 shifts at speeds according to the chart given in Section 17.
7 At 26 mph (42 kph) in 3rd gear, depress the accelerator to full throttle position. The car should accelerate in 3rd gear and should not downshift to 2nd.
8 At 30 mph (48 kph) in 3rd gear, depress the accelerator to the kickdown position when the transmission should downshift to 2nd gear.
9 At 18 mph (29 kph) in 3rd gear, depress the accelerator to the kickdown position; the transmission should downshift to 1st gear.
10 Stop and restart using full throttle depression, again check for correct 1-2 and 2-3 shifts according to the chart.
11 At 40 mph (64 kph) in 3rd gear, release the accelerator and select 'L'. Check for 3-2 downshift and engine braking. Check for 2-1 inhibited downshift and engine braking.
12 Stop, and with 'L' still engaged, release the brakes and, using full throttle, accelerate to 20 mph (32 kph). Check for no slip or clutch break-away noise and no upshifts.
13 Stop and select 'R'. Release the brakes and, using full throttle if possible, check for no slip or clutch break-away noise.
14 Stop on brakes facing downhill and select 'P'. Release the brakes and check that the car is held. Re-apply the brakes before moving the selector out of 'L'. Repeat this test with the car facing up-hill.
15 Check that the selector is trapped by the gate in the 'P' position.
16 At 30 mph (48 kph) in 3rd gear, D1, coast to a stop. Check roll out shifts for quality and speed.
17 Check the front pump by revving the engine between idle and 2000 rpm with the selector in 'N'. A high pitched whine indicates a noisy front pump, a restricted suction line or a dirty oil screen.
18 At or slightly above idle speed in neutral, a gear whine indicates dragging front clutch plates. A tendency for the car to creep in neutral is a further indication of dragging front clutch plates.

Pressure tests

Before carrying out these tests you must first make sure that the throttle kickdown cable is correctly adjusted as described in Section 10 and also that the engine idle speed is correct.

1 With a pressure gauge connected to the pressure take off point (see Section 10 paragraph 4) check and record pressures which should be within the limits quoted in the following table:-

Note: Figures given in table are normal for transmission temperatures from 150° to 185°F. only (65.5°C to 85°C).

Selector Position	Control Pressure Idle rpm	Control Pressure Stall rpm
D2	50 – 60	150 – 185
D1	50 – 60	150 – 185
L	50 – 60	150 – 185
R	50 – 60	190 – 210
N	55 – 60	—

2 Low pressure indicates leakage in the circuit tested. Low pressure in all selector positions is an indication of leakage, a faulty pump or incorrect pressure regulation.

3 High pressures, in all selector positions, indicate faulty pressure regulation, incorrect cable adjustment or stuck valves.

Converter diagnosis

If the general vehicle performance is below standard, check the stall speed as described later.

Inability to start on steep gradients combined with poor acceleration from rest is an indication that the converter stator one-way clutch is slipping or that the stator support is fractured, this will permit the stator to rotate in an opposite direction to the turbine and so prevent torque multiplication. Check the stall speed and if it is more than 650 rpm below normal the converter assembly must be replaced.

Below standard acceleration in top gear at speeds above 30 mph (48 kph) combined with a substantially reduced maximum speed, indicates that the stator one-way clutch has locked in the engaged position. The stator cannot rotate with the turbine and impeller and so the fluid flywheel phase of the converter performance cannot occur. This condition will be associated with severe overheating of the transmission but stall speed will be normal.

A stall speed substantially higher than normal indicates that the converter is not receiving its required fluid supply or that slip is occurring in the clutches of the gearbox. The torque converter is a sealed unit and, therefore, cannot be dismantled, any faults that arise will entail replacement of the complete unit.

Stall speed test

This test provides a rapid check on the correct functioning of the converter as well as the gearbox. The stall speed is the maximum speed at which the engine can drive the torque converter impeller whilst the turbine is held stationary.

As the stall speed is dependant on both engine and toque converter characteristics, it will vary with the condition of the engine and this must be taken into account in order to correctly interpret a low stall speed.

Allow the engine and transmission to attain their normal working temperature, set the handbrake, chock the wheels and apply the footbrake. Start the engine, select L or R and then fully depress the accelerator. Take a quick reading on the revolution counter and release the accelerator.

To avoid overheating, the duration of each stall test must not exceed 10 seconds.

Stall speed engine rpm	Condition indicated
Under 1,000	Stator free wheel slip
1,600 - 1,700	Normal
Over 2,100	Slip in the transmission gearbox

Fault diagnosis through road test

The following is a comprehensive list of faults which may be shown up when carrying out the road test procedure; the letters and figures shown opposite each fault are the actions required to cure it and details of these will be found in the Key. It is realised that in many cases you will not have the facilities to do the necessary remedial work, nevertheless details are given in order to give you some idea of what rectification entails.

18 Automatic transmission - fault diagnosis chart

Engagement	In Car	On Bench
Harsh	B. D. c. d.	2, 4
Delayed	A, C, D, E, F, a, c, d	b
None	A, C, a, c, d	b, 9, 10, 11, 13
No forward	A, C, a, c, d	B, 1, 4, 7
No reverse	A, C, F, a, c, j, k, h	b, 2, 3, 6
Jumps in forward	C, D, E, F	4, 7, 8
Jumps in reverse	C, D, E	2
No neutral	C, c	2

Upshifts		
No. 1–2	C, E, a, c, d, f, g, h, j	b, 5, 17
No. 2–3	C, a, c, d, f, g, h, k, l	b, 3, 17
Shift points too high	B, C, c, d, f, g, h, j, k, l	b
Shift points too low	B, c, f, g, h, l	B

Upshift Quality		
1–2 slips or runs up	A, B, C, E, a, c, d, f, g, k	b, 1, 5
2–3 slips or runs up	C, a, c, d, f, g, h, k, l	b, 3, 5
1–2 harsh	B, C, E, c, d, f, g, h	1, 7, 8
2–3 harsh	B, C, E, s, d, f	4
1–2 ties up or grabs	F, c	4, 7, 8
2–3 ties up or grabs	E, F, C	4

Downshifts		
No. 2–1	B, C, c, h, j	7
No. 3–2	B, c, h, k	4
Shift points too high	B, C, c, f, h, j, k, l	b
Shift points too low	B, C, c, f, h, j, k, l	b

Downshift Quality		
2–1 slides		7
3–2 slides	B, C, E, a, c, d, f, g	b, 3, 5
2–1 harsh		b, 1, 7
3–2 harsh	B, E, c, d, f, g, 5	3, 4, 5

Chapter 6/Gearbox

Reverse
   Slips or chatters   A, B, F, d, c, g   b, 2, 3, 6

Line Pressure
   Low idle pressure   A, C, D, a, c, d   b, 11
   High idle pressure   B, c, d, e, f, g
   Low stall pressure   A, B, a, c, d, f, g, h   b, 11
   High stall pressure   B, c, d, f, g

Stall Speed
   Too low (200 rpm or more)   13
   Too high (200 rpm or more)   A, B, C, F, a, c, d, f   b, 1, 3, 6, 7, 9, 13

Others
   No push starts   A, C, E, F, c   12
   Transmission overheats   E, F, e   1, 2, 3, 4, 5, 6, 13, 18
   Poor acceleration   13
   Noisy in neutral   m   2, 4
   Noisy in park   m   14
   Noisy in all gears   m   2, 4, 14, 16
   Noisy during coast (30–20 mph)   16, 19
   Park brake does not hold   C, 15   15

KEY TO THE FAULT DIAGNOSIS CHART

1 Preliminary checks in car
   A Low fluid level
   B Throttle cable incorrectly assembled or adjusted
   C Manual linkage incorrectly assembled or adjusted
   D Engine idle speed
   E Front band adjustment
   F Rear band adjustment

2 Hydraulic faults
   a) Oil tubes missing or broken
   b) Sealing rings missing or broken
   c) Valve body screws missing or not correctly tightened
   d) Primary valve sticking
   e) Secondary valve sticking
   f) Throttle valve sticking
   g) Compensator or modulator valve sticking
   h) Governor valve sticking, leaking or incorrectly assembled
   i) Orifice control valve sticking
   j) 1–2 shift valve sticking
   k) 2–3 shift valve sticking
   l) 2–3 shift valve plunger sticking
   m) Regulator

3 Mechanical faults
   1 Front clutch slipping due to worn plates or faulty parts
   2 Front clutch seized or plates distorted
   3 Rear clutch slipping due to worn or faulty parts
   4 Rear clutch seized or plates distorted
   5 Front band slipping due to faulty servo, broken or worn band
   6 Rear band slipping due to faulty servo, broken or worn band
   7 One-way clutch slipping or incorrectly installed
   8 One-way clutch seized
   9 Broken input shaft
   10 Front pump drive tangs on converter hub broken
   11 Front pump worn
   12 Rear pump worn or drive key broken
   13 Converter blading and/or one-way clutch failed
   14 Front pump
   15 Parking linkage
   16 Planetary assembly
   17 Fluid distributor sleeve in output shaft
   18 Oil cooler connections
   19 Rear pump

Chapter 7 Propeller shaft

Contents

General description ... 1	Universal joints - dismantling ... 6
Propeller shaft - lubrication ... 3	Universal joints - general ... 5
Propeller shaft - removal and replacement ... 2	Universal joints - reassembly ... 7
Sliding joint - dismantling, overhaul and reassembly ... 4	

1 General description

Drive is transmitted from the gearbox or automatic transmission unit by means of a finely balanced tubular propeller shaft incorporating Hardy Spicer needle roller bearing universal joints front and rear to allow for vertical movement of the rear suspension assemby and for movement of the complete power unit on its rubber mountings. Fore and aft movement of the rear suspension is catered for by a sliding spline at the front of the propeller shaft assembly.

The yoke flange of the universal joint is fitted to the gearbox mainshaft flange by four bolts which, in some cases, may be secured with plain washers, nuts and split pins and in others, more commonly, with plain washers and self locking nuts. A similar arrangement is used to secure the rear yoke flange to the pinion flange on the rear axle.

Depending on the age of your car, it may be fitted with one of three possible types of propeller shaft but they are all interchangeable. Fig. 7.1 illustrates an early type of shaft in which the journal assembly is provided with a grease nipple for lubrication of needle roller bearings and in this type the splined shaft is protected by a dust cap which screws onto the end of the sleeve yoke with a cork washer acting as an internal seal. The needle roller bearings and splined shaft of the other two types are of the 'sealed' for life pattern, that is to say the journal assembly is not provided with a grease nipple and no lubrication of the bearings or splines is required in service. These two shafts differ only in the spline sealing arrangements, one is fitted with an external rubber gaiter held in position on the sleeve yoke fits, sealing is provided by a rubber gaiter which fits inside the tube and over the shank of the sleeve yoke where it is retained by a clip.

2 Propeller shaft - removal and replacement

To allow removal (not disconnection) of the propeller shaft from the car, it is necessary to first remove either the engine or the rear suspension as you will find that this is much quicker and simpler operation than taking out the engine. Full instructions for removal of the rear suspension assembly will be found in Chapter 11.
1 Refer to Chapter 1, Section 5 and follow the operations covered in paragraphs 5,6,7, 8 and 13 to disconnect the propeller shaft flange yoke from the gearbox flange.
2 Remove the rear suspension as described in Chapter 11.
3 Withdraw the propeller shaft from the rear of the propeller shaft tunnel.
4 Refitting is the reverse of the removal procedure but the brake system will have to be bled after refitting the rear suspension, instructions for this work are given in Chapter 9.

3 Propeller shaft - lubrication

1 Later model cars are fitted with 'sealed for life' universal joints and sliding spline which do not require periodic lubrication. These items can be identified by lack of grease nipples at the universal joint journal assembly and at the sliding spline.
2 The propeller shaft of early model cars should be lubricated at the 2500 mile (4,000 km) servicing. This work can be done with the propeller shaft in situ.
3 The grease nipple for the front universal joint and the sliding spline is only accessible from inside the car through an aperture in the left hand vertical face of the gearbox cowl as shown in Fig.7.2

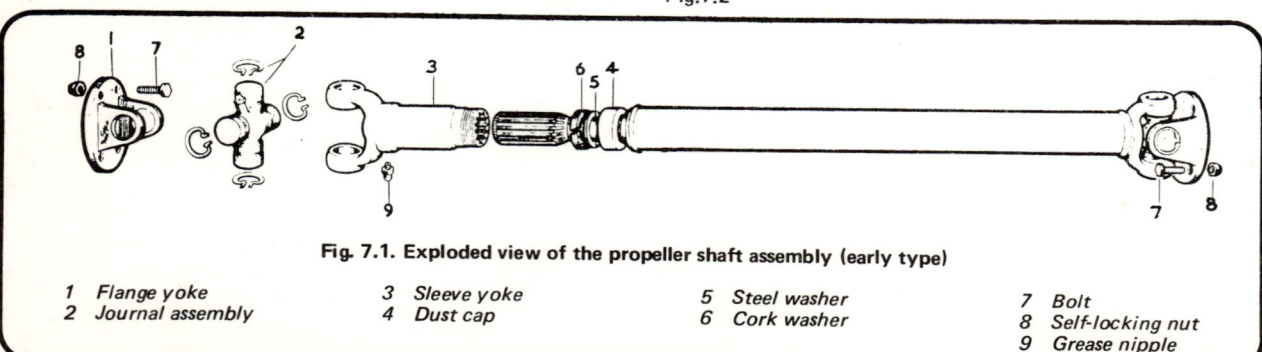

Fig. 7.1. Exploded view of the propeller shaft assembly (early type)

1 Flange yoke	3 Sleeve yoke	5 Steel washer	7 Bolt
2 Journal assembly	4 Dust cap	6 Cork washer	8 Self-locking nut
			9 Grease nipple

Chapter 7/Propeller shaft

4 Roll back the front carpet to expose the two snap fasteners which retain the gearbox cowl trim panel to the floor. Undo the fasteners and raise the panel.
5 Remove the rear metal or rubber aperture cover now exposed, look at the universal joint and sliding spline grease nipples and then move the car as necessary to bring them to a convenient position for application of the grease gun. Pump in grease until it appears at the journal side of the needle rollers, two or three strokes should suffice for the sliding spline.
6 The grease nipple for the rear end of the propeller shaft is accessible from underneath the car.
7 We recommend the use of Castrol LM grease for lubrication of the propeller shaft.

5 The male and female splines can bow be cleaned as may be necessary to allow close examination.
6 Check that the shaft and yoke assembly slide freely together, investigate any sign of stiffness which may be due to tightness in the splines resulting from burring, rectify as may be required but be careful not to reduce the overall width of the spline or splines.
7 Visually examine the splines for wear and if they appear to be satisfactory, assemble the yoke to the shaft and check that there is no rotary movement in excess of 0.004" (0.1 mm.) between the two items. If there is any wear it will be necessary to replace the assembly.
8 Reassembly is the reverse of the dismantling procedure. However, first grease the splines and then assemble the sleeve yoke on the splines so that the fixed yoke is aligned with the fixed yoke on the shaft, arrows are stamped on the two parts to facilitate alignment (see Fig.7.3).

5 Universal joints - general

1 Wear in the needle roller bearings is characterised by judder and vibration in the transmission on over-run, 'clonks' on taking up drive, and in extreme cases of lack of lubrication, metallic squeaking and ultimately grating and shrieking sounds as the bearings break up.
2 It is easy to check if the needle roller bearings are worn with the propeller shaft in position in the car. Refer to Chapter 1 and follow the instructions to give access to the front universal joint and gearbox flange. Use some means to prevent movement in the joint, follow the same procedure for the rear universal joint.
3 Any movement between the shaft and the front or rear couplings is indicative of bearing failure and/or wear in the spider, in this event the old bearings will have to be discarded and replaced by a new universal joint assembly.

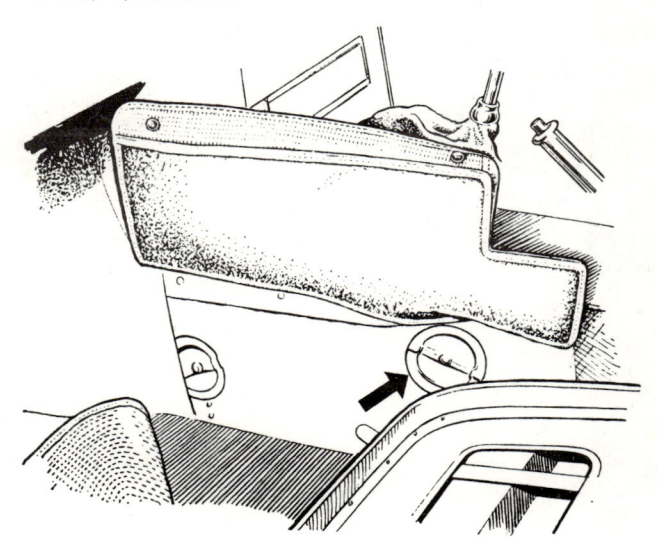

Fig. 7.2. Access hole in gearbox cowl for lubrication of front universal joint and the sliding spline

6 Universal joints - dismantling

1 Remove the propeller shaft, in the manner described in Section 2.
2 Thoroughly clean all dirt from the rings and the top of the bearing surfaces.
3 Remove all the snap rings by pinching with a pair of pliers and at the same time prising out with a screwdriver. If a ring proves difficult to remove it may be because it is jammed by the end of the bearing race so lightly tap the end of the race to relieve the pressure.
4 Hold the joint in the hand and with a hide faced or other type of soft hammer, tap the yoke lug as shown in Fig. 7.4. This will cause the top bearing to work outwards until it can finally be removed with the fingers (Fig.7.5). If the top bearing proves

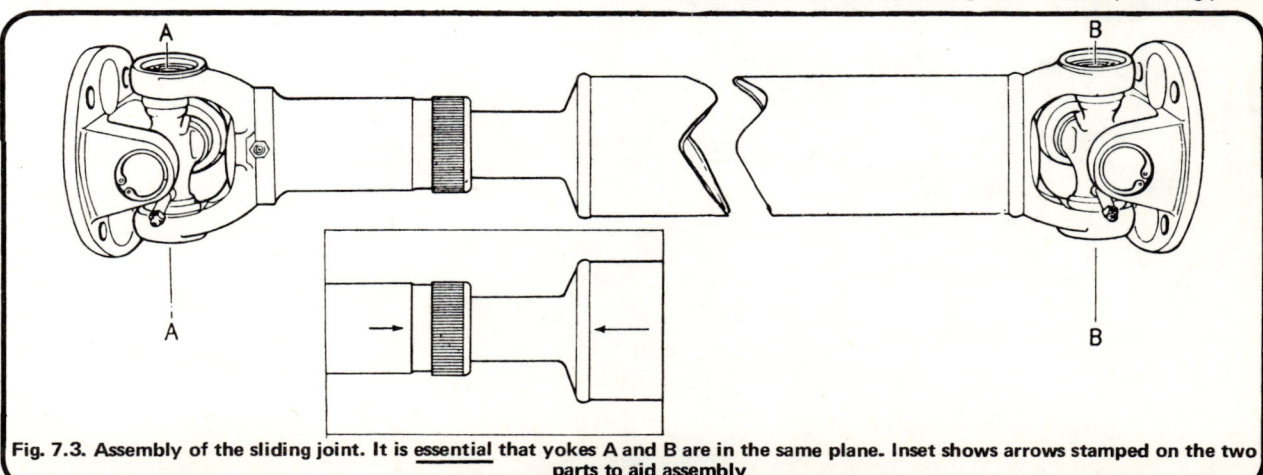

4 Sliding joint - dismantling, overhaul and reassembly

1 Refer to Section 2 and remove the propeller shaft assembly.
2 Where appropriate, roll back the rubber rings attaching the gaiter to the shaft and then slide off the sleeve yoke assembly.
3 Remove the gaiter from the sleeve yoke by detaching the steel ring and then slide the gaiter off the yoke.
4 Your car may be an older model fitted with a knurled dust cap instead of a rubber gaiter. Unscrew the cap to separate the sleeve yoke assembly.

Fig. 7.3. Assembly of the sliding joint. It is <u>essential</u> that yokes A and B are in the same plane. Inset shows arrows stamped on the two parts to aid assembly

obstinate it can be tapped out from inside (Fig.7.6) with a small diameter punch or piece of bar but be careful you do not damage the bearing if this is not being replaced.
5   Repeat the above procedure for the opposite bearing.
6   The splined sleeve yoke or flange can now be separated from the shaft yoke as depicted in Fig.7.7.
7   Rest the two trunnions which are now exposed, on wood blocks and tap the yoke with a soft nosed hammer to remove the two remaining races.
8   It is now advisable to look carefully at the yoke cross holes. It is a very rare event but these holes have been known to wear to a certain degree of ovality. If this has occurred the defective item will have to be replaced and if it is a fixed yoke on a shaft the complete assembly will have to be renewed.

7   Universal joint - reassembly

1   On early model cars it is advisable to fit new cork gaskets and gasket retainers on the spider using a tubular drift as illustrated in Fig.7.8.
2   It is a good tip to smear the wall of the race with vaseline to keep the rollers in position in the housing for assembly.
3   Insert the spider in the yoke holes, place the bearings in position and then lightly tap it home using a soft flat faced drift slightly smaller than the yoke hole in diameter.
4   Repeat the above for the opposite bearing in the yoke.
5   Fit new snap rings to the bearings and ensure they correctly located in the grooves.
6   Now place the mating yoke in position on the spider and fit the bearings and snap rings in the same manner as described above. It is essential that the sliding joint is refitted with its fixed yoke in lin with the fixed yoke at the end of the propeller shaft; arrows are stamped on the two items to facilitate alignment (see Fig.7.3).
7   Make sure that the joint moves freely in all directions, if it appears to bind tap lightly with a wooden mallet to relieve any pressure of the bearings on the end of the spider.
8   Refit the propeller shaft to the car as described in Section 2.

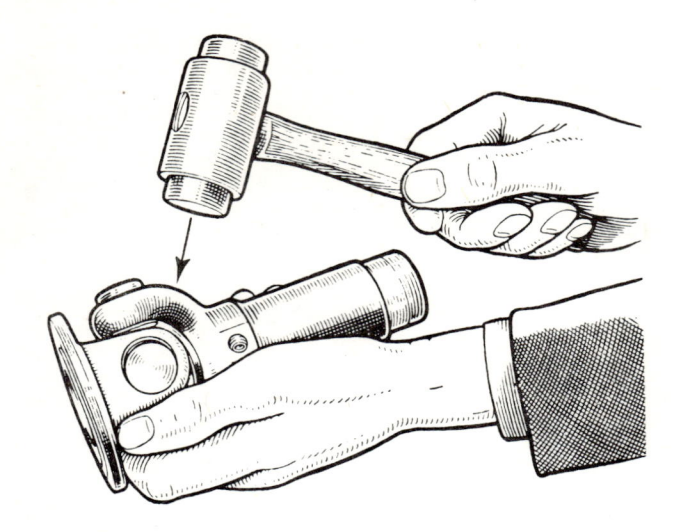

Fig. 7.4. Removing a universal joint bearing

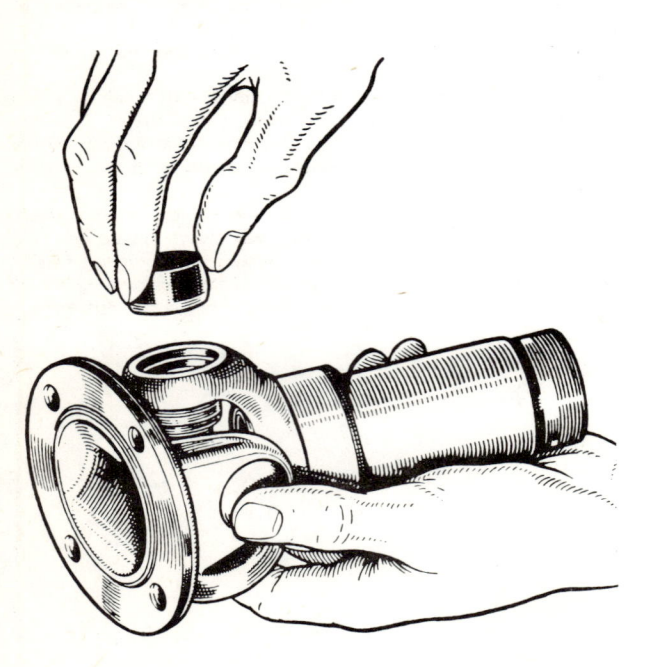

Fig. 7.5. Withdrawing the bearing

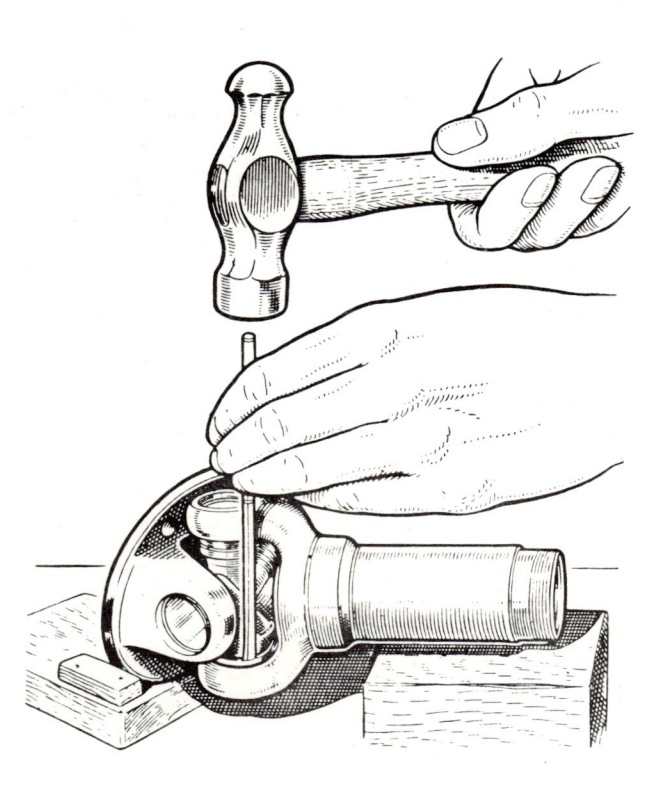

Fig. 7.6. Tapping out a bearing

Fig. 7.7. Separating the yokes

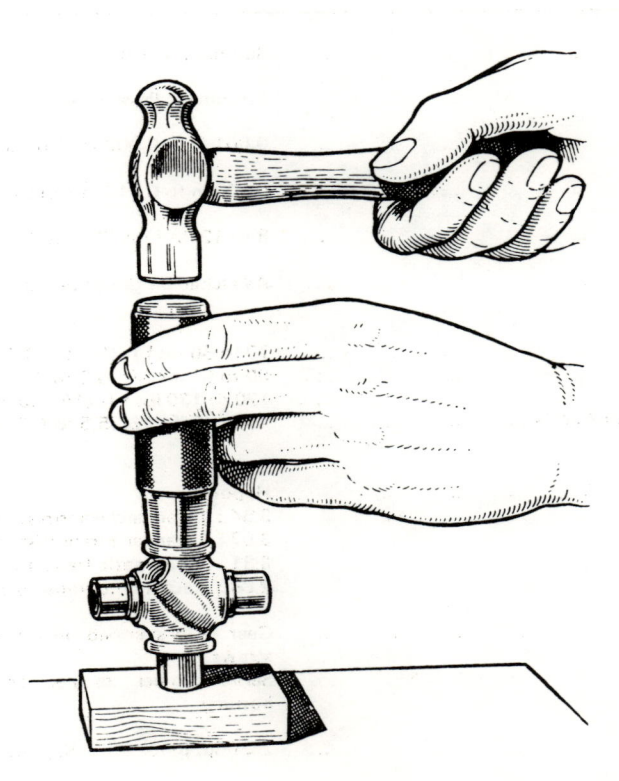

Fig. 7.8. Replacing a gasket

Chapter 8 Rear axle

Contents

Axle unit - removal and replacement ... 2	General description ... 1
Differential unit - dismantling ... 10	Half shaft - removal and replacement with rear suspension installed ... 3
Differential unit - reassembly ... 11	
Drive shafts - removal, checking of endfloat and refitting ... 7	Half shaft universal joints - dismantling and reassembly ... 5
Final drive unit - checking for wear ... 8	Half shaft universal joints - examination and check for wear ... 4
Final drive unit - dismantling ... 9	Rear hub - removal, replacement of bearings and oil seals, checking of endfloat and refitting ... 6
Final drive unit - reassembly ... 12	

Specifications

Type	Salisbury 4.H.U.
Differential type	Thornton "Powr-Lok"
Output shaft endfloat	0.001 in to 0.003 in (0.02 to 0.07 mm)
Differential bearing preload	0.006 in to 0.010 in (0.15 to 0.25 mm) shim allowance
Pinion bearing preload	8 to 12 lbs f ins (0.09 to 0.14 kg f m)
Backlash	As etched on the drive gear - minimum 0.004 in (0.10 mm)

Tightening torques:

Drive gear bolts	70 to 80 lbs f ft (9.7 to 11.1 kg f m)
Differential bearing cap bolts	60 to 65 lbs f ft (8.3 to 9.0 kg f m)
Pinion nut	120 to 130 lbs f ft (16.6 to 18.0 kg f m)
Thornton "Powr-Lok" differential bolts	40 to 45 lbs f ft (5.5 to 6.2 kg f m)

Final drive:

Type	Hypoid
Ratios	3.54 : 1 standard transmission USA and Canada
	3.07 : 1 standard transmission all other countries
	3.31 : 1 automatic transmission (2 + 2 only) USA and Canada
	2.88 : 1 automatic transmission (2 + 2 only) all other countries
Axle markings	Gear ratio stamped on a tag attached by one of the rear cover screws
	Serial number stamped on the underside of the gear carrier housing
Lubricant capacity	2¾ Imp. pints, 3¼ U.S. pints, 1.6 litres

Special tools

The following special tools are required for the efficient overhaul of the axle, alternatives, where suitable, are suggested in the text of this Chapter, but your attention is drawn to the note at the end of this Section.

Description of special tools:	Tool No.
Multi-purpose hand press used in conjunction with the following adaptors:	SL 14
Pinion bearing inner races - removing and replacing	SL 11 P/AB—2
Differential bearing - removing	SL 11 D/A—5
Differential bearing replacing - universal handle	SL 2 DB

Chapter 8/Rear axle

Used with adaptor	SL 2 D/B–2
Main tool and ring	SL 12
Used in conjunction with the following adaptors:	
Pinion bearing outer races - removing and replacing	SL 12 AB–4
Hub endfloat special collar	J.15
Hub endfloat dial test indicator	J.13
Pinion cone setting gauge	SL 3 P
Rear hub extractor	J.7
Pinion oil seal installation collar	SL 4 P/B

NOTE: Full servicing instructions for the rear axle are given in this Chapter but in view of the intricate adjustments and the number of special tools required, we advise that you take advantage of the factory reconditioning scheme and obtain a replacement axle on an exchange basis, if the need arises.

1 General description

All models are fitted with a Salisbury 4.HU type rear axle unit (Figs.8.1 and 8.2), it is mounted independently from the hubs and road wheels and is connected to them by short drive shafts with universal joints at each end which are coupled to the axle drive shafts which also provide a mounting for the discs of the inboard mounted disc brakes. A limited slip Thornton "Powr-Lok" differential is used and an exploded view of this item is given in Fig.8.3. A limited slip differential differs from the conventional type in that on slippery surfaces it will not allow the wheel with the lesser traction to spin, gain momentum and so swerve the car when a dry surface is regained. During cornering, the differential applies the major driving force to the inside rear wheel thus improving stability and cornering, and, under conditions of poor traction, it enables the wheel with the better traction to apply the major driving force to the road. Bumps do not adversely affect wheel action when the wheels are controlled by a limited slip differential as the free wheel does not spin and gain momentum and thus there is no sudden wheel stoppage to cause tyre scuffing or to cause the car to swerve. Because of this, the engine must **never** be run with the car in gear and with one wheel jacked up off the ground otherwise, owing to the action of the differential, the car may drive itself off the jack or stand. If it is desired to run the transmission with the car stationary, **both** wheels must be clear of the ground.

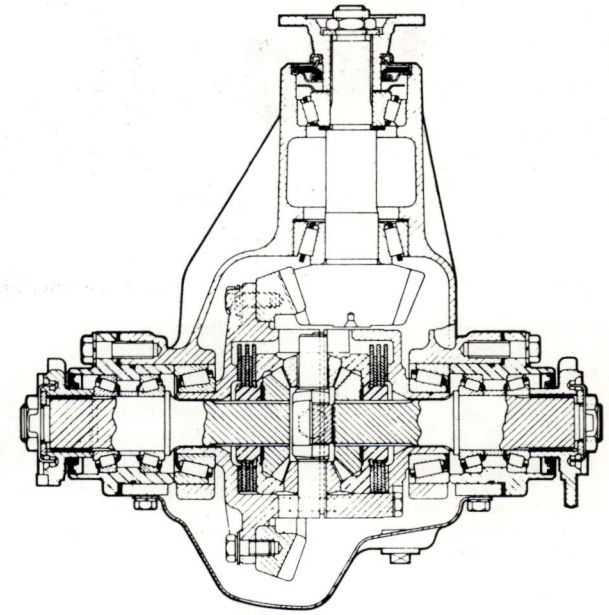

Fig. 8.1. Sectioned view of the axle unit

2 Axle unit - removal and replacement

1 Follow the instructions given in Chapter 11 for removal of the rear suspension assembly.
2 For ease of working, invert the suspension assembly on a bench. First remove the 14 bolts which secure the tie plate.
3 Remove the tie plate and disconnect the two shock absorber and coil spring units at each side.
4 Remove the four self locking nuts which secure the half shaft universal joints to the brake disc and axle output shaft flange at each side.
5 Withdraw the half shafts from the bolts taking careful note of the number and position of the shims as these control wheel camber angle.
6 Remove one self-locking nut from the inner wishbone fulcrum shafts and drift out the shafts; make sure that your drift is kept central on the shaft or the thread will be damaged.
7 Remove the hub (section 6) halfshaft, wishbone and radius arm assembly at each side and keep them identified for replacement on the side from which they were removed.
8 Take out the two bolts which hold the handbrake compensator linkage and withdraw the compensator.
9 Disconnect the hydraulic pipes at the brake calipers and blank off the unions and pipes to prevent the ingress of dirt.
10 Now turn the suspension over.
11 Break the locking wire to the four differential carrier mounting bolts (Fig. 8.4). Remove the bolts and then lift off the cross beam from the differential carrier by tilting forward over the nose of the pinion as illustrated in Fig. 8.5.

12 Replacement is the reverse of the above sequence, it is important, however, owing to heat dissipation from the brake discs, that you make sure the locknuts fitted on the output flange shaft studs are of the metal and not nylon locking type.
13 The self locking nut of the inner wishbone fulcrum shaft should be tightened to a torque of 55 lb.f.ft (7.6 kg.f.m) and the four differential carrier mounting bolts on the top of the cross beam to a torque of 75 lbs.f.ft (10.4 kg.f.m).

3 Half shaft - removal and refitting with rear suspension installed.

1 Follow the instructions given in Chapter 11 for the removal of the lower wishbone outer fulcrum shaft.
2 Remove the split pin from the castellated nut which you will see inside the rear hub. Prevent the half shaft from turning and then remove the nut and plain washer.
3 Using an extractor (Tool No.J.7) withdraw the hub and the hub carrier from the splined end of the half shaft (Fig. 8.6). The end float spacer, the inner oil seal and the sealing ring may be retained within the hub carrier, if not, collect them and make sure they are not mixed with the items from the other half shaft assembly if this is also being removed at the same time. Early cars were fitted with shims in addition to the spacer, these must be kept as a set for the assembly from which they were removed.
4 Now take off the forward shock absorber and spring unit in the manner described in Chapter 11.

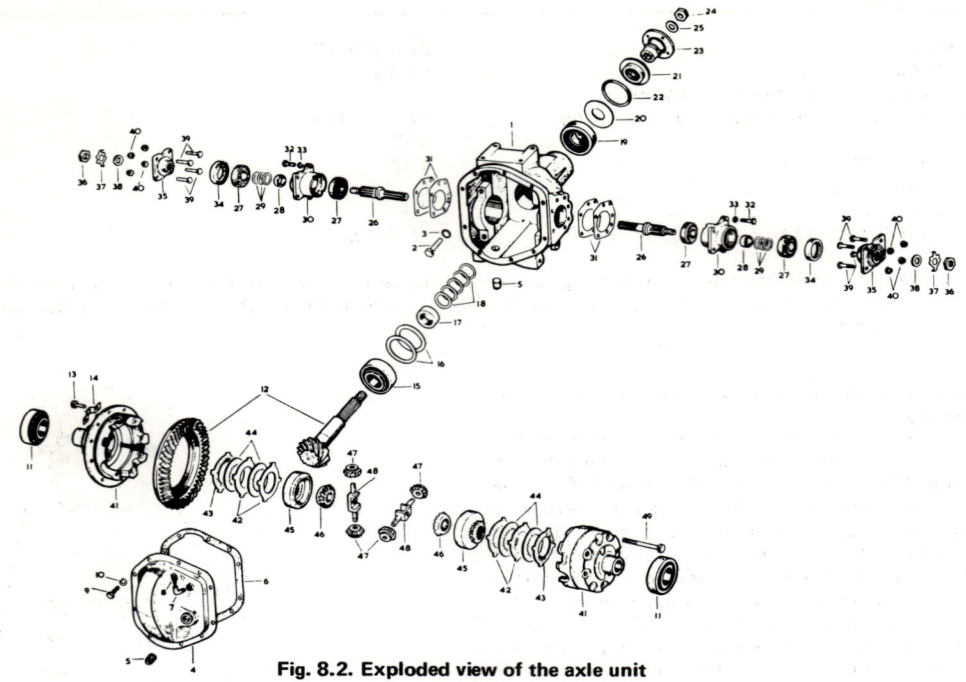

Fig. 8.2. Exploded view of the axle unit

1	Gear carrier	13	Setscrew	25	Washer
2	Setscrew	14	Locking plate	26	Drive shaft
3	Lockwasher	15	Roller bearing	27	Roller bearing
4	Cover	16	Shim	28	Distance piece
5	Plug	17	Distance piece	29	Shim
6	Gasket	18	Shim	30	Bearing housing
7	Elbow	19	Roller bearing	31	Shim
8	Breather	20	Oil thrower	32	Bolt
9	Setscrew	21	Oil seal	33	Spring washer
10	Spring washer	22	Gasket	34	Oil seal
11	Roller bearing	23	Companion flange	35	Flange
12	Crown wheel and pinion	24	Nut	36	Nut

- 37 Tab washer
- 38 Washer
- 39 Bolt
- 40 Self locking nut
- 41 Differential case
- 42 Flat friction plate
- 43 Dished friction plate
- 44 Friction plate
- 45 Side gear ring
- 46 Side gear
- 47 Pinion mate gear
- 48 Shaft
- 49 Bolt

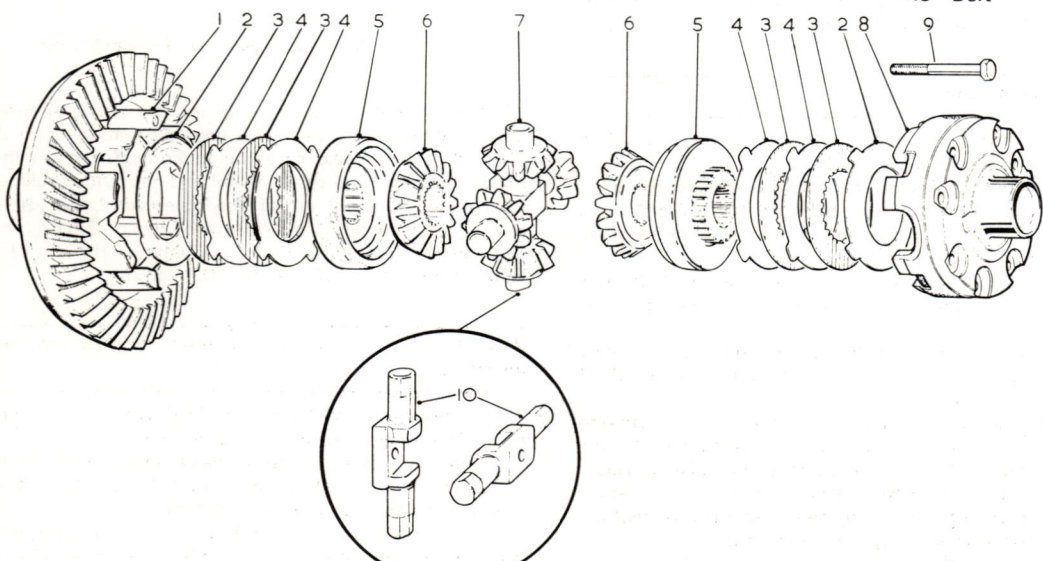

Fig. 8.3. Exploded view of the Thornton "Powr-Lok" differential

1 Differential casing - flange half
2 Dished clutch friction plate
3 Clutch friction disc
4 Clutch friction plate
5 Side gear ring
6 Bevel side gear
7 Bevel pinion mate gear assembly
8 Differential case - bottom half
9 Differential cas - bolt
10 Pinion mate cross shaft

Chapter 8/Rear axle

5 Remove the four self locking nuts which secure the half shaft inner universal joint to the axle output shaft flange and the inboard brake disc.
6 Withdraw the half shaft from the bolts (Fig. 8.7) and take note of the number of shims, these control the camber of the rear wheel and must be kept, and replaced, as a set.
7 When refitting, first replace the correct set and number of camber shims.
8 Place the half shaft inner universal joint over the four bolts and fit the locknuts, use metal type, not nylon, locknuts.
9 Refit the forward shock absorber and the coil spring.
10 Fit the end float spacer, and shims if applicable, a new inner oil seal and the sealing ring on to the splined end of the half shaft. Bring the hub carrier into line with the half shaft and then align the shaft so that the split pin hole is in line with the hole in the hub as shown in Fig. 8.8.
11 Locate the splines and then feed the splined shaft into the hub. When the threaded end emerges sufficiently, fit the plain washer and the castellated nut and draw the shaft into position by tightening the nut to a torque of 140 lbs.f.ft (19.3 kg.f.m). Fit the split pin.
12 Refit the lower wishbone outer fulcrum shaft, as described in Chapter 11.
13 If a new half shaft has been fitted you will have to check and adjust the camber of the wheels. Instructions for this will be found in Chapter 11.

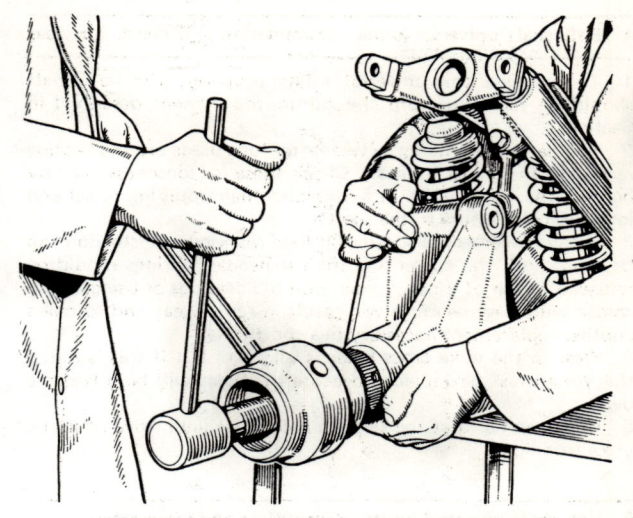

Fig. 8.6. Withdrawing a rear hub and carrier

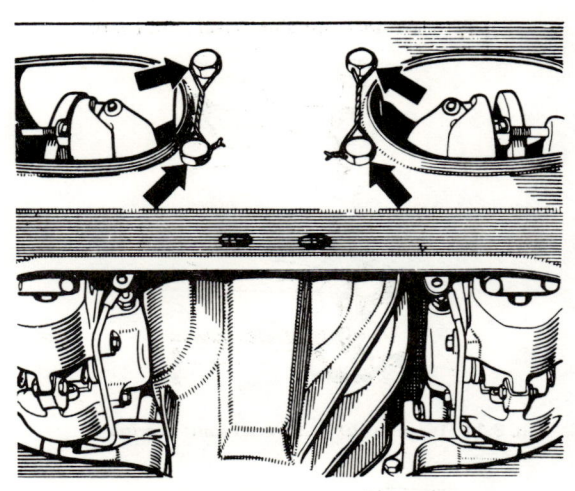

Fig. 8.4. The differential carrier mounting bolts

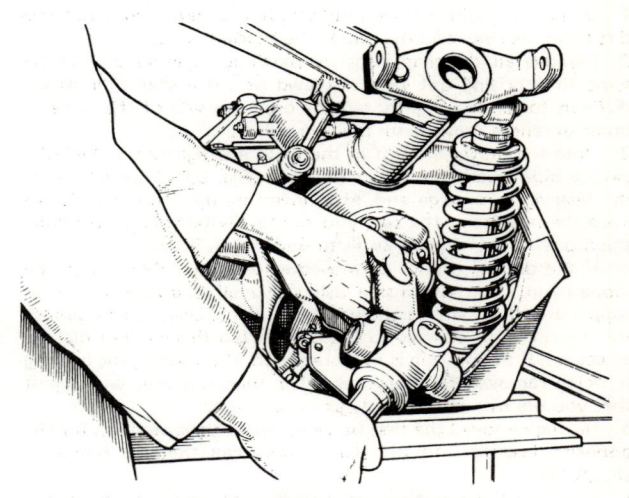

Fig. 8.7. Withdrawing a half shaft

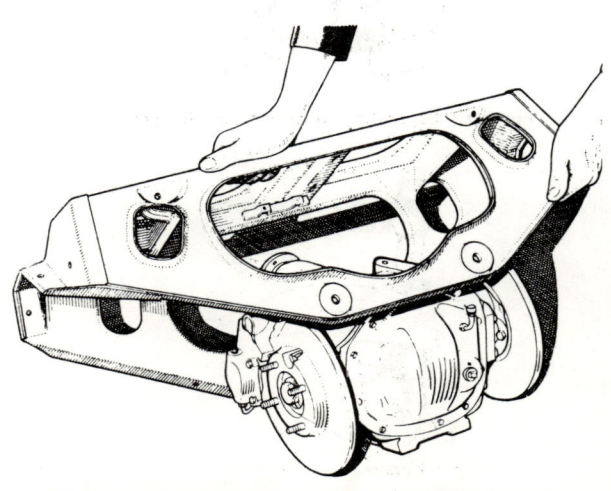

Fig. 8.5. Removing the cross beam from the axle unit

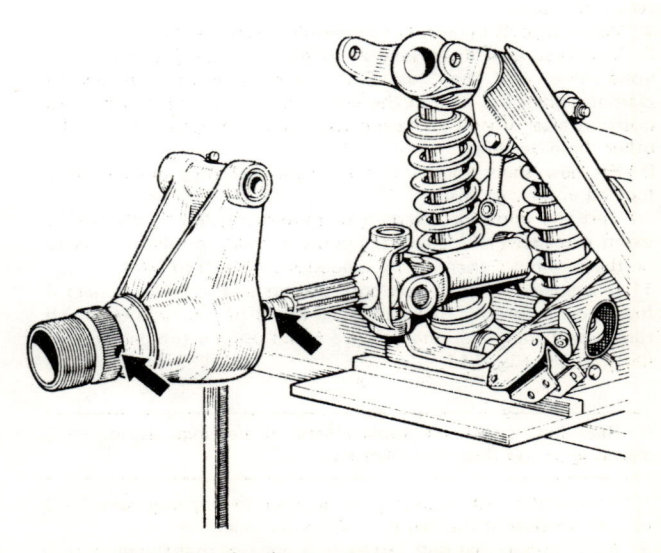

Fig. 8.8. Align split pin hole and access hole

Chapter 8/Rear axle

4 Half shaft universal joints - examination and check for wear

1 To examine the universal joints properly, the half shaft should be removed from the car in the manner described in Section 3.
2 The parts most likely to wear after long usage are the bearing races and the spider journals. Check these for looseness in fit by holding the outer yoke in a vice and then applying radial and longitudinal pressure to the shaft.
3 If any movement is noted or if load marking or distortion can be seen, both the spider and the the needle bearings should be renewed as use of a new spider with old bearings or using an old spider with new bearings will result in rapid wear and so make another replacement necessary in a short time.
4 Wear in the yoke cross holes is very rare but if this is noted the yoke must be replaced because the holes will have become oval.
5 It is essential that the bearing races are a light drive fit in the yoke trunnion.

5 Half shaft universal joints - dismantling and reassembly

1 Do not start dismantling until you have removed all paint and dirt from the rings and the top of the bearing races.
2 Remove all the snap rings by pinching together and at the same time prising out with a screwdriver. If a snap ring proves difficult to remove, lightly tap the opposite side of the yoke in order to relieve pressure on the ring.
3 Hold the universal joint in the hand and lightly tap the yoke with a hide faced hammer as shown in Fig. 8.9. This will cause the bearing to emerge and, after inverting the joint in order to keep the needles in the cap, you will finally be able to withdraw the cap with your fingers as illustrated in Fig. 8.10.
4 If the bearing cannot be moved as described above, it can be tapped out from the inside using a small diameter bar after supporting the shaft as shown in Fig. 8.11. Using this method, there is a risk of damage to the bearing but this will not matter, of course, if the bearing is to be renewed. If however, the bearing is being removed for purposes other than renewal, we suggest that you use this method only as a last resort.
5 Having removed the first bearing, repeat the operation for the opposite bearing and the flange yoke can then be removed, (Fig.8.12).
6 Now rest the two exposed trunnions on a wooden block and tap the yoke with a hide faced hammer to remove the two remaining races.
7 Wash all parts in petrol and carefully examine them.
8 The first task in reassembly is to insert the journal in the yoke holes and then, using a soft drift slightly smaller in diameter than the hole in the yoke, gently tap the bearing into position after liberally coating with grease. Repeat this for the other three bearings.
9 Fit new snap rings and make sure that they are correctly located in their grooves.
10 Test the joint for freedom of movement; tap lightly with a wooden mallet or hide faced hammer in order to relieve pressure on the bearings if there is any tendency for them to bind.
11 It is advisable to fit new cork gaskets and gasket retainers if the spider assembly fitted to your car is of the type to which these are fitted. The retainers are fitted using a tubular drift as illustrated in Fig. 8.13.

6 Rear hub - removal, replacement of bearings and oil seal, checking of end float and refitting.

1 Follow the instructions given in Section 3, paragraphs 1 - 3 for the removal of the rear hub and hub carrier.
2 Now support the hub carrier in a press so that the inner hub bearing is at the top. Press out the hub, as shown in Fig. 8.14. remove the hub (Fig. 8.15). If the inner oil seal, its seating ring and the spacer (and shims if fitted) are in place in the carrier,

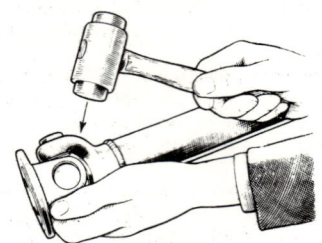

Fig. 8.9. Removing a bearing

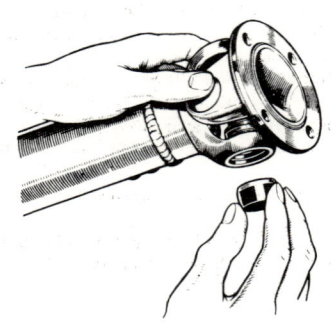

Fig. 8.10. Withdrawing a universal joint bearing

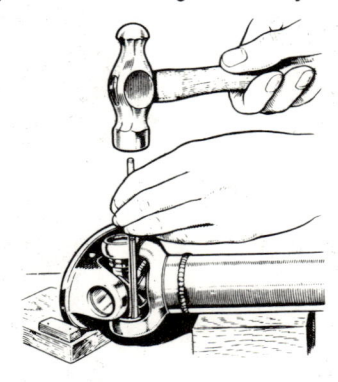

Fig. 8.11. Tapping out a universal joint bearing

Fig. 8.12. Separating the universal joint yokes

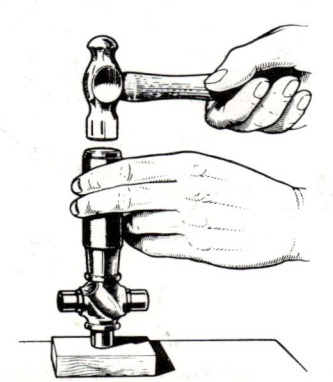

Fig. 8.13. Replacing a gasket retainer

remove them but keep them together for reassembly; discard the oil seal, however.
3 Prise out the outer oil seal and its seating ring; discard the oil seal.
4 Remove the inner bearing inner race.
5 Drift out the outer races of the inner and outer bearings if these are tightly held in position.
6 Use a suitable extractor to withdraw the outer bearing inner race.
7 Press the new inner and outer bearing outer races into the hub carrier making sure that they seat correctly in their recesses.
8 Hold the hub carrier so that the outer bearing will be at the top; place the outer bearing inner race in position, and now press the outer oil seal into its recess.
9 Place the outer oil seal seating ring over the hub and then press the hub into position in the outer bearing inner race, until it is fully home.
10 The next task is to adjust the end float of the hub bearing.
11 Hold the hub and the hub carrier vertically in a vice with the inner end of the hub uppermost.
12 Place the inner bearing inner race on the hub followed by the Special Collar of Tool No. J 15 and press the race onto the hub until the inner face is flush with the special collar as shown in Fig 8.16.

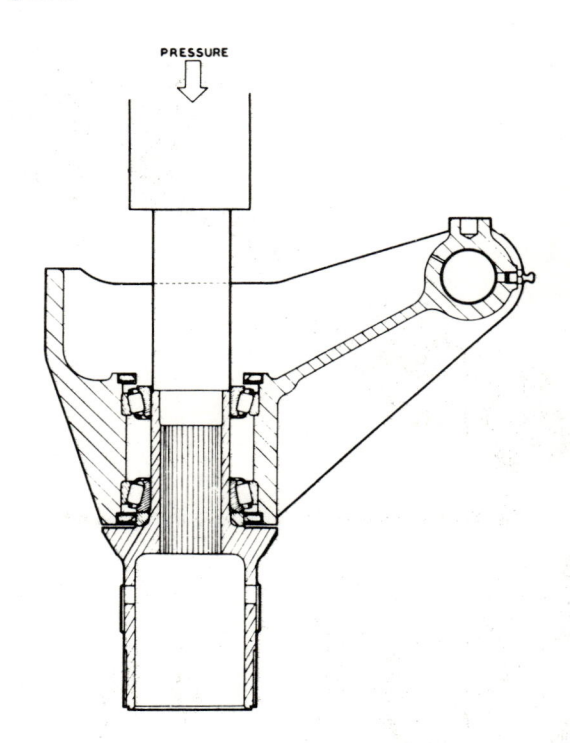

Fig. 8.14. Pressing a hub from the carrier

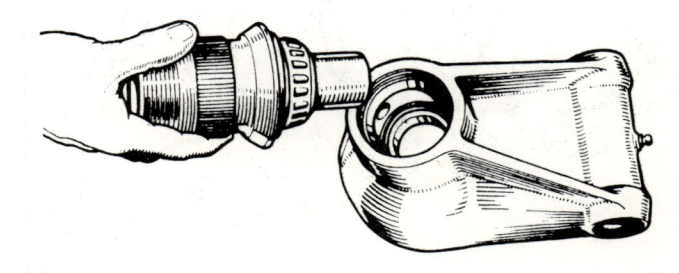

Fig. 8.15. A hub removed from its carrier

13 Now mount a Dial Test Indicator (DTI) on the carrier as shown in Fig. 8.17. Tap the hub inwards and then zerp the DTI on the end of the hub. Lever the hub outwards and record the movement.
14 Remove the Special Collar and then fit a spacer (if necessary) to give an end float of between 0.002" and 0.006' (0.051 to 0.152 mm). Spacers are supplied in thicknesses of 0.109 to 0.0151" in steps of 0.003" (0.076 mm) and are lettered 'A' (smallest) to 'R' (largest) omitting letters 'I' and 'N' as given in the following table:

Letter	Thickness ins.	mm
A	0.109	2.77
B	0.112	2.85
C	0.115	2.92
D	0.118	3.00
E	0.121	3.07
F	0.124	3.15
G	0.127	3.23
H	0.130	3.30
J	0.133	3.38
K	0.136	3.45
L	0.139	3.53
M	0.142	3.61
P	0.145	3.68
Q	0.148	3.75
R	0.151	3.87

15 For example, assume that the end float measured with the collar in position is 0.25" (0.64 mm). Take the mean permissible end float as 0.004" (0.10 mm) and subtract this from the measured end float giving 0.021" (0.53 mm). The Special Collar is 0.150" (3.81 mm) thick so the thickness of the spacer to be fitted will be 0.150-0.021" i.e.0.129" (3.28 mm). The nearest spacer in thickness to this is 0.130" (3.30 mm) so fit a letter 'H' spacer in place of the Special Collar.
16 Fit the inner oil seal and its seating ring.
17 Fit the half shaft splined end to the hub and tighten the castellated nut as described in Section 3 paragraph 11 but before fitting the split pin, carry out a further check to prove correctness of the hub end float.
18 Finally fit the split pin to secure the castellated nut and follow the instructions given in Section 3 for refitting the assembly to the car.

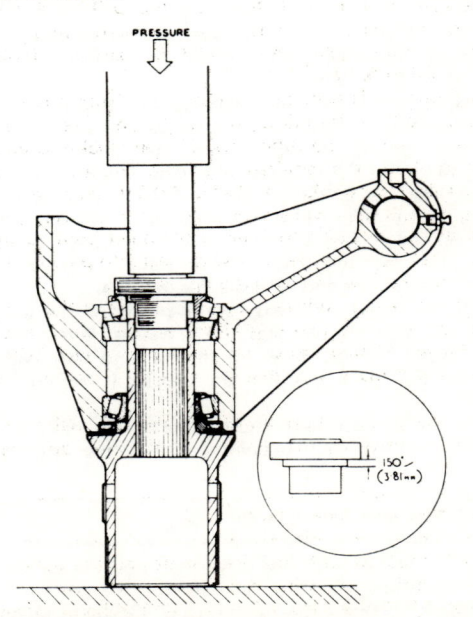

Fig. 8.16. Pressing in the hub inner bearing inner race using the special collar of Tool J.15

7 Drive shafts - removal, checking of end float, and refitting

1 This is a job best done with the final drive unit removed from the car; instructions for this will be found in Section 8.
2 An exploded view of the drive shaft drive assembly is given in Fig. 8.18.
3 Clean all dirt from the area of the drive shafts.
4 Remove the brake caliper and disc but be careful to note the position and number of the small shims which are located between the caliper and the carrier. These must be replaced in their original positions.
5 Unscrew the five bolts securing the drive shaft bearings housings and then, using a hide faced hammer, carefully tap the shaft out of the carrier (Fig. 8.19). Note the adjusting shims under the bearing housing and keep them together as a set.
6 The first task in dismantling the shaft assembly is to open the tab washer securing the nut at the outer end of the shaft and then remove the nut, the tab washer and the plain washer.
7 Next press the drive shaft with the inner bearing inner race the spacing collar and endfloat shims in position, through the flange and bearing housing.
8 If the bearings are to be replaced, remove the end float shims and spacing collar and keep them together as a set. Use a suitable extractor to withdraw the inner bearing inner race from the shaft.
9 Drift out the inner bearing outer race and, using a suitable sized tube on the outer race, press out the complete outer bearing and the oil seal.
10 When reassembling, first press in the new inner and outer bearing races until they are fully home in their recesses. **The races must be fitted so that the bearings will be opposed.**
11 Press the inner bearing inner race onto the shaft until it is fully home against the shoulder and so that the race is fitted the correct way round.
12 Fit the spacing collar and the original end float shims.
13 Fit the drive shaft into the bearing housing and place the bearing inner race on the shaft from the opposite end. Do not fit the oil seal yet.
14 Place the plain washer on the threaded end of the drive shaft followed by a new tab washer. Fit the nut and tighten down but do not lock the tab washer.
15 The end float of the shaft must now be checked and this is done by mounting a Dial Test Indicator (DTI) on the carrier, tap the shaft inwards and then zero the DTI on a convenient point on the shaft. Lever the shaft outwards and record the amount of movement, this should be between 0.001" and 0.003" (0.25 to 0.76 mm).
16 If adjustment is necessary, remove the flange nut and the tab and plain washers. Withdraw the flange and the outer bearing inner race. Remove the shims; check their thickness and add or remove to obtain the required end float. The shims are available in three thickness, 0.003", 0.005" and 0.010. Adding a shim will increase the end float whilst removing a shim will decrease it.
17 When the correct end float is obtained, replace the outer bearing inner race and press a new oil seal into position until is is flush with the casing and with the lips inwards.
18 Refit the flange followed by the plain and the tab washers making sure that the two tags on the washers locate in the holes in the flange. Tighten down on the nut and lock with one or more tabs but make sure that the tabs lie as flat on the nut as possible.
19 Refit the brake caliper and the brake disc making sure that the shims are correctly fitted between the caliper and the carrier.

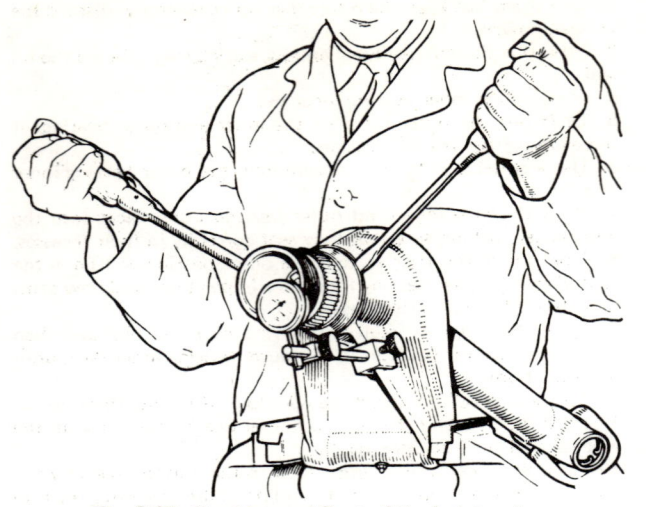

Fig. 8.17. Checking end float of the hub bearing

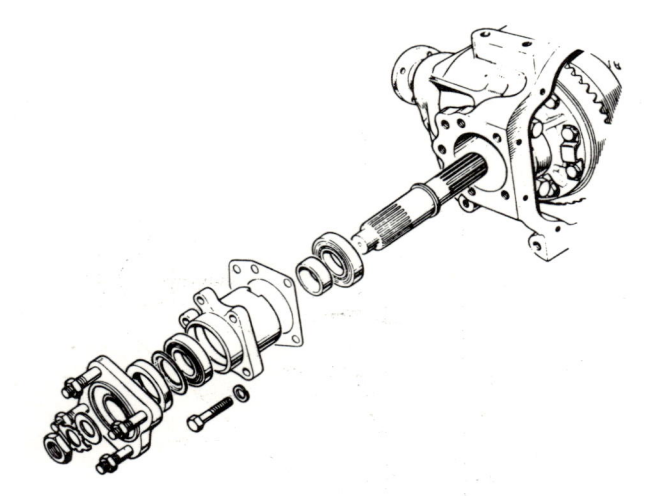

Fig. 8.18. Exploded view of a drive shaft assembly

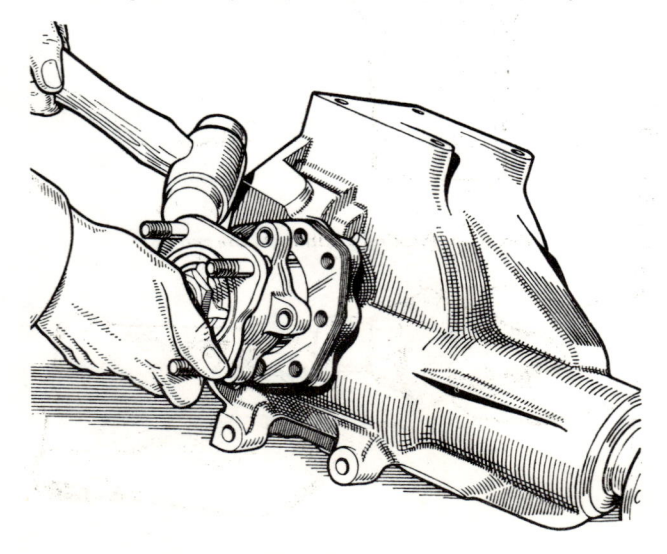

Fig. 8.19. Removing a drive shaft

8 Final drive unit - checking for wear

Wear is checked with one drive shaft and the pinion locked, in this condition the other drive shaft must not turn radially more than ¾" (19mm) measured on a 6" (152mm) radius. If this amount is exceeded, the final drive unit should be stripped and adjusted as described in Sections 9 and 12. A check on the end

Chapter 8/Rear axle

float of the drive shafts as set out in Section 7 paragraph 15 should be made also.

9 Final drive unit - dismantling

1 Follow the instructions given in Section 2 and remove the final drive unit.
2 Knock up the locking tabs and then unscrew the brake caliper mounting bolts (locking wire is used instead of tab washers on some early model cars).
3 Remove the caliper but be careful to note the number and position of the small shims which are located between the caliper and the carrier. These must be replaced in their original positions.
4 Remove the brake disc.
5 Thoroughly clean the exterior of the gear carrier

6 Drain the lubricant from the gear carrier and then remove the rear cover.
7 Flush out the unit and carefully inspect the parts for wear or damage in situ so that appropriate corrective action can be taken later.
8 Remove the drive shafts in the manner described in Section 7. Remove the two bolts which hold each differential bearing cap and then withdraw the differential unit.
10 The pinion can now be removed. First remove the nut and washer.
11 Using a suitable puller, withdraw the universal joint companion flange.
12 Now **press** the pinion out of the outer bearing, take note of the shims and keep these intact as a set. It is important that the pinion is pressed and not driven out otherwise damage to the outer bearing will result.
13 Remove the pinion oil seal to gether with the oil slinger and the outer bearing cone.
14 If the outer bearing is to be replaced, it can be removed using Tool No. SL 12 as shown in Fig. 8.20. If this tool is not available and the bearing cup is to be scrapped, it is possible to drive it out - you will find that the shoulder locating the bearing is recessed to allow this.
15 The inner bearing outer race can be removed in the same manner but take care of the shims fitted between the bearing cup and the housing abutment face.
16 The inner bearing may be driven out but if it is not to be renewed, use Tool no. SL. 14 with Adaptor SL.11 P/AB2 as shown in Fig.8.21.

10 Differential unit - dismantling

1 An exploded view of the differential unit is given in Fig. 8.3 and reference to this should be made during the following operations:
2 Before starting work, look for mating marks on the differential case, these will probably be similar to those depicted in Fig. 8.22 Make your own marks, if none can be seen, because it is important that the two halves of the casing are reassembled in their original position in relation to each other.
3 The crownwheel may be left in position if desired but, if it is to be removed, first mark its position in relation to the differential casing and then remove the ten securing bolts and the five lock straps. Carefully tap the crown wheel off the casing and place it where it will not be damaged. A word of warning - the crownwheel and the pinion are supplied as a matched assembly; they are lapped together at production and **must** be kept as a set.
4 Withdraw the differential bearings, if neccessary, using Tool No. SL.14 with Adaptor SL.11 D/A5 as depicted in Fig. 8.23.
5 Remove the eight bolts (9) securing the two halves of the casing.
6 Split the casing and take out the clutch discs(3) and the plates (2 and 4) from one side.
7 Remove the differential side gear ring (5).

8 Take out the pinion side gear (6) and the pinion mate cross shafts (7) complete with the pinion mate gears.
9 Separate the cross shafts (10).
10 Remove the remaining side gear and the side gear ring.
11 Extract the remaining clutch discs and plates.

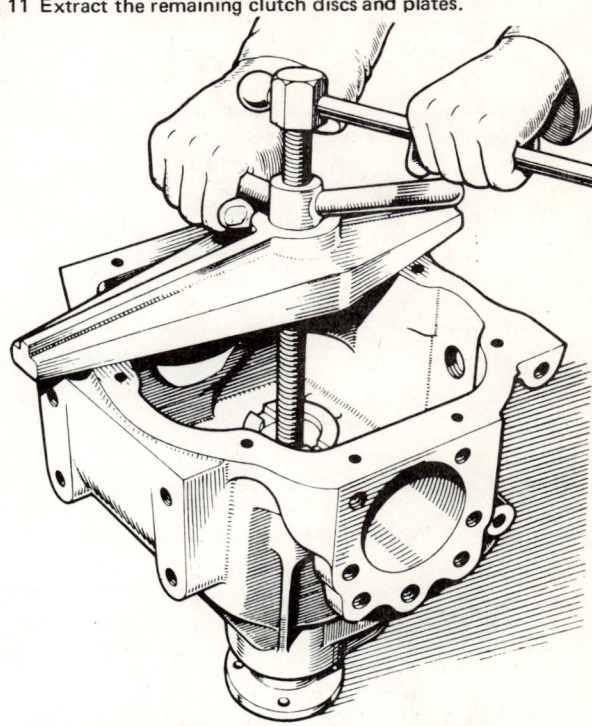

Fig. 8.20. Withdrawing the pinion inner bearing outer race

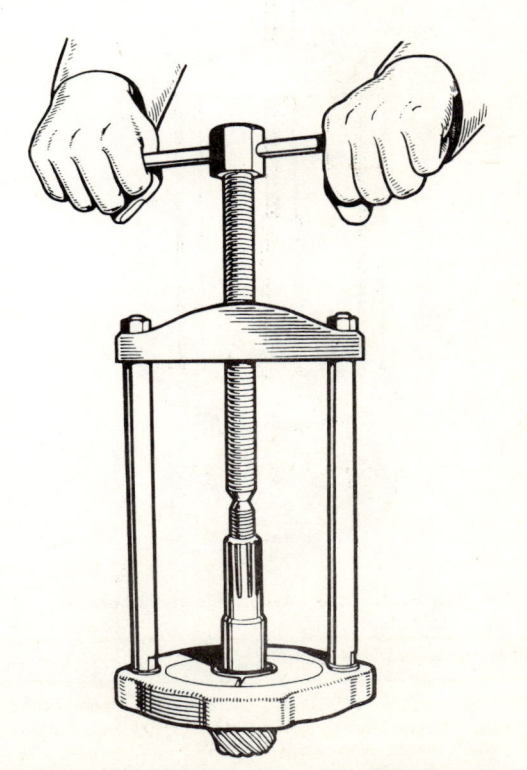

Fig. 8.21. Withdrawing the pinion inner bearing inner race

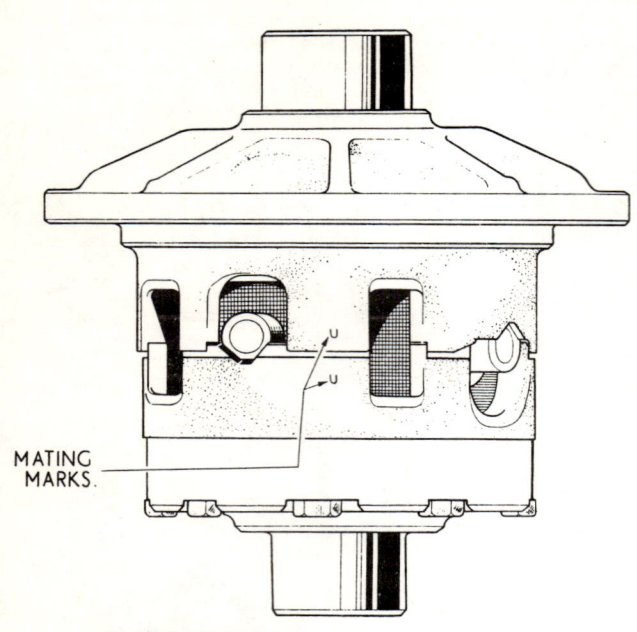

Fig. 8.22. Mating marks on the differential case

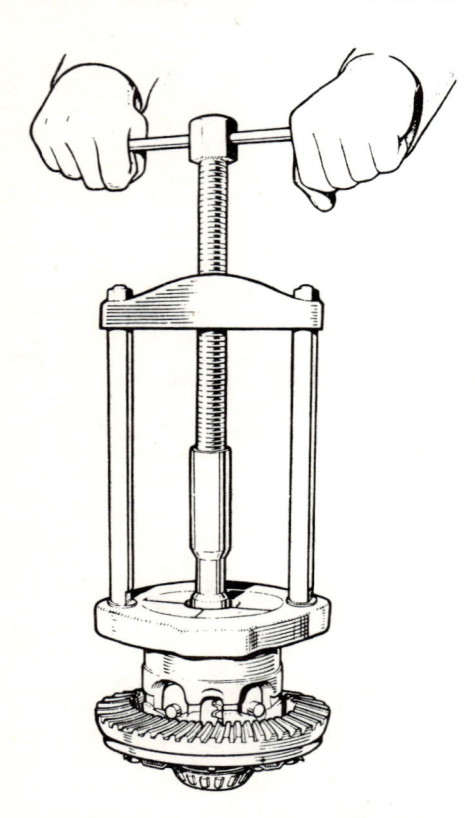

Fig. 8.23. Withdrawing a differential bearing

11 Differential unit - reassembly

1 Fit a dished clutch friction plate to the flange half of the casing so that the convex side is against the casing. Follow up by fitting the two clutch friction discs and friction plates alternately as shown in Fig. 8.3.
2 Fit the side gear ring so that the serrations on the gear mate with the serrations on the two friction discs.
3 Place one of the side gears into the recess in the side gear ring so that the splines in both are aligned.
4 Fit the pinion mate cross shafts together, assemble the gears and then offer the assembly to the flange half of the casing making sure that the ramps on the cross shafts coincide with the mating ramps in the case.
5 Build up with the remaining side gear and side gear ring so that the splines are aligned followed by the remaining clutch plates and discs.
6 Offer the bottom half of the casing to the flange half, with the alignment marks coinciding (Fig.8.22), and now position the tongues of the clutch friction plates so that they align with the grooves in the differential case.
7 Secure the two halves of the casing together with the eight bolts but do not tighten them at this stage.
8 Check the alignment of the splines in the side gear rings and side gears by inserting the two drive shafts, with the shafts in position (Fig. 8.24), tighten the eight bolts to a torque of 35 to 45 lbs.f.ft (4.8 to 6.2 kg.f.m). If the bolts are tightened without the shafts in position you will find it difficult, if not impossible, to enter them later on.
9 Replace the differential bearings, if applicable, using Tool No. SL 2DB and Adaptor No. SL 2D/B-2.
10 Check that the crownwheel bears the same serial number as the pinion and then, if applicable, assemble it to the casing so that your mating marks are aligned. Place the locking straps in position and secure the crownwheel with the ten bolts. It is advisable to use new lock straps.

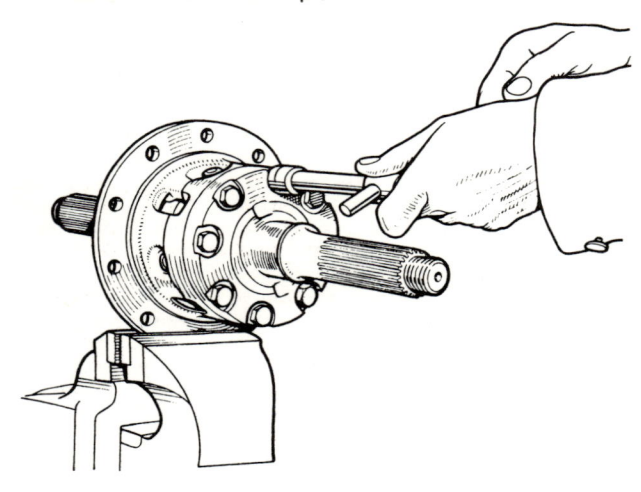

Fig. 8.24. Tightening the differential casing bolts with the drive shafts in position

12 Final drive unit - reassembly

1 Refit the pinion outer bearing outer race to the gear carrier using Tool No. SL.12.
2 Fit the pinion inner bearing outer race, with the original adjusting shims, using Tool SL.12 and Adaptor SL.12 AB-4 as shown in Fig. 8.25.
3 Press the inner bearing inner race onto the pinion using an arbor press and a piece of tube of such diameter that it will contact the inner race only and not the roller retainer.
4 The correct pinion setting must now be obtained and this is a job which calls for extreme accuracy. pinion setting must now be obtained and this is a job which calls for extreme accuracy.
5 The correct setting for a particular pinion is marked on the ground face of the pinion as shown in Fig. 8.26. The serial number at the top is also marked on the crownwheel and is for matching purposes (see paragraph 3 of Section 9). The letter to the left is a production code letter and for our purposes can be

Chapter 8/Rear axle

ignored. The letter and figure on the right refer to the tolerance on offset or pinion drop shown as 'A' in Fig. 8.27. The number at the bottom gives the cone setting distance for this pinion and it may be a Zero (0), or Plus (+), or Minus (-) quantity. When correctly adjusted, a pinion marked Zero will be at the cone setting distance, shown as 'B' in Fig. 8.27, from the centre line of the gear to the face on the small end of the pinion. A pinion marked '-2' as shown should be adjusted to dimension 'B' minus 0.002" (0.051 mm) whereas a cone marked, say '+2' must be set to dimension 'B' plus 0.002" (0.051 mm).
6 First place the pinion, with the inner bearing cone assembled, into the gear carrier and then turn the carrier over and support the pinion with a suitable block of wood.
7 Install the pinion bearing spacer if one was originally fitted to the assembly.
8 Install the original outer bearing shims on the pinion shank and make sure that they seat on the spacer or on the shoulder of the shank according to the construction of the particular unit.
9 Fit the pinion outer bearing inner race, the companion flange and its washer and nut. Tighten down on the nut. The oil slinger and the oil seal assembly are left out at this stage.
10 Now refer to Fig. 8.28 and mount a Dial Test Indicator (DTI) on the bracket of Tool I 3P as shown. The next stop is to zero the DTI on the 4HA setting of the gauge block and this is done with the gauge block and the bracket, with DTI. resting on a surface plate.
11 Mount the bracket and tool as shown in Fig. 8.28 and take a reading on the differential bore. Move the bracket and DTI so that the spindle of the DTI is indeed at the lowest point of the bore and this will be indicated by the lowest reading obtained. The minimum DTI reading shows the actual deviation of the pinion setting from the zero cone setting, record the direction of any such deviation as well as the magnitude.
12 If the pinion setting recorded does not match that stamped of the the face of the pinion, dismantle the pinion assembly and remove the pinion inner bearing outer race. Add or remove shims as required from the pack locating the bearing outer race. Shims are available in thicknesses of 0.003", 0.005", and 0.010" (0.076, 0.127 and 0.254 mm).
13 Refit the pinion assembly and again check the pinion setting.
14 When the correct setting has been obtained, check the pinion bearing preload. There should be a slight drag or resistance to turning with no end play of the pinion (the correct torque figure is given in Specifications at the beginning of this Chapter).

A low preload torque will result in excessive deflection of the pinion under load whilst a high torque will lead to failure of the bearings. The preload can be adjusted by removing (to increase), or adding (to decrease), shims between the outer bearing inner race and the pinion shank spacer or shoulder as the case may be but **do not** touch the shims behind the inner bearing outer race as this will upset all the work you have done in adjusting the pinion setting.
15 The oil seal assembly and the oil slinger can be fitted at this stage but they are usually left out until the differential assembly has been installed and adjusted.
16 Check that the differential bearing surfaces in the carrier and that the differential bearing caps are clean.
17 Install the differential assembly and secure in position with the caps and bolts ensuring that the numerals marked on the carrier face, and on the caps, correspond as shown in Fig.8.29. Tighten the bolts to the torque figure given in Specifications.
18 Refer to Fig.8.30 and mount the DTI as shown so that the indicator button is against the back face of the crownwheel.
19 Turn the pinion by hand and check the run-out of the wheel. This should not exceed 0.005" (0.13 mm) and if it does, remove the differential assembly, take off the crown wheel and clean the locating surfaces. Remove any burrs which may be present and then refit the assembly and recheck.
20 The next task is to adjust for backlash between the crownwheel and the pinion. The amount of backlash to be allowed for a particular crownwheel/pinion set is etched on the back of the crownwheel and is given as a figure representing the backlash to be allowed in thousandths of an inch.

Fig. 8.25. Fitting the pinion inner bearing outer race

Fig. 8.26. Typical markings on the ground face of the pinion

21 Slacken the differential cap bolts until the bearings are lightly held.
22 Install the drive shafts but do not fit any shims between the shaft bearing housings and the differential carrier. Secure each bearing housing with three bolts, evenly spaced, and tighten down evenly until the shaft is felt to contact the differential unit. Now bring the crownwheel fully into mesh with the pinion by slackening the bolts of one drive shaft bearing housings and tightening on the other side. Make sure that the crownwheel is not bearing on the pinion to the extent of effecting pinion preload.
23 Set up a DTI on the differential carrier as shown in Fig.8.31. so that the indicator button is bearing against one of the teeth of the crownwheel as nearly as possible in line with the direction of tooth travel. Record the reading of the DTI.
24 By tightening the drive shaft bearing bolts on one side and

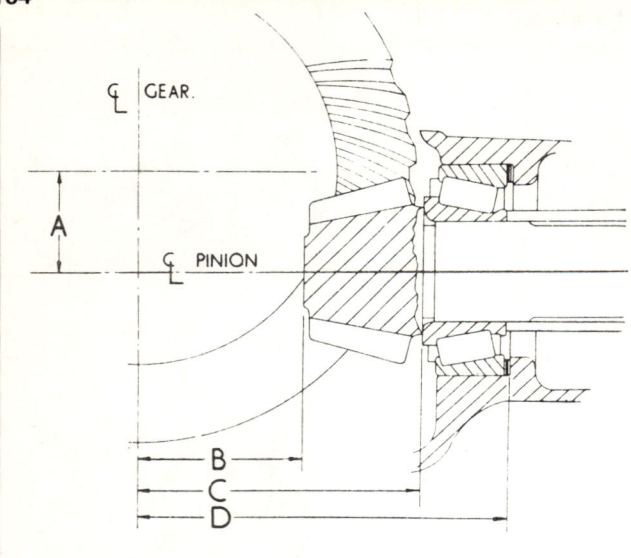

Fig. 8.27. Pinion setting distances

A	Pinion drop	1.5 in. (38.1 mm)
B	Zero cone setting	2.625 in. (66.67 mm)
C	Mounting distance	4.312 in. (108.52 mm)
D	Centre line to bearing housing	5.495 in. (139.57 mm) to 5.505 in. (139.83 mm)

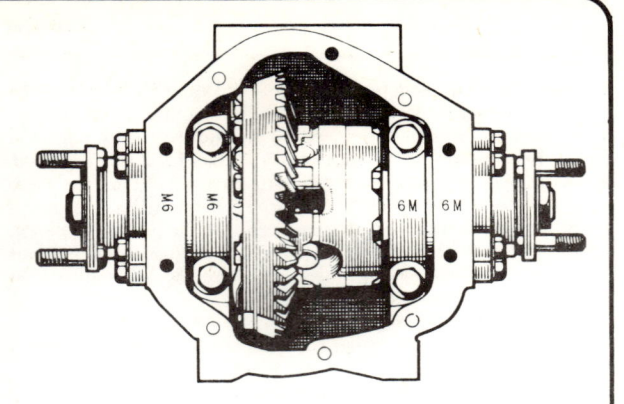

Fig. 8.29. Differential bearing cap markings

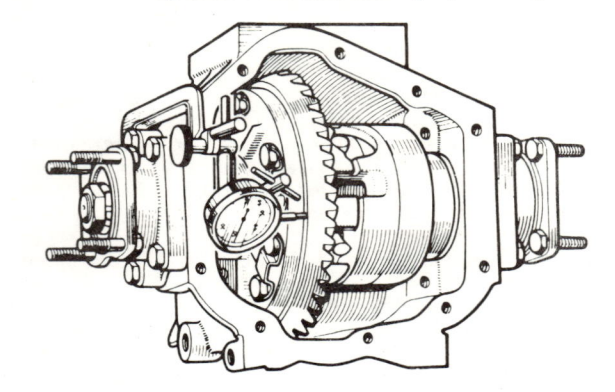

Fig. 8.30. Checking run-out of the crown wheel

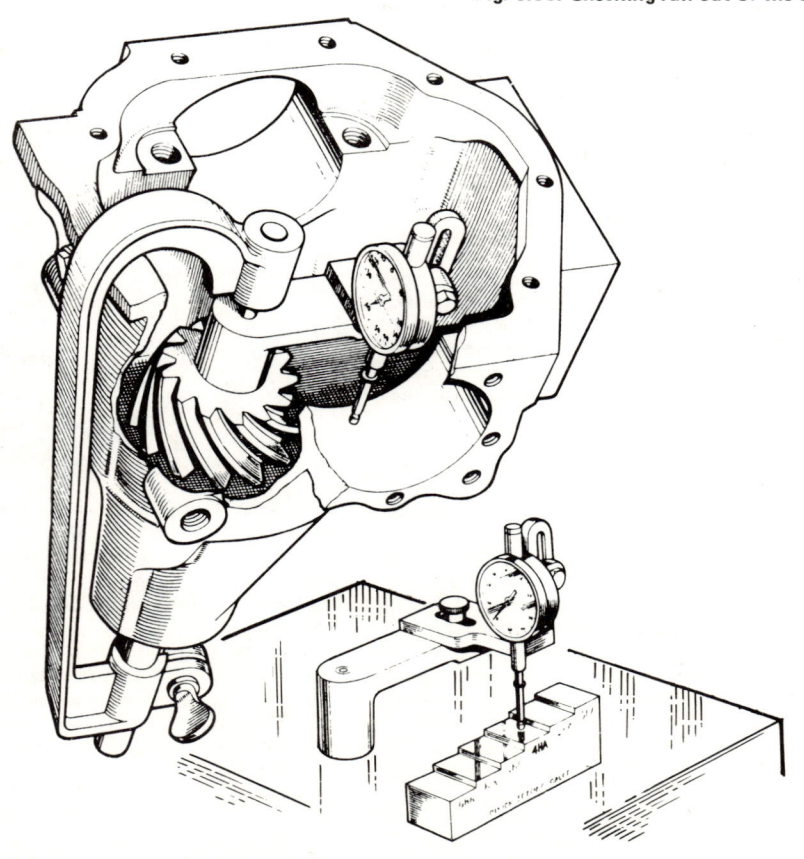

Fig. 8.28. Checking the pinion cone setting (inset illustrates zeroing the DTI on the 4HA dimension of the gauge block)

Chapter 8/Rear axle

slackening on the other, move the differential assembly away from the pinion to the amount of the backlash etched on the crownwheel.

25 When satisfied with the setting, tighten down on the differential bearing cap bolts to the torque given in Specifications at the beginning of this Chapter.

26 When the correct backlash has been obtained, check the gap on each side of the carrier between the drive gear bearing housing and the carrier using feeler gauges (Fig. 8.31). Check all round the housing to make sure that the gap is equal in positions. Now subtract 0.003" (0.076 mm) from the recorded gap to give the correct preload and then install shims on each side to the requisite amount. Shims are available in thicknesses of 0.003", 0.005", 0.010" and 0.030" (0.076, 0.127, 0.254 and 0.762 mm). For example, assume the backlash etched on the gear is 0.007" (0.178 mm) and when this figure is obtained as described above it is found that the gap found at one side is 0.054" (1.37 mm) and at the other side it is 0.046" (1.17 mm). The amount of shims required is 0.054" - 0.003" i.e. 0.051" (1.30 mm) at one side and 0.046" - 0.003" i.e. 0.043" (1.09 mm) at the other.

27 Finally secure the drive shafts in position with all bolts and tighten down to the torque given in Specifications.

28 When satisfied with all adjustments, mark a number of the crownwheel teeth very sparingly with a marking compound (engineer's blue is suitable) and move the painted teeth into mesh with the pinion until a good impression of tooth contact is obtained. Refer to Fig. 8.32 and take remedial action as may be necessary.

29 All the necessary adjustments have now been effected and final assembly can commence.

30 Remove the pinion nut, washer and withdraw the companion flange.

31 Install the pinion oil slinger and then fit the oil seal assembly using Tool SL 4P/B as shown in Fig.8.33. Place the oil seal with the dust excluder flange uppermost (do not forget the oil seal gasket as used with the metal case type seal on later models), fit the installation collar of the tool and then fit and tighten down the pinion nut and washer to drive the assembly home. Remove the nut, washer and installation collar.

32 Fit the companion flange, the washer and the nut. Tighten the nut to the correct torque loading (see Specifications) and lock.

33 Place a new rear cover gasket in position and then fit the cover and tighten down on the setbolts but do not forget to replace the final drive ratio and 'Powr-Lok' tags which are attached by two of the bolts.

34 Replace the drain plug and make sure that it is secure. Fill the unit with the appropriate quantity hypoid lubricants. Fit and tighten the filler plug.

35 Check for oil leaks at the cover, the pinion oil seal and where the differential cap bolts break through.

36 Replace the brake disc and caliper making sure that the caliper is correctly centralised by refitting the correct pack of shims (see Chapter 9 Braking system). Fit new tab washers, tighten the mounting bolts to a torque of 55 lb.f.ft (7.6 kf.f.m) and lock with the tabs.

37 Refit the half shafts (Section 3), refit the assembly to the cross beam and re-install in the car in the manner described in Section 2.

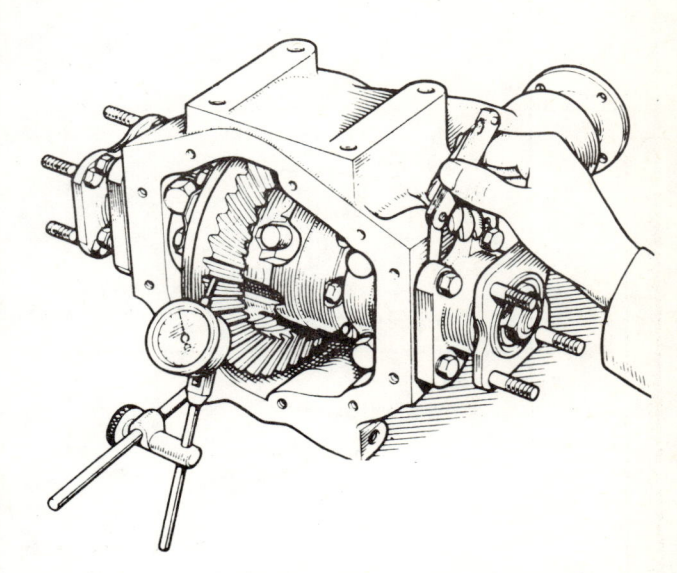

Fig. 8.31. Checking backlash and the position of the drive shafts

Fig. 8.33. Fitting the pinion oil seal

	TOOTH CONTACT (*DRIVE GEAR*)	CONDITION	REMEDY
A	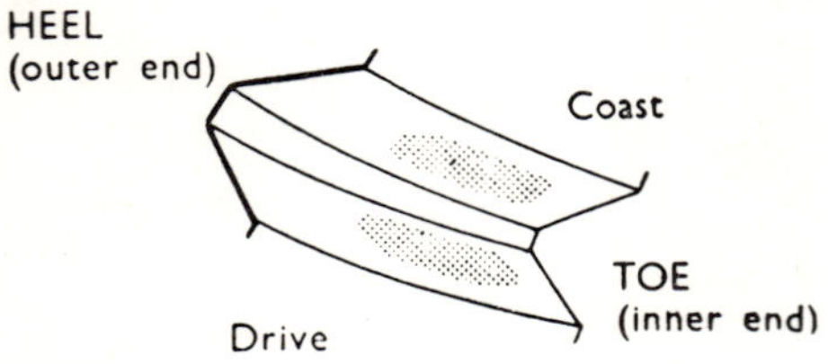 HEEL (outer end), Coast, TOE (inner end), Drive	**IDEAL TOOTH CONTACT** Evenly spread over profile, nearer toe than heel.	o ― o
B	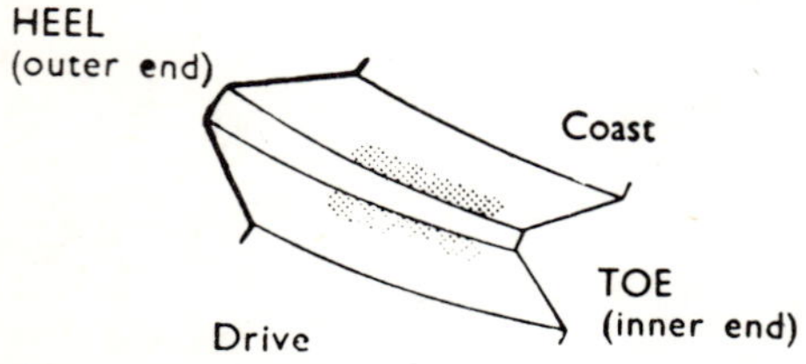 HEEL (outer end), Coast, TOE (inner end), Drive	**HIGH TOOTH CONTACT** Heavy on the top of the drive gear tooth profile.	Move the DRIVE PINION DEEPER INTO MESH. *i.e.,* REDUCE the pinion cone setting.
C	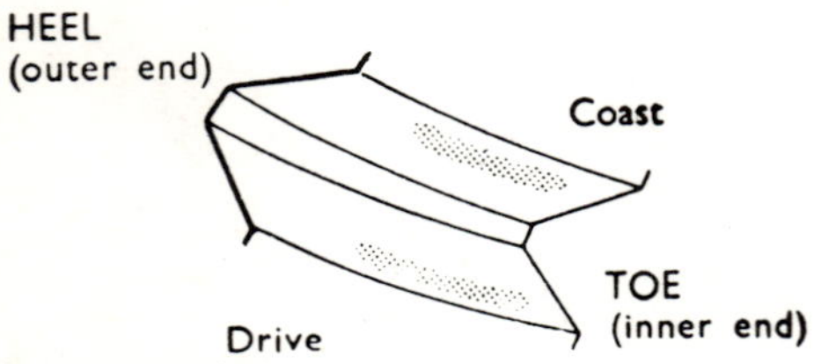 HEEL (outer end), Coast, TOE (inner end), Drive	**LOW TOOTH CONTACT** Heavy in the root of the drive gear tooth profile.	Move the DRIVE PINION OUT OF MESH. *i.e.,* INCREASE the pinion cone setting.
D	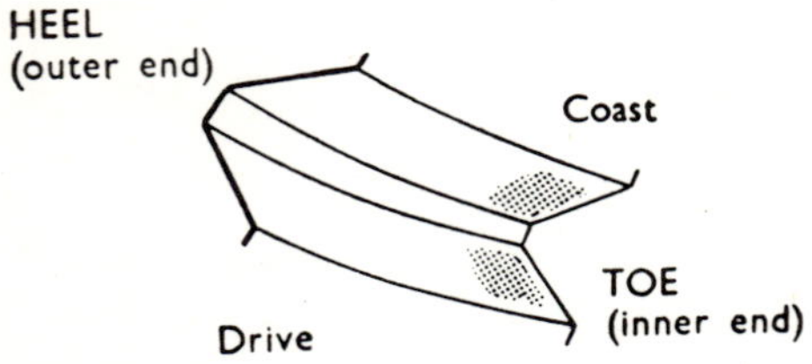 HEEL (outer end), Coast, TOE (inner end), Drive	**TOE CONTACT** Hard on the small end of the drive gear tooth.	Move the DRIVE GEAR OUT OF MESH. *i.e.,* INCREASE backlash.
E	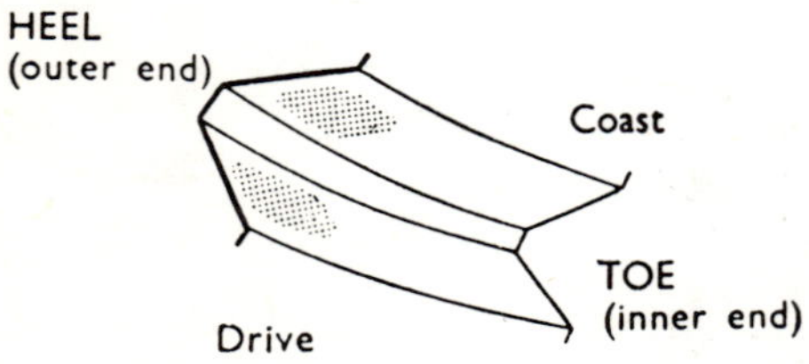 HEEL (outer end), Coast, TOE (inner end), Drive	**HEEL CONTACT** Hard on the large end of the drive gear tooth.	Move the DRIVE GEAR INTO MESH. *i.e.,* DECREASE backlash *but* maintain minimum backlash as given in "Data"

Fig. 8.32. Tooth contact indication

Chapter 9 Braking system

Contents

Bellows type vacuum servo, 3.8 litre models - general ... 12	mantling and reassembly ... 30
Bellows type vacuum servo unit - servicing ... 16	Handbrake friction pad carriers, non-self-adjusting type - dismantling and reassembly ... 29
Bleeding the hydraulic system ... 2	Handbrake friction pads - renewing ... 28
Brake/clutch pedal box assembly, 4.2 litre models - removal and refitting ... 25	Handbrake - general ... 26
Brake/clutch pedal box and vacuum servo unit, 3.8 litre models - removal and refitting ... 23	Lockheed dual-line servo system, 4.2 litre models - general... 17
Brake fluid level and handbrake warning light - setting ... 33	Lockheed dual-line servo system overhaul - general ... 18
Brake linkage, 3.8 litre models - dismantling and reassembly 24	Master cylinders, 3.8 litre models - dismantling and reassembly ... 11
Brake overhaul - general ... 7	Master cylinders, 3.8 litre models - removal and refitting ... 10
Brake piston seals (early type) - renewal ... 8	Master cylinder and reaction valve - dismantling and reassembly ... 22
Brake piston seals (later type) - renewal ... 9	
Fault diagnosis ... 34	Master cylinder and reaction valve - removal and refitting ... 21
Friction pads - removal and refitting ... 3	Rear brake caliper - removal and refitting ... 5
Front brake caliper - removal and refitting ... 4	Remote servo and slave cylinder - dismantling and reassembly ... 20
Front and rear brake discs - removal and refitting ... 6	
General description ... 1	Remote servo and slave cylinder - removal and refitting ... 19
Handbrake adjustment ... 32	Vacuum reservoir and check valve - removal and refitting ... 15
Handbrake cable - removal and refitting ... 31	Vacuum servo unit - dismantling and reassembly ... 14
Handbrake friction pad carriers - removal ... 27	Vacuum servo unit - removal and refitting ... 13
Handbrake friction pad carriers, self-adjusting type - dis-	

Specifications

	3.8	4.2
Make ...	Dunlop	
Type ...	Bridge type caliper with quick change pads	
Brake disc diameter		
Front ...	11" (27.9 cm)	
Rear ...	10" (25.4 cm)	
Master cylinder bore diameter ...	5/8" (15.87 mm)	7/8" (22.23 mm)
Brake cylinder bore diameter		
Front ...	2.1/8" (53.97 mm)	
Rear ...	1¾" (44.45 mm)	
Servo unit type ...	Dunlop bellows type vacuum	Lockheed dual-line
Main friction pad material ...	Mintex M.59 (early cars M.40 or M.33)	Mintex M.59
Handbrake friction pad material	Mintex M.34	

1 General description

The front wheel brake units consist of a disc mounted on the front wheel hubs and a braking unit rigidly attached to each of the front suspension members. The rear wheel discs are mounted between the half shafts and the drive shafts and are attached by the same bolts which secure the half shafts to the drive shafts,

the brake units are attached to the differential case. An example of a typical disc and brake unit assembly is shown at Fig.9.1.
The brake unit consists of a caliper which straddles the disc and houses a pair of friction pads located between a keep plate bolted to the caliper bridge and two support plates accommodated in slots in the caliper jaw. Cylinder blocks are bolted to each outer face of the caliper and contain a piston which is keyed to each friction pad. One cylinder block is connected to

Chapter 9/Braking system

the hydraulic braking system and interconnection between the blocks is provided by a rigid pipe. Wear on the friction pads is automatically taken up by a self adjusting mechanism built into each piston assembly and this maintains a "brake-off" working clearance of 0.008" to 0.010" (0.20 to 0.25 mm) between the pads and the disc throughout the life of the pads. This self adjusting mechanism is illustrated in Fig.9.2 which shows the set up for early model 3.8 litre cars whilst Fig.9.3 illustrates the mechanism as fitted to later model 3.8 litre and all 4.2 litre cars.

The front and rear brakes are hydraulically independent of each other, this is achieved in the case of 3.8 litre models by the use of two master cylinders, both of which are operated simultaneously by the brake pedal, one cylinder feeds the front brakes and the other the rear brakes only. Both cylinders have their own fluid reservoir as depicted in Fig.9.4 for right hand drive models where the reservoir to the right, and nearest the centre line of the car, supplies the rear brakes and the one in the centre feeds the front brakes. The layout of reservoirs for left hand drive models is shown at Fig.9.5, they are mounted on the front frame assembly adjacent to the exhaust manifold, here the forward reservoir serves the rear brakes and the centre one the front brakes. Both reservoirs are provided with an electrical switch connected to a warning light on the front facia panel to give an indication when fluid in the reservoir is dangerously low, in addition, each reservoir has an indicator pin located between the two switch terminals in the reservoir cap. This indicator is used by pressing down and then allowing it to return to its normal position, if it can then be lifted by the thumb and forefinger the reservoir needs topping up.

Independent hydraulic operation of the front and rear brakes of 4.2 litre models is provided by the Lockheed dual-line servo braking system as illustrated in Figs.9.21 and 9.22. This system utilises the same reservoir layout as already described for 3.8 litre models.

The handbrake, which operates on the rear brake discs only, is mechanically operated and is entirely independent of the hydraulic rear brake assembly.

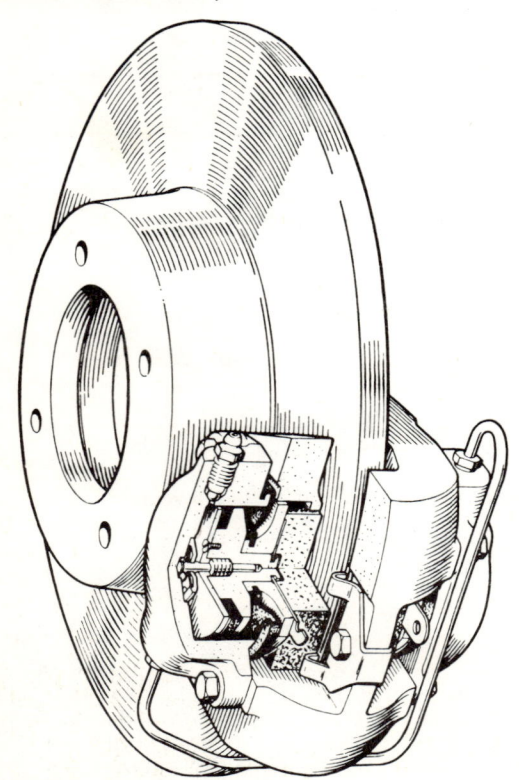

Fig. 9.1. Sectioned view of a disc brake (front brake assembly illustrated)

2 Bleeding the hydraulic system

Whenever the brake system has been overhauled to the extent of disconnecting a hydraulic union, or the level of fluid in a reservoir has become too low, air will have entered the system and bleeding (expelling air) will be necessary. During the following operations, the level of fluid in the reservoir should not be allowed to fall below half full, otherwise air will be drawn into the system again; if the exterior of the reservoir is kept clean it will be easy to see the fluid level. The recommended brake fluid is Castrol/Girling Universal Brake and Clutch Fluid, this conforms to SAE 70 R3 and, where this is not available, only fluid guaranteed to conform to that specification should be used.
1 Obtain a clean and dry glass jar, a length of plastic tubing of suitable diameter to fit tightly over the bleed screw, and a supply of hydraulic fluid.
2 Fill the appropriate master cylinder reservoir and the bottom inch of the jar with hydraulic fluid. Take great care not to allow any fluid to come into contact with the paintwork as it acts as a solvent and will damage the finish.
3 Place the car over a pit, or alternatively, jack up each wheel in turn to be worked on - commencing with the nearside rear brake. Ensure that the wheels are firmly chocked and that the raised wheel is firmly supported, preferably on an axle stand. Release the handbrake.
4 The location of the bleed screw (one per assembly) is shown at Fig.9.6. Remove the rubber dust cap, if fitted, clean around the screw and then slide one end of the plastic tube over the screw and insert the other end in the jar of fluid.
5 Use a suitable open ended spanner and unscrew the bleed screw about half a turn.
6 Have an assistant to depress the brake pedal slowly downwards and upwards through its full stroke.
7 Watch the flow of fluid in the jar and, when the air bubbles cease to emerge, with the next down stroke of the pedal hold it at the bottom of its stroke and then tighten the bleed screw.
8 An illustration of a typical set-up for bleeding the brakes is shown in Fig.9.7.
9 Repeat the above operation to bleed all the brakes.
10 If after all brakes have been bled, the brake pedal still feels spongy, this is an indication that there is still air in the system or that a master cylinder is faulty and should be overhauled as described later in this Chapter.
11 Finally check and top up the reservoirs with fresh hydraulic fluid. Never reuse old brake fluid.

3 Friction pads - removal and refitting

1 The minimum permissible thickness of friction pads, including the backing plates, is 0.25" (7 mm) after which the pads complete with backing plates should be renewed.
2 A friction pad assembly is shown in Fig.9.8.
3 To ensure maximum braking efficiency, it is advisable to renew the pads all round but in any event change all pads to the front or rear wheels as the case may be otherwise there is a distinct possibility of getting uneven braking with the result that the car will pull to one side or the other when the brakes are applied.
4 To change the front pads, jack up the car, make sure that it is firmly supported (use of an axle stand is recommended) and then remove the front wheel on the side that is to be worked on. In the case of the rear wheels, place the car over a pit, on a ramp, or raise the rear of the car to give access to the brake assemblies. If you raise the car, do make sure that it is safely supported before crawling underneath.
5 Remove the nut, washer and bolt securing the keep plate and withdraw the plate.
6 Due to the self adjusting feature of the brakes (later models) the pad will be very close to the disc, it is possible that the disc will have worn slightly in the area of contact of the pad throwing up a ridge on the periphery which will be accentuated by road

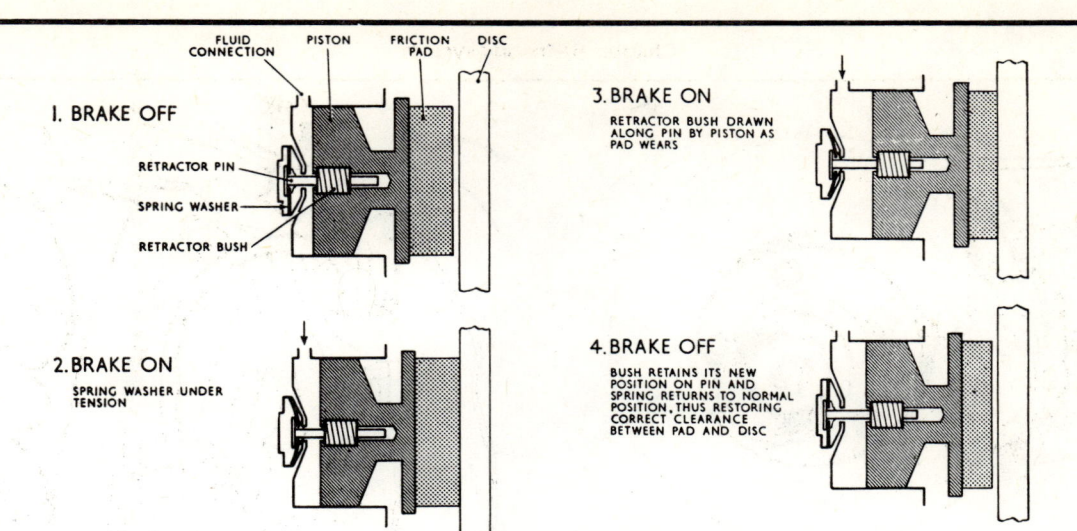

Fig. 9.2. Self adjusting mechanism (early type)

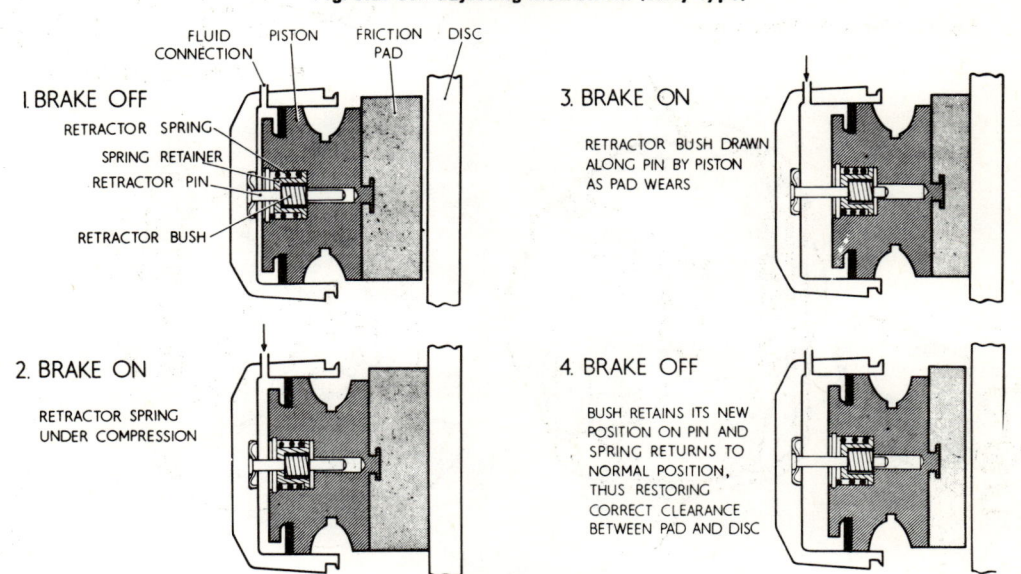

Fig. 9.3. Self adjusting mechanism (later type)

Fig. 9.4. Brake fluid reservoirs, right hand drive

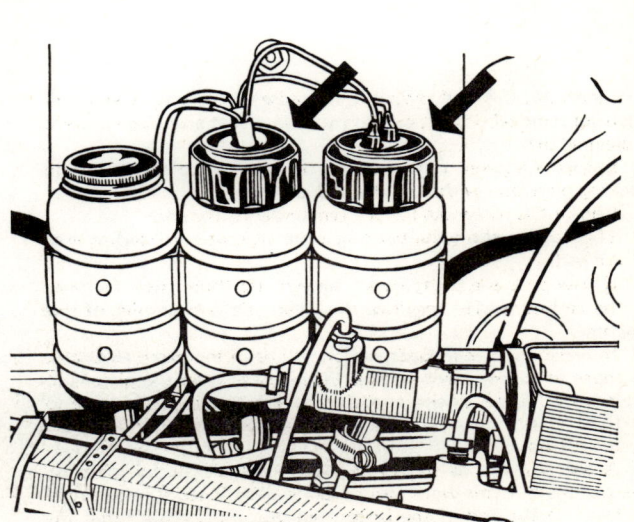

Fig. 9.5. Brake fluid reservoirs, left hand drive

Fig. 9.6. The brake bleed screw

Fig. 9.7. Bleeding the brakes

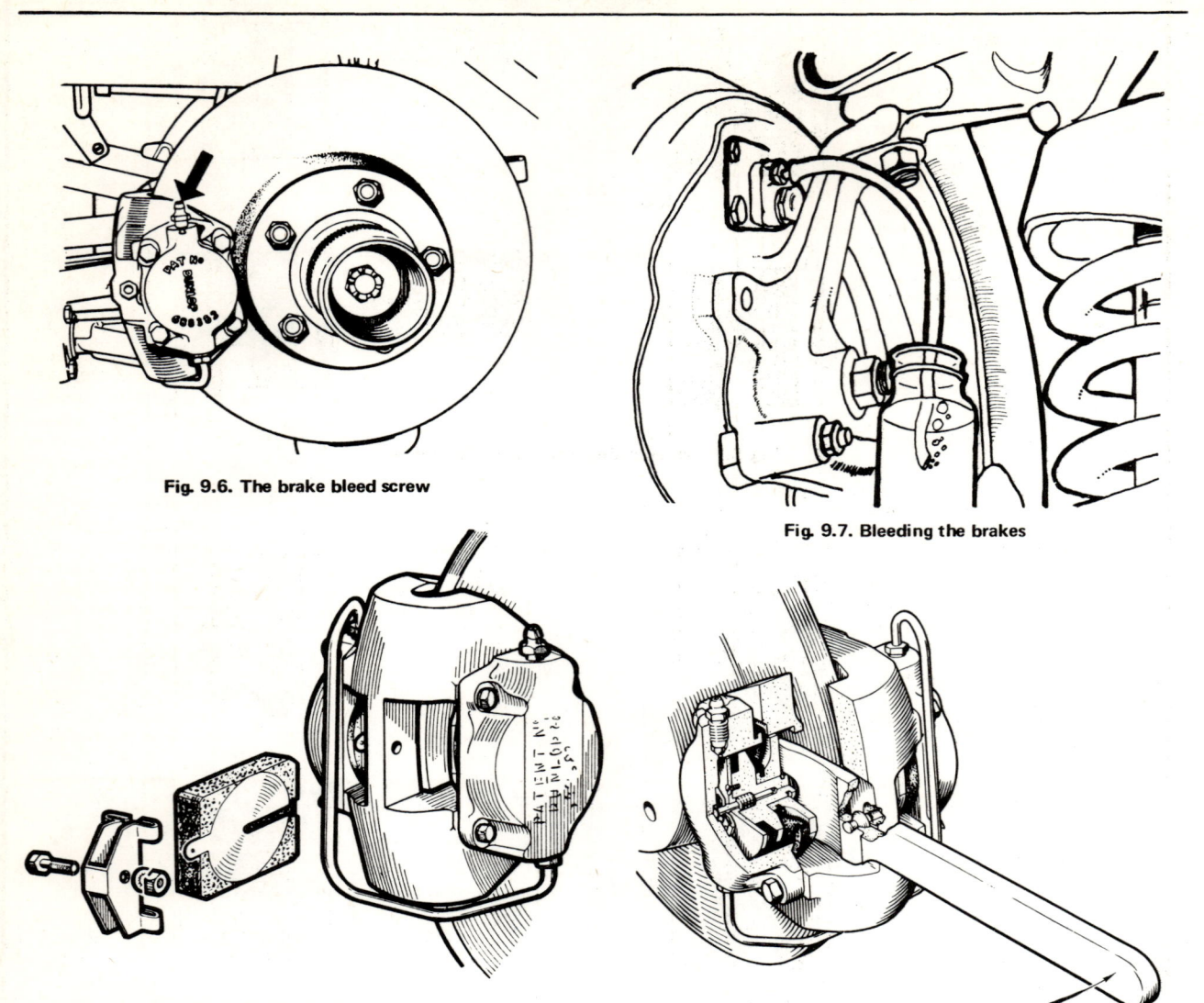

Fig. 9.8. A friction pad assembly

Fig. 9.9. Resetting a piston

dirt and rust, this will probably (because of the small clearance) obstruct removal of the pad - so make sure that any ridge present is cleaned off.

7 Engage a hooked instrument in the hole in the lug of the pad backing plate and withdraw the pad assembly.

8 Repeat the foregoing for the remainder of the pads.

9 Thoroughly clean the backing plate and the surrounding area of the pad.

10 If new pads are being fitted, always use those manufactured to the recommended specification given at the beginning of this Chapter.

11 In order to fit new pads, which will be of increased thickness to those removed, it will first be necessary to reset the pistons to their fully 'home' position. Before doing this it is advisable to partially empty the reservoir to accommodate the fluid displaced by the pistons.

12 A special tool, as shown in Fig.9.9 is available for resetting the pistons but this work can be done using a stout screwdriver to lever on the end of the piston but you must ensure that the piston is kept square in the bore, if it becomes tilted and is levered in that condition, it, or the cylinder bore, will be damaged.

13 Having moved the piston out of the way, insert the friction pad with its backing plate, replace the keep plate and secure with the nut and bolt.

14 Replace the road wheel, if applicable.

4 Front brake caliper - removal and refitting

1 An exploded view of the front brake caliper assembly is given in Fig.9.10.

2 Jack up the car and remove the road wheel. Support the car on a firmly based axle stand.

3 Disconnect the fluid feed pipe and immediately blank off the pipe and the caliper union to prevent the ingress of dirt.

4 Break the locking wire to the caliper mounting bolts and remove the bolts.

5 Withdraw the caliper assembly taking careful note of the number and position of shims fitted between the caliper and the mounting bracket.

6 Reassembly is the reverse of the above procedure. Refit the original shims and secure with the bolts but do not lock them at this stage.

Chapter 9/Braking system

7 Using feeler gauges, check the gap between each side of the caliper and the disc both at the bottom and at the top at each side of the disc. The difference should not exceed 0.010" (0.25 mm), if it does, fit shims between the caliper and the mounting plate to centralise the caliper body.
8 Lockwire the bolts, using soft iron locking wire, when you are satisfied with the location of the caliper.
9 If not already fitted, fit the bridge pipe connecting the two cylinder assemblies. It is essential that the bridge pipe is fitted with the hairpin bend to the inboard cylinder block. The pipe should carry a rubber identification sleeve marked "Inner Top" to aid assembly.
10 Connect the fluid supply pipe to the cylinder body.
11 Carry out the procedure outlined in Section 2 for bleeding the system.
12 Finally check for leaks at the unions and if all is satisfactory, replace the road wheel.

5 Rear brake caliper - removal and refitting

1 An exploded view of the rear brake caliper assembly as fitted to 3.8 litre (early and later models) and 4.2 litre cars is given in Figs.9.11 and 9.12 respectively. Although there are differences in these two assemblies, the removal and refitting procedure are similar.
2 The rear suspension must be removed in order to gain access to the rear calipers so first refer to Chapter 11 and follow the instructions given for this work.
3 Take out the split pin and remove the clevis pin which secures the compensator linkage to the handbrake operating lever.
4 Disconnect the hydraulic feed pipe and then immediately blank off the pipe and the union.
5 Remove the friction pads from the carrier in the manner described in Section 3.
6 Remove the front hydraulic damper and coil spring (see Chapter 11).
7 Remove the four steel type self locking nuts from the half shaft inner universal joint. Withdraw the joint from the bolts and allow the hub carrier to move outwards, support the carrier in this position. Take careful note of the shims between the universal joint flange and the brake disc, collect them and keep them in sets for replacement in the position from which they were removed. This is important because these shims control the camber angle of the rear wheels.
8 Knock back the locking tabs and remove the pivot bolts which secure the handbrake pad carriers to the caliper and retractor plate. Withdraw the handbrake pad carriers through the aperture at the rear of the crossmember.
9 Rotate the brake disc until the hole in the disc lines up with the caliper mounting bolts. Knock the locking tabs (locking wire is used on early models) and remove the bolts. Take note of the number of shims fitted over the mounting bolts and between the caliper and the adaptor plate (Fig.9.13).
10 The caliper can now be removed through the aperture at the front of the crossmember.
11 Refitting is the reverse of the removal sequence but be careful to ensure that the correct number of camber shims are refitted and that these are in their correct locations.
12 When you have refitted the halfshaft (using steel type self locking nuts) check the caliper for centralisation and adjust as necessary in the same manner as for the front caliper; described in Section 4, to obtain the clearance "A" in Fig.9.13.
13 Refit the fluid supply pipe and the bridge pipe if necessary, following the instructions given in Section 4, paragraph 9.
14 Reconnect the handbrake operating lever to the compensator linkage and use a new split pin to lock the clevis pin.
15 Reinstall the rear suspension assembly and then bleed the hydraulic system (see Section 2).

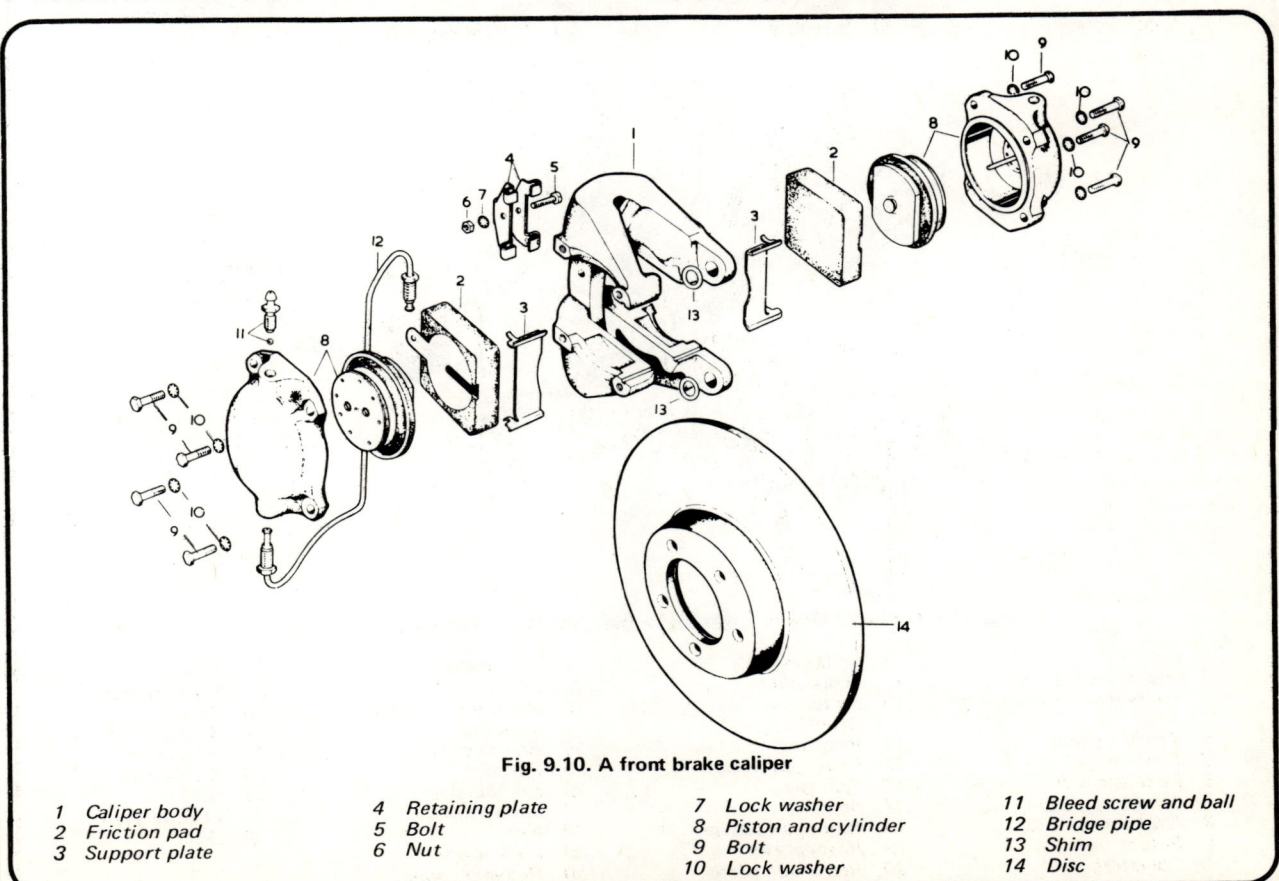

Fig. 9.10. A front brake caliper

1 Caliper body	4 Retaining plate	7 Lock washer	11 Bleed screw and ball
2 Friction pad	5 Bolt	8 Piston and cylinder	12 Bridge pipe
3 Support plate	6 Nut	9 Bolt	13 Shim
		10 Lock washer	14 Disc

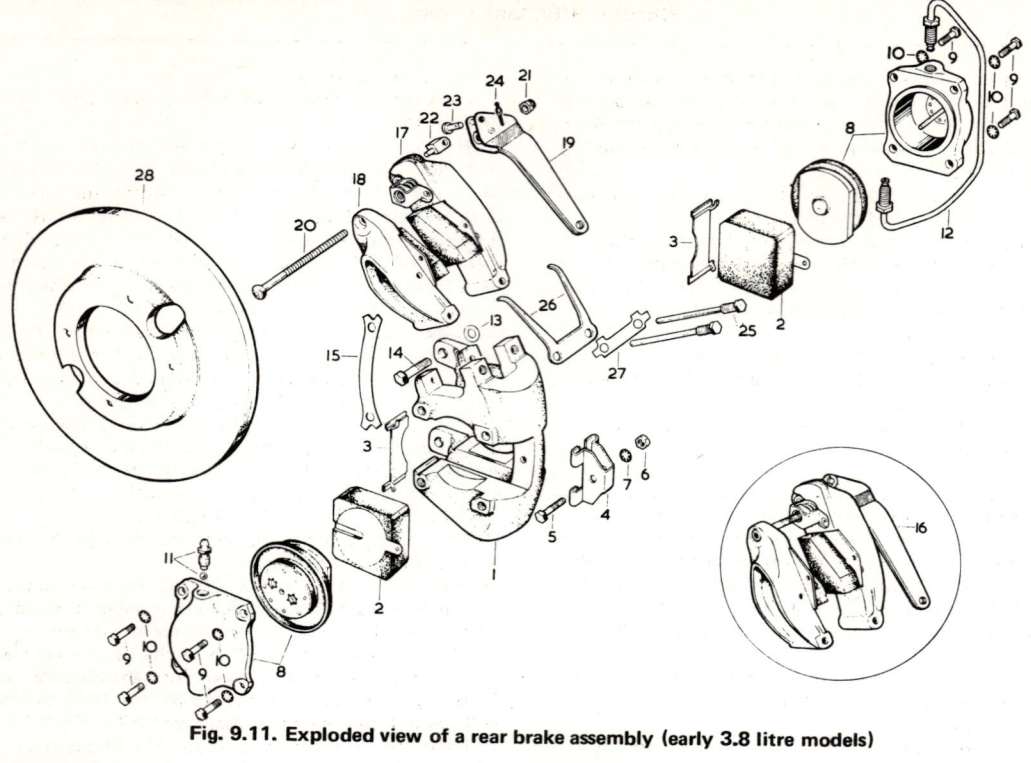

Fig. 9.11. Exploded view of a rear brake assembly (early 3.8 litre models)

1 Caliper body	8 Piston and cylinder	15 Tab washer	22 Pivot seat
2 Friction pad	9 Bolt	16 Handbrake assembly	23 Clevis pin
3 Support plate	10 Lock washer	17 Inner pad carrier	24 Split pin
4 Retaining plate	11 Bleed screw and ball	18 Outer pad carrier	25 Pivot bolt
5 Bolt	12 Bridge pipe	19 Operating lever	26 Retractor plate
6 Nut	13 Shim	20 Bolt	27 Tab washer
7 Lock washer	14 Setscrew	21 Self locking nut	28 Disc

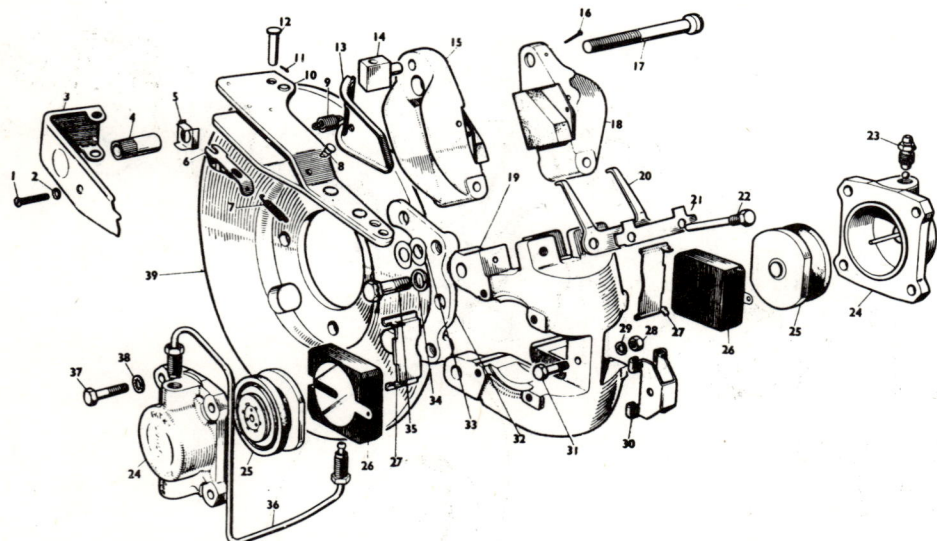

Fig. 9.12. Exploded view of a rear brake assembly, late 3.8 and 4.2 litre models

1 Bolt	11 Split pin	21 Tab washer	31 Bolt
2 Shakeproof washer	12 Hinge pin	22 Bolt	32 Locking plate
3 Protection cover assembly (rear)	13 Protection cover assembly (front)	23 Bleed screw and ball assembly	33 Shim
4 Adjusting nut	14 Pivot seat	24 Brake cylinder	34 Spring washer
5 Friction spring	15 Inner pad carrier	25 Piston	35 Setscrew
6 Pawl assembly	16 Split pin	26 Friction pad	36 Bridge pipe
7 Tension spring	17 Bolt	27 Support plate	37 Bolt
8 Anchor pin	18 Outer pad carrier	28 Nut	38 Lock washer
9 Return spring	19 Rear caliper	29 Lock washer	39 Disc
10 Operating lever	20 Retraction plate	30 Retaining plate	

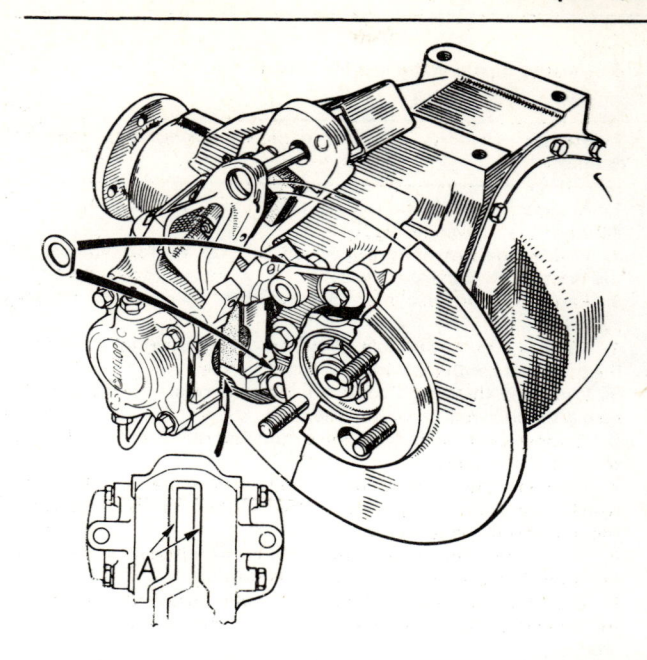

Fig. 9.13. Location of the shims fitted between the caliper and the adaptor plate

6 Front and rear brake discs - removal and refitting

1 To remove a front brake disc, jack up the car and remove the relevant wheel. Make sure that the car is safely supported before starting.
2 Follow the instructions given in Section 4 and remove the brake caliper. Now refer to Chapter 11 and remove the front hub in the manner described.
3 Remove the self locking nuts from the ten bolts securing the disc to the hub and then separate the disc and the hub.
4 Removal of the rear brake disc entails removal of the rear suspension; instructions for this will be found in Chapter 11.
5 Invert the suspension unit and remove the two hydraulic dampers and coil springs.
6 Follow the instructions given in Section 5 for the removal of the brake calipers.
7 Remove the steel type self locking nuts securing the half shaft inner universal joint. Tap the bolts back as far as possible and take note of the shims between the half shaft and the brake disc.
8 Lift the lower wishbone, hub carrier and halfshaft assembly upwards until the brake disc can be withdrawn from the mounting bolts.
9 Refitting of the brake discs is the reverse of the removal procedure. However, do not overlook the necessity to check and adjust the calipers for centralisation (Section 4), the replacement of the correct number of camber shims between the halfshaft and the rear brake disc, and finally the use of steel type self locking nuts for the halfshaft and rear brake disc securing bolts.
10 The last task is to bleed the system (Section 2) and top up the hydraulic fluid reservoirs as necessary.

7 Brake overhaul - general

The sections which follow deal with the dismantling and overhaul of brake hydraulic components and servo mechanisms. The importance of maintaining the braking system at top efficiency cannot be too highly stressed. To this end it is essential that certain elementary precautions are taken when handling the hydraulic assemblies.
1 Thorough cleanliness is a must at all times. Before starting to dismantle any unit, clean away all external dirt and then prepare a clean surface on the bench on which to work, we advise the use of a sheet of stout paper with which to cover the top of the bench.
2 Brake fluid should be used for cleaning internal components. The use of petrol, paraffin or solvents must be avoided at all costs as these will adversely affect rubber components, similarly, make sure your hands are not oily or greasy when handling rubber items.
3 All parts are manufactured with a high degree of precision so they must be handled with care, and should be carefully placed away from tools or other equipment likely to cause damage.
4 When it is not the intention to renew a rubber component, it must be thoroughly examined for serviceability. There must be no evidence of defects such as perishing, excessive swelling, cutting or twisting. If you are in any doubt as to the satisfactory condition of any item, discard it. As these rubber components are relatively cheap, our advice is to renew them as a matter of course.
5 The flexible pipes should show no signs of deterioration or damage. The bores should be cleaned with a jet of compressed air, under no circumstances should you attempt to clear a blockage by probing - as this may result in damage to the lining, and subsequent failure in service. Discard the pipe if a blockage cannot be cleared by an air blast.
6 When removing or refitting a flexible pipe, the end sleeve hexagon, "A" in Fig.9.14 should be held with a spanner to prevent the pipe from twisting whilst unscrewing the locknut "C" or the union nut "B". Excessive twisting of the pipe will cause weakness and possible early failure.

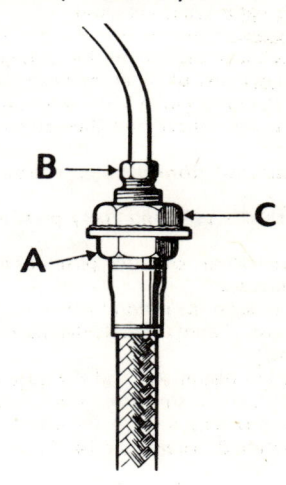

Fig. 9.14. Flexible pipe connections. Hold "A" when unscrewing or tightening "B" or "C"

8 Brake piston seals (early type) - renewal

1 Follow the instructions given in Section 4, or Section 5 if applicable, for removal of the brake caliper.
2 Remove the bolts securing the cylinder block to the caliper and withdraw the block.
3 Disengage the dust seal from the groove around the face of the cylinder block.
4 Connect the cylinder block to a source of fluid supply and apply pressure to push out the piston assembly. It is advisable to wrap a piece of cloth around the cylinder block so that the piston is caught.
5 Remove the screws securing the plate to the piston, lift off the plate and the piston seal. Withdraw the retractor bush from within the piston bore.
6 Cut-away and discard the dust seal.
7 Support the backing plate on a bush of sufficient diameter to just accommodate the piston. Using a piece of tube placed

Chapter 9/Braking system

against the end of the piston spigot and located around the shouldered head, press out the piston from the backing plate taking care not to damage the piston.
8 Engage the collar of the new dust seal with the lip on the backing plate, try to avoid excessive stretching of the seal.
9 Locate the backing plate on the piston spigot, support the piston and then press the backing plate fully home.
10 Insert the retractor bush into the bore of the piston.
11 Lightly lubricate the new piston seal with brake fluid and fit it to the piston face.
12 Attach the backing plate with the screws and then lock them by peening.
13 Make sure that the cylinder bore is clean and does not show any sign of damage.
14 Locate the piston assembly on the end of the retractor pin and, with the aid of a hand press, slowly apply an even pressure to the backing plate and press the assembly into the cylinder bore. Make sure that the piston assembly is in correct alignment in relation to the cylinder bore whilst pressing it home and also that the piston seal does not become twisted or trapped as it enters the bore.
15 Engage the outer rim of the dust seal in the groove around the face of the cylinder block and ensure that the two support plates are in position.
16 Reassemble the cylinder block to the caliper.
17 Follow the instructions given in Sections 4 or 5, as applicable for final refitting of the calipers.

9 Brake piston seals (later type) - renewal

The later type cylinder blocks can be distinguished by the letter "C" cast in the cylinder block at the inlet union hole.
1 Follow the instructions given in Section 4 or 5, as applicable, for removal of the brake caliper and then remove the friction pads (Section 3).
2 Disengage the dust seal from the groove around the face of the cylinder block.
3 Use a source of fluid supply and apply pressure to push out the piston assembly.
4 Using a blunt screwdriver, carefully push out and remove the piston seal and the dust seal.
5 It is not possible to strip the piston further.
6 Check that the piston and the cylinder bore are clean and show no signs of damage.
7 Lightly lubricate the piston seal and the dust seal with brake fluid and then place them on the piston using your fingers only.
8 From this point onwards follow the instructions for final assembly given in Section 8, paragraphs 14-17.

10 Master cylinders, 3.8 litre models - removal and refitting

1 Unscrew and remove the pipe unions from the end of the master cylinders, allow the fluid to drain from the reservoirs and then blank off the pipes and the unions to prevent the ingress of dirt.
2 Remove the two bolts and locknuts securing the top master cylinder flange.
3 Slacken the locknut on the top master cylinder push rod, unscrew the push rod from the yoke and remove the master cylinder.
4 Remove the two bolts and locknuts from the lower master cylinder flange.
5 Pull the cylinder forward as far as possible, remove the split pin and withdraw the clevis pin.
6 Remove the master cylinder.
7 Refitting is the reverse of the above sequence but it will be necessary to adjust the push rod on the top master cylinder so that there is 1/16" (1.58 mm) free play. This adjustment, through the balance lever, will give 1/32" (0.794 mm) free play to each master cylinder. When you are satisfied with this adjustment, tighten the locknut on the top master cylinder push rod.
8 Finally, top up and bleed the system (Section 2).

11 Master cylinders, 3.8 litre models - dismantling and reassembly

1 A sectioned view of a Dunlop type brake cylinder as fitted to 3.8 litre cars is given in Fig.9.15.
2 Follow the instructions given in Section 10 for removal of the master cylinder from the car.
3 Ease the dust excluder clear of the head of the master cylinder.
4 Using a pair of suitable pliers, remove the circlip to release the push rod complete with the dished washer.
5 Withdraw the piston complete with the return spring and the spring support and the valve assembly.
6 Remove both seals from the piston and remove the seal from the end of the valve.
7 Examine the bore of the cylinder for scores or abrasions which may cause leakage and if either of these faults are present, the unit should be renewed.
8 We strongly recommend that you renew all seals as a matter of course irrespective of their visual condition.
9 Lubricate the valve seal with hydraulic fluid and assemble it to the end of the valve making sure that the lip registers in the groove.
10 Lubricate the piston seals with hydraulic fluid and fit them in their grooves round the piston. Use your fingers only for this task.
11 Insert the piston into the spring support and assemble the return spring and the valve. Lubricate the piston with Girling Rubber Grease.
12 Smear the cylinder bore with hydraulic fluid and then slide the piston/valve assembly into the bore taking care not to damage or twist the seals.
13 Enter the push rod against the head of the piston and depress the piston sufficiently to allow the dished washer to seat on the shoulder at the head of the cylinder.
14 Fit the circlip and check that it is fully engaged with the groove.
15 Fill the dust excluder with Girling Rubber Grease, then refit it round the head of the cylinder.
16 Follow the instructions given in Section 10 and refit the master cylinder to the car.

12 Dunlop bellows type vacuum servo, 3.8 litre models - general

The power unit (Fig.9.16) consists of an air vacuum bellows which can expand or contract as the air pressure in the bellows is varied by the introduction of vacuum or atmosphere. One end of the assembly is connected to a bracket mounted on the bulkhead in the engine compartment and the other end is connected to the power lever which inturn is connected to the brake pedal assembly. A vacuum reservoir (Fig.9.19) is interposed between the vacuum servo and the inlet manifold of the engine with the object of providing a reserve of vacuum in the event of braking being required when the engine is not running. A vacuum check valve is fitted at the top end of the front face of the reservoir, with the topmost connection communicating with the inlet manifold whilst the second connection communicates directly with the vacuum port of the vacuum servo and so any reduction of pressure inside the reservoir is conveyed to the vacuum servo unit. Included in the inlet port of the check valve is a rubber spring loaded valve and when there is depression in the inlet manifold the valve is drawn away from its seat and thus the interior of the reservoir becomes exhausted. When the depression in the reservoir becomes equal to that of the inlet manifold, the valve spring will return the valve to its seat and so the highest possible degree of vacuum in the reservoir is maintained.

When the brake pedal is depressed, the air valve in the vacuum servo unit is closed and the vacuum valve opens, thus air

Chapter 9/Braking system 175

is evacuated out of the bellows causing them to contract, this excerts a pull on the power lever proportionate to the pedal pressure applied by the driver and so gives power assistance during application of the brakes.

Fig. 9.15. Sectioned view of a brake master cylinder, 3.8 litre models

Fig. 9.16. The Dunlop bellows type vacuum servo unit, 3.8 litre models

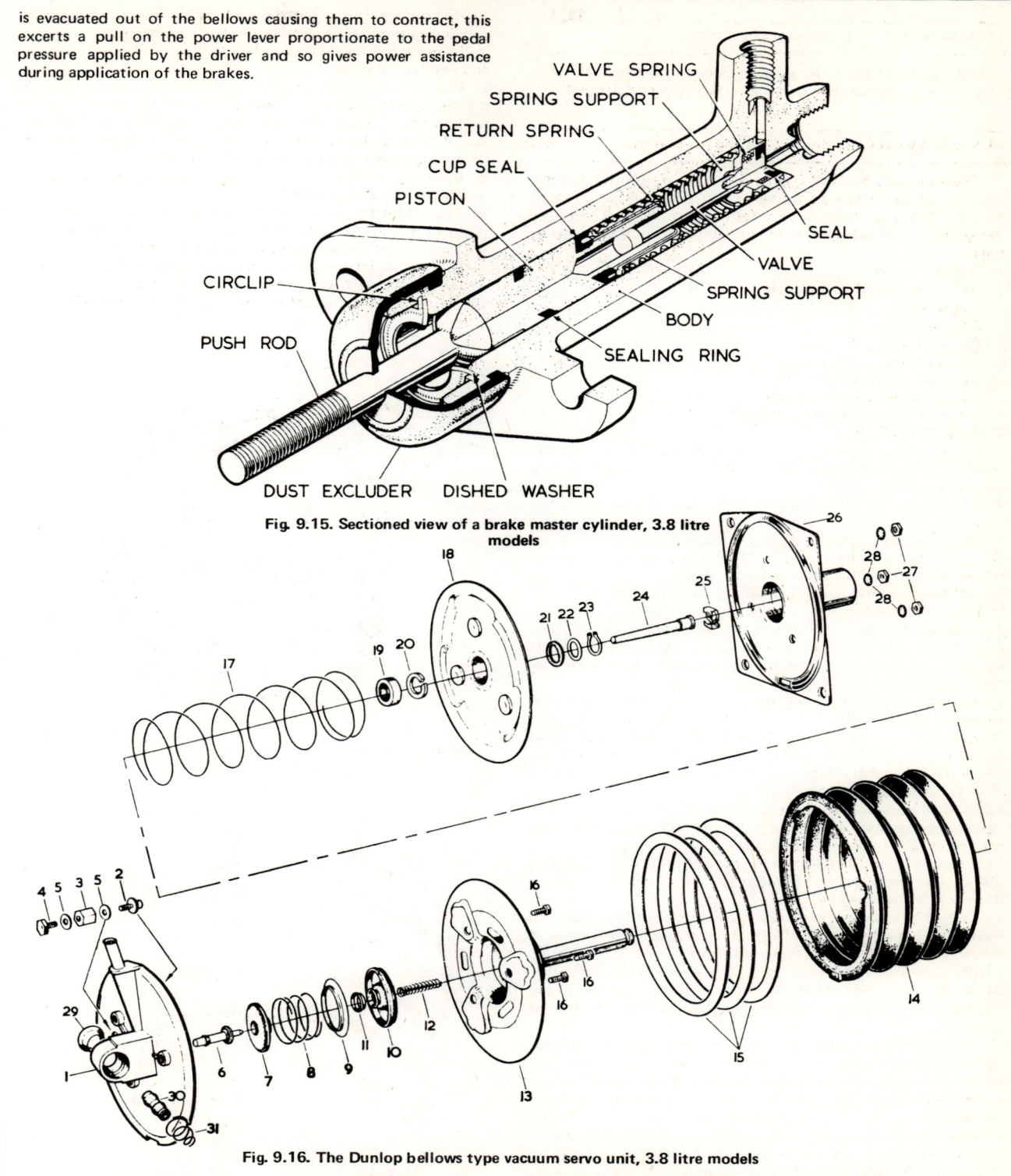

1	Valve housing	9	Balancing washer	17	Main return spring	25	Baffle
2	Nipple	10	Balancing diaphragm	18	Mounting hub	26	Mounting plate
3	Adaptor	11	Retainer	19	Seal	27	Nut
4	Plug	12	Control spring	20	Guide sleeve	28	Lock washer
5	Gasket	13	Retainer sleeve	21	Rubber buffer	29	Nylon bush
6	Air valve	14	Bellows	22	Stop washer	30	Eccentric bush
7	Vacuum valve	15	Support ring	23	Circlip	31	Spring
8	Return spring	16	Bolt	24	Air filter		

13 Vacuum servo unit - removal and refitting

Follow the instructions given in Section 23 until the servo unit can be withdrawn and replaced in the manner described.

14 Vacuum servo unit - dismantling and reassembly

The following parts should be renewed, irrespective of their visual condition, whenever the servo unit is dismantled so make sure of their availability before starting this work. The figure shown alongside each item is the identification number given in Fig.9.16.

Rubber washer	21
Air valve assembly comprising:	6
Air valve)	
Air valve buffer)	
Air valve cap)	
Air valve cup)	
Vacuum valve assembly	7
Vacuum balancing diaphragm	10

1 Remove the vacuum servo unit from the car.
2 Refer to Fig.9.16.
3 Clamp the servo unit in a soft jawed vice and take out the three setscrews, and shakeproof washers, attaching it to the mounting bracket (26).
4 Remove the air filter retaining baffle (25) and withdraw the air filter (24).
5 Hold the mounting hub (18) down against the return spring (17) and remove the circlip (23), the washer (22) and the rubber washer (21). Discard the rubber washer.
6 Holding the mounting hub down, ease off the lip of the bellows (14) from the hub.
7 Remove the mounting hub and the return spring.
8 Take out the three self locking screws to remove the guide sleeve (13) and bellows from the valve housing. Now withdraw the guide sleeve from the bellows.
9 Withdraw the air valve control spring (12), the valve balancing diaphragm (10), the vacuum valve spring (8), the vacuum valve assembly (7) and the air valve assembly (6) from the valve housing (1). Discard items 6,7 and 10 but first remove the valve balancing washer (9) and its retainer (11) from the valve balancing diaphragm.
10 Clean the mounting hub assembly with a dry cloth. Do not remove the leather seal as this will upset the silicon lubricant with which it is filled.
11 Clean all metal parts, except the mounting hub, but including the air filter in alcohol or some other oil free solvent and dry them with compressed air.
12 Clean the bellows, if necessary, by washing in a mild soap and water solution after taking off the three support rings. Rinse in clean water and dry with compressed air.
13 Inspect all parts for wear or damage and replace any part if its condition is at all doubtful. If the vacuum valve seat in the vacuum valve housing is damaged, the valve housing must be replaced.
14 Reassembly is the reverse of the dismantling procedure but when fitting the guide sleeve to the bellows the three bosses in the guide sleeve must line up with the three recesses in the bellows as shown in Fig.9.17. Also, when fitting the valve housing to the bellows, the boss on the bellows must line up with the recess in the valve housing and similarly the cutout in the mounting plate must line up with the boss on the bellows (Fig.9.18) when attaching it to the mounting hub.
15 When the unit is assembled, test the air valve by pressing the air valve cap down with the flat of a screwdriver. Two definite stages of movement should be felt and the valve should snap back smartly.

15 Vacuum reservoir and check valve (Fig.9.19) - removal and refitting

1 Remove the four drive screws and the two nuts and setscrews and detach the tray from below the vacuum reservoir.
2 Identify the two pipes to the check valve. Slacken the clips and remove the pipes.
3 Remove the four setscrews which hold the reservoir to the bulkhead and withdraw the reservoir from below the car.
4 It is best to leave the check valve in position in the reservoir, but, if it has to be removed, this can be done by unscrewing it - do not put undue pressure on the pipe unions.
5 Refitting of the reservoir is the reverse of the above but watch the following points:
a) The rubber hose from the vacuum servo unit attaches to the pipe of the check valve which has two grooves in its body. This is the pipe nearest the screwed connection of the valve.
b) The rubber hose from the inlet manifold attaches to the pipe of the check valve having two annular ribs in its body. This is the pipe moulded into the centre of the cap of the check valve.

Fig. 9.17. Lining up the bosses on the retainer sleeve with the recesses on the bellows

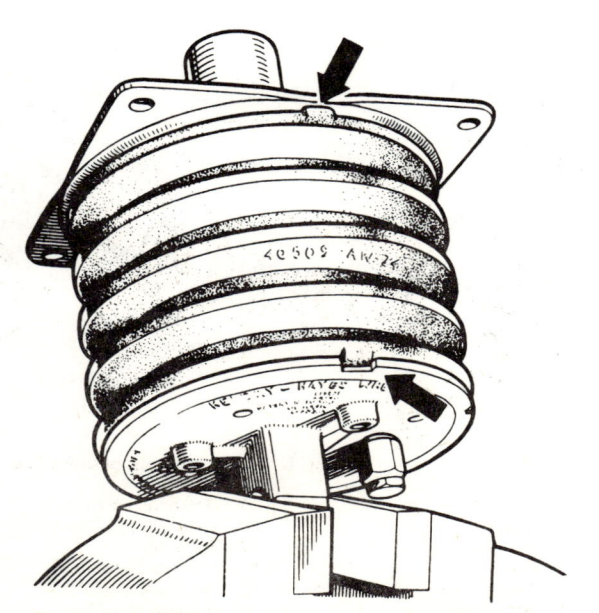

Fig. 9.18. Lining up the bosses on the bellows with the recesses on the mounting plate and valve housing

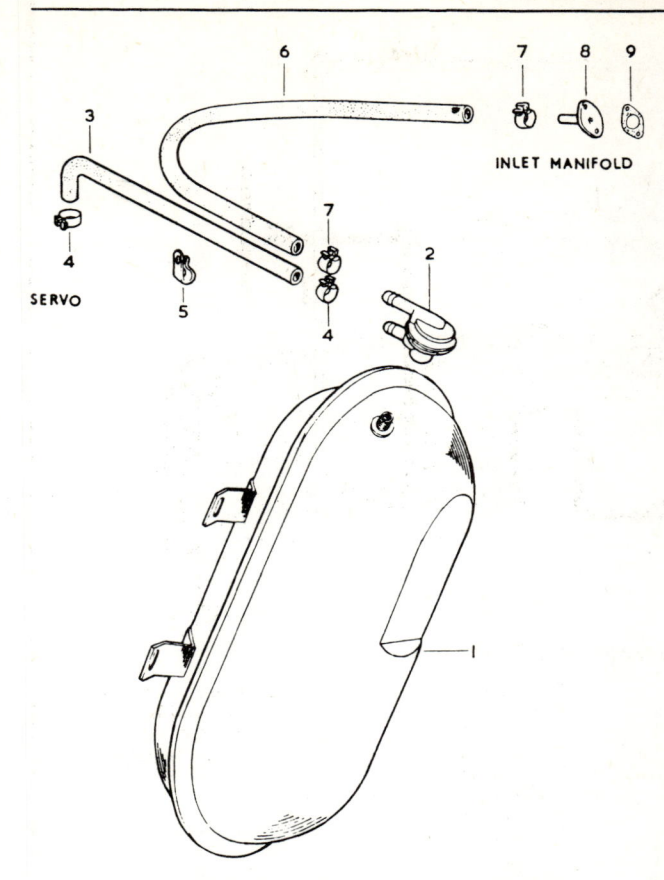

Fig. 9.19. The vacuum reservoir and check valve

1 Vacuum tank
2 Check valve
3 Hose
4 Clip
5 Clip
6 Hose
7 Clip
8 Adaptor
9 Gasket

16 Bellows type vacuum servo unit - servicing

This Section should be read in conjunction with Section 34, Fault Diagnosis which will be found at the end of the Chapter. Before doing any work on the servo unit, make sure that braking faults are not due to air in the system, worn friction pads, oil or grease on the discs or hydraulic leaks.

1 If excessive pressure is required at the brake pedal it is an indication of lack of power assistance. This may be due to a blocked, kinked or leaking vacuum line. Vacuum leaks may be accompanied by rough running of the engine symptomatic of a weak fuel mixture. However, to check, remove the rubber hose from the power unit and, with the engine running, check for vacuum. If no vacuum is discernible, check hoses for blockage, kinking, damage or loose connections. Finally check the valve unit in the vacuum reservoir and replace it if it is faulty. It may be possible to locate a vacuum leak by listening for a hissing sound when the engine is running.

2 Another possible cause is leaks in the unit. With the engine running have an assistant apply pressure on the brake pedal; and now listen for a hissing sound at the unit which, if heard, will probably entail removal and dismantling of the unit (as described in Sections 13 and 14) for rectification of the fault.

3 If the above tests prove satisfactory, the fault may lie in the valve adjusting eccentric being out of adjustment. To check, connect a vacuum gauge (capable of reading 0 to 30'' (0 to 76.20 cm) of mercury) to the union on the valve housing as shown in Fig.9.20. On early models the union will be found on the back of the mounting plate and on later models it is on the front of the auto-valve housing plate. Having made the connection, run the engine and apply normal full pressure to the brake pedal, the gauge should now register 20 inches (50.8 cm) of mercury. If no vacuum or only partial vacuum is registered, remove the return spring, slacken the locknut and apply a spanner to the hexagon of the eccentric bush (see inset at Fig.9.20) and turn until vacuum is obtained, this will be when the air valve is closed and the vacuum valve is fully open. Tighten the locknut, replace the return spring and check again. On releasing pressure from the brake pedal, the vacuum valve should close and the air valve open the gauge should register zero. The brakes should be free and if they are not: it is an indication that the vacuum valve is not completely closed, in which case, the eccentric must be adjusted in the opposite direction until the gauge registers zero and the brakes are free. If the adjustment is now correct, switch off the engine, remove the gauge and close the union. Do not adjust the eccentric more than necessary when carrying out the above work.

4 Slow return of the brake pedal is indicative of a choked air filter. The remedy in this case is to remove the servo unit from the car and, holding the pedal box lightly in a vice with soft jaws, collapse the bellows by hand. The end of the air inlet tube with circlip will now be exposed. remove the air intake baffle and lift out the filter. Clean the filter in alcohol or other oil free solvent and replace.

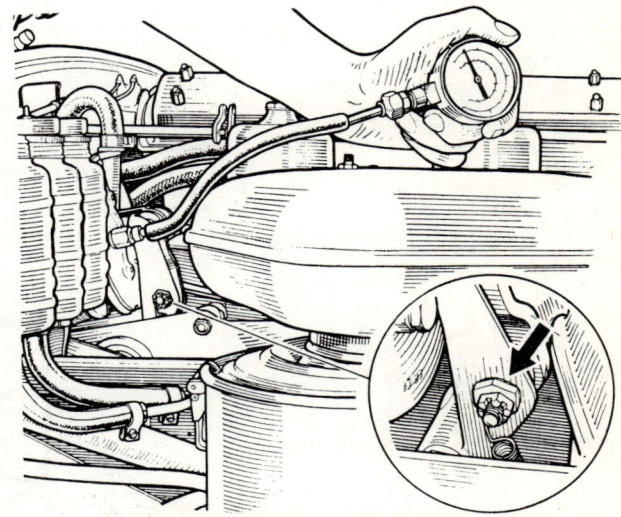

**Fig.9.20. Check the servo with a vacuum gauge
Inset shows the eccentric adjusting nut**

17 Lockheed dual-line servo system, 4.2 litre models - general

A schematic layout of this system as fitted to early and later model 4.2 litre cars is given in Figs.9.21 and 9.22. The dual-line servo system consists of an integral vacuum booster with tandem slave cylinder and a master cylinder combined with a booster reaction valve and two fluid reservoirs.

The master cylinder, Fig. 9.24 is of conventional design and consists of a single cast iron cylinder housing a piston sealed by a single hydraulic cup, the piston is deeply skirted to accommodate the operating push rod which is connected to the brake pedal. In the nose of the master cylinder is a smaller intermediate piston housed in its own bore, this piston is actuated by hydraulic pressure generated within the main chamber.

Mounted on the end of the master cylinder is a reaction valve which consists of pair of flow control valves which sequence the flow of air to the booster, both control valves are operated by the intermediate piston in the master cylinder. A flat plate is interposed between the main and the intermediate pistons and this enables the intermediate piston to function mechanically in the event of a hydraulic failure.

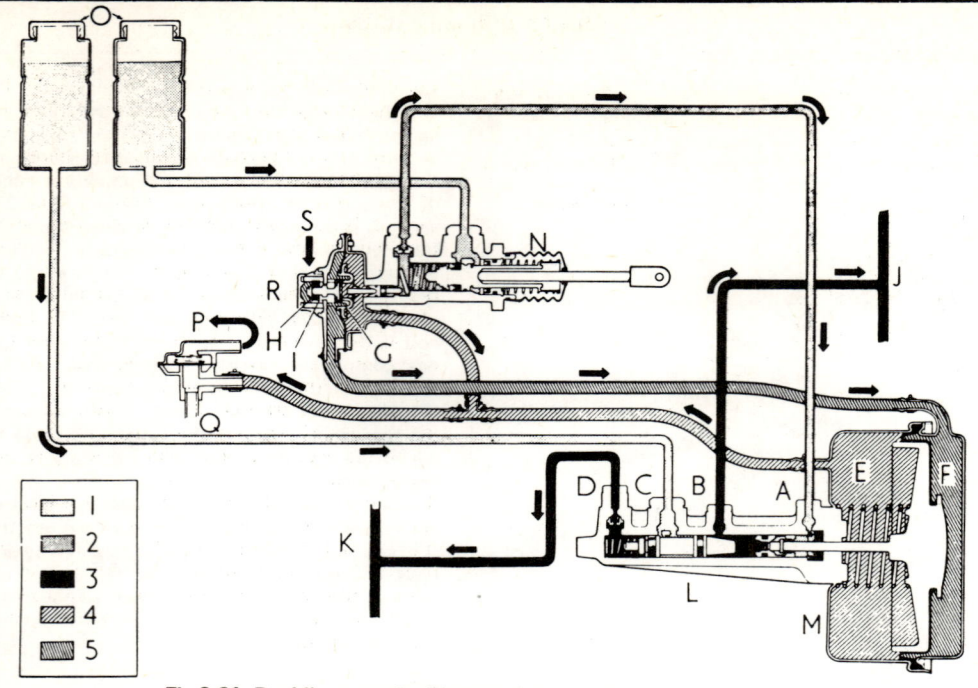

Fig.9.21. Dual-line servo braking system, early 4.2 litre cars

1 Fluid at feed pressure	A Primary chamber - slave cylinder	G Diaphragm	O Fluid reservoirs
2 Fluid at master cylinder delivery pressure	B Outlet port-rear brakes	H Filter	P To manifold
3 Fluid at system delivery pressure	C Inlet port - secondary piston	I Air control	Q To reservac
4 Vacuum	D Outlet port-front brakes	J To rear brakes	R Reaction valve
5 Air at atmospheric pressure	E Vacuum	K To front brakes	
	F Air pressure	L Tandem slave cylinder	
		M Vacuum cylinder	
		N Master cylinder	

Fig.9.22. Dual-line servo braking system, later 4.2 litre cars

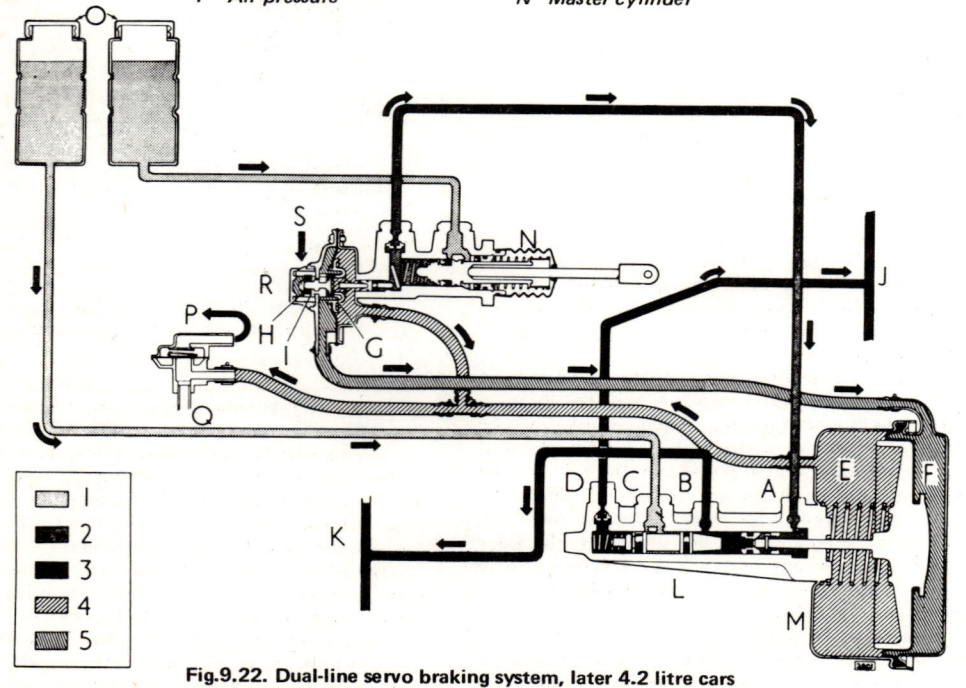

1 Outlet connection	9 Pushrod	17 Abutment plate	25 Seal
2 Gasket	10 Diaphragm support	18 Bearing	26 Retainer
3 Inlet connection	11 Diaphragm	19 Seal	27 Slave cylinder body
4 Piston	12 Key	20 Spacer	28 Spring
5 Pin	13 Cover	21 Cup	29 Trap valve
6 Retaining clip	14 Vacuum cylinder shell	22 Piston	30 Stop pin
7 Gasket	15 Screw	23 Cup	31 Gasket
8 Spring	16 Locking plate	24 Piston washer	

Chapter 9/Braking system

The booster portion of the integral booster and slave cylinder assembly, Fig.9.23, consists of a pressed steel tank housing a moulded phenolic resin piston and a rubber rolling diaphragm. A push rod, secured to the piston, extends through the forward face of the tank into the slave cylinder and provides the principal motive force for the tandem pistons.

The tandem slave cylinder, which is mounted on the forward face of the boost tank, consists of a single cast iron cylinder which houses two pistons in tandem each piston having its own inlet and outlet port. Either piston will operate independently in the event of a failure.

With reference to Figs.9,21 and 9.22, when the system is at rest both sides of the system are continuously exhausted by inlet manifold depression. When the brake pedal is depressed and moves the piston in the master cylinder, pressure is built up and fluid is forced out to the primary chamber of the slave cylinder (A). At the same time the intermediate piston in the end of the master cylinder closes the diaphragm valve (G) in the reaction valve and so isolates the vacuum (E) from the air pressure side (F) of the boost system. Further movement of the intermediate piston along its bore will crack the air control spool (I) in the reaction valve and thus admit air at atmospheric pressure to the rear of the boost cylinder piston. The air enters the system through a small cylindrical filter (H) on the reaction valve. The pressure imbalance so created will push the boost piston down the cylinder transmitting a linear force, through the push rod, to the primary piston of the slave cylinder. Forward movement of the primary piston, supplemented by the output of the master cylinder, transmits hydraulic pressure to the secondary piston (C) and fluid under pressure flows simultaneously from the two output ports (B and D) to the front and rear brakes.

This system has certain inbuilt safety factors. In the event of a failure in the pipe linking the master cylinder to the slave cylinder or in the pipe linking the master cylinder to the fluid reservoirs, the reaction valve will be actuated mechanically by the master cylinder piston providing the booster pressure to the front and rear brakes.

A failure in the fluid line coupling the slave cylinder to the front brakes will result in the slave cylinder secondary piston travelling to its fullest extent down the bore. This will have the effect of isolating the front brake line from the rest of the system and will permit normal fluid pressure to build up in the rear brake line.

If there is a fault in the rear brake line, the slave cylinder piston will travel along the bore until it contacts the other piston the two pistons will then travel along the bore together to apply the front brakes. On later cars this process is reversed as can be seen from Fig.9.22.

In the case of leaks in either the air or vacuum pipes both front and rear brakes may still be applied by the displacement of fluid at master cylinder pressure.

18 Lockhead dual-line servo system overhaul - general

1 The precautions, mentioned in Section 7, regarding the maintenance, cleanliness and avoidance of contamination (by oil or grease) of rubber components also apply when working on any part of the dual-line servo system.
2 Before starting work on the system it is advisable to obtain the appropriate repair kit which contains all the necessary rubber parts, required during overhaul. Three separate repair kits are available, namely:
a) Remote servo repair kit.
b) Reaction valve repair kit.
c) Master cylinder repair kit.
3 The vacuum non-return valve is a sealed unit and, if faulty, it must be replaced by a new assembly.
4 When a component is dismantled, all metal parts must be examined and particular attention paid to the following:
a) The reaction valve piston and bore.
b) The servo slave cylinder pistons and bore.
c) The master cylinder piston and bore.
d) The servo push rod stem.

19 Remote servo and slave cylinder - removal and refitting

1 Drain the hydraulic fluid from the system. The easiest, and cleanest, way to do this, is to follow the procedure as for bleeding the system (Section 2), but in this case carry on pumping the brake pedal until all fluid has been expelled. Discard the fluid.
2 Remove the trim from the floor recess panel on the left hand side of the car. The three nuts which secure the remote servo to the bulkhead will now be accessible.
3 Disconnect the four brake pipe unions and the two flexible hoses.
4 Remove the three nuts securing the servo to the bulkhead.
5 Remove the battery and the carrier bracket from the battery tray.
6 Remove the bolt securing the slave cylinder to the mounting bracket on the outer side member.
7 Withdraw the servo and slave cylinder.
8 Refitting the remote servo and slave cylinder is the reverse of the above procedure but the system must be refilled with fresh hydraulic fluid and bled, as described in Section 2.

20 Remote servo and slave cylinder - dismantling and reassembly

1 Having removed the servo slave cylinder as described in Section 19, support the servo slave cylinder in a vice, the vacuum cylinder shell uppermost, with specially formed wooden blocks placed either side of the cylinder and against the jaws of the vice.
2 Churchill Tool No J.31 is used to remove the cover. Secure the tool to the cover by fitting the three nuts and then turn the end cover in an anti-clockwise direction until the indents on the servo shell line up with the small radii around the periphery of the end cover. The cover can now be lifted off.
3 Refer to Fig.9.23.
4 Remove the diaphragm (11) from the groove in the diaphragm support (10). Remove the assembly from the vice, apply a gentle pressure to the diaphragm support and shake out the key (12).
5 Remove the diaphragm support and the return spring (8).
6 Open the tabs on the locking plate (16). Remove the three screws (15) and withdraw the locking plate, abutment plate (17) and servo shell (14) from the slave cylinder.
7 Take out the seal (19) and bearing (18) from the mouth of the slave cylinder. The push rod (9) together with the slave cylinder piston assembly can now be withdrawn.
8 The pushrod can be separated from the piston by sliding back the steel spring clip (6) around the piston and removing the pin (5).
9 There is no necessity to remove the cup (21) from the piston as a new piston complete with cup is contained in the repair kit.

Chapter 9/Braking system

10 Unscrew the fluid inlet connection (3) and extract the piston stop pin (30) from the base of the inlet port. Removal of the pin will be eased if you apply gentle pressure to the secondary piston (4).
11 Tap the open end of the slave cylinder on a block of wood to remove the secondary piston, together with the piston return spring (28), from the bore.
12 Remove the spring retainer (26) from the piston head extension and now work the piston seal (25) out of its groove and remove it with the piston washer (24). The plastic spring retainer (26) sometimes proves difficult to remove without damage, however, a new one is provided in the repair kit.
13 Remove the trap valve assembly by unscrewing the adaptor (1) from the fluid outlet port. If you remove the shim like clip from the body of the trap valve (29) make sure that it is not distorted in any way.
14 The first stage of reassembly is to replace the trap valve (29) complete with spring and clip into the outlet port and secure it by fitting the outlet adaptor (1) together with the copper gasket (31). Make sure that this gasket is in good condition before reusing it.
15 Lightly coat the four slave cylinder seals with brake fluid, which will act as a lubricant.
16 Locate the piston washer (24) over the piston head extension so that the convex face is towards the piston flange. Now, using the fingers only, assembly the two rubber seals (23 and 25) to the piston so that their concave faces oppose each other.
17 When both seals are in position, press the spring retainer (26) onto the piston head extension.
18 Fit the piston return spring (28) over the retainer and now offer the secondary piston assembly to the cylinder bore with the spring leading.
19 Press the piston assembly down the cylinder bore using a short length of brass bar (we advise against using a pencil or short piece of wood owing to the possibility of small pieces being broken off and causing trouble later on), until the drilled piston flange passes the piston stop pin hole.
20 Hold the secondary piston in that position until you have inserted the piston stop pin (30) into the fluid inlet port and have secured it by fitting the inlet adaptor (3). Make sure that the copper gasket (2) on the inlet adaptor is in good condition before reusing it.
21 Place the push rod (9) in the primary piston and, with the aid of a small screwdriver, compress the small spring within the piston so that the pin (5) can be inserted.
22 Check that the spring is indeed loaded between the heel of the piston and the pin as it is easy to pass the pin through the coils of the spring in error. If all is satisfactory, fit the pin retaining clip (6) by sliding it into position along the piston make sure that no corners are left standing proud after assembly.
23 Use your fingers only to fit a new cup (2) into the piston groove so that the lip of the cup (concave face) faces towards the piston head. Now assemble the piston into the slave cylinder bore.
24 Insert the spacer (20) followed by the gland seal (19) and the plastic bearing (18) into the slave cylinder counterbore and leave the bearing projecting slightly from the mouth of the bore.
25 Use the plastic bearing as a locating spigot and place the gasket (7) in position followed by the vacuum shell (14), the abutment plate (17) and the locking plate (16).
26 Insert the three securing screws (15) and tighten them to a torque of 150 to 170 lb f ins (1.7 to 1.9 kg f m). Lock the screws by bending up the tabs of the locking plate.
27 Locate the diaphragm support return spring (8) centrally inside the vacuum shell. Fit the diaphragm support (10) to the push rod and secure it by dropping the key (12) into the slot in the diaphragm support.
28 Stretch the rubber diaphragm (11) into position on the support and make sure that the bead on its inside diameter fits snugly into the groove in the support. If the surface of the rubber diaphragm appears wavy or crinkled it is an indication that it is not correctly seated. Assembly will be helped if you smear the outside edges of the diaphragm with brake fluid.
29 Position the end cover so that the hose connections will line up with the slave cylinder inlet and outlet ports on assembly and then secure the cover (using Tool No J31) by turning in a clockwise direction.

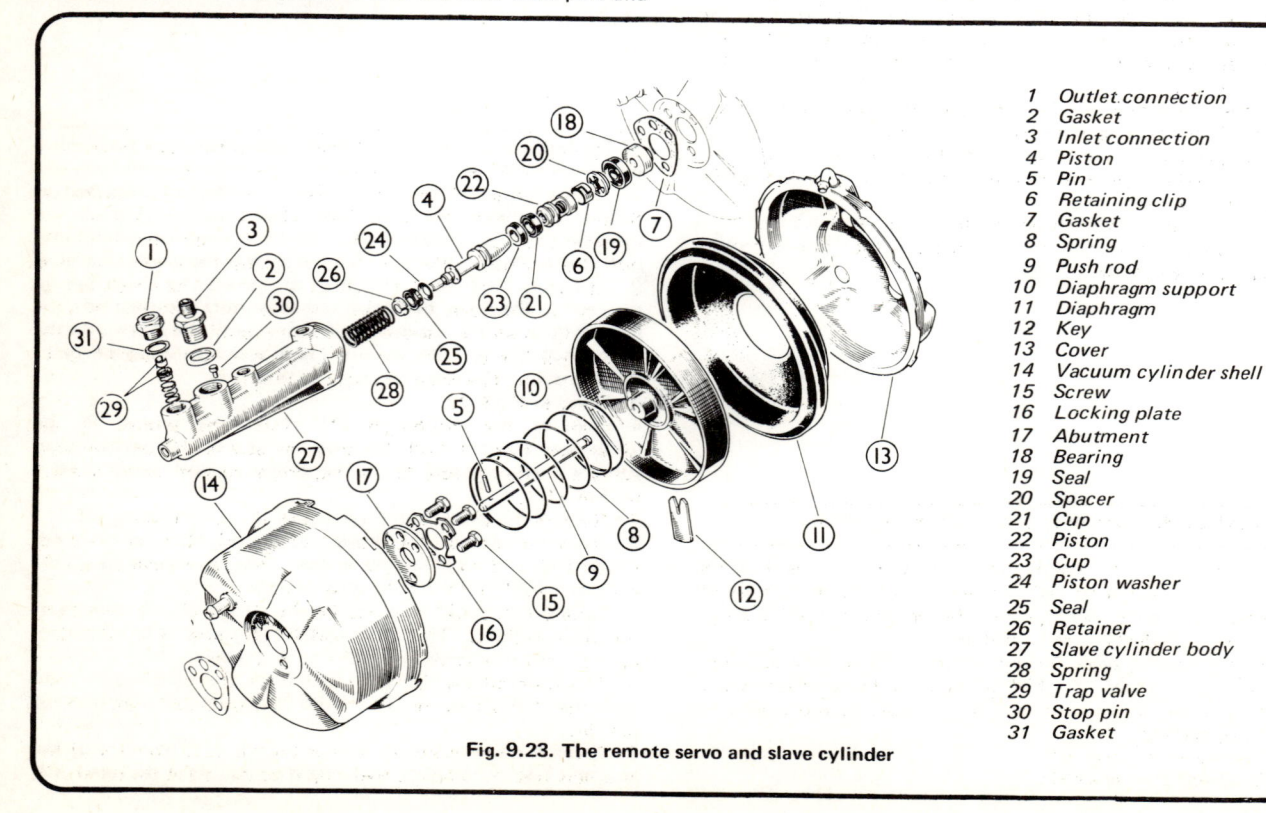

1 Outlet connection
2 Gasket
3 Inlet connection
4 Piston
5 Pin
6 Retaining clip
7 Gasket
8 Spring
9 Push rod
10 Diaphragm support
11 Diaphragm
12 Key
13 Cover
14 Vacuum cylinder shell
15 Screw
16 Locking plate
17 Abutment
18 Bearing
19 Seal
20 Spacer
21 Cup
22 Piston
23 Cup
24 Piston washer
25 Seal
26 Retainer
27 Slave cylinder body
28 Spring
29 Trap valve
30 Stop pin
31 Gasket

Fig. 9.23. The remote servo and slave cylinder

21 Master cylinder and reaction valve - removal and refitting

1 Drain the fluid from the system. The easiest, and cleanest, way of doing this is to follow the procedure for bleeding the system (Section 2) but in this case carry on pumping the brake pedal until all fluid has been expelled. Discard the fluid.
2 Disconnect the two hydraulic pipes to the master cylinder.
3 Disconnect the vacuum hose from the reaction valve.
4 Take out the split pin and then remove the clevis pin securing brake pedal to the master cylinder push rod. The clevis pin is accessible from inside the car.
5 For right hand drive cars, remove the top of the air cleaner and then, after referring to Section 22, carry out the operations to remove the reaction valve.
6 Remove the two nuts securing the master cylinder to the mounting and withdraw the master cylinder.
7 There is no need to remove the air cleaner on left hand drive cars and here the master cylinder and the reaction valve can be removed as a complete unit by taking off the two nuts securing the assembly to the mounting.
8 Refitting is the reverse of the above sequence but the system must be filled with fresh fluid and bled as described in Section 2.

22 Master cylinder and reaction valve - dismantling and reassembly.

1 Refer to Fig.9.24 whilst carrying out the following operations which assume that the reaction valve is still in position.
2 Unscrew the outlet adaptor (6) and extract the trap valve assembly (7) from the outlet port.
3 The inlet adaptor may be left in position but if you wish to take it out, remove the bolt (5), the banjo (9) and the two copper washers (10).
4 Remove the rubber boot (17), compress the piston return spring (16) and unwind the spirolox circlip (20) from the heel of the piston. The spring retainer (18) and the return spring can now be removed with the push rod (19).
5 Press the piston down the bore and remove the circlip (21) from the mouth of the cylinder bore.
6 The piston assembly complete with nylon bearings and rubber seals can now be withdrawn from the cylinder bore.
7 Remove the plastic bearing (22) complete with the O-ring (14), the secondary cup (13) and the rectangular section plastic bearing (12) from the piston by sliding them along the finely machined portion.
8 Remove the spring retainer (25) from the piston by holding the piston, head downwards, on a bench and then applying a downward force to the back face of the retainer with a slim open ended spanner. The retainer, being an interference fit on the piston head extension, may be damaged during this operation but this will not matter as a new retainer is included in the repair kit.
9 Withdraw the piston return spring (26), retainer (27), and lever (28) from the cylinder bore.
10 Now, to remove the reaction valve, first remove the filter cover (39) and collect the filter (36), the sorbo washer (37) and the spring (38).
11 Take out the five screws (43) holding the valve cover (42) to the valve housing (32). Withdraw the valve cover and collect the valve stem (40), the diaphragm (1) and its support (33).
12 Remove the two hexagon headed screws (2) and separate the valve housing (32) from the master cylinder.
13 The valve piston (31) can be removed by inserting a blunt instrument into the outlet port and easing the valve piston assembly along its bore until it can be removed by hand. Do not use pliers to extract the valve piston assembly because this may cause damage and subsequent malfunctioning.
14 Before assembling them; coat all rubber seals and plastic bearings with brake fluid, it is suggested, that this is left until immediately before assembly of a particular item in order to lessen the chance of picking up foreign matter. Do not lubricate the two valve rubbers.
15 The first stage in reassembly is to hold the master cylinder body at an angle of about 25° to the horizontal and then insert the lever (28), tab leading, into the cylinder bore; making sure, that the tab drops into the recessed portion at the bottom of the bore.
16 Place the piston washer (23) on the head of the piston, with the convex face towards the flange, together with a new main cup (24) and now press the plastic spring retainer (25) onto the extension of the piston head.
17 Drop the pressed steel spring retainer (27) into the bottom of the bore and follow up with the piston return spring (26). Check that the lever is correctly positioned.
18 Now press the piston assembly into the bore and locate the rectangular section plastic bearing (12), secondary cup (13), the bearing (22) together with the seal (14) in the mouth of the cylinder bore. Press this assembly down the bore to its fullest extent and fit the circlip to retain them.
19 Locate the other piston return spring (16) over the heel of the piston together with the spring retainer (18). Slide the retainer down the machined portion of the piston to compress the spring and then fit the spirolox circlip (20) into the groove in the heel of the piston.
20 Using your fingers only, stretch a new valve seal (29) and O-ring into position on the valve piston and insert the assembly into the valve box.
21 Secure the reaction valve housing to the body of the master cylinder with the two hexagon headed screws (2) and their spring washers. Use a new gasket between the valve housing and the body of the master cylinder. Tighten the screws to a torque of 160 to 180 lbs f ins (1,8 to 2 kg f m).
22 Stretch the reaction valve diaphragm onto the diaphragm support through the hole in the valve housing so that it engages the depression in the valve piston.
23 Using your fingers only, stretch the valve rubber, which is formed with the groove around its inside diameter, onto the valve stem flange. Insert the valve stem through the hole in the valve cover and secure it by placing the other valve rubber over the valve stem and then fitting the snap-on clip.
24 Now place the valve cover assembly into position on the valve housing; make sure, that all the holes line up, and that, the hose connections are in line with each other at the bottom of the unit. Fit the five self-tapping screws to secure the assembly.
25 Hold the master cylinder so that the reaction valve is uppermost and place the air filter together with the rubber washer into position with the small spring and the snap-on valve stem clip. Locate the air filter cover over the air filter and press it home firmly.
26 If you dismantled the trap valve, insert the small clip into the trap valve body (make sure that it does not become distorted) and locate the spring on the reduced diameter of the trap valve body. Assemble the trap valve complete, spring innermost, into the master cylinder outlet port.
27 Check that the copper gasket is in good condition (it is advisable to use a new one as a matter of course) and then screw the pipe adaptor into the outlet port. Repeat with the inlet port adaptor if this was removed.
28 The master cylinder push rod and the rubber boot can be fitted at this stage but you may find it more convenient to leave this for the time being and assemble them when refitting the assembly to the car.

23 Brake/clutch pedal box assembly and vacuum servo unit, 3.8 litre models - removal and refitting

1 On right hand drive models only, remove the air cleaner elbow and the carburettor trumpets. Slacken the rear carburettor float chamber banjo nut and push petrol feed pipe towards the float chamber. Remove the throttle rods from the bell crank above the servo bellows.
2 Now for both right and left hand drive models, disconnect the servo vacuum pipe.

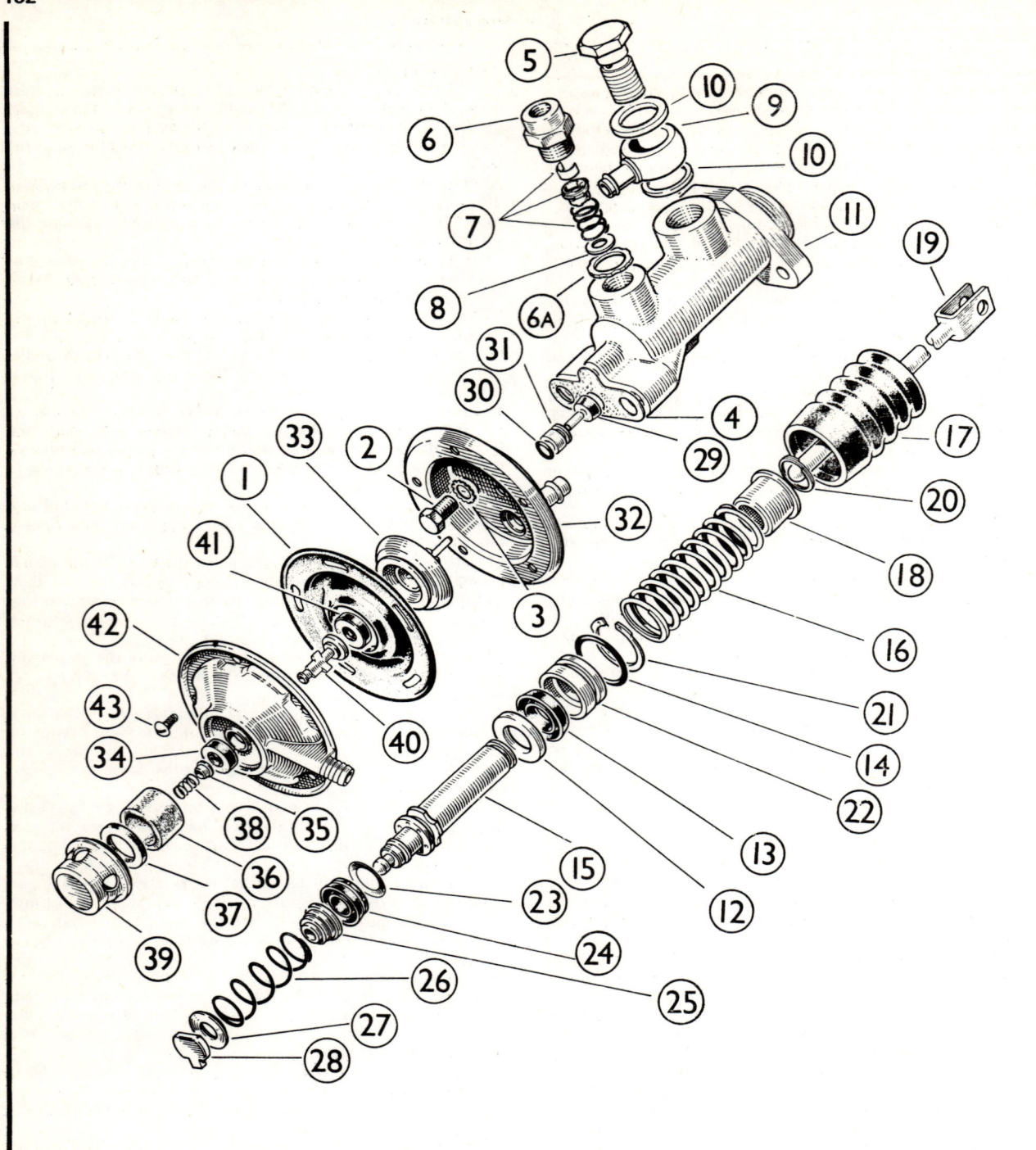

Fig.9.24. The master cylinder and reaction valve

1 Diaphragm	10 Copper gasket	27 Retainer	37 Sorbo washer
2 Screw	11 Body	28 Lever	38 Spring
3 Shakeproof washer	12 Bearing	29 Seal	39 Filter cover
4 Gasket	13 Secondary cup	30 Seal	40 Valve stem
5 Bolt	14 Seal	31 Piston	41 Valve rubber
6 Outlet adaptor	15 Piston	32 Valve housing	42 Valve cover
6a Copper gasket	23 Piston washer	33 Diaphragm support	43 Screw
7 Trap valve body	24 Main cup	34 Valve rubber	
8 Washer	25 Retainer	35 Valve cap	
9 Banjo	26 Spring	36 Filter	

Chapter 9/Braking system

3 Drain the brake and clutch fluid reservoirs by following the procedure for bleeding the systems, but in this case, pumping on the clutch and brake pedals until the reservoirs are empty. Discard the clutch and brake fluid.
4 Disconnect the fluid inlet pipes from the brake and clutch master cylinders and immediately blank off the pipes and the unions.
5 Identify the leads to the brake fluid reservoirs and then disconnect them. Remove the nuts from the clamp studs and remove the brake and clutch fluid reservoirs.
6 Disconnect the outlet pipes from the brake and clutch master cylinders and immediately blank off the pipes and the unions.
7 Remove the brake master cylinders as described in Section 10.
8 Move inside the car and take off the brake and clutch pedal pads by unscrewing the retaining nuts.
9 Remove the six self locking nuts and one plain nut and shakeproof washer which hold the servo assembly to the bulkhead.
10 Compress the servo bellows by hand, lift the servo assembly and at the same time turn it through about 90° clockwise to allow the pedals to pass through the hole in the bulkhead (Fig.9.25).
11 Work can now start on separating the vacuum servo unit.
12 Remove the bulkhead rubber seal.
13 Remove the four nuts and the one setscrew fastening the brake master cylinder mounting bracket.
14 Remove the self locking nut from the serrated pin and remove the conical spring and washer.
15 Remove the pinch bolt from the brake pedal shaft.
16 Remove the vacuum check point from the front valve housing.
17 Remove the brake master cylinder support bracket with the linkage and pedal shaft assembly from the pedal housing. Collect the fibre washer from between the brake and clutch pedals.
18 Remove the throttle bellcrank bracket on right hand drive models by removing the four self locking nuts.
19 Remove the brake vacuum servo assembly.
20 Refitting is the reverse of the above procedure but note the following points:
21 Do not overlook fitting the rubber seal over the exhausting tube between the servo bellows and the bulkhead.
22 When refitting the securing nuts inside the car, ensure that the plain nut and the shakeproof washer go on the short stud at the front centre.
23 When replacing the fluid reservoirs ensure that the brake fluid warning light wires are fitted with one feed wire (red/green) and one earth wire (black) to each reservoir cap.
24 Ensure that the petrol feed pipe is clear of the rear float chamber before tightening the banjo union nut.
25 Ensure that all clevis pins enter freely and without force.
26 Finally, refill the reservoirs with clean hydraulic fluid and bleed both the brake and clutch systems.

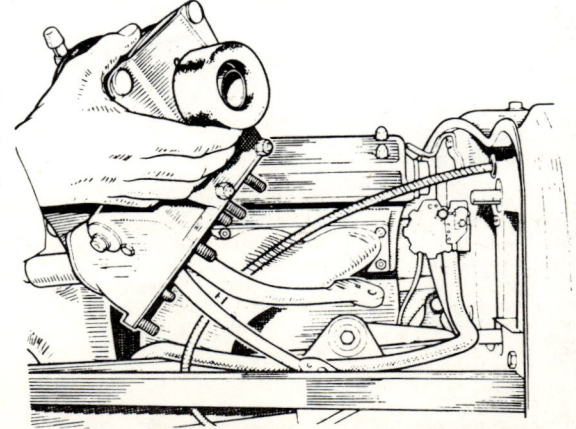

Fig.9.25. Removing the brake/clutch pedal box assembly

24 Brake linkage, 3.8 litre models - dismantling and reassembly

1 Remove the brake/clutch pedal box assembly as described in Section 23.
2 Refer to Fig.9.26.
3 Remove the bulkhead rubber seal (42).
4 Remove the four nuts and one setscrew fastening the brake master cylinder mounting bracket (35).
5 Remove the self locking nut from the serrated pin (21) and remove the conical spring (24) and the retaining washer.
6 Remove the pinch bolt from the brake pedal lever (1).
7 Remove the circlip (9) and the washer (10) from the pedal shaft.
8 Take out the vacuum check point union from the front valve housing.
9 Remove the brake master cylinder support bracket with linkages and pedal shaft assembly from the pedal housing (4). Collect the fibre washer from between the brake and clutch pedals.
10 Remove the throttle bellcrank bracket on right hand drive models.
11 Remove the four self locking nuts and take off the brake vacuum servo assembly (41).
12 Remove the setscrews and brass bush (5) from the pedal housing. Remove the split pin and withdraw the clevis pin.
13 Remove the clutch master cylinder and withdraw the clutch pedal. If the caged needle roller bearing (6) is faulty, it should be pressed out and replaced.
14 Remove the self locking nut and withdraw the bolt from the pivot bracket (32). Remove the brake master cylinder support bracket.
15 Remove the servo operating arm return spring (20). Remove the castellated nut (19) and the eccentric barrel nut (18). Remove the belleville washer (16), spacing collar (15), chamfered washer (14), and the rubber buffer (13).
16 Take off the power lever (11) from the pedal shaft and pin assembly (8). Check the nylon bushes (12) and renew them if necessary.
17 Remove the steel bush (23) and the nylon bush (22) and press out the serrated pin if necessary.
18 Remove the self locking nut and bolt attaching the two-way fork (25) to the pivot bracket and balance link; then dismantle.
19 Remove the upper brake master cylinder fork end (37) by removing the split pin and taking out the clevis pin.
20 Press out the spacing sleeves (30) and (33) from the compensator fork and lever. Check the nylon bushes (31) and (34) and renew if necessary.
21 Remove the grub screw (28) and press out the joint pin (26) from the two-way fork. Check and renew the nylon bushes (27), and (31), if necessary.
22 Remove the four plain nuts and shakeproof washers and take off the servo mounting bracket from the pedal block.
23 Reassembly is the reverse of the above but note the following points:
24 Ensure that all linkages are entirely free especially the balance link and the servo operating arm.
25 When replacing the pedal shaft assembly on the pedal lever make sure that the pedal pad is lined up with the clutch pedal pad and also that the brake pedal lever does not foul the pedal box on full stroke.
26 Do not forget to fit the fibre washer between the brake and clutch pedals.
27 After refitting the assembly to the car (Section 23) and after the hydraulic system has been refilled and bled, reset the air valve operation with the eccentric barrel nut in the manner described in paragraph 3 of Section 16.

25 Brake/clutch pedal box assembly, 4.2 litre models - removal and refitting

1 Drain the brake and clutch hydraulic reservoirs by following

the procedures for bleeding the system but in this case carry on pumping on the brake and clutch pedals until the reservoirs are empty.

2 For right hand drive models, remove the air cleaner elbow and the carburettor trumpets. Slacken the rear carburettor banjo nut and push the petrol feed pipe towards the float chamber.

3 Remove the servo vacuum pipe and clips.

4 Remove the fluid inlet pipes from the clutch and brake master cylinders. Immediately blank off the pipes and the unions.

5 Identify and then remove the brake fluid warning light wires from the caps of the brake fluid reservoirs. Remove the nuts from the clamp studs and then take off the brake and clutch reservoirs.

6 Remove the outlet pipes from the clutch and brake master cylinders. Immediately blank off the pipes and unions.

7 Remove the reaction valve assembly on right hand drive models by taking out the five screws securing the assembly to the valve housing and now the valve housing can be removed by unscrewing the two setscrews which secure it to the body of the master cylinder.

8 Also on right hand drive models, remove the throttle bell-crank bracket.

9 Move inside the car and take off the brake and clutch pedal pads by unscrewing the securing nuts.

10 Follow the instructions given in Chapter 12 for removal of the dash casing which, when removed, will expose the nuts securing the pedal box assembly. Remove the nuts together with the distance pieces and the brake pedal stop plate. Note that there are six self locking nuts and one plain nut with a shakeproof washer which goes on the centre bottom stud.

11 Remove the brake/clutch pedal box assembly by lifting outwards and at the same time turning through about 90° to allow the pedals to pass through the hole in the bulkhead as illustrated in Fig.9.25.

12 Refitting is the reverse of the removal sequence but watch the following points.

13 Make sure that you fit the plain nut and shakeproof washer on the bottom centre stud when securing the pedal box assembly inside the car.

14 Fit one feed wire (red/green) and one earth wire (black) to each of the brake fluid reservoir caps.

15 When tightening the banjo union nut (right hand drive models) ensure that the petrol feed pipe is clear of the rear float chamber.

16 Finally refill the brake and clutch systems with fresh hydraulic fluid and then bleed the systems.

26 Handbrake - general

An exploded view of the handbrake actuating mechanism is given in Fig.9.27 and of the pad carriers in Figs.9.11 and 9.12. It is a mechanical system which is entirely independent of the hydraulic foot braking system in that it has its own brake calipers, which operate on the rear brake disc only, and which are mounted on, and above, the rear foot brake calipers.

Each handbrake unit consists of two carriers, one each side of the brake disc and attached to the inside face of each carrier is a friction pad. The handbrakes of later models are self adjusting to compensate for friction pad wear but on early model cars, any wear of the pads which may occur has to be remedied by periodical adjustment for which an adjuster bolt is provided at the caliper. Adjustment for length of the handbrake cable is also provided.

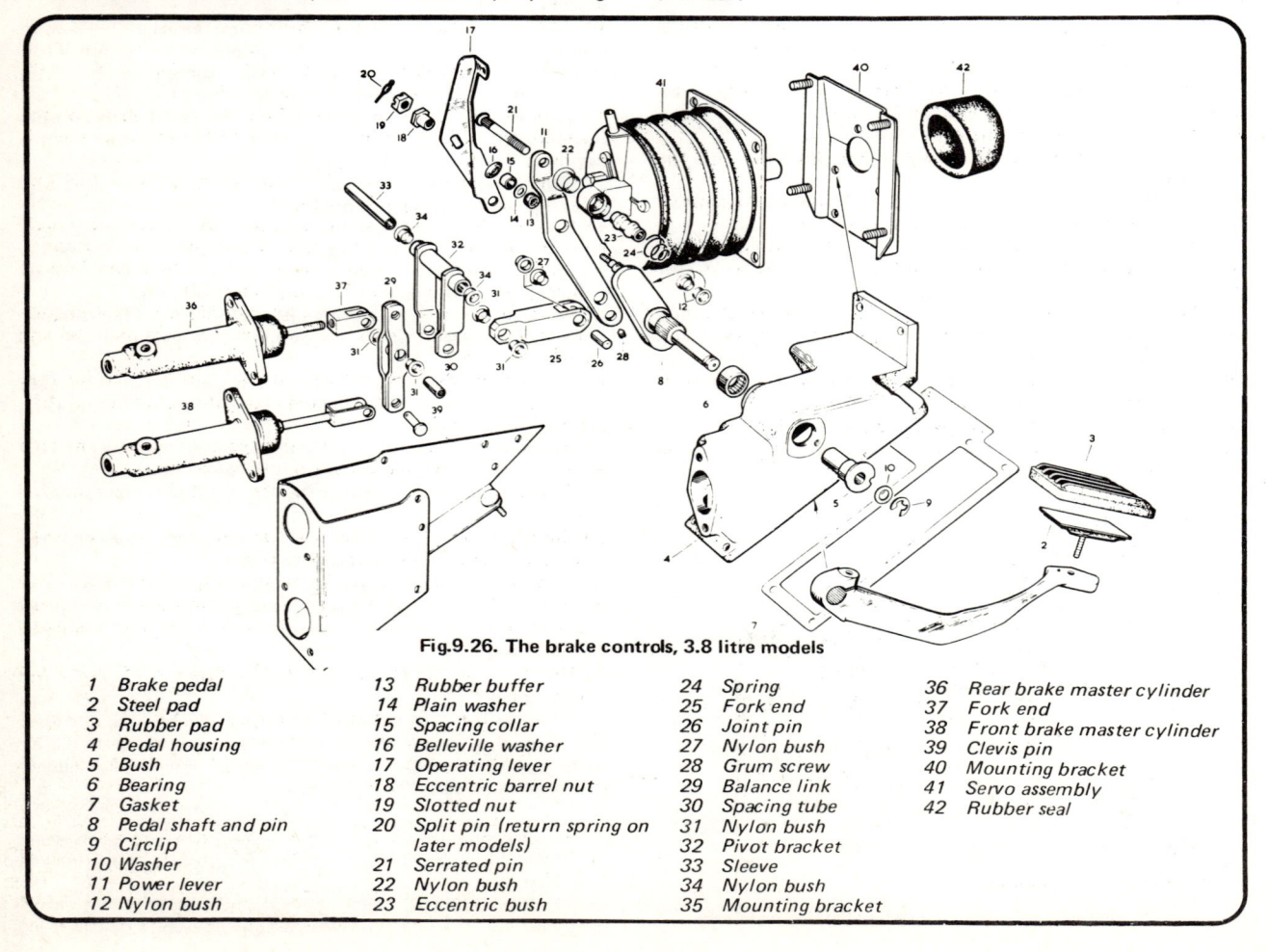

Fig.9.26. The brake controls, 3.8 litre models

1 Brake pedal	13 Rubber buffer	24 Spring	36 Rear brake master cylinder
2 Steel pad	14 Plain washer	25 Fork end	37 Fork end
3 Rubber pad	15 Spacing collar	26 Joint pin	38 Front brake master cylinder
4 Pedal housing	16 Belleville washer	27 Nylon bush	39 Clevis pin
5 Bush	17 Operating lever	28 Grub screw	40 Mounting bracket
6 Bearing	18 Eccentric barrel nut	29 Balance link	41 Servo assembly
7 Gasket	19 Slotted nut	30 Spacing tube	42 Rubber seal
8 Pedal shaft and pin	20 Split pin (return spring on later models)	31 Nylon bush	
9 Circlip	21 Serrated pin	32 Pivot bracket	
10 Washer	22 Nylon bush	33 Sleeve	
11 Power lever	23 Eccentric bush	34 Nylon bush	
12 Nylon bush		35 Mounting bracket	

Chapter 9/Braking system

A sectioned view of the self adjusting type of brake is given in Fig.9.28. When the brake lever in the car is operated, the lever (A) is moved away from the friction pad carrier (B) and draws the friction pads (F) together. If the running clearance between the friction pads and the brake disc is normal, the pawl (C) returns to its normal position when the lever is released. If the clearance is greater than it should be, the pawl will turn the ratchet nut (D) on the bolt thread and so draw the adjuster bolt (E) inwards thus bringing the friction pads closer to the brake disc until the normal running clearance is restored.

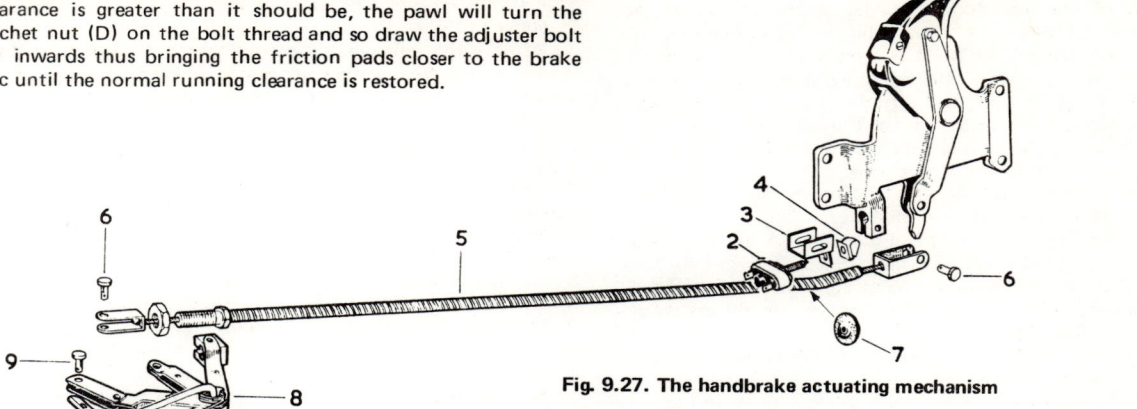

Fig. 9.27. The handbrake actuating mechanism

1 Handbrake lever assembly
2 Warning light switch
3 Mounting bracket
4 Spring striker
5 Handbrake cable
6 Clevis pin
7 Grommet
8 Compensator linkage
9 Clevis pin

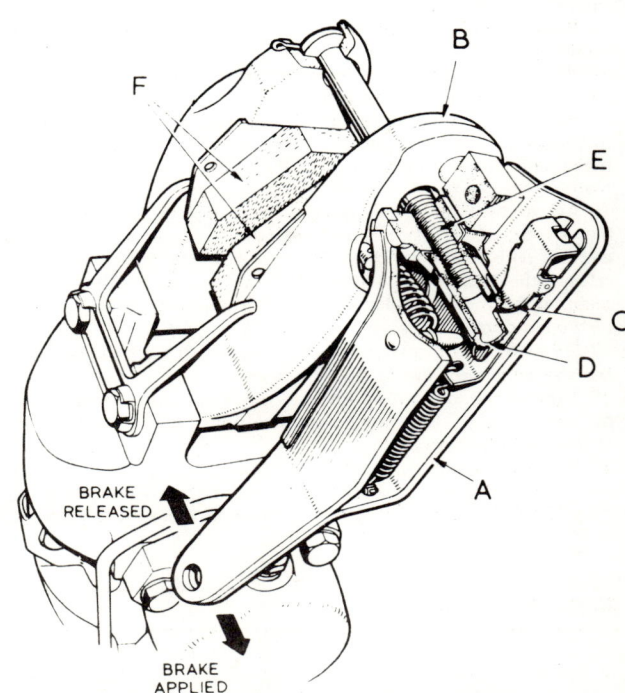

Fig. 9.28. Self adjusting handbrake mechanism

27 Handbrake friction pad carriers - removal and refitting

1 Place the car over a pit, or on a ramp, to give access to the rear suspension.
2 Take out the split pin followed by the clevis pin to disconnect the handbrake compensator linkage from the handbrake operating lever.
3 Knock up the tabs, unscrew the pivot bolts and withdraw them. Lift off the forked phosphor bronze retraction plate. It may well be that the shank of the pivot bolts will have rusted and so may prove difficult to remove, if this occurs the only advice we can give is to sparingly use penetrating oil and to work the bolts out using a mole wrench.
4 Remove the friction pad carriers from the caliper bridge by moving them rearwards around the disc and withdrawing from the rear of the suspension assembly.
5 Repeat the above for the carriers on the other side of the car.
6 Refitting is the reverse of the above sequence, but the handbrake will have to be adjusted and the way of doing this (in respect of the non-self adjusting type brake) will be found in Section 31.
7 When the self adjusting type of friction pad carriers have been refitted to the car, and before you make the connection to the compensator linkage, pull and release the operating lever repeatedly when the ratchet will be heard to click. Carry on doing this until the ratchet does not operate and this indicates that the correct clearance between the friction pads and the disc has been obtained. Connect the operating levers to the compensator linkage and check the functioning of the handbrake, it is possible, if the handbrake does not lock the wheels, that the handbrake cable requires adjustment; instructions for this will be found in Section 32.

28 Handbrake friction pads - renewing

1 Follow the instructions given in Section 27 for removal of the friction pad carriers.
2 Slacken the nut in the outer face of each carrier and remove the friction pad by pulling outwards with a hooked tool engaged in the hole at the rear of the friction pad securing plate.
3 Place the new friction pad over the securing plate and then enter them into the carrier so that the short face of the friction pad is upwards. Make sure that the securing plate slot is engaged with the head of the securing bolt.
4 Follow the instructions given in Section 27 for refitting the friction pad carriers to the car.

29 Handbrake friction pad carriers, non-self adjusting type - dismantling and reassembly

1 Separate the friction pad carriers by taking out the adjuster bolt but be careful to control the run of the self locking nut in the forked end of the operating lever.
2 Take out the split pin and withdraw the clevis pin to detach

the pivot seat from the forked end of the operating lever.
3 Do not attempt to remove the spring or the squared nut because, if either are damaged, the pad carrier will have to be renewed.
4 Before reassembling the friction pad carriers, make sure that the trunnion block has complete freedom of movement in the forked end of the operating lever. Also ensure that the pin of the pivot seat is a sliding fit in the drilling at the extreme end of the friction pad carrier, the pivot seat must also be a sliding fit between the forked ends of the operating lever. The clevis pin must be a sliding fit both through the eye of the pivot and through the holes in the forked ends of the operating lever.
5 Assemble the operating lever and the pivot seat but do not fit them to the inner pad carrier at this stage.
6 Pass the adjusting bolt through the outer pad carrier and screw it into the retaining nut and spring.
7 Fit the operating lever and pivot assembly to the inner pad carrier and screw in the adjusting bolt until it is flush with the outer face of the trunnion block. The spring should be preloaded by inserting the blade of a screwdriver between the retaining nut and the cage as illustrated in Fig.9.29.
8 Now adjust on the bolt until it again becomes flush with the outer face of the trunnion block.
9 Place the self locking nut on the trunnion block and screw the adjuster bolt into it until it is flush with the outer face of the nut. Withdraw the preloading screwdriver.

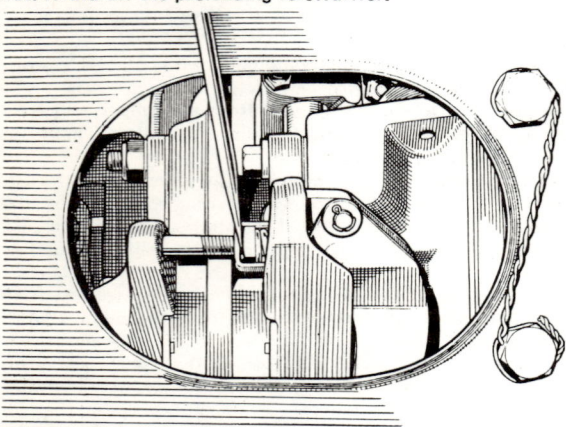

Fig. 9.29. Pre-loading the handbrake caliper return spring

30 Handbrake friction pad carriers, self adjusting type - dismantling and reassembly

1 Remove the cover securing bolt.
2 Take out the split pin and withdraw the pivot clevis pin.
3 Remove the dust cover and take out the split pin from the screwdriver slot in the adjusting bolt.
4 Unscrew the adjusting bolt from the ratchet nut and withdraw the nut and bolt.
5 Detach the pawl return spring and withdraw the pawl over the locating dowel.
6 Detach the operating lever return spring and then remove the operating lever and the lower cover plate.
7 Reassembly is the reverse of the above sequence.
8 After final assembly, and before the split pin is replaced in the slot of the adjusting bolt, screw the bolt in, or out until there is a distance of 7/16" (11.1 mm) between the friction pads, this represents the thickness of the brake disc plus 1/16" (1.5 mm).
9 Fit a new split pin to lock the adjuster bolt when the above adjustment is made.

31 Handbrake cable - removal and refitting

It is unlikely that the handbrake friction pads will wear to the extent of affecting the correct functioning of the handbrake. The most likely cause, after a long period of use, is for the cable to stretch and where this cannot be compensated by the adjustment described in Section 32 it will be necessary to fit a new cable.
1 Place the car over a pit or on a ramp.
2 Refer to Fig.9.27.
3 Remove the split pin and then withdraw the clevis pin from the inner cable fork at the compensator linkage.
4 Slacken the locknut and unscrew the outer cable from the retaining block.
5 Remove the spring holding the cable away from the propeller shaft.
6 Remove the four nuts holding the seats to the seat slides and remove the seats.
7 Remove the two screws, one each side of the radio control panel, which secures the ashtray.
8 Remove the two screws which secure each side of the radio control panel to the brackets under the instrument panel.
9 Withdraw the radio control panel casing.
10 Remove the setscrews securing the propeller shaft tunnel cover to the body.
11 Remove the gear lever knob and pull the handbrake into the fully 'ON' position. Now remove the propeller shaft tunnel by lifting at the rear and at the same time turning towards the nearside of the car and sliding it over the gear lever and the handbrake lever.
12 Remove the split pin and the clevis pin from the fork end at the handbrake lever.
13 Slacken the pinch bolt and remove the outer cable from the retaining block.
14 Remove the grommet and withdraw the cable from the rear end of the propeller shaft tunnel.
15 Refitting is the reverse of the above but when fitting the outer cable to the retaining block at the compensator linkage, make sure that the longer end of the block is facing towards the front of the car.
16 Adjust the cable by screwing in each adjuster bolt at the rear brake (non-self adjusting type) until the friction pads are in hard contact with the brake discs. Now for both types of brake, fully release the handbrake and, if applicable, slacken the locknut securing the threaded adaptor to the compensator at the rear end of the handbrake cable.
17 Screw out on the adaptor until there is no slack in the cable, but the cable must not be under tension.
18 Tighten the locknut and now, for the non-self adjusting type of brake, carry out the clearance adjustment of the pads described in Section 32.
19 Finally for both types of brake, check that the friction pads are not bearing on the discs when the handbrake lever is in the fully 'OFF' position and also that the wheels are locked when it is in the 'ON' position.

32 Handbrake adjustment

When the friction pad carriers of the non-self adjusting type of handbrake have been removed and refitted, or, if applicable, at the 2500 mile (4000 km) servicing of the car, they should be adjusted as follows:
1 Remove the carpet from the luggage compartment floor and then remove the seven screws holding the rear axle cover plate.
2 Insert a 0.004" (0.10 mm) feeler gauge between the face of one handbrake friction pad and the disc and then screw in the adjuster bolt, using the special key supplied in the tool kit (Fig.9.30), until the feeler gauge is just nipped. Withdraw the feeler and check the disc for free rotation. Repeat for the other side.
3 If, after carrying out the above adjustment, the handbrake does not function satisfactorily, the trouble may be due to maladjustment of the handbrake cable in which case refer to Section 31 and follow the adjustment procedure described.

Chapter 9/Braking system

4 On those models not provided with a rear axle cover plate, there is no alternative but to drop the rear suspension in order to make the above adjustment.

33 Brake fluid level and handbrake warning light - setting

This light is actuated by three switches, one in the cap of each fluid reservoir and by an interrupter switch operated by the handbrake lever. If the light does not glow when the ignition is switched on and the handbrake is applied and you are satisfied that the electrical circuit is in order, then the necessity for adjustment of the handbrake switch is indicated, this will also be the case if the light is not extinguished when the handbrake is 'OFF' and the reservoirs are full of fluid.

1 Refer to Section 31 and follow the instructions for removal of the propeller shaft tunnel.
2 Check the handbrake for full travel and make sure that the spring steel lever which operates the plunger of the switch is not distorted or misaligned.
3 Check the operation of the spring steel lever and the switch plunger with the handbrake in the 'ON' and 'OFF' position. Bend the lever, if necessary, to obtain correct operation. However, if this does not cure the trouble, recheck the electrical circuit starting with the bulb and paying particular attention to the switch.

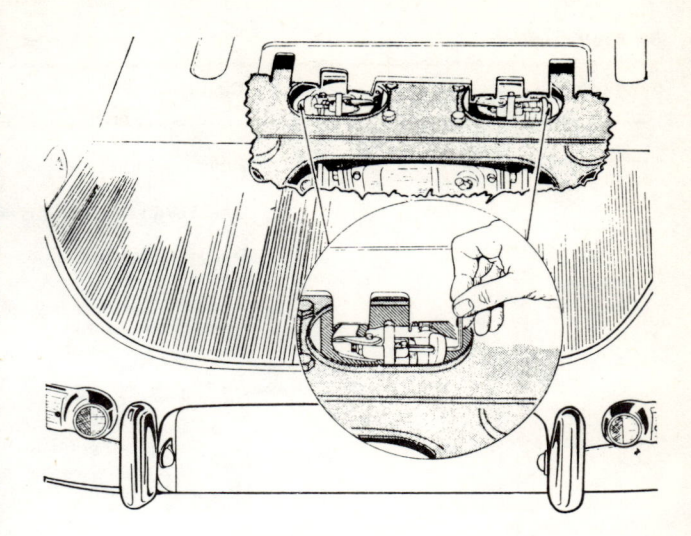

Fig. 9.30. Handbrake adjustment (non-self adjusting type)

Chapter 9/Braking system

34 Fault diagnosis

Symptom	Cause	Remedy
Brake travel excessive	Fluid level too low	Top up master cylinder reservoir and check for leaks.
	Wheel cylinder or caliper leaking	Dismantle wheel cylinder or caliper, clean, fit new rubbers and bleed brakes.
	Master cylinder leaking	Dismantle master cylinder, clean, fit new rubbers and bleed brakes.
	Brake flexible hose leaking	Examine and fit new hose. Bleed brakes.
	Brake line fractured	Replace with new pipe. Bleed brakes.
	Brake system unions loose	Check, tighten unions as necessary. Bleed brakes.
	Normal wear of linings	Fit replacement shoes or friction pads.
Brake pedal feels springy	New linings not yet bedded-in	Use brakes gently until springy feeling ceases.
	Brake discs badly worn, weak or cracked	Fit new brake drums or discs.
	Master cylinder securing nuts loose	Tighten nuts and make sure that spring washers are fitted.
Brake pedal feels spongy and soggy	Wheel cylinder or caliper leaking	Dismantle wheel cylinder or caliper, clean, fit new rubbers and bleed brakes.
	Master cylinder leaking	Dismantle master cylinder, clean, fit new rubbers and bleed brakes.
	Brake pipe line or flexible hose leaking	Fit new pipe line or hose. Bleed brakes.
	Unions in brake system loose	Examine for leaks and tighten. Bleed brakes.
Brakes uneven and pulling to one side	Friction pads or discs contaminated with oil or grease	Ascertain and rectify source. Clean pads and discs and fit new pads if necessary.
	Tyre pressures incorrect	Check and rectify.
	Brake disc loose	Check and tighten as necessary
	Brake pads fitted incorrectly	Remove and fit correct way round.
	Different types of linings fitted at each wheel	Fit linings of correct specification all round.
	Anchorages for front or rear suspension loose	Check and tighten as necessary. Ensure rubbers are not perished.
	Brake discs badly worn, cracked or distorted	Fit new brake discs.
Brakes tend to bind, drag or lock-on	Incorrect adjustment of brake pads	Check and rectify pad adjustment.
	Handbrake cable over-tightened	Adjust correctly.
	Master cylinder by-pass port choked	Dismantle and clean master cylinder. Bleed brakes.
	Wheel cylinder piston seized	Dismantle and rectify wheel cylinder. Bleed brakes.
	Blockage of port through which air enters servo valve assembly	Remove and clean servo unit. Bleed brakes.
Brakes fail to release	Handbrake over-adjusted	Check and adjust correctly.
	Master cylinder by-pass port choked	Dismantle master cylinder, clean and bleed brakes.
	Excessive friction between wheel cylinder seals and cylinder body	Dismantle wheel cylinder and rectify. Bleed brakes.

Chapter 10 Electrical system

Contents

Alternator - dismantling and reassembly ... 19	Pre-engaged starter motor M45G - bench examination, test and rectification ... 37
Alternator - inspection, bench test and repair ... 21	Pre-engaged starter motor M45G - dismantling and reassembly 35
Alternator output control unit model 4 TR - general ... 22	Revolution counter, early models - testing, removal and refitting ... 52
Alternator - precautions ... 17	
Alternator - removal and refitting ... 18	Revolution counter (tachometer) later models - removal and refitting ... 51
Alternator - testing in situ ... 20	
Ammeter, fuel water temperature and oil pressure gauges - removal and refitting ... 48	Speedometer - fault diagnosis and rectification ... 55
Battery - charging ... 5	Speedometer - removal and refitting ... 53
Battery - electrolyte replenishment ... 4	Speedometer drive cable - removal, examination and refitting ... 54
Battery - maintenance and inspection ... 3	
Battery - removal and replacement ... 2	Starter motor M45G - bench examination, test and rectification ... 36
Battery indicator - general ... 49	Starter motor M45G - dismantling and reassembly ... 34
Battery indicator - removal and refitting ... 50	Starter motor - general ... 31
Dynamo - dismantling ... 8	Starter motor - removal and refitting ... 32
Dynamo - examination ... 9	Starter motor - testing in position ... 33
Dynamo - general ... 6	Switches - removal and refitting ... 47
Dynamo - reassembly ... 10	The bi-metallic resistance instrumentation - fault finding ... 56
Dynamo - removal and refitting ... 7	The instrument panel - general ... 44
Dynamo - testing in position ... 11	Traffic hazard warning device - general ... 27
Electric clock, early models - adjustment, removal and refitting ... 46	Voltage regulator - checking continuity between the battery and regulator ... 13
Electric clock, later models - maintenance, removal and refitting ... 45	Voltage regulator RB 310 - cleaning, checking and adjustment 14
Flasher unit - fault tracing and rectification ... 26	Voltage regulator RB 340 - cleaning, checking and adjustment 15
Fitment of auxiliary lights - general ... 57	Voltage regulator, testing - general ... 12
Fitment of radio and cassette and cartridge players - general 58	Warning light control unit model 3AW - general ... 24
Fuse units model 4 FJ - general ... 25	Windscreen washer type 5SJ - dismantling, testing and reassembly ... 39
General description ... 1	
Horns - removal and refitting ... 28	Windscreen washer type 5SJ - testing in position ... 38
Horns and horn relay - adjustment, fault tracing and rectification ... 29	Windscreen wipers - fault diagnosis and rectification ... 43
Lamps - removal, adjustment and refitting ... 30	Windscreen wipers - general ... 40
Lucas 11 AC alternator - general ... 16	Windscreen wiper motor - removal and refitting ... 41
Output control unit model 4 TR - checking and adjusting ... 23	Windscreen wiper spindle housings - removal and refitting ... 42
	Wiring diagrams - colour coding ... 59

Specifications

System type ...	12 volt negative earth
Battery	
Type ...	Lucas FRV11/7A
Voltage ...	12
No. of plates per cell ...	11
Capacity, 10 hour rate ...	55 ampere hours
Capacity, 20 hour rate ...	60 ampere hours
Charging system	
3.8 litre - early models ...	Dynamo Type C45 PVS-6
	Current and voltage regulator - Type RB 310
- later models ...	Dynamo Type C42
	Current and voltage regulator - Type RB 340

Chapter 10/Electrical system

4.2 litre	Lucas 11 AC alternator
	Alternator output control unit - Model 4 TR
	Warning light control unit - Model 3AW

Fuse unit

3.8 litre	Qty. 4 Model 4FJ
4.2 litre	Qty. 3 Model 4FJ

Fuses

3.8 litre

	Amps.	Fuse No.
Headlamps:		
Main beam	35	1
Dip beam	35	2
Horns	50	3
Fuel pump	5	4
Side, tail, panel, number plate lamps	35	5
Horn relay-washer, radiator fan motor, stop lamps	35	6
Flasher-heater, wiper, choke, fuel, water, oil gauges	35	7
Head lamp flash, interior lamps, cigar lighter	35	8
Electrically heated backlight (optional extra)	15	—

4.2 litre

	Amps.	Fuse No.
Headlamps:		
Main beam	35	1
Dip beam	35	2
Horns	50	3
Spare	—	4
Side, panel, tail and number plate (not Germany) lamps	35	5
Horn relay, washer, radiator fan motor and stop lamps	35	6
Flashers, heater, wiper, choke, fuel, water and oil gauges	35	7
Headlamp flasher, interior lamps and cigar lighter	35	8
Heated backlight (when fitted)	15	In line
Radio, optional extras	5	In line
Traffic hazard warning system	35	In line

Horns

Early models	One type WT.618 low note
	One type ET.618 high note
Later models	One type 9H low note
	One type 9H high note

Lamps

LAMP	LUCAS BULB NUMBER	VOLTS	WATTS	APPLICATION
Head - 3.8 litre	416	12	60/40	Home and R.H. drive export
	417	12	60/40	L.H. drive export
	410	12	45/40	European continental
(Yellow)	411	12	45/40	France
	Sealed beam unit			U.S.A. and Canada
Head - 4.2 litre	Sealed beam unit	12	75/45	Home and R.H. drive export
			50/40	L.H. drive except Europe
	410	12	45/40	L.H. drive European
	411	12	45/40	France
			(Yellow)	

ALL MODELS:

Side	989	12	6	Not Belgium, Germany, Holland, Switzerland
Front and rear flashing indicators	382	12	21	
Reversing light	273	12	21	
Rear/brake	380	12	21/6	
Number plate illumination	207	12	6	
Interior lights	382	12	21	Open 2 seater
	989	12	6	Fixed head Coupe and 2 + 2
Map light	989	12	6	
Instrument illumination:				
Headlamp warning light				
Ignition warning light				
Fuel level warning light				
Handbrake/brake fluid warning light	987	12	2	

Chapter 10/Electrical system

Mixture control warning light
Electrically heated backlight indicator light
Traffic warning device indicator light

Switch indicator strip				
Flashing indicator warning light	281	12	2	
Automatic transmission selector quadrant	281	12	2	2 + 2
Headlamp pilot lamp	989	12	6	Belgium, Germany, Holland, Switzerland

Starter

3.8 litre ... ... Model M 45 G
   Lock torque ... ... 22 lbs. f. ft with 430 - 450 amps at 7.8 - 7.4 volts
   Torque at 1000 rpm ... ... 8.3 lbs. f. ft with 200 - 220 amps at 10.2 - 9.8 volts
   Light running current ... ... 45 amps at 5,800 - 6,800 rpm

4.2 litre ... ... Pre-engaged motor Model M 45 G
   Lock torque ... ... 22.6 lbs. f. ft with 465 amps at 7.6 volts
   Torque at 1000 rpm ... ... 9.6 lbs. f. ft with 240 amps at 9.7 volts
   Light running current ... ... 70 amps at 5,800 - 6,500 rpm

Windscreen washer ... ... Electrical Lucas Type 5SJ

Windscreen wiper ... ... Two speed, triple arm. Single Lucas Type DL.3 motor

1 General description

All models have 12 volt electrical systems in which the negative battery terminal is earthed. The major components of the system are a 12 volt battery located at the lower left-hand side of the engine compartment; a dynamo in the case of 3.8 litre and an alternator for 4.2 litre models mounted on the engine at the front left-hand side; a starter motor at the rear right-hand side of the engine, and a current and voltage regulator or alternator output control unit - according to model, in the engine compartment.

The battery supplies a steady current for the ignition, lighting and other electrical circuits and provides a reserve of electricity when the current consumed by the electrical equipment exceeds that being produced by a charging system.

Although full instructions for the periodic overhaul and minor servicing of the various electrical components are given in this Chapter it must be appreciated that rectification of major faults will require specialised knowledge and equipment and therefore, where such faults arise, the defective item should be removed and replaced with a serviceable item which can usually be obtained on an exchange basis.

Wiring diagrams covering the various models will be found at the end of the Chapter.

2 Battery - removal and replacement

1 Open the bonnet.
2 Disconnect the battery terminals. These may be attached by screws or, the earth terminal, by a pinch nut and bolt.
3 Unscrew the two wing nuts securing the battery retaining strap, remove the fixing rods and the strap.
4 Lift out the battery.
5 Refitting is the reverse of removal but before fitting the cable connectors, clean the terminals and lightly coat them with petroleum jelly.

3 Battery - maintenance and inspection

Note: Never use a naked light when examining a battery as an explosive mixture of oxygen and hydrogen is given off when it is on charge or when standing idle.

1 Normal weekly battery maintenance consists of checking the level of the electrolyte in the cells to ensure that the separators are covered by about ¼" of electrolyte. If the level has fallen, top up with distilled water only. Do not overfill the battery. If the battery is overfilled or any electrolyte is spilled, clean it away immediately as the electrolyte is extremely corrosive.
2 Keep the terminals clean and covered with petroleum jelly.
3 Clean the top of the battery regularly. Maintaining it in a clean and dry condition will help to prevent corrosion and will also ensure that the battery does not become partially discharged by leakage through dampness and dirt.
4 Inspect the battery securing nuts, the clamp plate, the tray and leads for corrosion which will show up as white fluffy deposits brittle to the touch. If corrosion is found, clean it off with ammonia and then paint over the clean metal with an anti-rust, anti-acid paint.
5 Inspect the battery case for cracks. If a crack is found, clean and plug it with one of the proprietary compounds now on the market. If leakage through the crack has been excessive it will be necessary to refill the cell concerned with fresh electrolyte. Cracks are commonly caused at the top of the battery case by pouring in distilled water in the middle of winter AFTER instead of BEFORE a run. This gives the water no chance to mix with the electrolyte and so it freezes and splits the battery case.
6 If very frequent topping up is necessary and it has been established that the case is not cracked, the fault is due to overcharging of the battery and indicates a need to check and reset the voltage regulator. (Refer to Sections 14, 15 or 22 as applicable).
7 If the battery persists in a low state of charge, first consider the conditions under which the battery is used. If it is subjected to long periods of discharge without suitable opportunities for recharging, a low state of charge must be expected. If, on the other hand, the battery remains in a low state of charge when the car is in regular use with reasonably long running periods each day then it may be that the dynamo or the voltage regulator are at fault. A heavy discharge tester can be used to determine whether or not the fault lies in the battery. Your local garage can easily carry out the test for you. The contact prongs of the tester are pressed against the exposed positive and negative terminals of each cell. A good cell will maintain a reading of 1.2 - 1.5 volts for at least six seconds. If, however, the reading falls off rapidly that particular cell is probably faulty and will especially if other cells are also at fault, account for the low state of charge of the battery.

Chapter 10/Electrical system

8 The specific gravity of the electrolyte in each cell should be checked periodically using a hydrometer but to avoid misleading readings do not take hydrometer readings immediately after topping up with distilled water. The readings given by each cell should be approximately the same and if one cell differs appreciably from the others, an internal fault in that cell is indicated.

9 The appearance of the electrolyte drawn into the hydrometer will given an indication of the state of the plates. If the electrolyte is very dirty or contains small particles in suspension, it is possible that the plates are in poor condition.

10 The specific gravity of the electrolyte varies with its temperature and so, for convenience in comparing specific gravities, this is always corrected to $60°$ F. The method of correction is as follows:-

For every $5°F$ below $60°F$ **deduct** 0.002 from the hydrometer reading to obtain the correct specific gravity at $60°F$.

For every $5°F$ above $60°F$ **add** 0.002 to the hydrometer reading.

11 The specific gravity of electrolyte corrected to $60°F$ is given in the following table:-

State of charge	U.K. and climates ordinarily below $80°F$ ($26.6°C$) Specific gravity corrected to $60°F$	Climates frequently over $80°F$ ($26.6°C$) Specific gravity corrected to $60°F$
Fully charged	1.270 – 1.290	1.210 – 1.230
About half charged	1.190 – 1.210	1.120 – 1.150
Completely discharged	1.110 – 1.130	1.050 – 1.070

4 Battery - electrolyte replenishment

1 If the battery is in a fully charged state and one of the cells maintains a specific gravity reading lower than the others and a check of each cell has been made with a discharge tester with satisfactory results, then it is likely that at some time electrolyte has been lost from the cell.

2 Top up the cell with a solution of 1 part sulphuric acid to 2.5 parts water. If the cell is already fully topped up, draw some electrolyte out of it with a pipette.

3 When mixing sulphuric acid and water **never add water to the acid**. Always pour the acid slowly onto the water in a glass container. **If water is added to the sulphuric acid it will explode.**

4 Recharge the battery after topping up with the electrolyte and recheck the hydrometer readings.

5 Battery charging

1 In winter time when a heavy demand is placed on the battery it is a good idea to have the battery charged occasionally from an external source at a rate of 1.5 amps can be safely used overnight. Special rapid boost charges which are claimed to restore the power of the battery in 1 to 2 hours are most dangerous unless they are thermostatically controlled as they can cause damage to the battery plates through overheating.

6 Dynamo - general

The model C.45 PVS-6 and the model C42 dynamos are of similar construction and the instructions which follow, cover both types. The main difference is that the bearing for the armature spindle in the commutator end bracket of the Model C42 is a porous bronze bush whereas in the Model C.45 PVS-6 it is a ball race. An exploded view of the Model C42 is given in Fig.10.1 and a view of the Model C.45 PVS-6 showing the end plate removed, and illustrating the foregoing difference, will be found at Fig. 10.2.

The dynamo is a shunt wound; two pole; two brush generator which is arranged to work, in the case of the C.45 PVS-6, in conjunction with a Lucas regulator unit Type RB.310 and in the case of the C42 with a regulator Type RB.340.

A fan integral with the driving pulley draws cooling air through the dynamo via the inlet and outlet holes provided in each end bracket.

The regulator unit governs the output of the dynamo and this is dependent on the state of charge of the battery. When the battery is in a low state of charge, the ouput from the dynamo is high, on the other hand, if the battery is fully charged the output is just sufficient to keep the battery in good condition. Provision is made for an increase in output to balance the current taken by the lighting and other electrical systems.

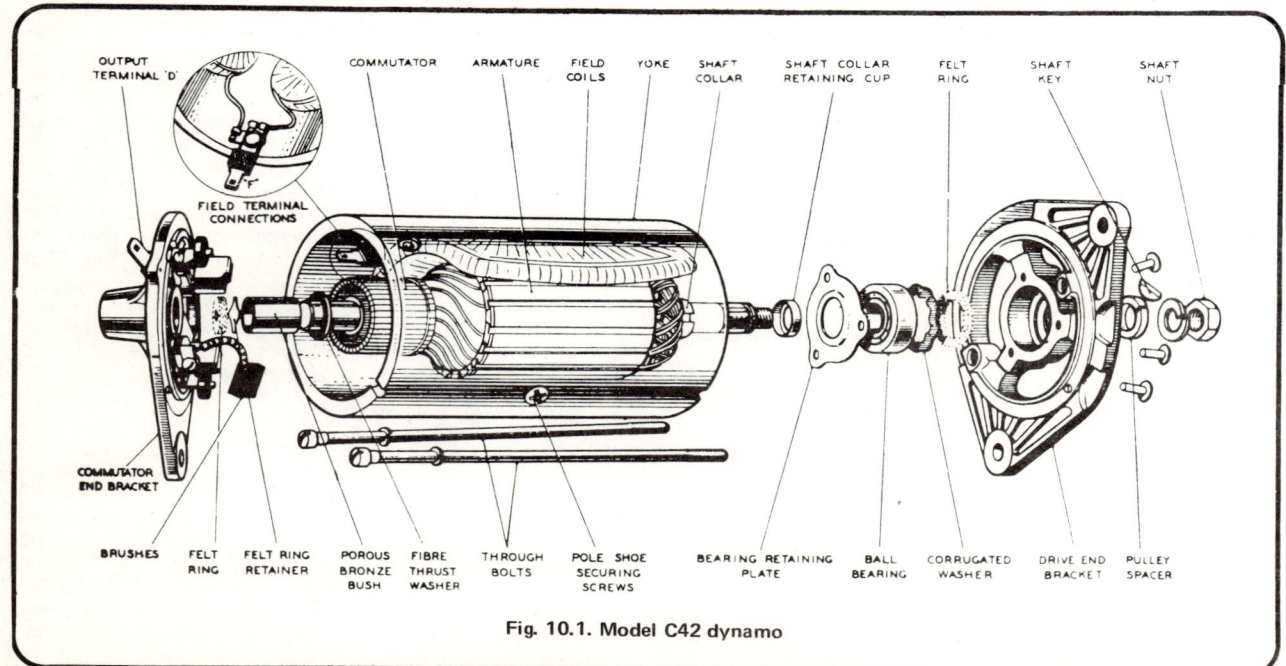

Fig. 10.1. Model C42 dynamo

Chapter 10/Electrical system

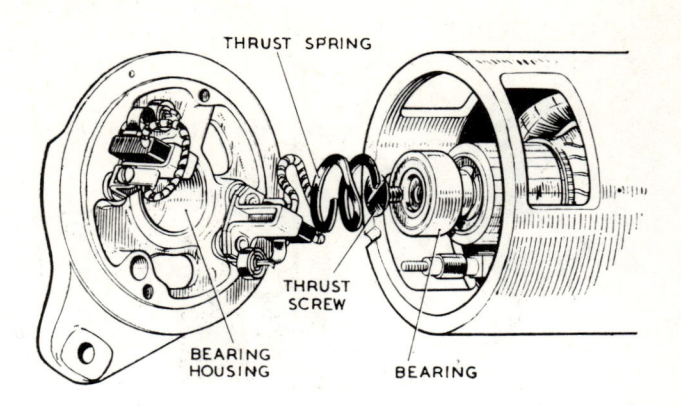

Fig. 10.2. Model C.45 PVS-6 dynamo - end plate removed

7 Dynamo - removal and refitting

1 Note the location of the two terminals at the rear of the dynamo. You will find that they are of different sizes.
2 Disconnect the cables.
3 Unscrew the nut and remove the bolt which secures the adjusting link to the dynamo.
4 Remove the nut and bolt securing each end plate to their mounting bracket. Take the weight of the dynamo as you remove the bolts.
5 Tilt the front of the dynamo downwards to release all tension on the belt and then work the belt off the pulley.
6 Lift the dynamo away from the car.
7 Refitting is the reverse of the removal.
8 When the dynamo is in position and before you finally tighten the mounting bolts, move the dynamo to a position where it is possible to depress the belt ½" (12 mm) at a point midway between the pump and dynamo pulleys.
9 Do not overtighten the belt as this will place a heavy load on the bearing and will cause early failure. But do not have the belt too slack, as this will cause a screeching noise on acceleration due to the belt slipping - rapid wear of the belt can be expected and due to the slipping whether noise is present or not the rate of the charge will be low.

8 Dynamo - dismantling

1 Follow the instructions for the removal of the dynamo from the car given in Section 7.
2 Remove the nut securing the driving pulley. Take off the pulley and collect the Woodruff key from the shaft.
3 Unscrew and remove the two through bolts.
4 Take off the commutator end bracket from the yoke (Fig. 10.1).
5 Remove the drive end bracket complete with the armature.
6 Lift the brush springs to one side and draw the brushes out of the brush holders. Note their positions if they are not being renewed because they will have 'bedded-in' and so should be replaced in their original positions. Unscrew the screws which hold the brush leads to the commutator end bracket, collect the lockwashers and then remove the brushes.
7 The bearings need not be removed or the armature shaft separated from the driving end bracket unless the bearings or the armature are to be renewed. If the items are to be separated support the driving end bracket and press out the shaft using a hand press.

9 Dynamo - examination

1 Fit the brushes to their respective holders and check them for freedom of movement. If movement is sluggish, remove them and ease the sides lightly polishing with a smooth file.
2 Measure the length of the brushes, the minimum permissible length is ¼" and brushes not meeting this must be renewed and bedded to the commutator.
3 Fit the commutator end bracket over the commutator, fit the brushes in their holders and, using a spring balance, test the brush spring tension. The tension of a new spring when the tension falls below that value. It should be noted that it is possible to examine the brushes and to check the spring tension without dismantling the later type of dynamo as this work can be done through the 'window' (see Fig. 10.3).
4 Clean the commutator with a petrol moistened cloth. A commutator in good condition will be smooth and free from pits or burned spots. Minor blemishes can be removed by polishing the commutator with fine glass paper whilst rotating the armature. It may be possible to rectify a badly worn commutator by mounting the armature in a lathe and taking a light cut with a very sharp tool at a high speed. But do not remove more metal than absolutely necessary.

After working in the lathe, polish the commutator with fine glass paper. Now undercut the insulators between the segments to a depth of 1/32" as illustrated in Fig. 10.4. A hacksaw blade ground to the thickness of the insulator is a handy tool for undercutting but make sure that the insulator is cut away squarely and cleanly as shown in Fig. 10.5.
5 Burnt commutator segments are indicative of an open circuited armature winding. If you have no armature testing facilties the only way it can be checked is by substitution.
6 The old armature is removed from the driving end bracket in the manner described in Section 8.
7 It is essential, when fitting the new armature, that the inner journal of the ball race is supported by a piece of mild steel tube of suitable diameter.
8 The resistance of the field coils can be checked without removing them from the yoke by using an ohmmeter connected between the field terminal and the yoke. The field resistance is 4.5 ohms. If a meter is not available, the field coils can be checked by connecting a 12 volt battery between the field terminal and the yoke with an ammeter in series.
9 The ammeter reading should be about 2.7 amperes, if a zero reading or 'infinity' on the ohmmeter is recorded, an open circuit in the field winding is indicated. If the current reading is about 2.7 amps or the meter reading is lower than 4.5 ohms it means that the insulation of one of the coils has broken down.
10 Replacement of the field coils involves the use of equipment not normally available to most home mechanics so our recommendation is that in the event of field coil failure, you obtain a replacement exchange dynamo.
11 Check the condition of the bearings. They must be changed when wear has reached the point of allowing visible side movement of the armature shaft. A bush bearing is fitted to the commutator end bracket and a ball bearing at the drive end bracket of the C42 type of dynamo, whilst the C45 PVS-6 has a ball race at each end.
12 To change a bearing bush, first obtain a replacement and then allow it to soak for 24 hours immersed in thin engine oil so that the pores of the brush are filled with lubricant.
13 Remove the old bearing bush from the commutator end bracket using an extractor or by screwing an 11/16" tap into the bush for a few turns and then using the tap as an extractor but make sure that the tap is screwed in squarely to avoid damage to the bracket and that side loads are not applied when pulling out the bush.
14 Examine the felt ring in the bearing housing and if it is in good condition it can be re-used.
15 Replace the felt ring and then press the new bearing bush into the housing using a shouldered and highly polished mandrel of the same diameter as the armature shaft, Fig. 10.6. Press in the bush until the end is flush with the inner face of the bracket.
16 To replace the ball race at the driving end, refer to Fig. 10.7 and drill out the rivets which secure the retaining plate to the end bracket and then remove the plate.

17 Press the bearing out of the end bracket and collect the corrugated washer, the felt washer and the oil retaining washer.
18 Pack the new bearing with high melting point grease.
19 Refit the oil retaining washer, the felt washer and the corrugated washer to the bearing housing in that order and then locate the bearing in the housing and press it home.
20 Assemble the bearing retaining plate and insert new rivets from the inside of the end bracket and then open them out with a punch to secure the plate.
21 The ball race fitted at the commutator end of the armature shaft as illustrated in Fig. 10.2 is secured to the shaft by a thrust screw. The bearing can be removed with an extractor after the thrust screw has been removed. When fitting a new bearing, first pack it with high melting point grease and then press it home against the shoulder on the shaft and finally secure it with the thrust screw.

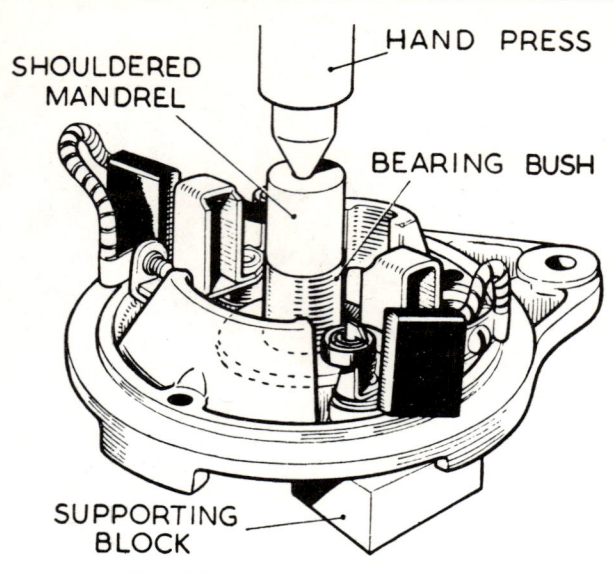

Fig. 10.6. Fitting a bearing bush

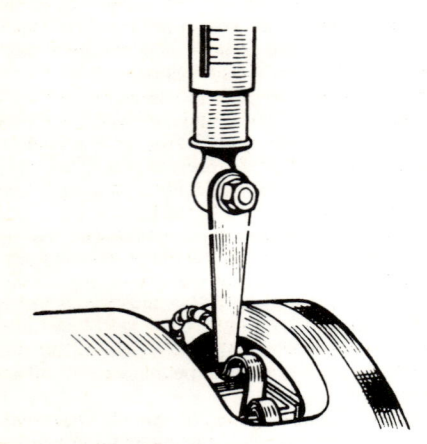

Fig. 10.3. Testing tension of the brush spring

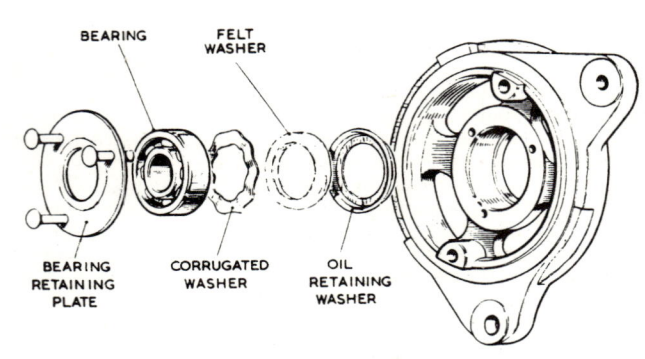

Fig. 10.7. The drive end bracket bearing assembly

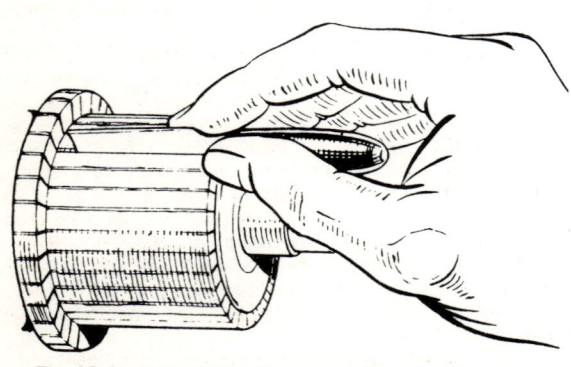

Fig. 10.4. Undercutting the commutator insulators

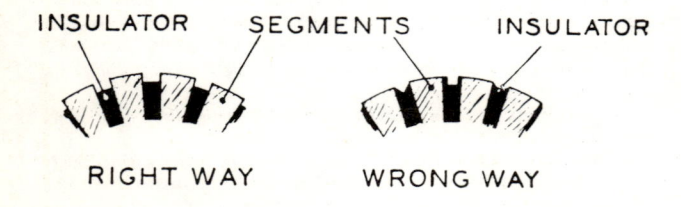

Fig. 10.5. Correctly and incorrectly cut insulators

10 Dynamo - reassembly

1 Fit the drive end bracket to the armature shaft and during this operation the inner journal of the race in the bracket must be supported. A piece of tube 4" in length and 1/8" thick and 11/16" internal diameter will be found suitable for this. The drive end bracket itself must not be used as a support.
2 Place the yoke over the armature to mate with the drive end bracket.
3 Assemble the brushes to the same brush holders from which they were removed (unless new brushes are being fitted) and secure their leads to the terminals with the screw and shakeproof washer. Lift up the brush springs and place them to one side so that they are not bearing on the brushes.
4 Make sure the brushes are clear of the commutator end of the holders and then fit the commutator end bracket to the armature shaft until the brush holders are partly over the commutator. Release the brush springs and then slide the commutator end bracket home against the yoke so that the projection on the bracket locates in the yoke.
5 After making sure that the mounting holes in the commutator end and the driving end brackets are correctly aligned, refit the two through bolts and tighten down.
6 Lubricate the commutator end bearing (bush type) by injecting a few drops of medium engine oil into the hole marked 'oil' at the end of the bearing housing (Fig. 10.8).

Chapter 10/Electrical system

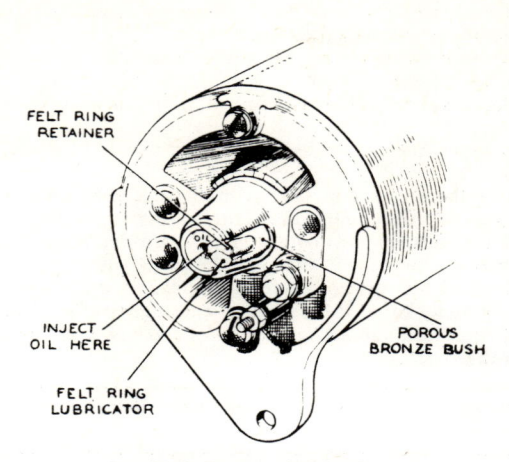

Fig. 10.8. Dynamo lubrication

11 Dynamo - testing in position

1 If it is noted, from ammeter readings during normal running, that there is no charge or low or intermittent charge, proceed as follows to determine the cause of the fault.
2 Check the fan belt and adjust as necessary.
3 If the dynamo or control box connections have been upset, check that they are connected correctly. The large and the small dynamo terminals should be connected to control box terminals 'D' and 'F' respectively.
4 Next switch off all lights and accessories (do not forget the rear window heater, if this is fitted and is connected to operate when the ignition is switched 'ON'), disconnect the cables from the dynamo terminals and connect the two terminals with a short length of wire.
5 Start the engine and allow it to run at normal idling speed. Now attach the negative lead of 0.20 volt moving coil voltmeter to one dynamo terminal and the positive lead to a good earth on the yoke.

6 Gradually increase engine speed and note the reading of the voltmeter which should rise rapidly and without fluctuation. Do not allow the reading to reach 20 volts and do not race the engine in an attempt to increase the voltage. A generator speed of about 1000 rpm is all that is required.
7 If the voltage does not rise rapidly and without fluctuation, an internal fault is indicated and the dynamo will have to be removed for detailed examination and test as indicated in Section 8 but before doing this, if a radio suppressor is fitted between the output terminal of the dynamo and earth, disconnect it and re-test the dynamo. If readings are now satisfactory, the capacitor is the cause of the fault.

12 Voltage regulator testing - general

The regulator should not need adjustment in service but if the battery fails to stay in a well charged condition, or if the ouput of the dynamo does not fall when the battery is fully charged, the setting should be checked and, if necessary, corrected. However, before suspecting a fault in the regulator, make sure that the fault is not due to a defective battery or to slipping of the dynamo belt.

Only a good quality **moving coil voltmeter** (0 - 20 volts) must be used when checking the regulator.

13 Voltage regulator - checking continuity between the battery and regulator.

1 If the dynamo and the battery are in good order, disconnect the cables from terminal blades 'B' (see Fig. 10.9) of the control box and connect them to the negative terminal of a moving coil voltmeter.
2 Connect the positive terminal of the voltmeter to a good earthing point and note the reading.
3 If the meter registers battery voltage (ie 12 volts), the wiring is in order and the regulator settings should be checked.
4 If there is no reading, reconnect the cables to terminal 'B' and examine the wiring between the battery, ammeter and control box for defects in the cables or loose connections.

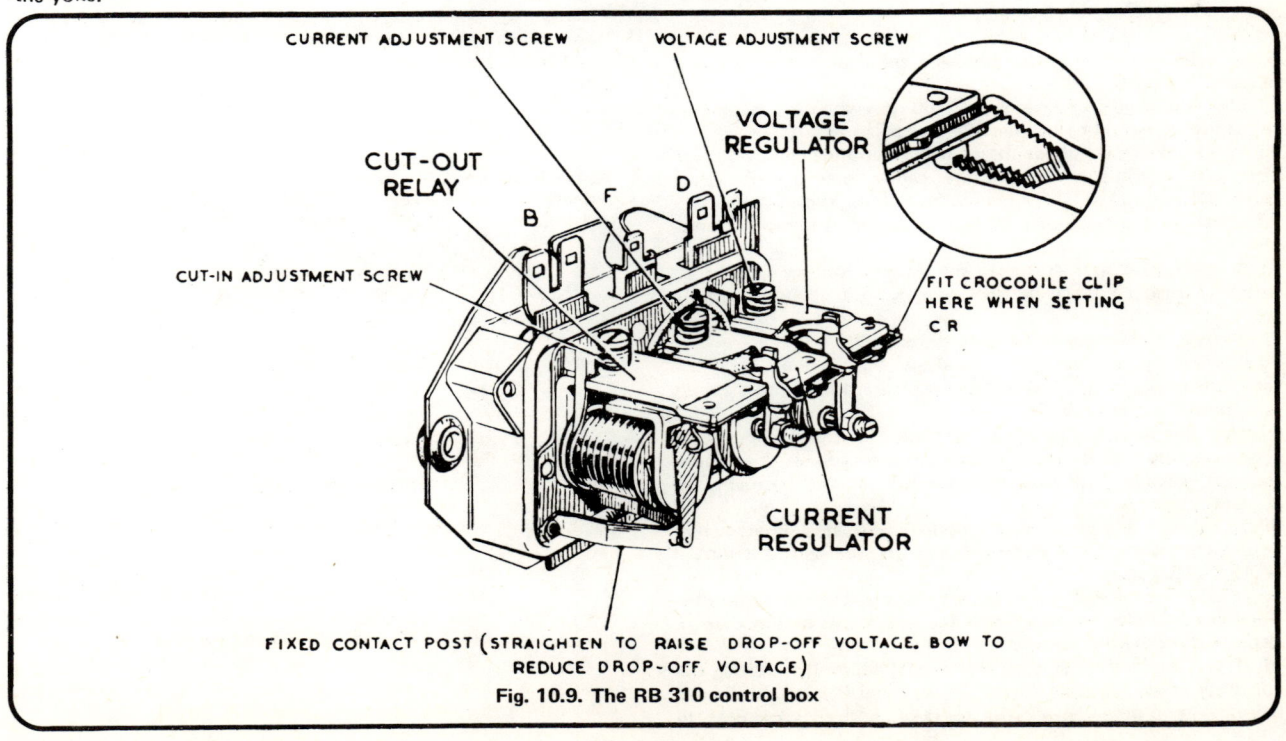

Fig. 10.9. The RB 310 control box

14 Voltage regulator RB 310 - cleaning, checking and adjustment

1 If it is found necessary to clean the regulator contacts, this can be done with a fine carborundum stone or fine emery cloth. The cut-out contacts should be cleaned with fine glass paper. All traces of metal dust or foreign matter must be removed after cleaning, by use of a clean cloth and methylated spirits.
2 Check continuity between the battery and the control box as described in Section 13.
3 To adjust the voltage regulator, disconnect the cable from terminal 'B' and join the ignition and battery feeds together using a jumper lead.
4 Connect the voltmeter between terminal 'D' of the box and a good earthing point.
5 In order to obtain accurate readings, the air temperature in the vicinity of the control box should be known as indicated below. Any adjustment that is found to be necessary should be completed within 30 seconds because heating of the shunt coil by the energising current may cause false settings to be made.
6 Start the engine and run up to 2,000 rpm when the open circuit voltage readings should be within the following limits:

Ambient temperature in vicinity of regulator	Voltage setting
10°C (50° F)	15.1 – 15.7
20°C (68° F)	14.9 – 15.5
30°C (86° F)	14.7 – 15.3
40°C (104°F)	14.5 – 15.1

7 If the voltmeter reading is outside the above limits, switch off the engine and turn the voltage regulator adjusting screw, adjacent to the 'D' terminal as shown in Fig. 10.9, clockwise to raise or anticlockwise to reduce the setting
8 Check your adjustment by repeating the operation at paragraph 6.
9 Now check and set the current regulator. To do this, the dynamo must be made to develop its full rated output. The regulator must be made inoperative by short circuiting the regulator contact with a crocodile clip placed between the insulated fixed contact bracket and the frame of the regulator as shown in Fig. 10.9.
10 Disconnect the leads to terminal 'B' and connect a 0.40 first grade moving coil ammeter between these and their terminal blades.
11 Start the engine and run it at 2,700 rpm when the reading on the ammeter should be between 24 and 26 amps.
12 If the reading is outside these limits, stop the engine and turn the current adjusting screw (the centre one of the three) clockwise to raise, or anticlockwise to decrease the setting.
13 Check your adjustment by repeating the operation at paragraph 11.
14 If the regulator is correctly set but the battery is still not being charged, it may be that the cut-out is not properly adjusted.
15 Partly withdraw the cable from terminal 'D'.
16 Connect your voltmeter between the exposed part of terminal 'D' and a good earth but take care not to short circuit terminal 'D' to the base.
17 Start the engine and slowly increase the speed at the same time watching the voltmeter closely. You should see the voltage increase steadily and then drop slightly at the moment the contacts close.
18 The cut-in voltage is that which is indicated immediately before the pointer of the voltmeter drops back and this should be 12.7 - 13.3 volts.
19 If you cannot determine with accuracy, the moment when the contacts close, try again with the headlamps switched on - a definite drop should be seen.
20 If the cut-in voltage is outside limits; adjust by rotating the adjusting screw adjacent to the 'B' terminal (Fig. 10.9) a small amount at a time - clockwise to raise, and anticlockwise to reduce the setting. Do all your settings as quickly as possible because of temperature rise effects.
21 Having checked and set the cut-in voltage, the next task is to check the drop-off voltage.
22 Disconnect the cables at terminal 'B' and then connect them together.
23 Connect your voltmeter between terminal 'B' and a good earth.
24 Start the engine and run up to above cut-in speed, reduce engine speed slowly and observe the voltmeter pointer. Opening of the contacts is indicated by the pointer dropping to zero and this should occur from a voltage of 9.5 - 11.0 volts.
25 The drop-off voltage is adjusted by altering the height of the fixed contact. To do this carefully straighten (see Fig. 10.9) the legs of the fixed contact post to raise, and bow them to reduce, the drop-off voltage.

15 Voltage regulator RB 340 - cleaning, checking and adjustment

1 The servicing procedures for this regulator, which is used in conjunction with the Type C42 dynamo, are identical to those described in Section 14 for the RB.310 regulator except that there are differences in setting procedures and voltage limits.
2 The regulator settings are adjusted by means of toothed cam adjusters as illustrated in Fig. 10.10 in place of the screws used with the RB.310 type. The best way of rotation the cams is to use a special setting tool of the type shown, as you will have a finer degree of control than using, say, a screwdriver. Clockwise rotation of the cam will increase, and anticlockwise rotation will decrease the setting.
3 Temperature and voltage setting for voltage regulator adjustment (paragraphs 2 - 8 of Section 14) are:

Ambient temperature	Open circuit voltage setting
10°C (50° F)	15.0 – 15.6
20°C (68° F)	14.8 – 15.4
30°C (86° F)	14.6 – 15.2
40°C (104°F)	14.4 – 15.0

4 Current regulator setting (paragraph 11 of Section 14) is 30 ± 1½ amperes.
5 Cut-in voltage (paragraph 18 of Section 14) is 12.6 - 13.4 volts and drop-off voltage (paragraph 24) is 9.25 - 11.25 volts.
6 A circuit diagram of the RB. 340 control box is given at Fig. 10.11

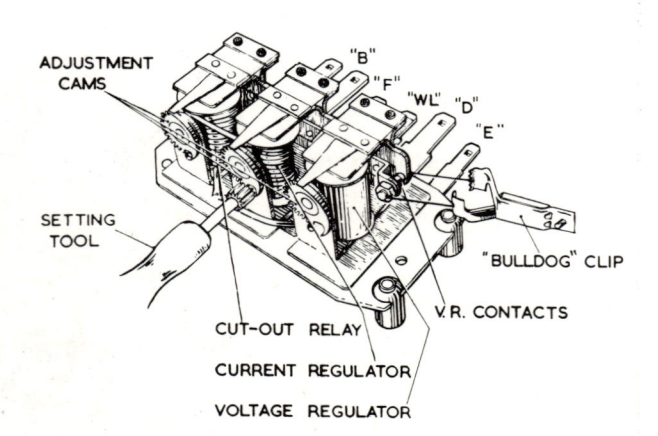

Fig. 10.10. The RB 340 control box

Chapter 10/Electrical system

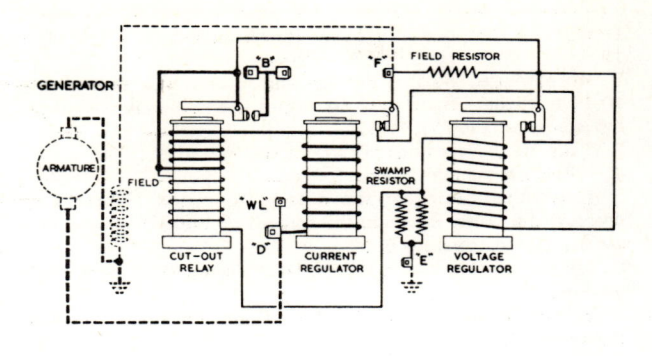

Fig. 10.11. Circuit diagram of the RB 340 control box

the AC output. Cooling of the diodes is provided by air flow through the alternator from a ventilating fan at the drive end.

The alternator is matched to a Model 4TR ouput control unit which controls the alternator field current and hence the terminal voltage. There is no need for a cut-out in the control unit as the diodes in the alternator prevent reverse currents passing through the stator when the machine is stationary or is generating less than the battery voltage. A separate current limiting device is not provided as the self limiting properties of the alternator effectively limit the output to a safe value.

The output control unit and the field windings of the alternator are isolated from the battery, when the engine is not running, by a separate pair of contacts in the ignition switch. On cars fitted with a steering column lock, the field windings are isolated by means of a relay which replaces the ignition switch control.

On later models a Lucas 3AW warning light control is built into the circuit.

16 Lucas 11 AC alternator - general

An exploded view of the Lucas 11 AC alternator is given in Fig. 10.12. The unit consists basically of a stationary output winding having built in rectification and a rotating field winding energised from the battery through a pair of slip rings.

The stator is housed between the slip ring end cover and the drive end bracket and consists of a 24 slot, 3 phase star connected winding on a ring shaped lamination pad.

The 8-pole rotor carries a field winding which is connected to two face type slip rings, it is supported by a ball race in the drive end bracket and on a needle roller bearing in the end cover of the slip ring, which also carries the brush gear for the field system. Two carbon brushes, one positive and one negative, bear against a pair of concentric brass slip rings which are carried on a moulded disc attached to the end of the rotor, the positive brush is always associated with the inner slip ring. Six silicone diodes, Fig. 10.13, are carried on the slip ring end cover and are connected in a 3 phase bridge circuit to provide rectification of

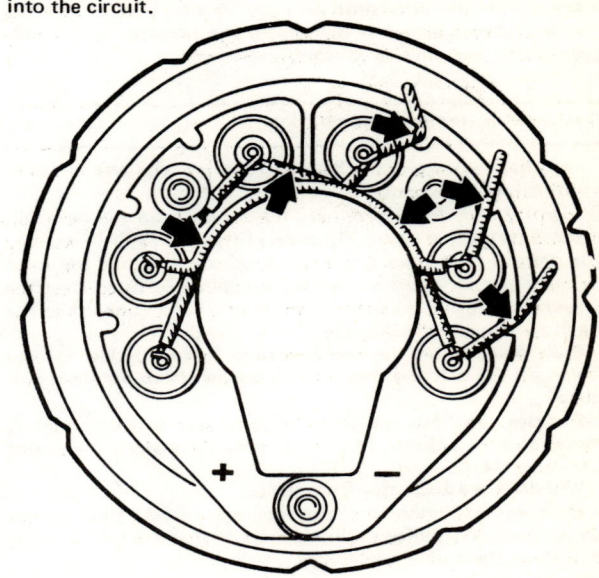

Fig. 10.13. The diodes and connections in the slip ring end cover

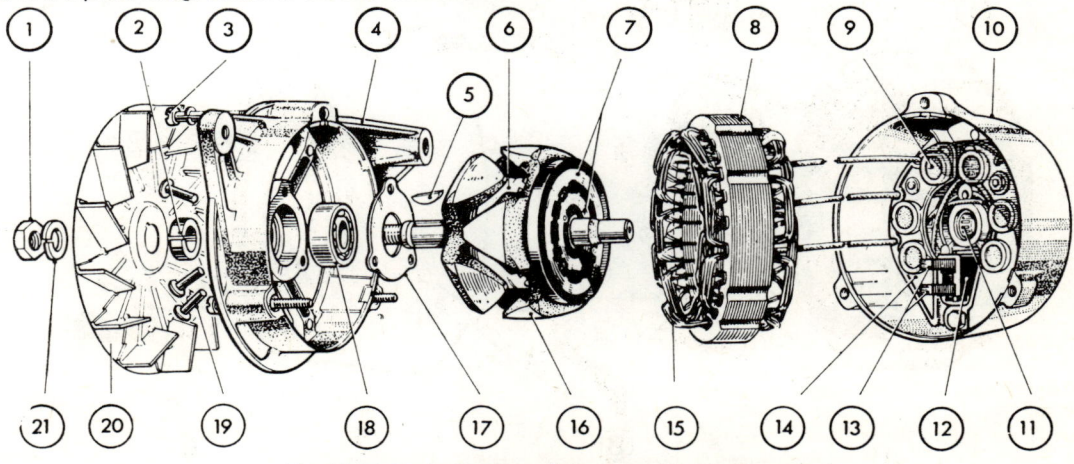

Fig. 10.12. Exploded view of the Lucas 11 AC alternator

1 Shaft nut
2 Bearing collar
3 Through fixing bolts (3)
4 Drive end bracket
5 Key
6 Rotor (field) winding
7 Slip rings
8 Stator laminations
9 Silicon diodes (6)
10 Slip ring end bracket
11 Needle roller bearing
12 Brush box moulding
13 Brushes
14 Diode heat sink
15 Stator windings
16 Rotor
17 Bearing retaining plate
18 Ball bearing
19 Bearing retaining plate rivets
20 Fan
21 Spring washer

17 Alternator - precautions

1 The car electrical system **must not** be checked using an ohmmeter having a hand driven generator until the transistors in the control box unit and the diode rectifiers in the alternator have been disconnected.
2 When you are replacing a battery or are using a slave battery to start the engine, take every precaution that you do not reverse the battery connections otherwise the diode rectifiers will be damaged. The battery **must** be connected **negative** earth.
3 When cables are disconnected at the alternator, **never** earth the brown/green cable because if this is earthed when the ignition is switched **on** the control unit and the wiring may be damaged. Similarly, **never** earth the main output cable of the alternator or its terminal as damage to the alternator or circuit may result.
4 **Never** run the alternator on open circuit, that is with the main lead disconnected as damage to the rectifier diodes may occur due to peak inverse voltages.

18 Alternator - removal and refitting

1 For the reasons given in Section 17, it is a wise precaution to first disconnect the battery.
2 Identify and then disconnect the cables from the terminals on the slip ring end cover. Those cars fitted with a 3AW warning light control unit have a positive lock connector in the main alternator output cable so making it impossible to connect the harness incorrectly. The cable to the control unit connects to the fourth terminal on the slip ring.
3 Push inwards on the spring loaded jockey pulley to ease tension on the belt and then remove the belt from the alternator pulley.
4 Remove the bolts securing the alternator to the mounting bracket and the adjuster link at the same time taking the weight of the alternator.
5 Withdraw the alternator from the car.
6 Refitting is the reverse of the removal procedure but do not release the jockey pulley until you are sure that the belt is sitting correctly in the pulley 'vee' tracks.

19 Alternator - dismantling and reassembly

1 Remove the alternator from the car as described in Section 18.
2 An exploded view of the slip ring end cover is given in Fig. 10.14 but where figures are quoted in the following paragraphs the reference is to Fig. 10.12.
3 Take off the shaft nut (1) and its spring washer (21).
4 Withdraw the fan (20).
5 Break the staking of the nuts to the bolts (3) and then unscrew the bolts. Examine the threads of the bolts and if they are damaged discard the bolts.
6 Mark the drive end bracket (4), the lamination pack when marking it.
7 Withdraw the drive end bracket (4) and the rotor (16) from the stator (8).
8 It is advisable not to separate the rotor from the drive end bracket unless the bearing is suspect or the rotor is to be replaced. In this event, first remove the shaft key (5) and the collar (2). Now use a hand press to remove the rotor from the drive end bracket.
9 Remove the terminal nuts and their washers and insulating pieces, the brush box screws and the hexagon headed setscrew from the slip ring end cover and the stator and diode heat sink assemblies can now be withdrawn.
10 Close up the retaining tonque at the root of each field terminal blade and withdraw the brush spring together with the terminal assemblies from the brushbox.
11 When reassembling, the procedure for which is the reverse of the above, tighten the brushbox fixing screws to a maximum torque of 10 lb. f. in (0.15 kg.f.m). Make sure that the drive end bracket, the lamination pack and the slip ring end bracket are reassembled to the marks you made before dismantling and then fit and tighten the through bolts evenly to a torque of 45 to 50 lb.f.in (0.158 to 0.576 kg.f.m) and then re-stake the nuts after tightening.

20 Alternator - testing in-situ

1 Disconnect the battery.

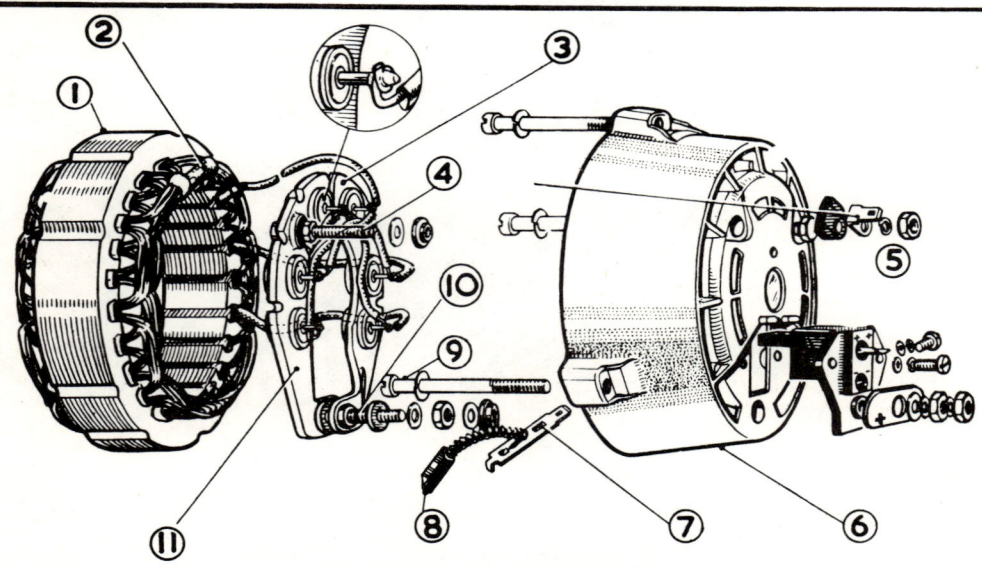

Fig. 10.14. Exploded view of the slip ring end cover

1 Stator	4 Warning light terminal 'AL'	7 Terminal blade retaining tongue
2 Star point	5 Field terminal (2)	8 Rotor slip ring brush (2)
3 Negative heat sink anode base diodes (black)	6 Slip ring end cover	9 Through bolts (3)
		10 Output terminal (+)
		11 Positive heat sink and cathode base diode (red)

2 Remove the two thumb screws and lower the instrument panel to give access to the ammeter. Disconnect the brown and the brown/white cables from the ammeter and connect them to a good quality moving coil ammeter capable of reading to at least 75 amps.

3 Disconnect the cables from the base of the control unit and then, as shown in Fig. 10.15, connect the black and brown green cables together by means of a short length of cable with Lucar tags attached. By so doing, you will connect the field winding of the alternator across the battery terminals and so bypass the output control unit.

4 Reconnect the battery.

5 Start the engine and slowly increase speed to about 2,000 rpm at which speed the ammeter reading should be approximately 40 amps. A low reading indicates either a faulty alternator or poor wiring connections in the circuit.

6 Check all connections; in particular the earth connections, and repeat the test.

7 If all connections are in order and the low reading persists, measure the resistance of the rotor coil by means of an ohmmeter connected between the field terminal blades with the external wiring disconnected. The resistance must approximate 3.8 ohms.

8 If you do not have an ohmmeter available, the rotor coil can be tested by connecting a 12 volt DC supply between the field terminals with an ammeter in series as shwon in Fig. 10.16. The ammeter reading should be approximately 3.2 amps.

9 An infinity reading on the ohmmeter or a zero reading on the ammeter indicates an open circuit in the field system, that is, in either the brush gear slip rings or in the winding. On the other hand, if the current reading is much above or the ohmmeter reading is much below the values quoted, the indication is of a short circuit in the rotor winding and this will entail replacement of the rotor slip ring assembly.

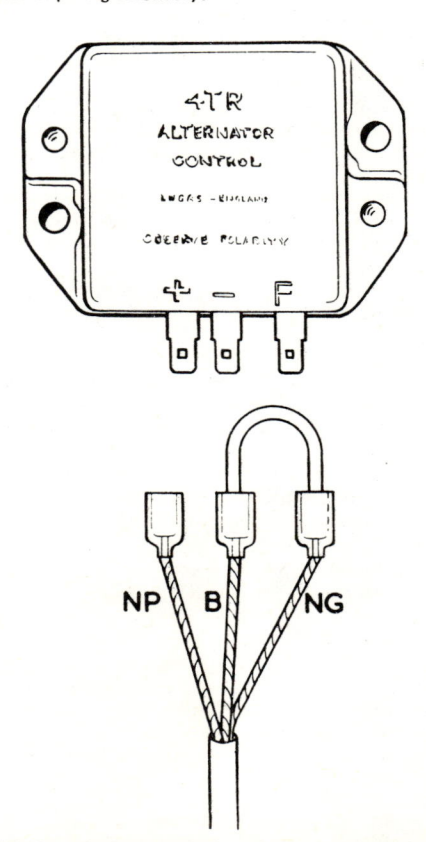

Fig. 10.15. Detach the connections from the base of the control unit

10 If the test at paragraph 5 resulted in a zero reading, stop the engine. Check that battery voltage is reaching the rotor windings by connecting a voltmeter between the cable ends attached to the alternator field terminals and then switch on the ignition. If no reading is obtained, a fault is indicated in either the field isolating contacts in the ignition switch or in the wiring associated with this circuit. However, when a steering column lock is fitted, field isolation is by means of a relay. Check each item in turn and rectify as necessary.

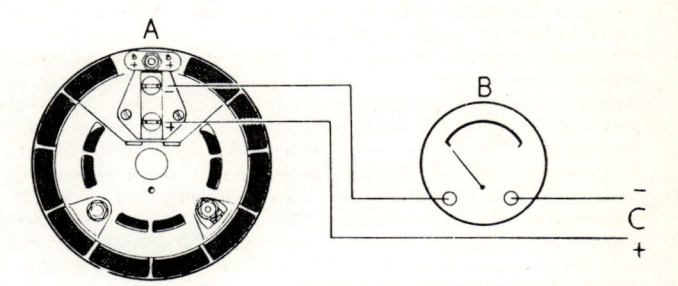

Fig. 10.16. Checking the rotor coil using a battery and ammeter

A Alternator   B Ammeter   C Battery

21 Alternator - inspection, bench test and repair

1 Follow the instructions given on Section 19 for dismantling the alternator.

2 Check the length of the brushes, the minimum acceptable length is 5/32" (3.97 mm). Discard them if they are at, or are approaching that dimension. New brushes are supplied complete with spring and terminal blade and have merely to be pushed in until the tongue registers; but, to be sure that the terminal is properly retained, carefully lever up the retaining tongue so that it makes an angle of 30° with the terminal blade.

3 See that the brushes are in good condition, that they are not chipped or showing any obvious defect.

4 Check that the brushes move freely in the holders. If they appear sluggish, clean the bearing sides with a cloth moistened with petrol and if this does not effect a cure, polish the sides with a smooth file but make sure that all trace of dust is removed before refitting to the holders.

5 If you have suitable equipment available, it is a good idea to check the brush spring pressures. With the brush spring compressed to 25/32" (19.84 mm) the brush spring pressure when iftted to the holder should be 4 - 5 ozs. (113 - 114 grms) and 7½ - 8½ ozs (212 - 242 grms) with the spring compressed to 13/32" (10.31 mm).

6 Examine the slip rings. The surfaces should be smooth and free from oil or other foreign matter. They can be cleaned with a cloth moistened in petrol but if there is any evidence of burning it is permissible to use fine glass paper; **Do not** use emery cloth or similar abrasive.

7 It is unlikely that the slip rings will be scored or pitted but discard them if they are in this condition. Do not try to rectify them by machining as any resultant eccentricity could adversely affect the high speed performance of the alternator.

8 Check the rotor windings by connecting an ohmmeter or 12 volt DC supply between the slip rings in the manner described in Section 20 paragraphs 7 and 8 where this test was then made with the brush gear in circuit. Reading should be the same as given in those paragraphs.

9 Test for defective insulation between each of the slip rings and one of the rotor poles using a mains low wattage test lamp for the purpose. If the lamp lights it indicates that the coil is earthing and the rotor/slip ring assembly must be replaced.

10 Unsolder the three stator cables from the heat sink assembly (Fig. 10.14) but take care not to overheat the diodes.

11 Identify the three cables as 'A', 'B', and 'C' and this gives

three pairs of cables 'AB', 'AC' and 'BC' for testing the stator windings.

12 Measure the voltage drop across each of these pairs of cables in turn whilst passing a 20 amp current between the cable ends. The voltage drop should be about 4.3 volts in each of the three measurements. If any reading is outside of that value, the stator must be replaced.

13 Using a mains test lamp, check for defects in the insulation between the stator coils and lamination pack, by connecting the lamp between any of the three cable ends and the lamination pack. If the lamp lights, the stator coils are earthing and a replacement stator must be fitted.

14 Check the diodes before you re-solder the cables to the heat sink assembly.

15 Check each diode by connecting it in series with a 1.5 watt test bulb across a 12 volt DC supply and then reverse the connections. The lamp should light and current should flow in one direction only. If the lamp lights when current is flowing in either direction or it does not light at all, the diode is defective.

16 Diodes are not individually replaceable but are supplied already pressed into the appropriate heat sink portion which, as already described, consists of two mutually insulated assemblies one of positive and the other of negative polarity. The positive assembly carries three cathode base diodes marked in black.

17 When soldering the interconnections, use 'M' grade 45/55 tin/lead solder. Be careful not to overheat the diodes or bend their pins. It is advisable to lightly grip the diode pins with a pair of long nosed pliers when soldering so that the pliers will absorb excess heat.

18 After soldering, the connections must be neatly arranged around the heat sinks as shown in Fig.10.13 and tacked down with 'MMM' EC 1022 adhesive where shown in the figure in order to ensure adequate clearance for the rotor.

19 The stator connections must pass through the appropriate notches at the edge of the heat sink, these connections can now be re-made.

20 If the bearings permit excessive side movement of the rotor shaft they must be renewed. The needle roller bearing in the slip ring end cover is not replaceable; it is supplied, complete with the end cover.

21 The drive end ball bearing is removed by first filing away the heads of the retaining plate rivets and then punching out the rivets.

22 Next, press the bearing out of the housing.

23 Pack the new bearing with high melting point grease, locate it in the housing and press it home.

24 Place the bearing retaining plate in position and secure with three new rivets.

22 Alternator output control unit model 4TR - general

The model 4TR electronic Ouput Control Unit is similar in action to the vibrating contact type of voltage control units such as the RB 310 and RB 340 but switching is achieved by transistors instead of vibrating contacts. A Zener diode provides voltage control in place of the coil and tension spring system of those units and no cut-out is required as the diodes incorporated in the alternator prevent reverse currents flowing. Similarly, no current regulator is required as the self regulating properties of the alternator limit the output current to a safe value.

The control unit is isolated from the battery, when the engine is switched off, by a special double pole ignition switch or, on those cars fitted with a steering column lock, by a relay which replaces the ignition switch control.

Extreme care must be taken at all times not to reverse the battery connection as this will damage the semi-conductor devices in the control unit as well as those in the alternator.

A circuit diagram of the unit is given in Fig. 10.17.

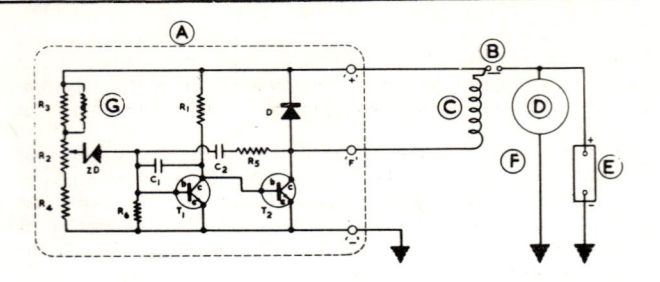

Fig. 10.17. Circuit diagram - Model 4TR control unit

A Control unit
B Field isolating device
C Rotor field winding
D Alternator
E 12-volt battery
F Stator winding (rectified) output
G Thermistor

23 Output control unit model 4TR - checking and adjusting

Check and adjustment of the control unit can only be carried out satisfactorily if the alternator and associated circuits are correct (see Sections 20 and 21) and a well charged battery is in use.

1 Leave the existing connections to the alternator and the control unit undisturbed but move the tags sufficiently on the control unit to allow connection of a high quality voltmeter between the positive and negative terminals. It is advisable to use a voltmeter of the supressed zero type capable of reading to 12 - 15 volts.

2 Run the alternator, whether testing in situ or on the bench, at charging speed for eight minutes.

3 Switch on an electrical load of approximately 2 amps (side and tail lights will give this loading).

4 Now run the engine at 1500 rpm when the voltmeter should show a reading of 13.9 to 14.3 volts with an ambient temperature of 20 to 26º (68 to 78ºF).

5 If the reading is outside the above limits, but has risen to some degree above battery terminal voltage before finally reaching a steady value, the unit can be adjusted to bring it within limits.

6 If the voltmeter reading remains steady at battery terminal voltage or if, on the other hand, it increases in an uncontrolled manner, the control unit is faulty and as it cannot be serviced it must be replaced.

7 To adjust the control unit, first stop the engine and then take out the control unit mounting screws, leave the voltmeter connected.

8 Lift the control unit away from the bulkhead and if you look at the rear you will find a blob of sealing compound which conceals a potentiometer adjuster; the location of which is shown at Fig. 10.18 'A'. Carefully chip away the sealing compound.

9 Check that the voltmeter is still firmly connected between positive and negative terminals of the control unit.

10 Start the engine and again run it at 1500 rpm. Turn the potentiometer by means of the slot, in a clockwise direction to increase, and in an anticlockwise direction, to decrease the setting. The potentiometer should be turned only a small amount at a time as you will find that a small movement will make a considerable difference to the voltage reading.

11 When the voltage reading is within limits (see paragraph 4), stop the engine. Now check the setting by restarting the engine and again checking at 1500 rpm with a load of 2 amps.

12 If the reading are correct, remove the voltmeter and then refit the control unit after sealing the potentiometer.

24 Warning light control unit model 3AW - general

This device which is fitted to later model 4.2 litre cars, is connected to the centre point of one of the pairs of diodes in the

Chapter 10/Electrical system

alternator and operates in conjunction with the ignition warning light to give an indication that the alternator is charging.

The unit is mounted on the bulkhead adjacent to the control box and is similar in appearance to the flasher unit but has a distinctive green label. The internal components consist of an electrolytic capacitor, a resistor and a silicone diode mounted on a base with three terminals.

The unit is sealed so no servicing or adjustment is possible.

If the unit appears faulty, first make sure that the remainder of the charging circuit is serviceable and then check by substitution. However, a faulty diode or an intermittent or open circuit in the alternator can cause excessive voltages to be applied to the warning light so before fitting a new replacement unit first check that the voltage between the alternator 'AL' terminal and earth is between 7 and 7.5 volts with the engine running at 1500 rpm. If a higher voltage is registered, check that all charging circuit connections are clean and tight and then, if necessary, check the alternator rectifier diodes.

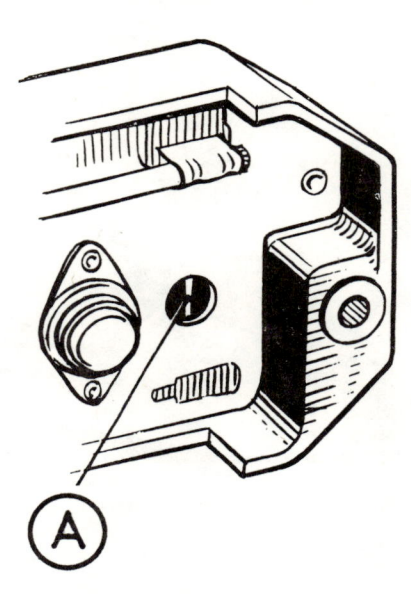

Fig. 10.18. The poteniometer adjuster

25 Fuse units model 4 FJ - general

The fuse units (illustrated at Fig. 10.19) are located behind the instrument panel and access to them is gained by undoing the instrument panel securing thumbscrews and then hinging the panel downwards. The circuits controlled by individual fuses (see Specifications at the beginning of this Chapter) are shown on the indicator panel and it is essential that if a fuse blows it is replaced by one of the same value after tracing and correcting the fault.

Spare fuses are carried and are retained in position in their holders by small spring clips. It is prudent to replace a spare fuse as soon as possible.

26 Flasher unit - fault tracing and rectification

The flasher unit is housed in a cylindrical container plugged into a base block which is part of the main wiring harness and is located on the bulkhead behind the facia on the driver's side as shown in Fig.10.20.

Electrical contact is made by means of three blades at the base of the unit and as these are offset the possibility of making a wrong connection is prevented.

A circuit diagram of the flasher unit is given at Fig. 10.21, in the event of faults arising in the system, proceed as follows:
1 Check the bulbs for broken filaments.
2 Refer to the wiring diagram at Fig. 10.21 and check all circuit connections.
3 Remove the facia panel.
4 Switch on the ignition and check with a voltmeter that terminal 'B' on the flasher unit is at 12 volts with respect to earth.
5 Connect terminals 'B' and 'L' of the flasher unit together and now operate the direction-indicator switch, if the unit operates it is defective and must be replaced.
6 If the foregoing tests prove satisfactory, it must be the switch which is at fault and this is best tested by substitution.

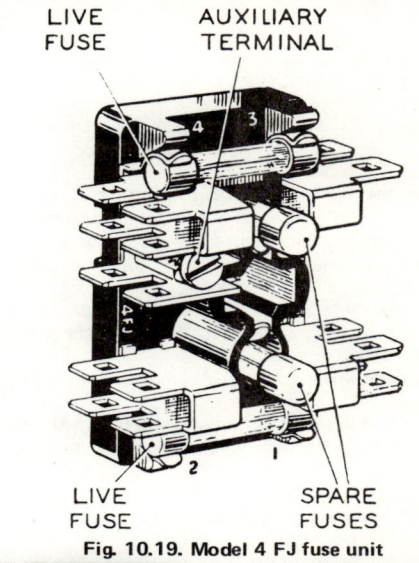

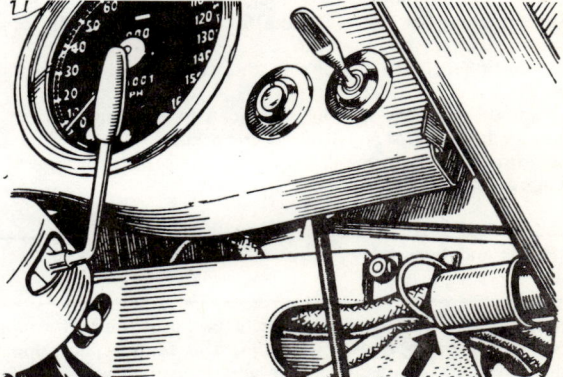

Fig. 10.19. Model 4 FJ fuse unit

Fig. 10.20. Showing the location of the flasher unit

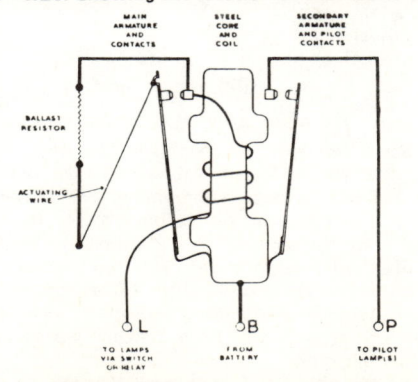

Fig. 10.21. Flasher unit - circuit diagram

27 Traffic hazard warning device - general

This system operates in conjunction with the four flashing indicator lamps and is controlled by a dash panel switch to cause all indicator lamps to flash simultaneously. A warning lamp is built into the circuit to indicate that the system is in operation. The circuit is protected by a separate 35 amp in-line fuse.

The flasher unit is located and is similar in appearance to the flashing turn indicators, however, it has a different internal circuit so a correct replacement must be fitted in the event of failure.

Failure of one or more bulbs will not, as in the case of the flashing turn indicators, prevent the remaining bulbs from operating.

28 Horns - removal and refitting

To remove the WT.618 horns from 3.8 litre and early 4.2 litre models:
1 Disconnect the battery.
2 Take out the six screws securing the headlamp rim and remove the rim, the rubber seal and the headlamp glass. You may find that the headlamp rim screws are corroded in position, in which case, the use of a penetrating oil is advised otherwise you may shear them when unscrewing.
3 Take out the three screws securing the headlight duct to the diaphragm and withdraw the duct forwards through the glass aperture. The horn is now exposed.
4 Take out the central screw and remove the dome cover from the horn.
5 Identify and then remove the two cables from the tags.
6 Remove the two securing nuts and bolts and then withdraw the horn.

To remove the Type 9H horns from later models:
7 Disconnect the battery.
8 The horns are mounted on brackets attached to the subframe lower crossmember. It will make for easier work if the car is raised to give access to the horn.
9 Identify and then disconnect the cables from the terminal tags.
10 Remove the nuts securing the horn to the captive bracket mounting bolts and then withdraw the horn.
11 Refitment of both the WT 618 and 9H horns is the reverse of the above procedures.

29 Horns and horn relay - adjustment, fault tracing and rectification

1 In the event of a horn failing to sound or if the performance becomes uncertain, make sure that the fault is not due to external causes before you make any adjustments. Carry out a check on the following:
a) Battery connections.
b) Loose horn fixing bolts.
c) Loose or broken connections in the horn circuit.
d) Check that fuse No. 3 (50 amps) and fuse No.6 (35 amps) have not blown.
e) Faulty relay. The location of the horn relay is shown in Fig. 10.22. Use a test lamp to check that current is available at terminal 'C2' (brown/purple cable) and then switch on the ignition and test for current at terminal 'W1' (green cable). If satisfactory, remove the purple/black cable from terminal 'W2' and earth the terminal to a clean part of the frame, the relay coils should operate and close the contacts. The next step is to remove the brown/purple cable from terminal 'C2' and check for continuity using an earthed test lamp; carry out this check with the horn button depressed and the ignition switched on.
Failure of the above tests will mean that the relay is faulty and must be replaced.
2 To adjust the WT.618 horn, follow the instructions in Section 27 to remove the domed cover of the horn. Now refer to Fig. 10.23 and connect a 0 - 20 amp first grade moving coil ammeter in series with the horn. Slacken the contact locknut and adjust the contact until the horn will pass 13 - 15 amps at 12 volts. Tighten the locknut and refit the domed cover.
3 To adjust the 9H horn, remove it from the car as described in Section 27. Fig. 10.24 illustrating the horn shows two screws 'A' and 'B'; screw 'A' located adjacent to the horn terminal is provided to take up wear in the moving parts of the horn, it does not have any affect on the pitch of the horn note. The screw 'B', the centre slotted core, must on no account be disturbed. Connect a moving coil ammeter, in series with the horn supply feed, it is advisable to protect the ammeter from overload by connecting an on/off switch in parallel with its terminals and to keep the switch on except when taking readings, that is when the horn is sounding. Turn screw 'A' clockwise until the horn operates within the limits of 6.5 to 7 amps.

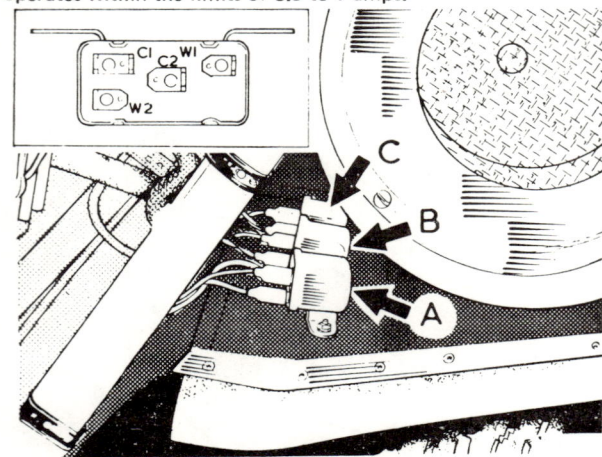

Fig. 10.22. Location of the horn relay

A   Horn relay
B   Alternator/ignition relay
C   Air conditioning equipment relay (when fitted)
      (Inset shows the connections)

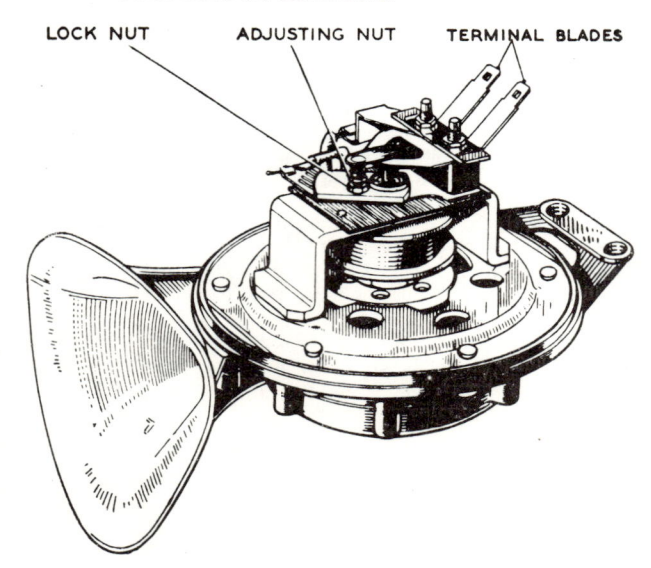

Fig. 10.23. The WT 618 horn (dome removed)

Chapter 10/Electrical system

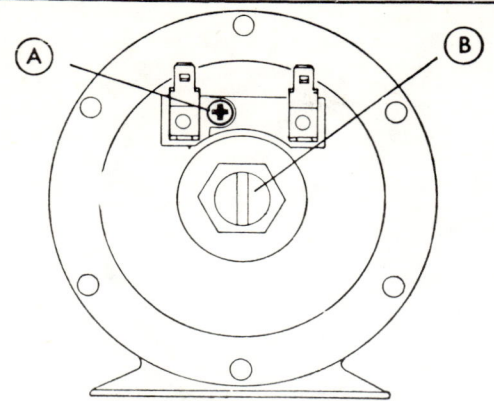

Fig. 10.24. The Lucas Type 9H horn

A Contact breaker adjustment screw
B Slotted centre core (do not disturb)

30 Lamps - removal, adjustment and refitting

1 To guard against accidental blowing of a fuse it is always a wise precaution to disconnect the battery before breaking any connections which may leave you with bare cable ends.
2 Except where otherwise stated, the refitment of components is the reverse of the removal sequences following.
3 To remove the **non-sealed beam type of headlamp** the components of which are illustrated at Fig. 10.25, first take out the six screws holding the glass headlamp cover retaining ring to the wing of the car. You may find the screws have corroded into position and are difficult to remove, in this case we suggest that you treat them with penetrating oil.
4 Take off the ring followed by the rubber ring and the glass cover.
5 Slacken the three cross headed screws which hold the headlamp glass and reflector unit rim. Turn the rim in an anti-clockwise direction and remove it. Do not confuse the retaining screws with screws 'A' and 'B' shown in Fig. 10.26 as these screws are for adjustment of the headlamp. There is no need to completely remove the retaining screws.
6 The light unit can now be withdrawn.
7 Pull off the plug with cables attached from the unit. The bulb retaining spring clips can now be released to remove the bulb.
8 Replace with a bulb of the correct type (see Specifications at the beginning of this Chapter), note that a groove in the bulb plate must register with a raised portion on the bulb retainer.
9 Ro replace the **sealed beam type of headlamp** (Fig.10.27), first prise off the headlamp rim, this is retained by spring clips and care should be taken not to scratch the surround.
10 Remove the three cross headed screws which secure the retaining ring and then remove the ring. Do not confuse these screws with the headlamp setting screws 'A' and 'B' shown in Fig. 10.28.
11 Withdraw the sealed beam unit and unplug the adaptor.
12 The headlamps should be set so that when the car is carrying its normal load, the driving beams are parallel with each other and with the road when on full beam. However, although the headlamp can be set approximately, at home by adjustment on screw 'A' in Figs. 10.26 and 10.28 to alter the vertical trim and on screw 'B' to alter the horizontal position, this is a job which should be left to a garage having the necessary beam setting equipment in order to get the best results from your lights.
13 To remove the **sidelamp** take out the three screws which retain the lamp glass and remove the glass (Fig. 10.29).
14 The sidelamp bulb is the inner one of the two exposed, the other is the **flasher** bulb. Both bulbs are removed by pressing in and turning anticlockwise.
15 The **rear/brake/rear flasher** bulbs are removed by taking out the two screws retaining the lamp glass and then removing the glass (Fig. 10.30).
16 The rear/brake bulb is the inner one of the two bulbs now exposed and the other is the rear flasher bulb.
17 To remove the **number plate** lamp bulb, take out the screw retaining the rim and glass and detach the glass rim and its gasket. Remove the bulb by pressing inwards and turning anticlockwise.

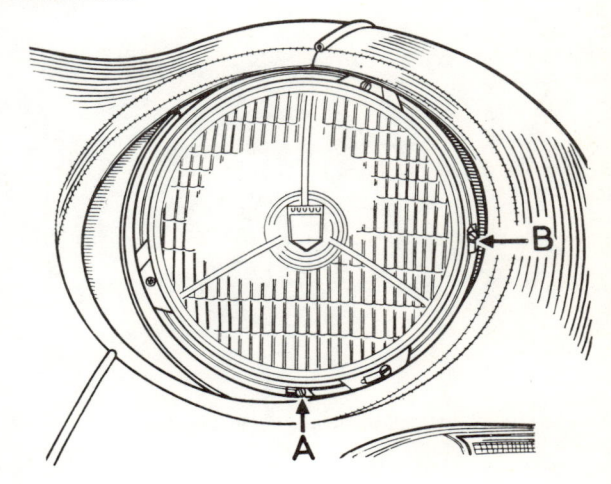

Fig. 10.26. Headlamp adjustment screws - non sealed beam type

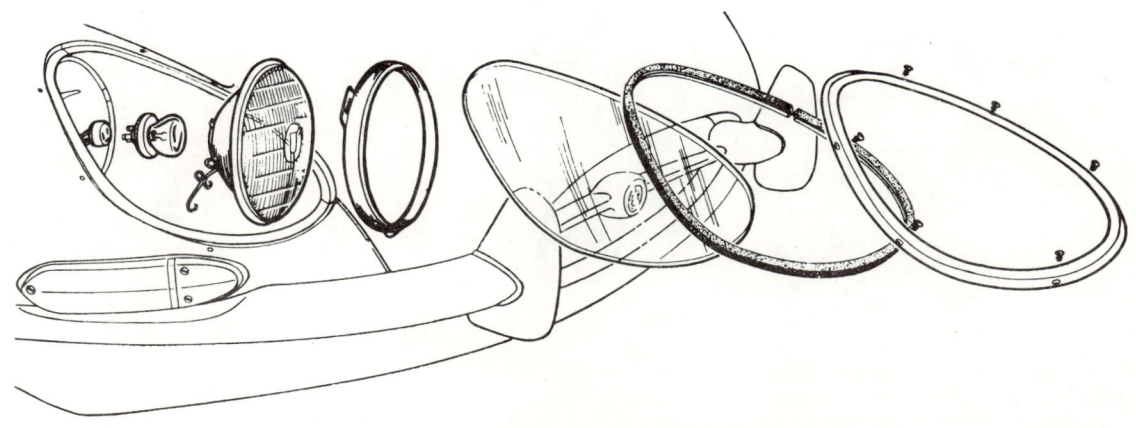

Fig. 10.25. Headlamp removal - non-sealed beam type

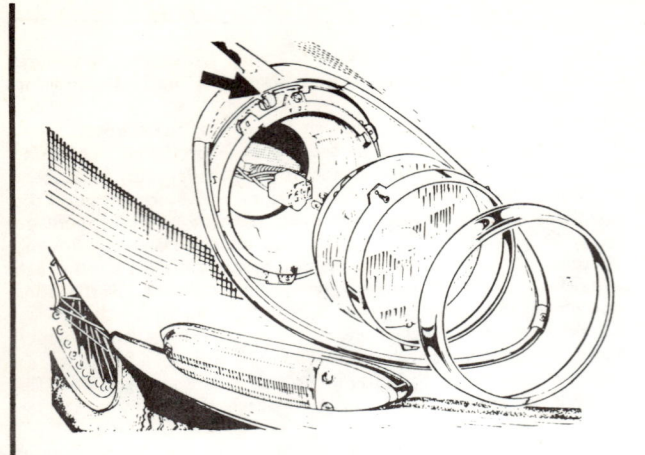

Fig. 10.27. The sealed beam headlamp (the arrow indicates one of the spring clips which retain the rim)

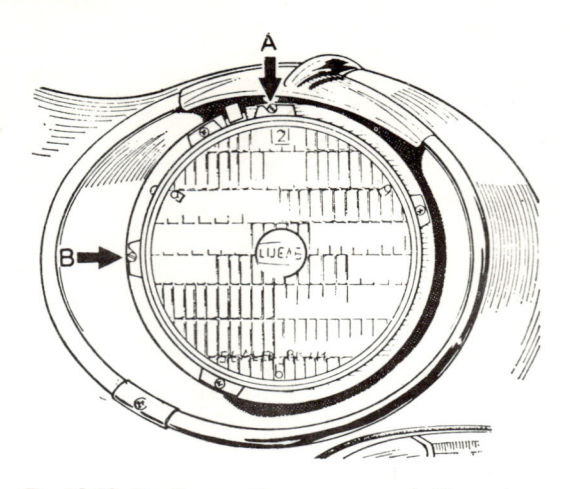

Fig. 10.28. Headlamp setting screws — sealed beam type

A for adjustment in vertical plane
B for adjustment in horizontal plane

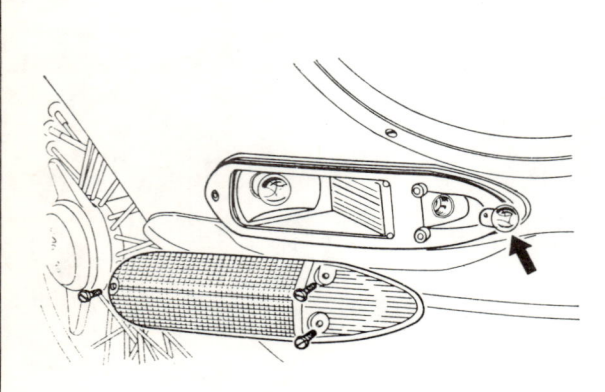

Fig. 10.29. Removing a side lamp bulb

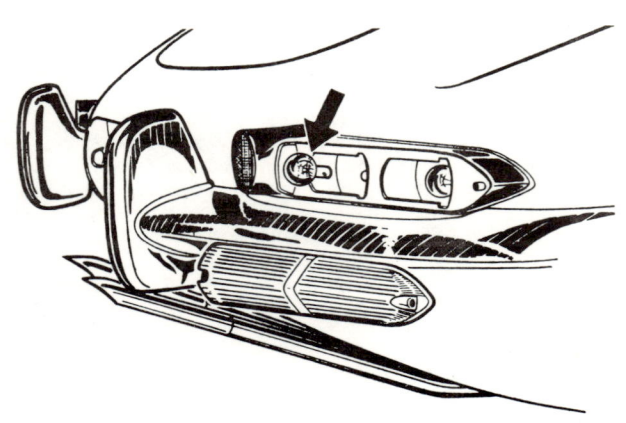

Fig. 10.30. Rear brake/rear flasher bulb removal

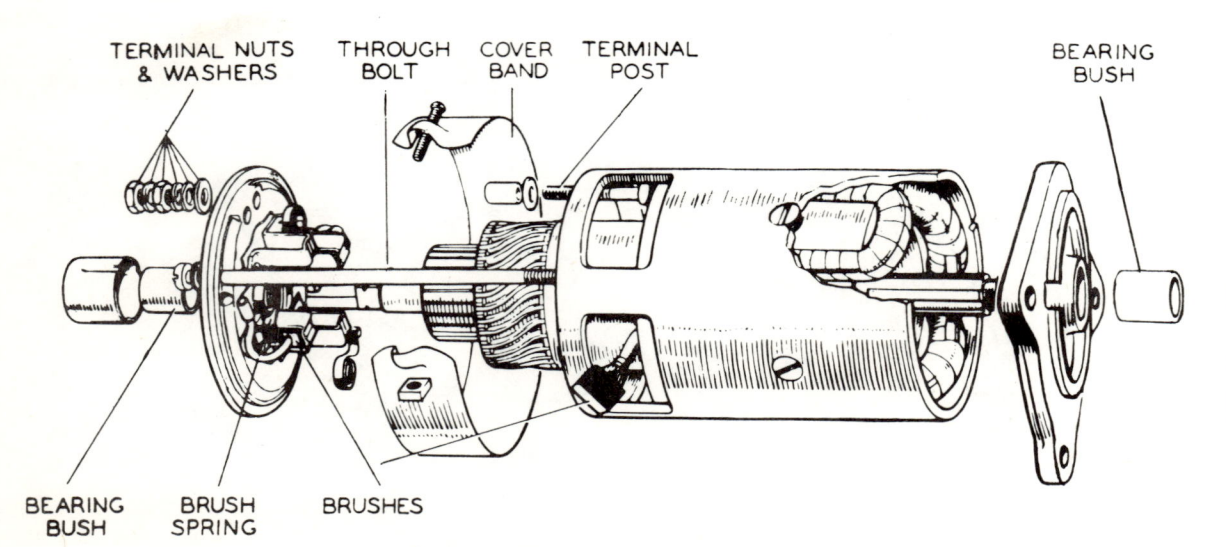

Fig. 10.31. The model M45G starter motor

Chapter 10/Electrical system

31 Starter motor - general

The model M45 G starter motor (Fig. 10.31) fitted to 3.8 litre models is a four pole four brush machine which has an extended shaft carrying the starter drive. It is of similar construction to the dynamo except that heavier gauge copper wire is used in the armature and field coils. The field coils are series parallel connected between the field terminal and the brushes. The internal connections are shown in Fig. 10.32.

The pre-engaged starter motor as fitted to 4.2 litre models is illustrated at Fig. 10.33, the purpose of this type of motor is to prevent premature pinion ejection. The starter motor is connected to the battery only after the pinion has been meshed with the ring gear on the flywheel and this is effected by solenoid operated linkage system.

After the engine has started, the current is automatically switched off before the pinion is retracted, when this happens the armature, which is spinning, is brought rapidly to rest by a braking device which consists of a pair of moulded shoes driven by a cross peg on the armature shaft and spring loaded against a steel ring insert in the commutator end bracket. Thus, the possibility of damaged teeth resulting from attempts being made to re-engage a rotating pinions is minimised.

A bridge shaped bracket is secured to the front end of the machine by the through bolts and this bracket carries the main battery input and solenoid winding terminals, short extension cables being connected between these and the corresponding solenoid terminals.

The starter solenoid is an electro-magnetic actuator which is mounted pick-a-back fashion on the yoke of the motor. It contains a soft iron plunger which is linked to the engaging lever, the starter switch contacts and a coil consisting of a heavy gauge pull-in or series winding and a lighter gauge hold-on winding. The starter solenoid is sealed in a roll-over outer case and cannot be dismantled.

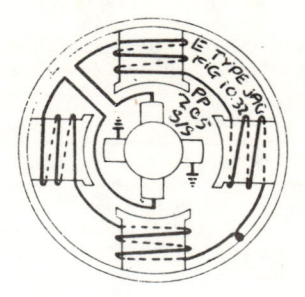

Fig. 10.32. The internal connections of the M45G starter motor

32 Starter motor - removal and refitting

1 Place the car over a pit or on a ramp or raise the car to give access to the starter motor from underneath. If the car is raised on a jack; do make sure that it is safely supported before crawling underneath.
2 Disconnect the battery.
3 Disconnect the cable from the terminal at the end of the starter motor or the cables from the actuating solenoid as the case may be.
4 Slacken the clips to the two rubber hose pipes from the brake servo vacuum located on the bulkhead above the starter motor. Identify the pipes and then disconnect them.
5 Remove the four nuts and their washers which secure the servo vacuum tank to the bulkhead and remove it.

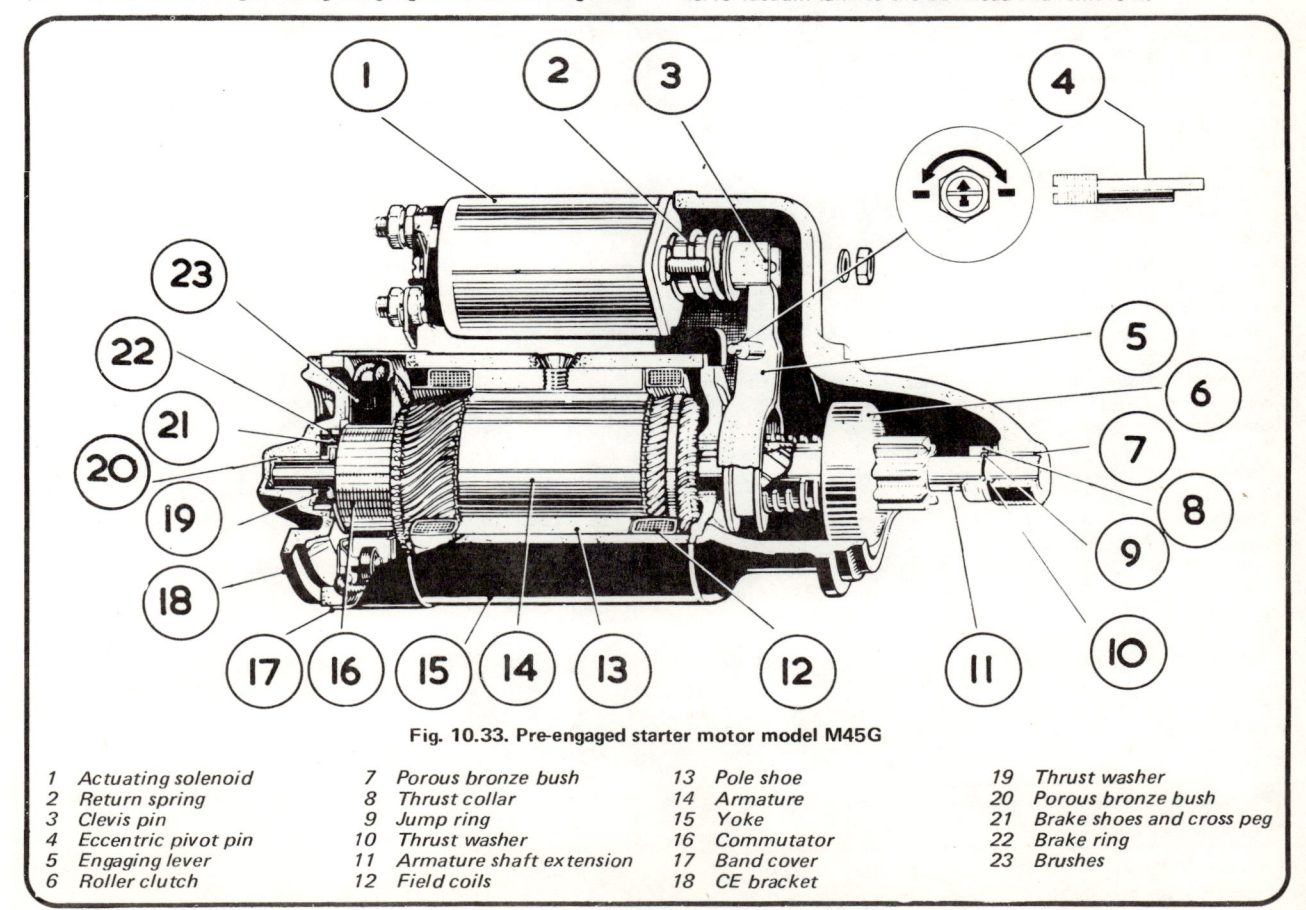

Fig. 10.33. Pre-engaged starter motor model M45G

1 Actuating solenoid	7 Porous bronze bush	13 Pole shoe
2 Return spring	8 Thrust collar	14 Armature
3 Clevis pin	9 Jump ring	15 Yoke
4 Eccentric pivot pin	10 Thrust washer	16 Commutator
5 Engaging lever	11 Armature shaft extension	17 Band cover
6 Roller clutch	12 Field coils	18 CE bracket
19 Thrust washer		
20 Porous bronze bush		
21 Brake shoes and cross peg		
22 Brake ring		
23 Brushes		

Chapter 10/Electrical system

6 Remove the nuts from motor securing bolts. Take the weight of the motor, withdraw the bolts and then withdraw the motor. If the motor is of the pre-engaged type, collect the spigot plate and the dowels.
7 Refitting is the reverse of the foregoing sequence but you must ensure that the vacuum tank hoses are refitted to the correct unions.

33 Starter motor - testing in position

The following tests cover both types of starter motor.
1 Switch on the lights and the ignition. Ensure that the selector lever of automatic models is in the 'P' or 'N' position.
2 Operate the starter control, if the lights go dim but the motor is not heard to operate, it shows that current is flowing through the motor windings but that, for some reason, the armature is not rotating. It may be that the pinion has not disengaged from the starter ring on the flywheel in which case the motor will have to be removed from the car for examination.
3 If, when the starter is operated, the lamps retain their full brilliance and the motor does not operate, check the circuit for continuity starting with the battery connections (especially the earth connection), then look carefully at the engine earth connection followed by the connections to the motor and the starter switch. If it is established that voltage is getting to the motor when the switch is operated, an internal fault in the motor is indicated.
4 Sluggish or a slow action of the motor is usually caused by a loose connection resulting in high resistance in the circuit, check as described in paragraph 3.
5 If the motor is heard to operate but it does not turn the engine, a fault in the drive is indicated which will involve removal of the motor for rectification.

34 Starter motor M45G - dismantling and reassembly

1 Before completely dismantling the motor to check for a fault, first make sure that the brushes are not the cause of the trouble. This can be done by slackening the nut and bolt securing the 'window' cover band, slide the band clear of the 'windows' and the brushes can now be lifted out for examination Fig. 10.34. It is also possible to check the weight of the brush spring (correct tension 30-40 ozs) using a spring balance through the window, Fig. 10.35.
2 Leave the brushes out of their holders if they are satisfactory and further dismantling is necessary.
3 Remove the nuts from the terminal post at the commutator end bracket.
4 Unscrew the two through bolts, and making sure that the brushes do not foul the yoke, remove the commutator end bracket from the yoke.
5 Relate the brushes, if they are serviceable, to their respective holders and then undo the lead securing screws and remove the brushes. (Fig. 10.36).
6 Withdraw the driving end bracket, complete with the armature and drive, from the yoke.
7 Now refer to Fig. 10.37 for dismantling of the drive.
8 Take out the split pin from the shaft nut (B), hold the squared end of the shaft with a spanner and then unscrew the nut.
9 Take off the main spring (C) followed by the remainder of the components.
10 Reassembly of the starter motor and drive is a reversal of the above procedure.

35 Pre-engaged starter motor M45G - dismantling and reassembly

1 Before completely dismantling the motor to check for a fault, first make sure that the brushes are not the cause of the

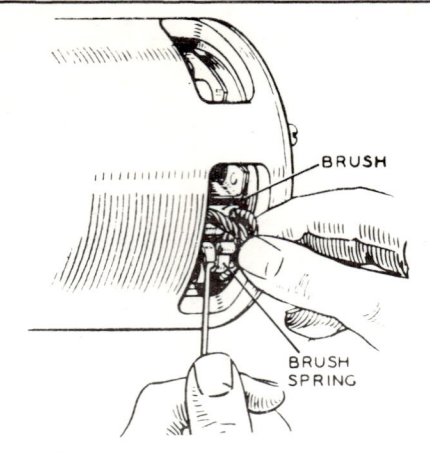

Fig. 10.34. Checking the brush gear

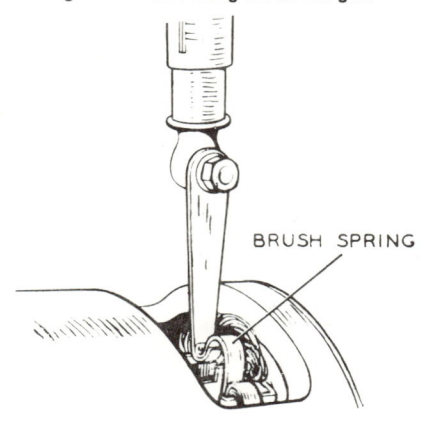

Fig. 10.35. Checking the tension of the brush spring

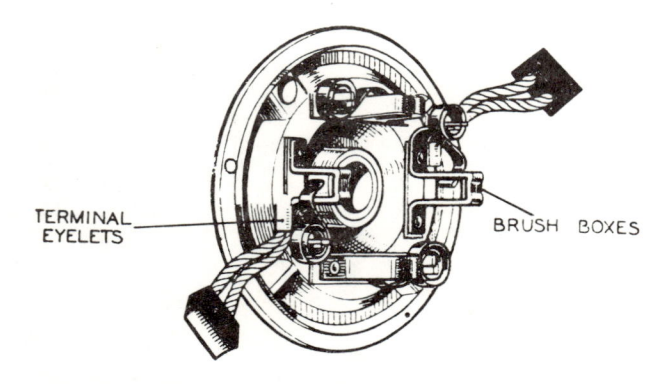

Fig. 10.36. The commutator end bracket brush connections

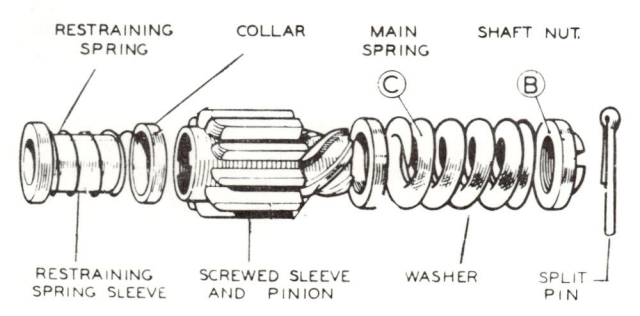

Fig. 10.37. The starter drive assembly

Chapter 10/Electrical system

trouble. This can be done by slackening the nut and bolt securing the window cover band and then sliding the band clear of the windows. The brushes can now be lifted out for examination. They should be free in their holders, undamaged, a minimum length of 5/16" (7.94 mm) and the spring tension, on a new brush, should be 52 ozs. (1.47 kg.). (See Fig.10.35).

2 Disconnect the copper link between the lower solenoid terminal and the yoke of the motor.

3 Take off the two solenoid unit securing nuts, detach the extension cables and withdraw the solenoid from the drive end bracket. Be careful when disengaging the solenoid plunger from the starter drive engagement lever.

4 If you have not already done so, slacken the nut and bolt securing the cover band, slide the band clear of the windows, lift off the brush springs and then remove the brushes from the holders. (See Fig. 10.34).

5 Unscrew and remove the two through bolts from the commutator end bracket.

6 The commutator end bracket and yoke can now be removed from the intermediate and drive end bracket.

7 Remove the rubber seal from the drive end bracket.

8 Slacken the nut on the starter drive engagement lever eccentric pivot pin and then unscrew and remove the pin.

9 Separate the drive end bracket from the armature and intermediate bracket assembly.

10 Using a mild steel tube of suitable bore, remove the thrust washer from the end of the armature shaft extension.

11 Prise the jump ring out of its groove and then slide the drive assembly and the intermediate bracket from the shaft.

12 Prise out the jump ring and separate the operating bush and engagement spring.

13 Reassembly is the reverse of the above sequence but when refitting the commutator end bracket, ensure that the moulded brake shoes seat squarely and then turn them so that the ends of the cross peg in the armature shaft engage correctly with the slots in the shoes.

14 Tightening torque figures are:

Nuts on solenoid copper terminals - 20 lb.f.in. (0.23 kg.f.m.)
Solenoid fixing bolts - 4.5 lb.f.ft. (0.62 kg.f.m.)
Starter motor through bolts - 8.0 lb.f.ft. (0.83 kg.f.m.)

15 After final assembly it will be necessary to check for correct movement of the pinion, and for opening and closing of the starter switch contacts.

16 To check pinion movement, refer to Fig. 10.38 and connect the solenoid Lucar terminal to a **6 volt** battery as shown (do not use a 12 volt supply otherwise the armature will turn).

17 Measure the distance between the pinion and the thrust washer on the extension of the armature shaft when the pinion is lightly pressed towards the armature. This dimension should be 0.005" to 0.015" (0.13 to 0.38 mm).

18 If the setting is outside of those limits, disconnect the battery.

19 Now adjust the setting by first slackening the nut on the eccentric pivot pin (inset at Fig. 10.38) and then turn the pin clockwise to decrease and anticlockwise to increase the gap. The head of the arrow stamped on the end of the pin ashould be set only between th limits of the arrows cast in the drive end bracket.

20 Tighten the pin securing nut and then reconnect the battery and re-check the setting.

21 Having correctly set the pinion travel, the next task is to check the opening and closing of the starter switch contacts.

22 Refer to Fig. 10.39, remove the copper link connecting the solenoid 'STA' with the starter motor terminal.

23 Connect a switched 10 volts DC supply between the solenoid Lucar terminal and the large terminal 'STA'. **Do not** close the switch at this stage.

24 Connect a separately energised test lamp across the solenoid main terminals.

25 Use a tool, an open ended spanner is ideal with its jaws passing over the armature shaft extension, to act as a stop and so restrict the pinion travel to its out of mesh clearance which is normally 1/8" (3.17 mm).

26 Now energise the shunt winding using a 10 volt DC supply and then close the switch in the series winding circuit. The solenoid contacts should close fully and remain closed, a steady light from the test lamp will indicate that this is so.

27 Switch off and remove the stop from the pinion.

28 Switch on again and hold the pinion assembly in the fully engaged position.

29 Switch off and the test lamp should go out indicating that the contacts have opened.

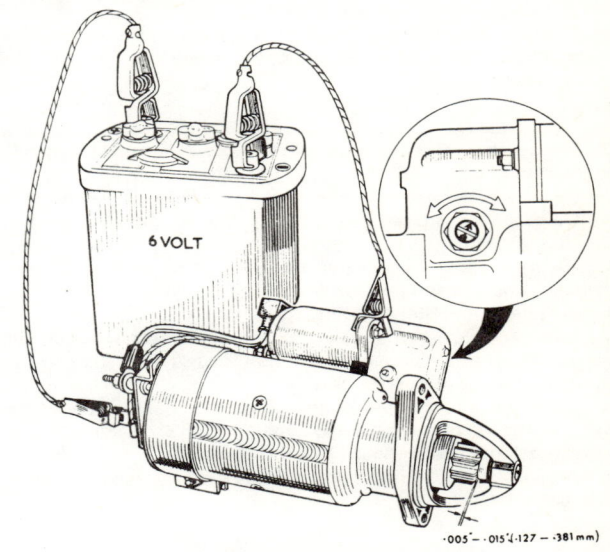

Fig. 10.38. Setting pinion movement

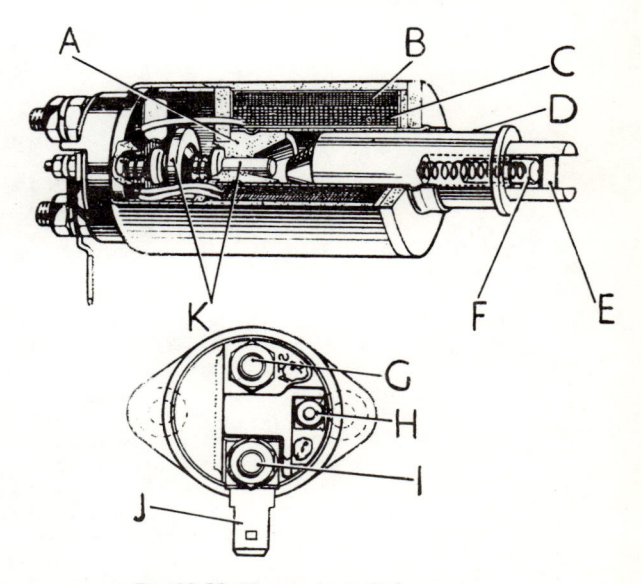

Fig. 10.39. The starter switch

A Core
B Shunt winding
C Series winding
D Plunger
E Clevis pin
F 'Lost motion' device
G Starter terminal
H Solenoid terminal
I Battery terminal
J Accessories terminal
K Spindle and moving contact assembly

36 Starter motor M45G - bench examination, test and rectification

1 Except for the following points, follow the instructions given for the dynamo in Section 9:
a) The minimum permissible length of the brushes is 5/16".
b) The acceptable tension of the brush springs is 30-40 ozs.
c) The insulators between the commutator segments **must not be undercut**.
d) No attempt should be made to machine the armature core or to true a distorted armature shaft.
e) If either the screwed sleeve or the pinion of the drive are worn or damaged, they must be replaced as a pair, not separately.

37 Pre-engaged starter motor M45G - bench examination, test and rectification

1 Remove the starter motor from the car as described in Section 32.
2 Securely clamp the motor in a vice and, using a 12 volt battery, check the light running current which should be 70 amps at between 5800 and 6500 rpm.
3 Look for excessive sparking at the commutator during the above test. If this is present, check that the brushes are clean and are free to move in their holders.
4 Carry out a torque test using a constant voltage supply of 7.6 volts at the starter terminal and compare with the values given in Specifications at the beginning of this Chapter.
5 An indication of the fault or faults which may show up on the above tests is:

Symptom	Probable Fault
1 Speed, torque and current consumption correct.	Assume motor to be in normal operating condition.
2 Speed, torque and current consumption low.	High resistance in brush gear, e.g., faulty connections, dirty or burned commutator causing poor brush contact.
3 Speed and torque, low current consumption high.	Tight or worn bearings, bent shaft, insufficient end play, armature fouling a pole shoe, or cracked spigot on drive end bracket. Short circuited armature, earthed armature or field coils.
4 Speed and current consumption high, torque low.	Short circuited windings in field coils.
5 Armature does not rotate, high current consumption.	Open circuited armature, field coils or solenoid unit. If the commutator is badly burned, there may be poor contact between brushes and commutator.
6 Armature does not rotate, high current consumption.	Earthed field winding or short circuit solenoid unit. Armature physically prevented from rotating
7 Excessive brush movement causing arcing at commutator.	Low brush spring tension or out-of-round commutator. "Thrown" or high segment on commutator.
8 Excessive arcing at the commutator.	Defective armature windings, sticking brushes or dirty commutator.
9 Excessive noise when engaged.	Pinion does not engage fully before solenoid main contacts are closed. Check pinion movement as detailed under Setting Pinion Movement.
10 Pinion engaged but starter motor not rotating.	Pinion movement excessive. Solenoid main contacts not closing. Check pinion movement as detailed under Setting Pinion Movement.

6 Check the brushes for damage and make sure they are free in their holders. If movement is sluggish they may be wiped with a cloth moistened in petrol or, if this does not cure the trouble, they may be lightly rubbed down with a smooth file. Make sure that all dust is removed before refitting to the holders.
7 The brush connectors are soldered to terminal tags. Two are connected to the brush boxes and two to the free ends of the field coils. Unsolder the connections and resolder those of the new brushes in place.
8 The brushes are pre-formed so no bedding in to the commutator is necessary.
9 Check the commutator which should be burnished and free from pits or burned spots.
10 The commutator may be cleaned with a cloth moistened with petrol but if this is ineffective, mount the armature in a lathe and polish the commutator with fine glass paper and finish by removing all dust with an air blast.
11 If the commutator cannot be rectified by the above it is permissible to mount it between centres in a lathe and to take a very light skim with a sharp tool at high speed. The commutator must not be reduced below a diameter of 1.17/32" (38.89 mm).
12 **The insulators between the commutator segments must not be undercut.**
13 Examine the armature, if the conductors are found to be lifted from the commutator risers it is an indication of over-speeding and careful checking of the clutch assembly (see paragraph 24) is called for.
14 If the armature core has been fouling the pole faces it shows that either the bearings are worn or that the shaft is distorted. A damaged armature must be replaced, no attempt should be made to machine the core or to straighten a distorted shaft.
15 Connect a 110 volt AC test lamp between a commutator segment and the armature shaft to check armature insulation. The lamp should not light when connected between any segment and the shaft. If a short circuit is suspected, check that it is not due to foreign matter and if the fault cannot be traced, fit a replacement armature.
16 Carry out a field coil continuity test by connecting a 12 volt test lamp between the terminal on the yoke and each brush with the armature removed from the yoke. Ensure that both brushes and their flexible connectors are clear of the yoke. If the lamp does not light, an open circuit in the field coils is indicated. Replace the complete starter as replacement of the field coils requires the use of special equipment.
17 Connect a 110 volt AC test lamp between the terminal post and a clean part of the yoke to test for insulation. If the lamp lights, the field coils are earthed to the yoke and must be replaced.
18 Now use the test lamp to check the insulation of the pair of brush boxes on the commutator end bracket by connecting the lamp between the box and the bracket. If the lamp lights the insulation is faulty and the end bracket will have to be replaced.
19 Examine the porous bronze bush bearings in the commutator and drive end brackets and the indented bronze bearing in the intermediate bracket. If they are damaged or worn to the extent of allowing excessive side play of the armature shaft, they must be replaced.
20 The bearings in the intermediate and drive end brackets can be pressed out, but the best way to remove the bearing in the commutator end bracket is to use a 9/16" (14.29 mm) tap. Screw the tap into the bearing and then use it to withdraw the bush.
21 Allow a new porous bronze bearing bush to soak in clean engine oil for 24 hours before fitting. Press the bearing into position using a shouldered and highly polished mandrel approximately 0.0005" (0.013 mm) greater in diameter than the shaft which is to fit the bearing.
22 **Do not** ream the porous bearings after fitting as this will destroy its self-lubricating properties.
23 Press the intermediate bearing bush into position and then lubricate the bearing surface with Rocol 'Molypad' molybdenised non-creep, or similar oil.

Chapter 10/Electrical system

24 Examine the roller clutch drive assembly (Fig. 10.40). If it is in good condition it should take up the drive instantaneously in one direction and should rotate easily and smoothly in the other. There should be no roughness or tendency to bind in movement around or along the shaft splines. Similarly, the operating bush must be free to slide smoothly along the driving sleeve when the engagement spring is compressed. Make sure that the trunnion blocks pivot freely on the pegs of the engaging lever.
25 Finally, lubricate all moving parts of the roller clutch drive Castrol LM grease.

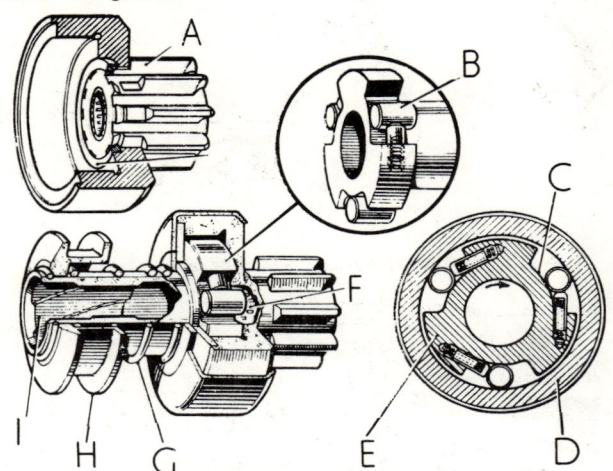

Fig. 10.40. The roller clutch drive components

A Alternative construction (pinion pressed and clear-ringed into driven member)
B Spring loaded rollers
C Cam tracks
D Driven member (with pinion)
E Driving member
F Bush
G Engagement spring
H Operating bush
I Driving sleeve

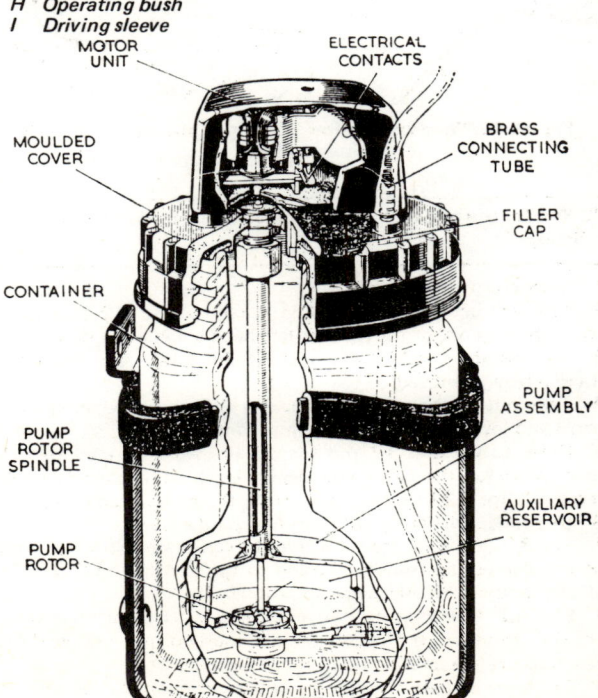

Fig. 10.41. Type 5SJ windscreen washer assembly

38 Windscreen washer type 5SJ - testing in position

The Type 5SJ windscreen washer is illustrated in Fig. 10.41.
1 Check the polarity as indicated on the moulding housing and then connect a direct current voltmeter to the motor terminals.
2 Switch on the ignition and then operate the washer switch at the same time observing the voltmeter. If a low or zero voltage is indicated, check the fuse, the switch and external connections and rectify as necessary.
3 If the voltmeter gives a reverse reading, transpose the connections to the motor.
4 If supply voltage is indicated at the terminals but the unit fails to operate, an open circuit winding or a fault in the brush gear can be suspected and the motor should be dismantled and tested as outlined in Section 39.
5 If the motor can be heard to operate but does not move freely, connect a suitable DC ammeter in series with the motor and operate the switch. If the current reading exceeds 2 amps, remove the motor and check that the pump impeller shaft rotates freely. If the shaft is difficult to turn, the water pump unit will have to be replaced. If the shaft turns freely, the fault is in the motor which will have to be dismantled for inspection.

39 Windscreen washer type 5SJ - dismantling, testing and re-assembly

1 Disconnect the external tube and the electrical connections and remove the cover from the bottle.
2 Remove the self tapping screw securing the motor to the cover and lift off the motor unit but be careful not to lose the coupling which connects the armature coupling to the pump spindle coupling. In some models it will be found that this coupling is a piece of split tube and cases have been experienced where this slips, and although the motor operates, the pump spindle is not turned. If this occurs, squeeze the tube with a pair of pliers to tighten it on the pump and armature spindles.
3 Remove the armature coupling from the armature shaft by holding the shaft with a pair of pointed pliers and, using a second pair of pliers, draw off the armature coupling.
4 Remove the screws from the bearing plate and take off the plate and the rubber gasket.
5 Take out the two screws holding the terminals and the terminal nuts and brushes can now be removed and the armature withdrawn, but take care not to lose the bearing washer which is loosely fitted to the armature shaft.
6 It is advised that you do not disturb the pole assembly unless this is absolutely necessary. If it has to be removed, take careful note of its position in relation to the motor housing. The narrower pole piece is adjacent to the terminal locations. Also take note of the position of the pole clamping member which, when fitted ocrrectly, locate on both poles, if it is not fitted correctly pressure will be applied to one pole piece only.
7 If the motor has been overheated or if any part of the housing is damaged, there is no alternative to replacement.
8 Examine the armature, if it is damaged or if the windings are loose or badly discoloured, fit a new armature.
9 Clean the commutator with a non-fluffy cloth moistened in petrol and if it is badly discoloured, polish it with very fine glass paper.
10 Check the resistance of the armature winding using an ohmmeter. The resistance should be 2.8 - 3.1 ohms.
11 Examine the brushes, if they are less than 1/16" (1.59 mm) in length they should be replaced.
12 Reassembly of the unit is the reverse of the dismantling procedure but watch the following points:-
a) Fill the bearing recess in the motor with Rocal Molypad molybdenised grease and be sure to remove any excess from the face of the bearing boss.
b) See that the pole piece assemblies are secure and that they are firmly located on the circular spigot and are the right way round.

c) Make sure that the brushes bear firmly against the commutator.
d) Before replacing the motor on the cover, be sure that the armature coupling is pushed fully home and that the intermediate coupling is in place.

40 Windscreen wiper - general

The windscreen wiper assembly (Fig. 10.42) consists of a two speed motor coupled by connecting rods to three wiper spindle bearings. In early models a control cable is attached to the centre spindle bearing mechanism for adjustment of the parking switch, a knurled adjusting knob is attached to the cable and is accessible in the engine compartment on the bulkhead as shown in Fig. 10.43.

Turning this control will raise or lower the parking limits of the wiper arms.

Adjustment of the parked position on later models is controlled by the location of the parking switch carrier plate mounted on the gear housing. Refer to Fig. 10.44, unscrew the three drive screws shown, sufficiently to release the tension on the clamping plate and now rotate the switch carrier plate in the direction of the arrows on the plate. Only a slight movement should be necessary after which, tighten the screws and recheck.

Your adjustments should be such that the blades do not park below the lower edge of the windscreen glass.

41 Windscreen wiper motor - removal and refitting

1 Disconnect the battery.
2 Remove the spring clip holding the throttle pedal link rod to the bellcrank lever and then withdraw the rod.
3 Mark the position of the carrier bracket on the bulkhead, take out the two setscrews and detach the bracket.
4 Remove the plastic strap from the motor, identify the cables and then disconnect them from the snap connector block.
5 Remove the two thumb screws in the top of the instrument panel and lower the panel to give access to the ball joint on the centre wiper motor spindle bearing. Disconnect the ball joint.
6 Remove the four setscrews attaching the motor to the bulkhead and withdraw the motor with the link rod attached. Collect the plate and seal behind it.
7 It is essential that the length of the link rod is not altered whilst you have the motor off the car, otherwise, the wiper arms will be placed out of phase with each other.
8 Refitting is the reverse of the above sequence, the main point to watch is that the lever is central in its bearing when refitting the throttle bell crank carrier bracket. Adjustment for this is provided by means of the two slotted holes in the bracket.

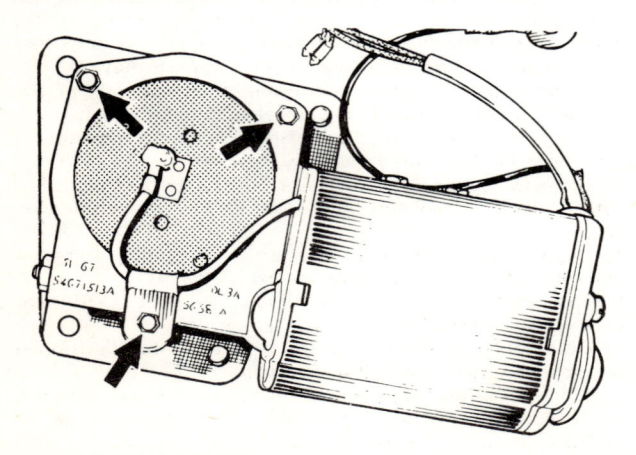

Fig. 10.44. Windscreen wiper parking adjustment (later models)

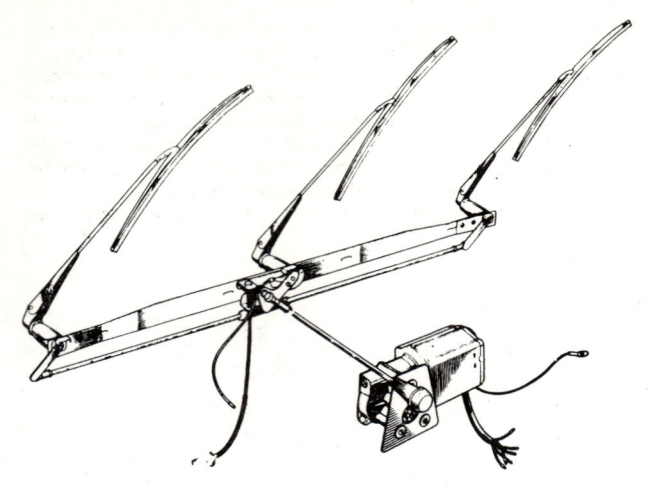

Fig. 10.42. The DL3 windscreen wiper motor and linkage

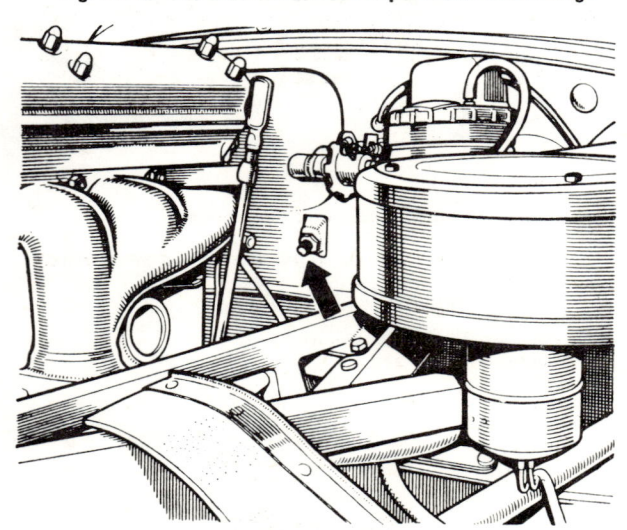

Fig. 10.43. Windscreen wiper parking adjusting screw (early models)

42 Windscreen wiper spindle housings (Fig. 10.45) - removal and refitting.

1 Disconnect the battery.
2 The instructions contained in paragraphs 3 - 9 (inclusive) cover the removal of the right or left hand spindle housing.
3 Depress the catch and withdraw the wiper arm from the spindle being removed.
4 Take out the two thumb screws at the top of the instrument panel and lower the panel to the horizontal position.
5 Refer to Chapter 12 and for 3.8 litre cars remove the side facia panel or the glove box depending on the housing to be removed. For 4.2 litre cars, remove the screen rail assembly.
6 Disconnect the ball joint from the spindle lever.
7 For 3.2 litre cars, from outside the car unscrew the large nut which secures the spindle housing to the scuttle and then take off the chrome distance piece and rubber seal.
8 For 4.2 litre models, the next step after paragraph 6 is, working from inside the car, remove the two nuts and washers which secure the housing bracket to the base plate.
9 For both models, the spindle housing can now be withdrawn from inside the car.
10 The link rod can be separated from the housing by removing the spring retainer and then withdrawing the pivot pin. Do

Chapter 10/Electrical system

not alter the length of the link rod otherwise the wiper blades will be made out of phase with each other.

11 The following paragraphs cover the removal of the centre spindle housing.

12 Withdraw the wiper arm from the spindle and then follow the instructions at paragraph 5.

13 Disconnect the ball joints from the two outer spindle cranks.

14 On early models, disconnect the two cables attached to the parking switch and then remove the nut attaching the parking switch control to the engine side of the bulkhead and withdraw the control from inside the car.

15 For all models, unscrew the large nut securing the housing to the scuttle and take off the chrome distance piece and rubber seal.

16 Withdraw the housing from inside the car.

17 Remove the spring retainer and withdraw the pivot pin with the attached outer link rods.

18 Refitting of all housings is the reverse of the removal sequence but you must ensure, if both outer housings were removed that the spindle with the longer crank is refitted to the driver's side.

43 Windscreen wipers - fault diagnosis and rectification

Poor performance of the wipers may not necessarily be due to a faulty motor, it can be electrical or mechanical in origin and may be due to low voltage at the motor from poor connections or discharged battery; it could also be due to excessive loading on the wiper blades or the spindles binding in the housing.

1 If the cause of the fault is not mechanical and no fault can be found in the supply to the motor, refer to the circuit diagram at Fig. 10.46 and proceed as follows.

2 Use a first grade moving coil voltmeter and measure the voltage between the motor supply terminal (green cable) and a good earth. The reading should be 11.5 volts with the wiper working normally. If a low reading is obtained, recheck the cables and connections and if these are in order suspect the switch but this can only be checked by substitution.

3 If the voltage is correct, remove the windscreen wiper arms and blades and then measure the light running current with an ammeter connected in series with the supply cable as follows.

4 Check the 'fast' speed current by using a fully charged 12 volt battery and connecting the green cable on the wiper motor to the negative terminal. Join the yellow and red cables together and then connect them to the positive battery terminal. Connect the blue and white cables together and check the cycles per minute of the wiper spindle. This should be 58 - 68 cycles per minute. Check the reading on the ammeter.

5 To check the 'slow' speed current. Connect the green cable to the negative battery terminal. Join the brown and red cables together and connect them to the positive battery terminal. Connect the blue and white cables together and now check the cycles per minute of the wiper spindle. It should be making 44-48 cycles per minute. Record the ammeter reading.

6 The light running current must not exceed 3.0 - 3.7 amps at 'slow' speed or 2.2 - 2.9 amps at 'fast' speed. If the current is in excess of these values the motor is unserviceable and must be changed.

44 The instrument panel - general

The instrument panel fitted to later cars differs from that on earlier models in that:-

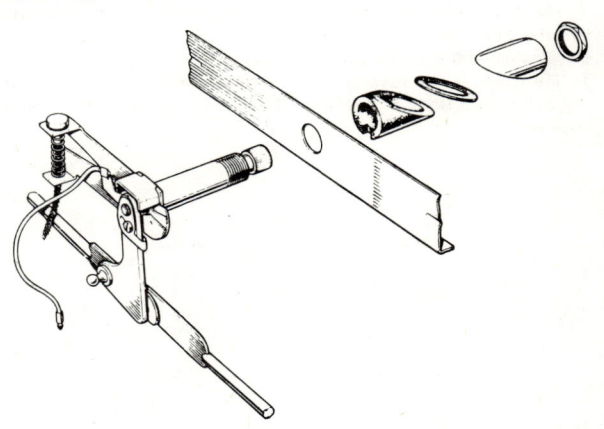

Fig. 10.45. A spindle housing assembly

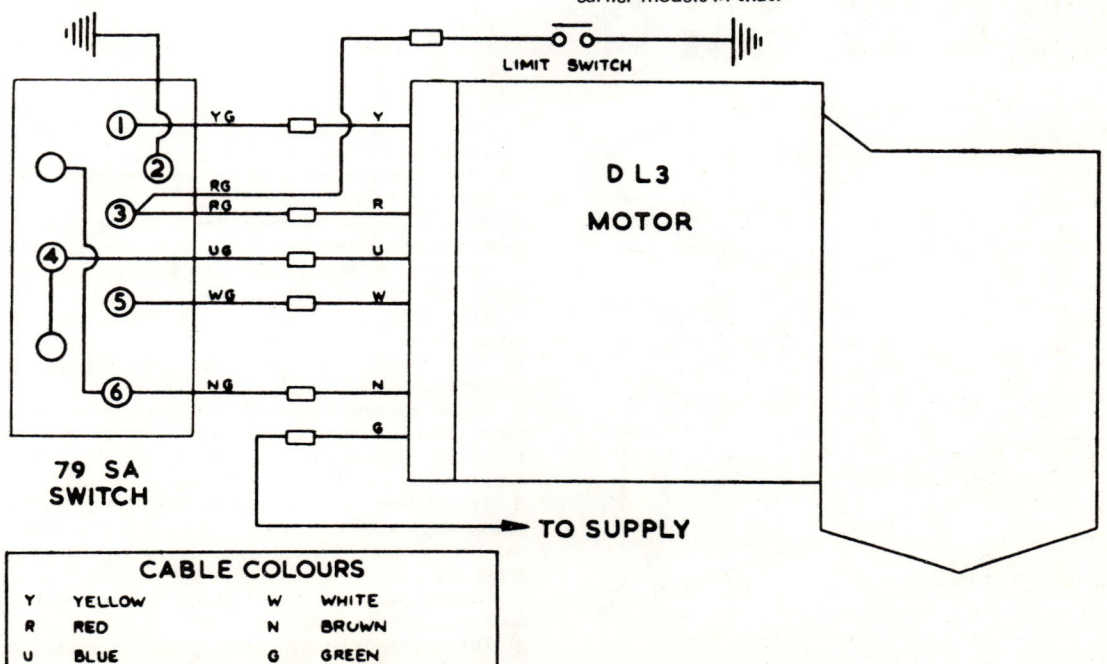

Fig. 10.46. Wiring connections - windscreen wiper switch to motor

Chapter 10/Electrical system

a) Rocker switches replace tumbler switches.
b) The clock is now seperate to the revolution counter.
c) A battery indicator replaces the ammeter.
d) The ignition/starter switch is a combined instrument and is now mounted on a sub-panel in place of the two separate items mounted on the instrument panel.

45 Electric clock, later models - maintenance, removal and refitting

The clock, as shown in Fig. 10.47 is fitted in the centre of the instrument panel and is a fully transistorised item powered by a mercury cell housed in a plastic holder attached to the back of the clock. Adjustment is provided by a small knurled knob 'A' (Fig. 10.47) for setting the hands and this action automatically starts the clock. A small slotted screw 'B' is for time keeping regulation, turn the screw towards the '+' sign if the clock is gaining and towards the '-' if losing. Moving the indicator scale through one division will alter the timekeeping by five minutes per week.

The window of the clock is a plastic moulding and should only be cleaned with a damp cloth or chamois leather, oil, petrol or cleaning fluids are harmful and should bot be used.

The lift of the mercury cell is about 18 months after which it should be renewed to ensure pefect timekeeping.

1 To replace the mercury cell battery (shown at Fig. 10.48), lower the instrument panel, lever the battery out of the holder and discard it. Press the new battery into position and replace the instrument panel.

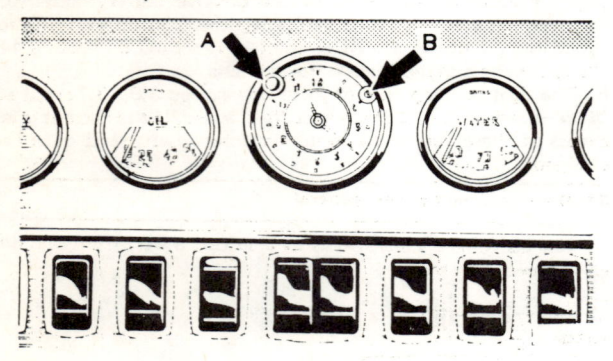

Fig. 10.47. Electric clock - later models

A hand setting   B time regulator

Fig. 10.48. Changing the clock battery

2 To remove the clock, lower the instrument panel and withdraw the illumination bulb holder from the back of the clock.
3 Remove the two nuts passing through the clamp strap at the back of the clock.
4 Withdraw the clock complete with the battery holder from the instrument panel.
5 Refitting is the reverse of the above.

46 Electric clock, early models - adjustment, removal and refitting

1 Adjustment is effected by means of a small screw at the back of the clock as shown in Fig. 10.49. If the clock is gaining, turn the screw towards the '—' sign, and towards the '+' sign if losing.
2 To remove the clock, first disconnect the battery.
3 Now refer to Section 52 and follow the instructions for removal of the revolution counter.
4 Disconnect the cable at the snap connector.
5 Remove the two nuts holding the clock to the revolution counter and withdraw the clock.
6 Slacken the knurled nut and remove the flexible setting drive.
7 Refitting is the reverse of the removal sequence again following the instructions in Section 52 for refitment of the revolution counter.

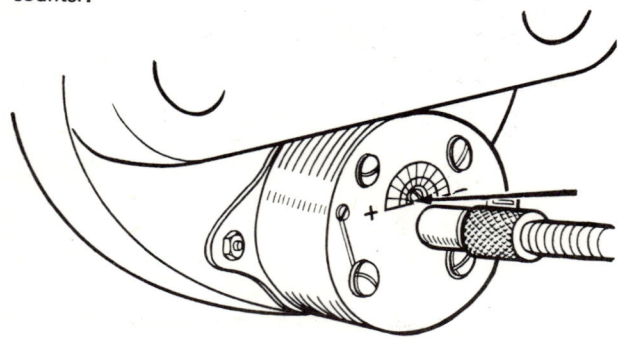

Fig. 10.49. Clock adjustment (early models)

47 Switches - removal and refitting

1 Disconnect the battery.
2 Remove the two thumb screws at the top of the instrument panel and then lower the panel to the horizontal position.
3 Identify the leads to the switch being removed and then disconnect them from the lucar tags.
4 For tumbler type switches, hold the switch lever in a horizontal position and unscrew the chromium ring from the face of the instrument panel. Withdraw the switch from the hidden face of the panel.
5 For **rocker type** switches, press in on the two locking tabs at the bottom and top faces of the body of the switch and push the switch through the aperture. The location of the locking tabs is shown in Fig. 10.50.
6 The **ignition switch** fitted to early models can be removed by unscrewing the chrome ring at the front of the instrument panel and then withdrawing the switch from the hidden face of the panel.
7 Remove the **starter push switch** by taking off the nut on the hidden face of the panel.
8 The Lucas 47SA combined ignition and starter switch is mounted on a bracket attached to the steering column or on the air conditioning control bracket when this system is fitted (see Fig. 10.51). Remove the locking ring and withdraw the switch

Chapter 10/Electrical system

through the bracket with the brass locknut and wave washer.
9   In each case, refitting is the reverse of the removal procedure.

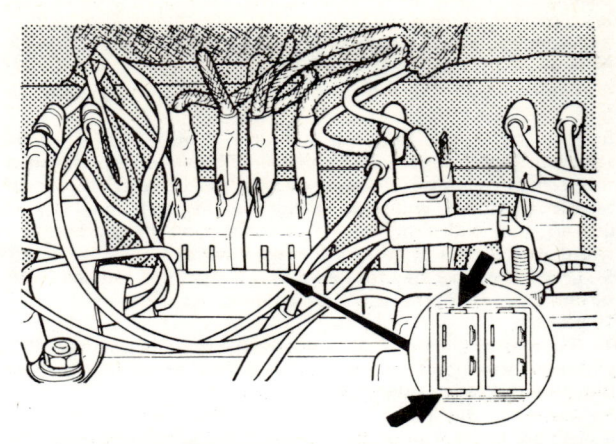

Fig. 10.50. The locking tabs of the rocker type switches

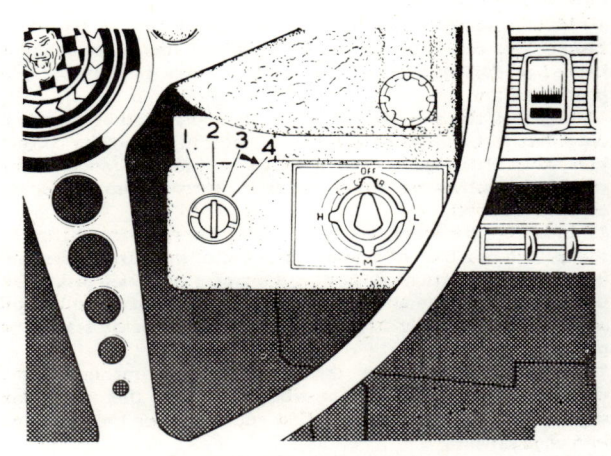

Fig. 10.51. The Lucas 47A combined ignition and starter switch - location when air conditioning fitted

1   Auxiliaries   3   Ignition "ON"
2   Ignition "OFF"   4   Starter

48 Ammeter, fuel, water temperature and oil pressure gauges - removal and refitting

1   Disconnect the battery.
2   Hinge the instrument panel downwards. Withdraw the illumination bulb holder from the instrument, identify and detach the leads.
3   Remove the two knurled nuts holding the instrument to the 'U' clamp and then withdraw the instrument through the front face of the panel.
4   Refitting is the reverse of the above procedure but in addition to making sure that the leads are fitted in their correct positions you must see that the 'U' piece does not foul any terminal or bulb holder.

49 Battery indicator - general

The instrument is, in effect, a voltmeter with a specially calibrated dial to indicate the condition of the battery. It does not register the charging rate of the alternator.

50 Battery indicator - removal and refitting

1   Disconnect the battery.
2   Remove the thumb screws at the top of the instrument panel and lower the panel to the horizontal position.
3   Identify the leads to the instrument and then remove the illumination bulb holder and detach the leads.
4   Remove the two nuts and clamp strap and withdraw the instrument forward through the panel.
5   Refitting is the reverse of the above.

51 Revolution counter (tachometer) later models - removal and refitting

This is an impulse tachometer instrument having transistors and a printed circuit. The pulse lead (white) is wired in circuit with the S/W termianl on the ignition coil and the ignition switch. Its performance is not affected by the setting of the distributor contact points, by corrosion of the spark plug points or gap settings.
Connection to the back of instrument is by means of a plug and socket as shown in Fig. 10.52. The contacts are offset to prevent incorrect connection.
1   Disconnect the battery.
2   Refer to Chapter 12 and follow the instructions for the removal of the screen rail facia.
3   Remove the two knurled nuts and retaining pieces and the earth lead at the back of the instrument.
4   Withdraw the instrument from the front face of the facia and remove the illumination bulb holders.
5   Pinch together the prongs of the plastic retaining clip and withdraw the plug and socket assembly.
6   Separate the plug from the socket and withdraw the instrument from the car. **Do not** detach the green and white cables connected to the plug and the instrument.
7   Refitting is the reverse of the above. Ensure that the plug and socket assembly is locked by the retaining clip.

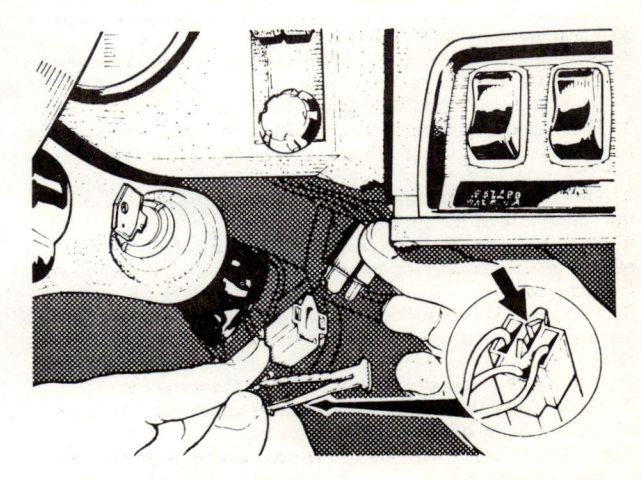

Fig. 10.52. Tachometer plug and socket assembly
(Inset shows the clip in its fitted position)

52 Revolution counter, early models - testing, removal and refitting

This instrument is of the electrical type operated from an AC generator fitted to the rear of the right hand camshaft on the engine.
1   If the instrument appears faulty, connect the leads of an AC

voltmeter across the terminals of the generator at the rear of the right hand camshaft.
2 Run the engine and check for voltage. It can be assumed that, as a rough guide, there will be one volt output per 100 engine revolutions.
3 If current is registered, reconnect the leads to the generator and then disconnect the leads at the rear of the instrument and connect them to the voltmeter. Again run the engine, and if current is registered it can be assumed that the instrument is at fault and must be replaced.
4 The first step in removing the revolution counter is to disconnect the battery.
5 Remove the two knurled nuts, the retaining pieces and the earth lead from the back of the instrument and withdraw it into the car.
6 Identify the leads to the instrument and to the clock and then disconnect them and the lead to the illumination lamp.
7 The instrument may now be removed from the car.
8 The clock can be removed from the instrument as described in Section 46.
9 If no reading is obtained during the check at pargraph 2 above, it means that the generator is faulty and must be replaced.
10 Having disconnected the leads from the Lucar tags on the generator, take out the three Allen securing screws and the plate washer and then withdraw the generator in a rearward direction at the same time taking not of the position of the tongued driving spindle.
11 Refit the generator in the reverse manner to the above after making sure that the drive is in the same position as when removed. If you do this, and provided the engine has not been turned whilst the generator was not in position, there should be no difficulty in seating the generator. However, do not use force to position it, check, using a mirror, the position of the slot in the camshaft and position the drive on the generator accordingly.
12 Refit the revolution counter head in the reverse manner to the removal procedure.

53 Speedometer - removal and refitting

1 Disconnect the battery.
2 Raise the steering to the highest position.
3 Remove the two knurled nuts holding the speedometer at the back of the facia panel and then remove the two retaining pieces and the earth lead.
4 Unscrew the knurled sleeve nut securing the flexible drive to the rear centre of the speedometer.
5 Pull the instrument forward into the car, identify and then remove the three warning lamps and the two illumination lamps from the back.
6 Unscrew the knurled sleeve nut securing the flexible odometer trip setting drive to the speedometer and remove the drive.
7 The speedometer can now be removed from the car.
8 Refitting is the reverse of the above sequence. The location of the warning and illumination lamps is:
   Two instrument illumination lamps - apertures at the side of the instrument.
   Headlamp warning light - right hand bottom aperture.
   Fuel warning light - centre bottom aperture.
   Ignition warning light - left hand bottom aperture.

54 Speedometer drive cable - removal, examination and refitting

1 Refer to Section 53 and follow the instructions to the point of removal of the speedometer to give access to the knurled drive nut at the rear of the instrument. Unscrew the nut to disconnect the drive.
2 The outer flexible drive casing must now be disconnected at the right angle drive attachment on the gearbox. You will find full instructions, with accompanying photographs, on the method of gaining access to the drive in Chapter 1 in the Section dealing with engine removal - work inside the car. Disconnect the drive from the connector.
3 Release the cable from the securing clips and withdraw it from inside the car.
4 Working from the instrument end, withdraw the inner flex from the outer casing.
5 It is not often that the complete drive will have to be removed. If faults arise which can be attributed to the inner flex, this can be removed by disconnecting the drive at the speedometer and then withdrawing the flex. If the inner flex is broken, you will also have to disconnect at the gearbox and withdraw the piece of broken flex from that end.
6 Check the inner flex for kinks by rolling it on a flat and clean surface any kinks present will be seen and felt as depicted in Fig. 10.53.
7 Examine the inner flex generally for wear or damage. Pay particular attention to the squared drive ends of the flex, if these show signs of wear at the corners, change the flex.
8 Assemble the inner to the outer flex and measure the projection of the inner flex beyond the outer casing at the instrument end. This should be 3/8", as shown in Fig. 10.54, to ensure correct engagement in the instrument and point of drive.
9 Now check that the inner flex rotates concentrically in the outer casing as shown in Fig. 10.55
10 Make sure that the inner flex is not binding in the outer casing. This condition cannot be satisfied if the outer casing is kinked or damaged in any way so first make sure that it is in good condition.
11 If either the speedometer drive complete or the inner flex only, has to be renewed make sure that you obtain the correct item when ordering, by quoting the make, year and model of the car. If alternatives are shown for your car, state also the length of drive required.
12 Lubricate the inner flex before assembly prior to refitting to the car. Place a blob of grease on the end of the outer cables and insert the flex through it so that the grease is carried inside. Do not use an excessive amount of grease and **do not** use oil. If oil appears in the flexible drive during use, suspect the oil seal at the gearbox drive end.
13 Refit the speedometer drive in the reverse manner to the removal sequence. Where possible, withdraw the inner flex slightly and connect the outer first and then slide the inner flex into engagement.
14 The condition of the flexible cable affects the performance of the speedometer to a very great extent. Poor installation will show up as apparent faults in the speedometer so, when refitting, pay particular attention to the following:
a) The run of flexible drive must be smooth. The minimum permissible bend radius is 6" and there must be no bend within 2" of the connections.
b) Avoid sharp bends at the clips but make sure that the drive is adequately secured, do not allow it to flap freely.
c) Make sure you do not crush the flex by overtightening on a clip.
d) Ensure that the outer flex connections are secure and are fully home, you may have to clean out the threads to make sure of this. The connections should be finger tight only.

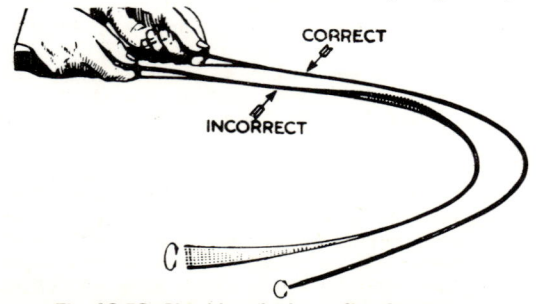

Fig. 10.53. Checking the inner flex for kinks

Chapter 10/Electrical system

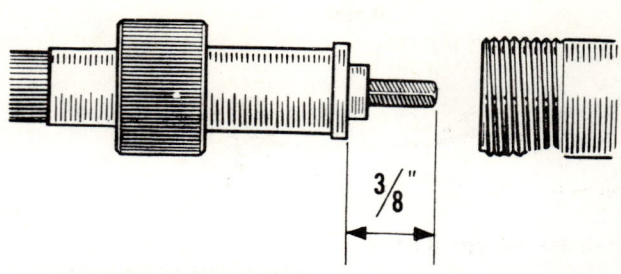

Fig. 10.54. Showing protrusion of the inner flex from the outer cable

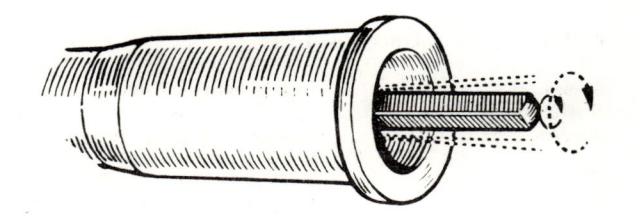

Fig. 10.55. Checking concentricity of the inner flex

55 Speedometer - fault diagnosis and rectification

The performance of the speedometer is dependent on the serviceability of the flexible drive so first make sure that the drive is in good condition and that the requirements of Section 53 are fulfilled.

1 **Instrument not operating**, incorrect connection of the flexible drive or insufficient engagement of the inner shaft. Broken or damaged inner flexible shaft. Check these possible faults by removal and examination as described in Section 54, if found satisfactory it is the instrument which is at fault.

2 **Instrument inaccurate**, check tyre pressures, inaccuracy can also be caused by badly worn tyres or the use of non-standard tyres. If you do not know the history of the car, check the code number (Fig.10.56) and make sure by enquiry of your local Jaguar dealer that the correct speedometer is fitted. It may be that a non-standard rear axle has been fitted or the drive ratio in the gearbox is non-standard. You can check this by removing the flexible drive from the speedometer and then fitting a cardboard counter on the inner flex as shown in Fig.10.57. Now measure the radius of the rear wheel, that is from the centre of the hub to the ground. Make a chalk mark at any position on a rear wheel and using this as a datum, push the car forward in a straight line for six turns of the wheel and at the same time have an assistant note the number of turns made by the cardboard counter. Now apply the figures obtained in the formula:

$$\frac{1680N}{R} = \text{Turns per mile number}$$

Where N = the number of turns made by the counter for six turns of the rear wheels
R = the radius of the rear wheels measured in inches

The turns per mile number is given on the front face of the speedometer as shown in Fig.10.58.

The turns per mile number should correspond within 25 either way of the number obtained from your calculation. If it does not, investigate the possible fitment of non-standard item.

3 **Pointer waver**, this can be caused by an oiled-up instrument resulting from failure of the oil seal at the drive end of the shaft. Replace the oil seal, clean and lubricate the flexible drive and fit a replacement speedometer.

The inner flexible drive may not be engaging fully. Check this against Fig.10.54 and then make sure that the connections were correct and if all is in order the fault is probably due to damage to the inner flexible drive.

Ensure that the outer casing is not kinked, that the bends are not too sharp and that it is not being crushed by a clip.

If the above are in order, the fault probably lies in the instrument.

4 **Noisy operation**, this, as illustrated in Fig.10.59, may take a number of forms. If the fault appears to be in the drive, look for damage to the outer casing, overtightness of clips or incorrect run. Check lubrication of the inner drive and if the fault persists the only remedy may be fitment of a new drive assembly. Noise in the instrument itself, which will usually be accompanied by inaccuracies, will entail fitment of a new instrument.

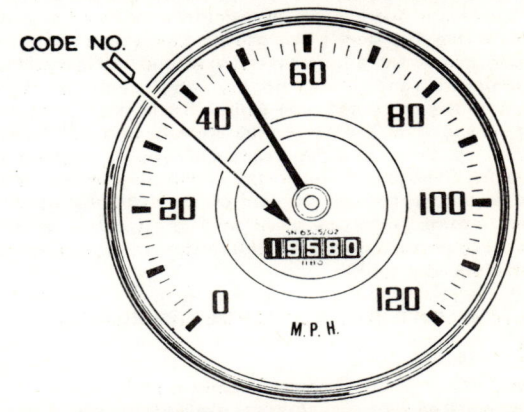

Fig. 10.56. Showing the location of the code number on the face of the speedometer

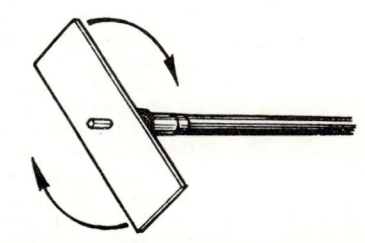

Fig. 10.57. Fitment of a cardboard counter to check rear axle and gearing

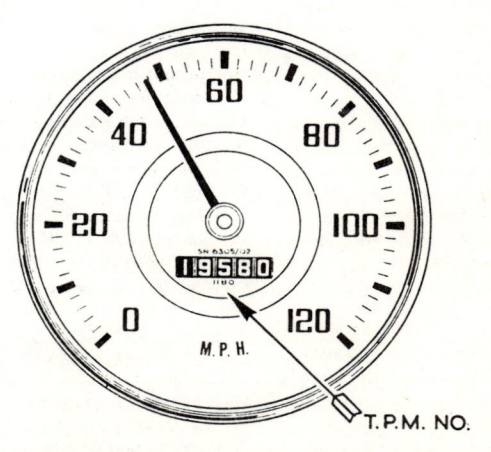

Fig. 10.58. Showing the location of the TPM number

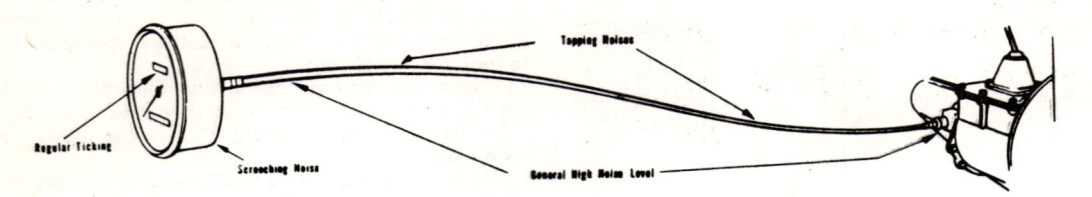

Fig. 10.59. Showing apparent source and type of noise

56 The bi-metallic resistance instrumentation - fault finding

The engine temperature, engine oil pressure and petrol tank content gauges are operated by transmitters mounted on the engine or in the fuel tank. The gauge units operate on the thermal principle having a heater winding wound on a bi-metallic strip. The engine temperature and the petrol contents transmitters are of the resistance type whilst the oil pressure gauge fitted in the head of the engine oil filter is a voltage compensated pressure unit consisting of a diaphragm, a bi-metallic strip with a heater winding and a pair of contacts. A circuit diagram for these units and a combined wiring diagram of the fuel contents and water temperature gauges are given in Figs. 10.60 to 10.63. (inclusive). The possible causes of faults arising in the gauges are given in the following tables:

ENGINE TEMPERATURE AND PETROL TANK GAUGE FAULTS

Symptom	Unit possibly at fault
Instrument panel gauge showing a 'zero' reading	Voltage regulator — Check that output voltage at terminal 'I' is 10 volts. Instrument panel gauge — Check for continuity between the gauge terminals with the leads disconnected. Transmitter unit in petrol tank or engine unit — Check for continuity between the terminal and the case with lead disconnected. Wiring — Check for continuity between the gauge, the transmitter and the voltage regulator, also that the transmitter unit is earthed.
Instrument panel gauge showing a high/low reading when ignition switched on	Voltage regulator — Check output voltage at terminal 'I' is 10 volts. Instrument panel gauge — Check by substituting another instrument panel gauge. Transmitter unit in petrol tank or engine — Check by substituting another transmitter unit in petrol tank or engine unit. Wiring — Check for leak to earth.
Instrument panel gauge showing a high reading and overheating	Voltage regulator — Check output voltage at terminal 'I' is 10 volts. Wiring — Check for short circuits on wiring to each transmitter unit.
Instrument panel gauge showing an intermittent reading	Voltage regulator — Check by substituting another voltage regulator. Instrument panel gauge — Check by substituting another instrument panel gauge. Transmitter unit in petrol tank or engine unit — Check by substituting another transmitter unit in petrol tank or engine unit. Wiring — Check terminals for security, earthing and wiring continuity.

OIL PRESSURE GAUGE FAULTS

Symptom	Unit possibly at fault
Instrument panel gauge showing a 'zero' reading	Wiring — Check for continuity between the gauge and the transmitter unit and that the latter is earthed. Instrument panel gauge — Check for continuity between the gauge terminals with leads disconnected. If satisfactory replace the transmitter unit.
Instrument panel gauge showing a reading with ignition switched on but engine not running	Transmitter unit on oil filter head — Check by substituting another transmitter unit.
Instrument panel gauge showing a high reading and overheating	Transmitter unit on oil filter head — Check by substituting another transmitter unit.
Instrument panel gauge showing a below 'zero' reading with ignition switched off	Instrument panel gauge — Check by substituting another instrument panel gauge.

NOTE: The instrument panel gauges **must never** be checked by short circuiting the transmitter units to earth.

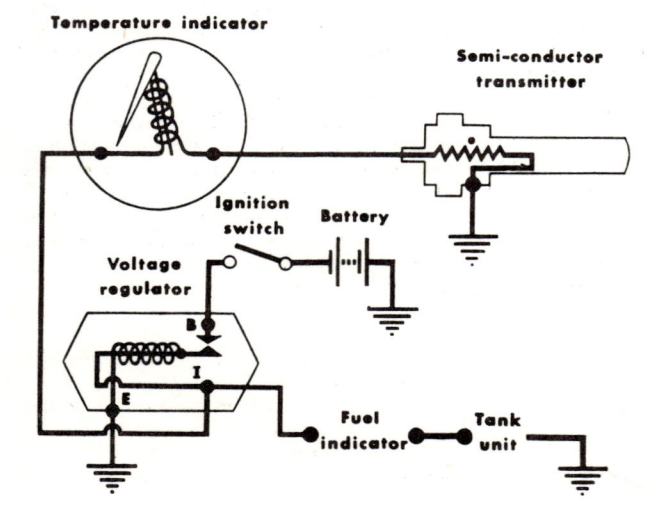

Fig. 10.60. Combined wiring diagram of fuel and water temperature gauges with voltage regulator

Chapter 10/Electrical system

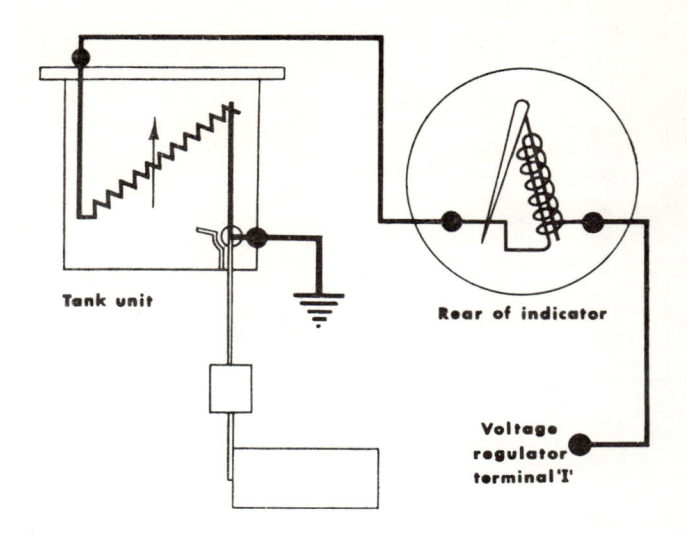

Fig. 10.61. Circuit diagram for fuel contents gauge

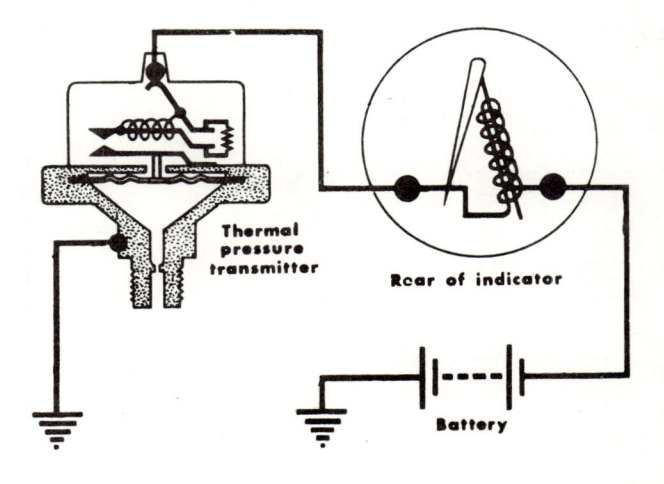

Fig. 10.62. Engine oil pressure gauge circuit

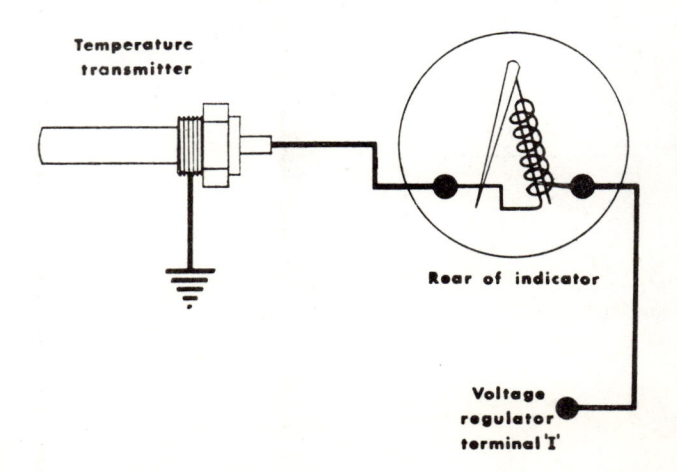

Fig. 10.63. Water temperature gauge circuit

57 Fitment of auxiliary lights - general

The first point to be considered when deciding to fit auxiliary lighting is (if the lights are external and are intended to supplement the existing headlamps) will the proposed position of the light and its subsequent use, contravene traffic regulations. It is well worthwhile making enquiries beforehand because excuses for some minor contravention will not be well received at a later date.

Having decided on the position of a light, the next thing to be considered is connecting it into the existing system and deciding whether or not you wish the light to be operative when the ignition is switched off, or only when the ignition is switched on. Here a study of the appropriate wiring diagram for your car is called for but if you have difficulty in picking up a suitable circuit go to the junction box and you will find that most connections into the circuit have spare connectors and use of a test lamp with the ignition switched on and off, will show up a connection which will meet requirements. Trace this from its identification number in the wiring diagram and then make sure that the lamp you are fitting is not going to overload the circuit. Do not try to get round this by fitting a heavier fuse as, in case of trouble, you may find that an existing item of electrical equipment is irreparably damaged, or wiring is burnt out. Under no circumstances must you connect direct to the battery, the wiring must go through a fuse unit.

You will probably wish to have the light switched separately to existing services; suitable switches, complete with mounting brackets can be bought quite cheaply and look neat when fitted. In the case, say, of a reversing light, or of a heated rear window it is a wise precaution to have an indicating light in circuit with the switch to show you that the accessory is switched on.

Fitting a light to the outside of the car will mean drilling the bodywork. Before doing this be absolutely certain that the position you have chosen is correct, centre punch the position of the hole and then use a pilot drill to avoid the possibility of a larger drill running off centre and damaging the paintwork. It is advisable to paint the bare metal to protect against subsequent corrosion.

58 Fitment of radios

The actual fitment of a radio and loudspeaker presents no problem as space for these is already provided for in the manufacture of the car. However, the question of whether or not a radio will fit the car without modification is one point to consider when buying the radio. The main point to bear in mind when buying a radio or a tape player is that it must be a **negative earth** otherwise you will be faced with extra expense in the purchase of equipment to reverse polarity.

Three radios, designed for use in the 'E' Type, are available as manufacturer's supply, these have Long and Medium, Medium, Medium and Short wavebands only.

The input for the radio must be from a fused supply with additional protection provided by a 3 amp fuse in the circuit. This is built into a special quick release connector in the radio suppy line.

You will need to suppress the following with condensers as follows:

Coil	-	1 mfd condenser
Petrol pump	-	1 mfd condenser
Alternator/dynamo	-	1 mfd condenser
Clock	-	1 mfd condenser
Oil pressure transmitter	-	0.25 mfd condenser

The position for the aerial is a matter of choice but experience has shown that the best place is on the body just forward of the nearside windscreen pillar, in this position there is minimum blockage of forward vision and least electrical interference. When fitting the aerial, use a pilot drill after centre punching the position for the hole, in order to prevent the possibility of a larger diameter drill running off centre and damaging the paintwork. Treat the bare metal edge with paint to reduce the chance of subsequent corrosion. Make sure that the aerial is well mounted on its rubbers and that there is no chance of its earthing against the body. Feed the lead to the radio using an existing grommet (if available) but if you have to drill and pass it through the body at any point fit a grommet to protect the lead. Make sure that the lead will not foul any existing equipment.

Much of the foregoing also applies to the fitment of a tape player but here positioning is a matter of choice and may well call for the manufacture of supporting brackets for the cassette and the speakers. Make sure that the cassette is fed from a fused supply and is also protected with its own fuse.

59 Wiring diagrams - colour coding

Wiring diagrams for the various models are given in Figs. 10.64 to 10.67 (inclusive). The cable colour code used in these diagrams is given below, where a cable is shown as having two code letters, the first letter denotes the main colour and the second gives the tracer colour.

B	-	Black	S	-	Slate
U	-	Blue	W	-	White
N	-	Brown	Y	-	Yellow
R	-	Red	D	-	Dark
P	-	Purple	L	-	Light
G	-	Green	M	-	Medium

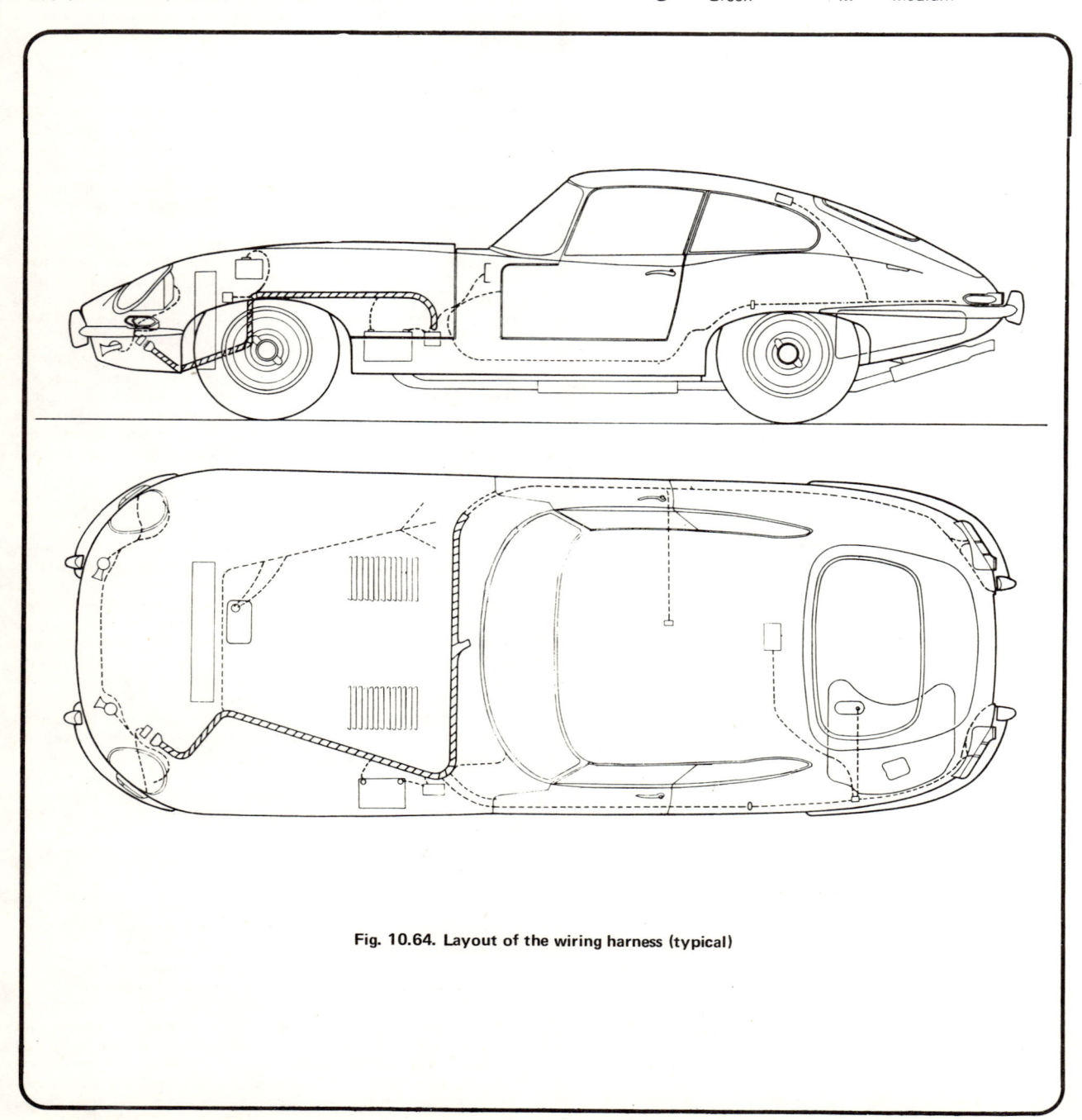

Fig. 10.64. Layout of the wiring harness (typical)

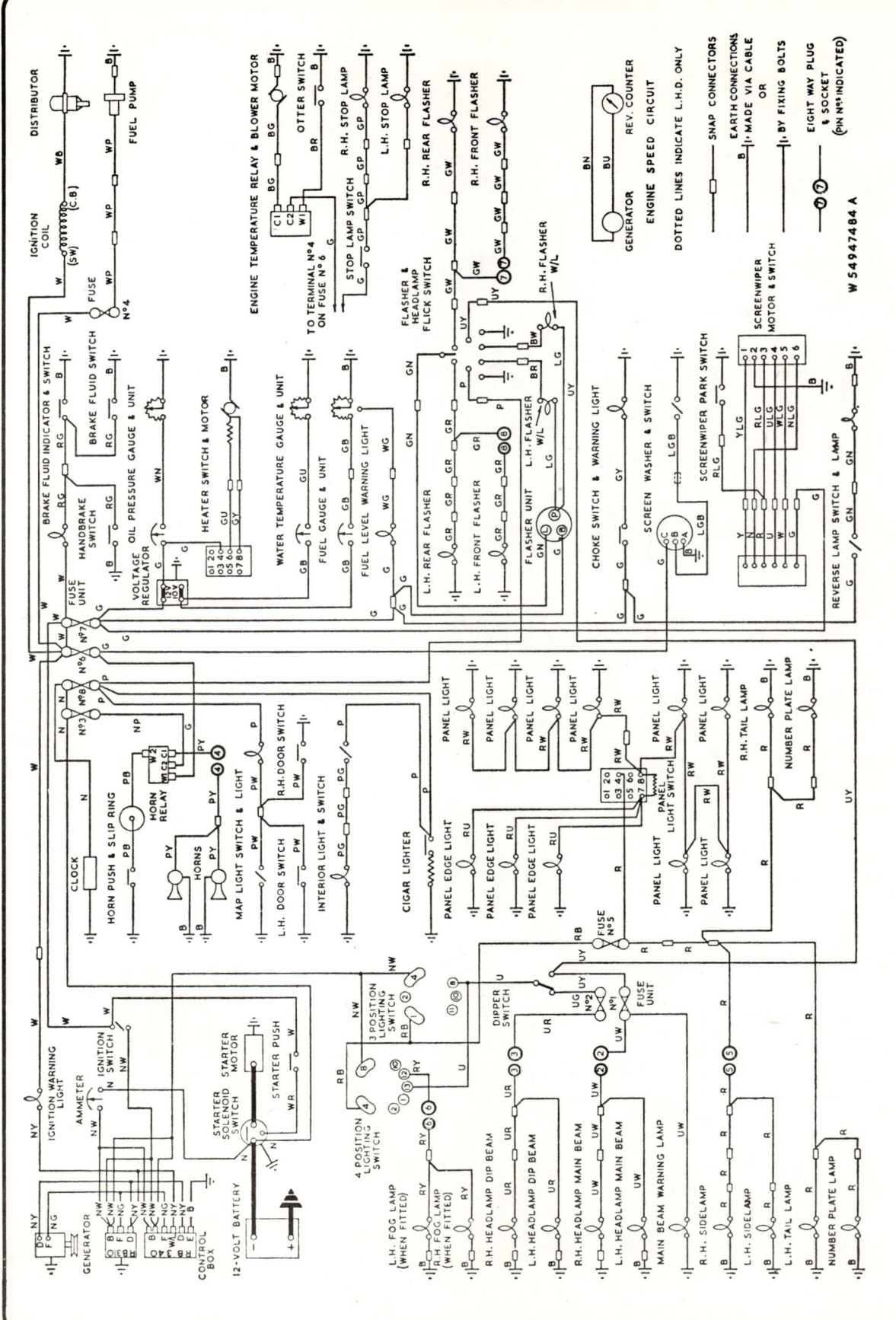

Fig. 10.65. Wiring diagram - 3.8 litre models

Fig. 10.66. Wiring diagram - 4.2 litre models from chassis numbers IE.50001 (RH drive) and
IE.75001 (LH drive) to:-
IE.15979 - Open 2 seater   )
IE.34582 - Fixed head Coupe) LH drive - USA only
IE.16009 - Open 2 seater   )
IE.34751 - Fixed head Coupe) LH drive - other than USA
IE.77709 - 2 + 2   )

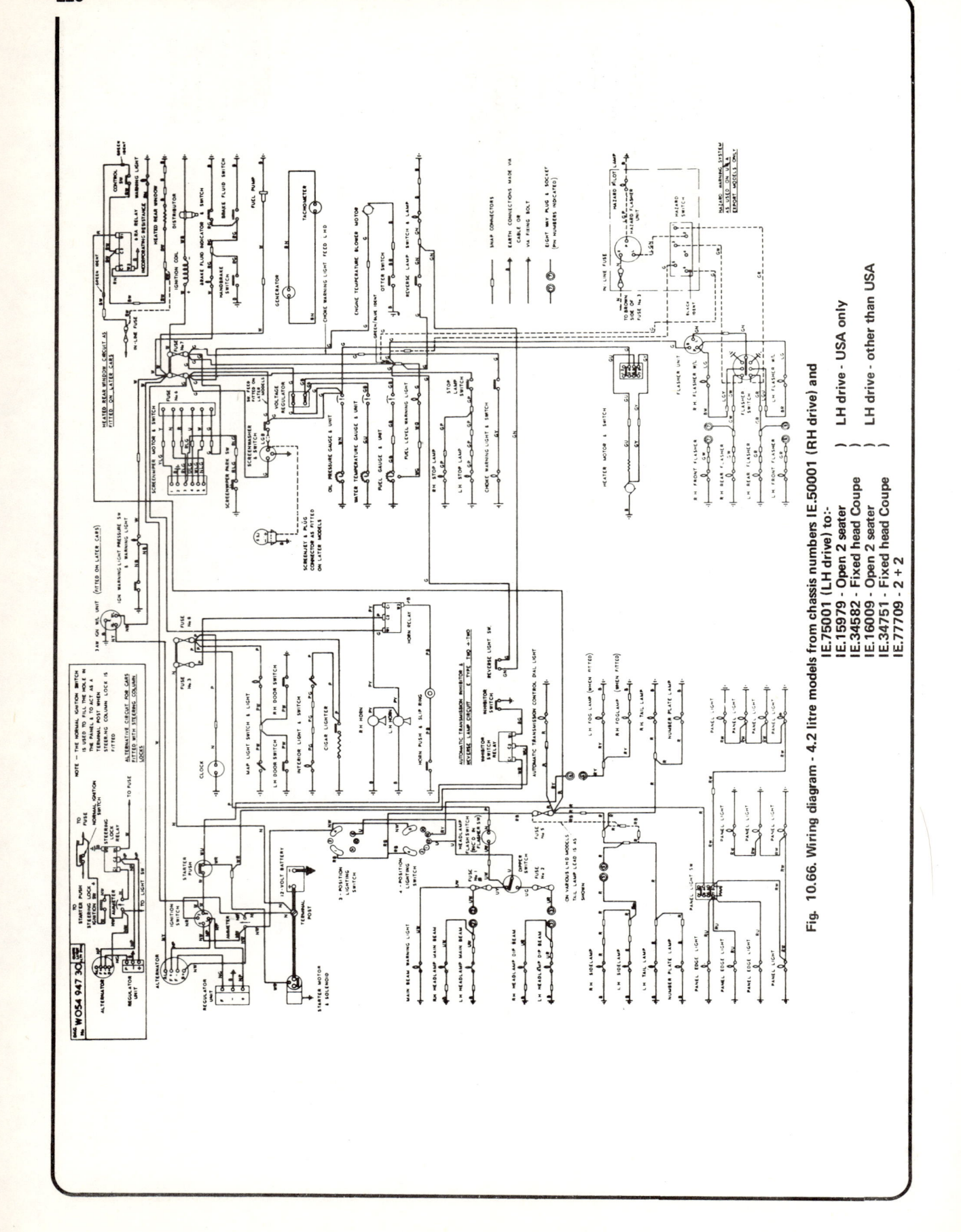

Fig. 10.67. Wiring diagram - 4.2 litre models with the following and subsequent chassis numbers:-

RH drive
Open 2 Seater IE.2037

Fixed Head Coupe IE.21786

2+2 IE.51197

LH drive
IE.15980 USA only
IE.16010 other than USA

IE.34583 USA only
IE.34752 other than USA
IE.77709

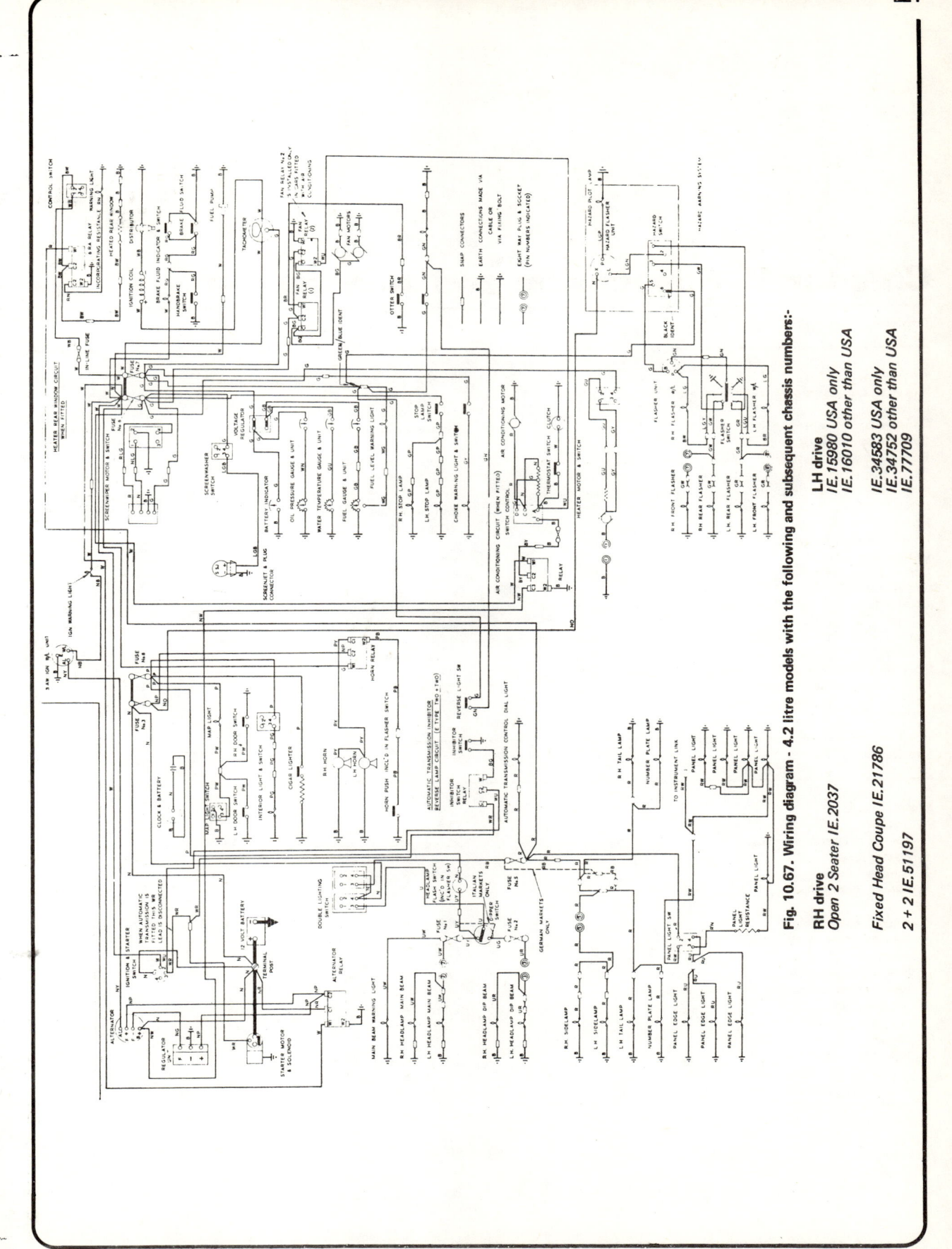

Chapter 11 Suspension and steering

Contents

Accidental damage ... 31	Rear wheel camber angle - adjustment ... 19
Anti-roll bar - removal, renewing the link arm bushes and refitting ... 8	Road spring and hydraulic damper assembly - dismantling and reassembly ... 15
Camber angle - adjustment ... 11	Road spring and hydraulic damper assembly - removal and refitting ... 14
Castor angle - adjustment ... 10	Steering arm - removal and refitting ... 29
Front hydraulic dampers - removal and refitting ... 7	Steering wheel - removal and refitting ... 20
Front suspension assembly - general ... 2	Stub axle carrier and lower wishbone ball joint - removal and refitting ... 5
Front suspension components - accidental damage ... 12	Torsion bar - adjustment ... 9
Front suspension lower wishbone - removal and refitting ... 4	Upper and lower steering columns, collapsible type - general ... 24
Front wheel alignment ... 30	Upper steering column (collapsible type) - removal and refitting ... 25
Front wheel hubs - removal and refitting ... 6	Upper steering column (conventional type) - dismantling and reassembly ... 22
General description ... 1	Upper steering column (conventional type) - removal and refitting ... 21
Upper steering column (collapsible type) - dismantling and reassembly ... 26	Upper wishbone, fulcrum shaft and ball pin - removal and refitting ... 3
Lower steering column (conventional type) - removal and refitting ... 23	Wishbone - removal and refitting ... 17
Rack and pinion assembly - dismantling, examination and reassembly ... 28	Wishbone outer pivot bearing - adjustment ... 18
Rack and pinion housing - removal and refitting ... 27	
Radius arm - removal and refitting ... 16	
Rear suspension - removal and refitting ... 13	

Specifications

Front suspension:
  Type ... Independent torsion bars
  Dampers (shock absorbers) ... Telescopic hydraulic
  Castor angle ... $2° \pm \frac{1}{2}°$ positive
  Camber angle ... $\frac{1}{4}° \pm \frac{1}{2}°$ positive
  Swivel inclination ... $4°$

Rear suspension:
  Dampers (shock absorbers) ... Telescopic hydraulic
  Track ... 50¼ inch
  Rear wheel camber ... $\frac{3}{4}° \pm \frac{1}{4}°$ negative
  Road wheel movement from mid laden position:
    Full bump ... 3.1/8 inch
    Full rebound ... 3.1/8 inch

  Rear road spring: 3.8 litre (early models) | Late 3.8 litre and 4.2 litre cars
    Free length (approx) ... 10.1 inch (25.65 cm) | 10.5 inch (26.67 cm)
    Number of coils (approx) ... 9.3/8 | 10
    Wire diameter ... 0.432 inch (11.0 mm)
    Colour ... — | Red

Steering:
  Type ... Rack and pinion
  Number of turns - lock to lock ... 2½
  Turning circle ... 37 ft (11.27 m)
  Diameter of steering wheel ... 16 inch (40.5 cm)
  Front wheel alignment ... 1/16 inch to 1/8 inch (1.6 to 3.2 mm) toe-in

Chapter 11/Suspension and steering

1 General description

The front suspension assemblies fitted to all models are of similar construction and differ only in respect of the coil springs. The assembly is illustrated in Fig.11.1.

The front suspension units are comprised of the upper and lower wishbone to which are attached the stub axle carriers, torsion bars and hydraulic dampers. The front end of the torsion bar is attached to the lower wishbone whilst the rear end is secured to brackets on the chassis frame. Each torsion bar is controlled by a hydraulic damper the top of which is attached to brackets on the forward chassis assembly, the bottom being bolted to the lower wishbone. The upper wishbone is a one piece forging and is secured to a threaded fulcrum shaft by means of pinch bolts through clamps formed on the wishbone inner mounting, the fulcrum shaft itself is mounted on two rubber/steel bonded bushes. The outer ends of the upper wishbone carry the wishbone ball joint and this is secured to the hub carrier by the tapered shank of a ball pin and a locknut. The lower wishbone, a two piece assembly, is secured at the inner ends to the fulcrum shaft on rubber/steel bonded bushes whilst the outer end is secured to the lower wishbone ball joint, as for the upper wishbone assembly, by the tapered shank of a ball pin and locknut. The anti-roll bar fitted between the lower wishbones is attached to the chassis front member by rubber mounted brackets. The wheel hubs run on two tapered roller bearings with the inner races fitted on a shaft located in a tapered hole in the stub axle carrier.

A view of one side of the rear suspension is given in Fig.11.2. Two tubular links locate the wheels in a transverse plane, one of these is the half shaft which has universal joints at each end and the other link is attached to, and pivots on, the wheel carrier and the cross beam adjacent to the differential casing, the pivot bearings of this lower link are widely spaced in order to give maximum rigidity in a longitudinal plane. The suspension utilises four coil springs each enclosing a telescopic hydraulic damper, two springs being mounted each side of the differential casing. The complete assembly is carried in a fabricated steel cross beam which is attached to the body of the car on four rubber mounting blocks and is located by radius arms which pivot on rubber bushes mounted on each side of the car between the lower link and a mounting point on the body. An anti-roll bar is fitted between the two lower wishbones and is attached to the underframe side members by rubber insulated brackets.

The steering gear is of the rack and pinion type in which motion is transmitted from the inner steering column through the pinion to the steering rack which carries tie rods, attached by ball joints enclosed in rubber bellows, connected to the steering arms at the stub axle carriers. The steering rack assembly forms a complete unit attached to the front crossmember of the forward chassis frame between the engine and the radiator. The steering column engages the splined end of the pinion shaft to which it is secured by a clamp bolt.

With effect from the chassis numbers quoted below, the upper and lower steering columns are of the collapsible type designed to comply with USA safety regulations. It should be noted that this feature is not built into any right hand drive cars.

Model	Chassis No.	Remarks
Open 2 seater	1E.15980	USA only
Open 2 seater	1E.16010	Other than USA
Fixed head coupe	1E.34583	USA only
Fixed head coupe	1E.34752	Other than USA
2 + 2	1E.77709	

2 Front suspension assembly - general

It is not advisable to remove the front suspension assemblies, illustrated at Fig.11.3, as complete units. The various components should be removed and replaced as separate items.

3 Upper wishbone, fulcrum shaft and ball pin - removal and refitting

1 Apply the handbrake and chock the rear wheels.
2 Slacken, but do not remove, the hub cap to the front wheel of the side being worked on. The hub caps are marked "RIGHT (OFF) SIDE" and "LEFT (NEAR) SIDE" and the direction of rotation for removal is clockwise for the right hand and anti-clockwise for the left hand side.
3 Place a jack under the lower wishbone fulcrum support bracket and raise the car until the wheel is clear of the ground. For safety, place a stand under the wishbone fulcrum rear support bracket.
4 Having satisfied yourself that the car is safely supported, remove the road wheel. **Do not** place the jack or stand under the forward frame cross tubes.
5 Refer to Fig.11.4.
6 Remove the self locking nut (11) and the washer (10) securing the upper wishbone ball pin (6) to the stub axle carrier. Tie up the axle carrier to the frame member to prevent the brake hose being strained.
7 Tap the stub axle smartly with a hammer in the area of the ball pin to break the grip of the taper, and so separate the ball pin from the stub axle carrier.
8 Remove the two bolts, nuts and lockwashers securing the fulcrum shaft rear bracket to the chassis frame. Collect the stiffener plate located behind the two nuts on the inner face of the frame member.
9 Remove the three setscrews and lockwashers retaining the front bracket of the fulcrum shaft.
10 Lift the wishbone assembly away from the car but take careful note of the shims (12 and 13) fitted behind the fulcrum shaft carrier brackets.
11 Take out the split pins and then unscrew the nuts retaining the carrier brackets to the fulcrum shaft. Withdraw the brackets and the rubber bushes.
12 Remove the circlip (1) followed by the top cover (2), the shim (3), the spring (4) and the ball pin socket (5).
13 Now remove the ring (8) securing the rubber gaiter (9) to the upper wishbone and remove the gaiter and the insert between it and the upper wishbone.
14 Take out the ball pin (6).
15 If the rubber/steel bushes in the fulcrum shaft brackets (Fig.11.5) have to be renewed, they can be drifted or pressed out of the brackets.
16 To remove the fulcrum shaft, slacken the two clamp screws locking the shaft to the wishbone. Turn the shaft in a clockwise direction, looking from the rear, until the threaded portion is clear of the wishbone. The shaft can now be withdrawn through the wishbone arms.
17 Reassemble the rubber/steel bushes to the fulcrum shaft brackets by pressing into position until the bush protrudes an equal amount each side of the bracket. You will find that fitting the bushes will be made easier if you lubricate them with a solution of 12 parts of water to one part of liquid soap.
18 Reassemble the fulcrum shaft in the reverse manner to removal and when it is correctly located, lock it in position by tightening the clamp screws.
19 Before completing the build up of the ball joint it is necessary adjust the ball pin for correct working clearance. This is done by first making sure that all parts are clean. Now assemble the joint, less the spring and the lower gaiter and retaining ring, but adding additional shims between the cover plate and the ball pin socket until the ball pin is tight in its socket when the circlip is fitted. Shims are currently produced in 0.004" (0.10 mm) thicknesses. Now remove shims to the value of 0.004" (0.10 mm) and reassemble the ball joint complete with the spring, it should now be possible to move the ball pin by hand. Finally lubricate the joint with the recommended grade of grease.
Note: Shims should not be added to take up excessive wear in the ball pin and socket. If these parts are badly worn they should

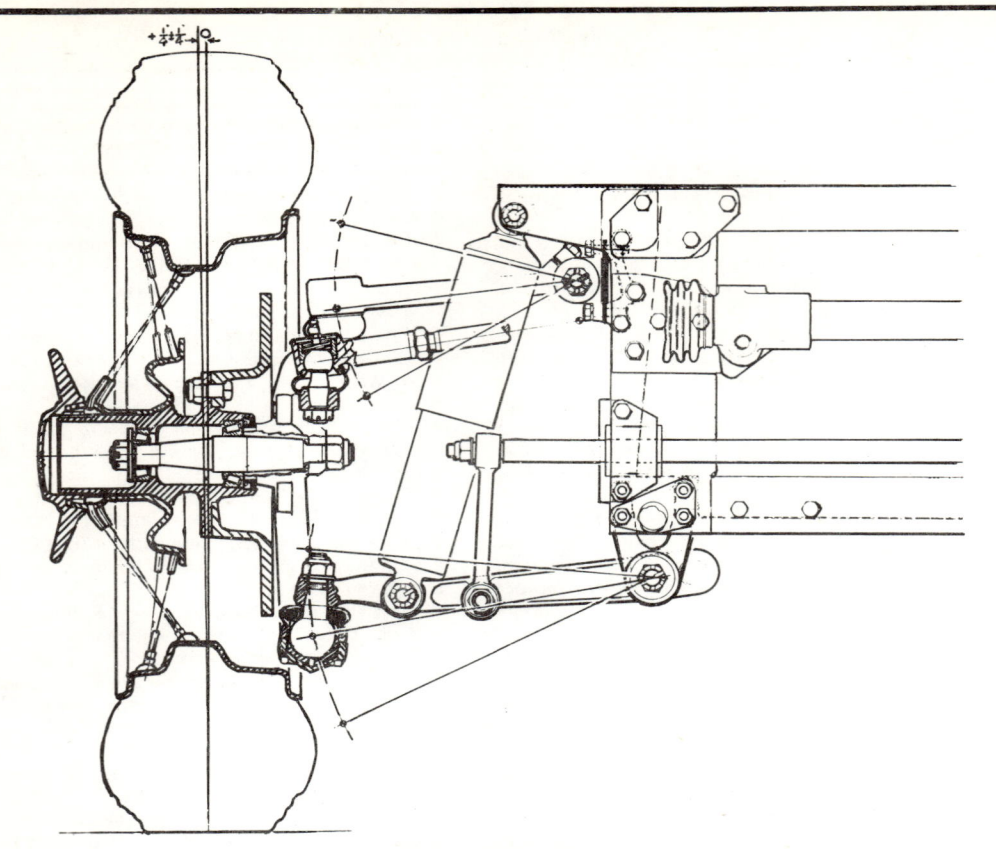

Fig. 11.1. The front suspension assembly

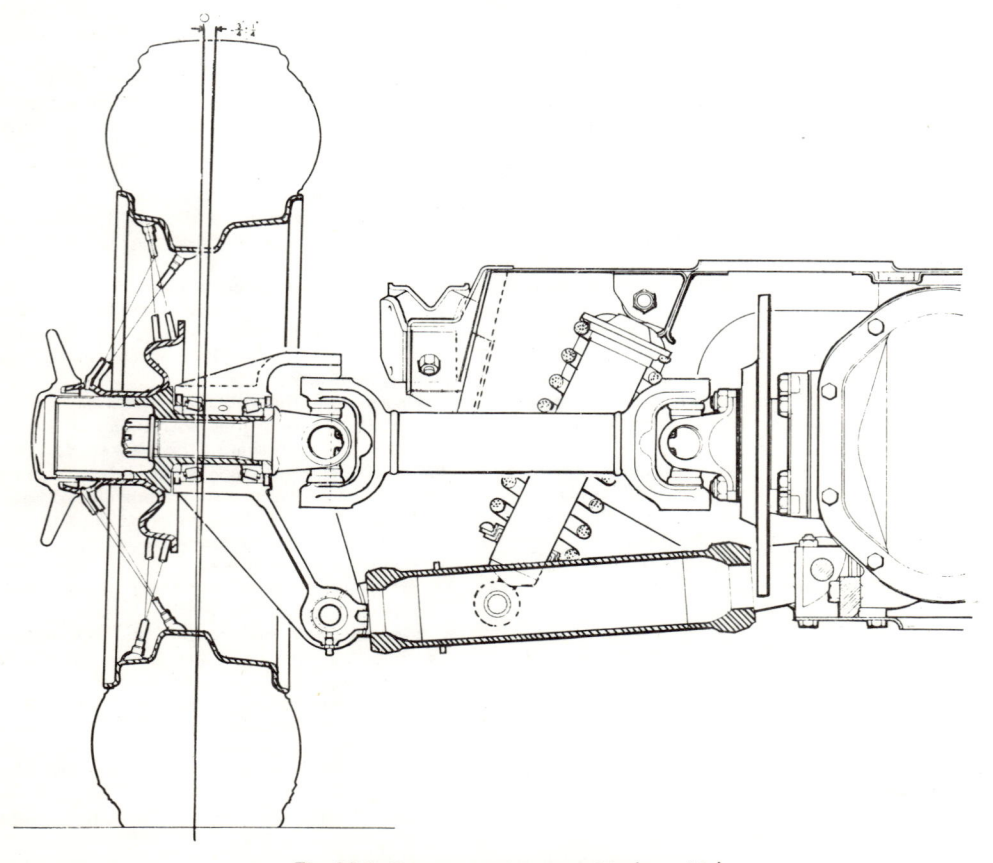

Fig. 11.2. Rear suspension assembly (one side)

Chapter 11/Suspension and steering

be replaced. When testing for movement of the ball in its socket, move the ball only in a longitudinal direction, it must not be allowed to come into hard contact with the sides of the socket.
20 Refit the rubber gaiter and insert, and secure them in position by the retaining ring. Renew the gaiter if the condition of the existing item is at all doubtful.
21 From this point onwards, refitting of the wishbone assembly is the reverse of the removal sequence but ensure that the correct number of shims are refitted between the fulcrum shaft brackets and the chassis frame. **Do not** tighten the slotted nuts at each end of the fulcrum shaft until the full weight of the car is on the suspension otherwise undue torsional loading of the rubber bushes will result with possible premature failure.
22 Finally, refer to Sections 10 and 11 and check the castor and camber angles.

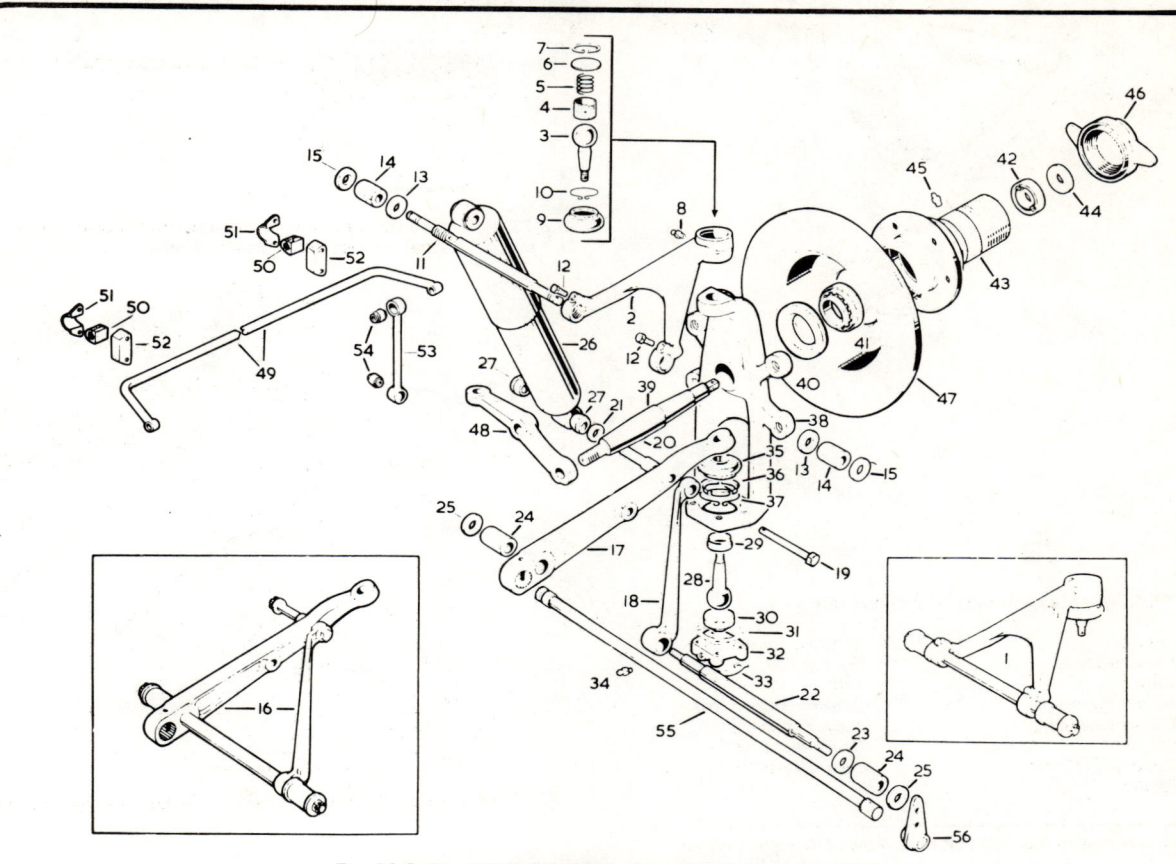

Fig. 11.3. Exploded view of the front suspension assembly

1	Upper wishbone assembly (right-hand)	29	Ball pin spigot
2	Upper wishbone (right-hand)	30	Morganite socket
3	Upper wishbone ball pin	31	Shims
4	Ball pin socket	32	Lower ball pin cap
5	Spring	33	Tab washers
6	Top cover	34	Grease nipple
7	Circlip	35	Rubber gaiter
8	Grease nipple	36	Gaiter retainer
9	Rubber gaiter	37	Clip
10	Clip	38	Stub axle carrier
11	Upper wishbone fulcrum shaft	39	Stub axle
12	Pinch bolt	40	Oil seal
13	Distance washer	41	Inner bearing
14	Rubber bush (upper wishbone)	42	Outer bearing
15	Special washer	43	Front hub (right-hand)
16	Lower wishbone assembly (right-hand)	44	"D" washer
17	Lower wishbone lever (right-hand front)	45	Grease nipple
18	Lower wishbone lever (right-hand rear)	46	Hub cap
19	Bolt	47	Brake disc
20	Sleeve	48	Steering arm
21	Washer	49	Anti-roll bar
22	Lower wishbone fulcrum shaft	50	Rubber bush
23	Distance washer	51	Bracket
24	Rubber bush (lower wishbone)	52	Distance piece
25	Special washer	53	Anti-roll bar link
26	Shock absorber (front)	54	Rubber bush
27	Shock absorber (bottom bush)	55	Torsion bar
28	Lower wishbone ball pin	56	Bracket - torsion bar (rear end)

Chapter 11/Suspension and steering

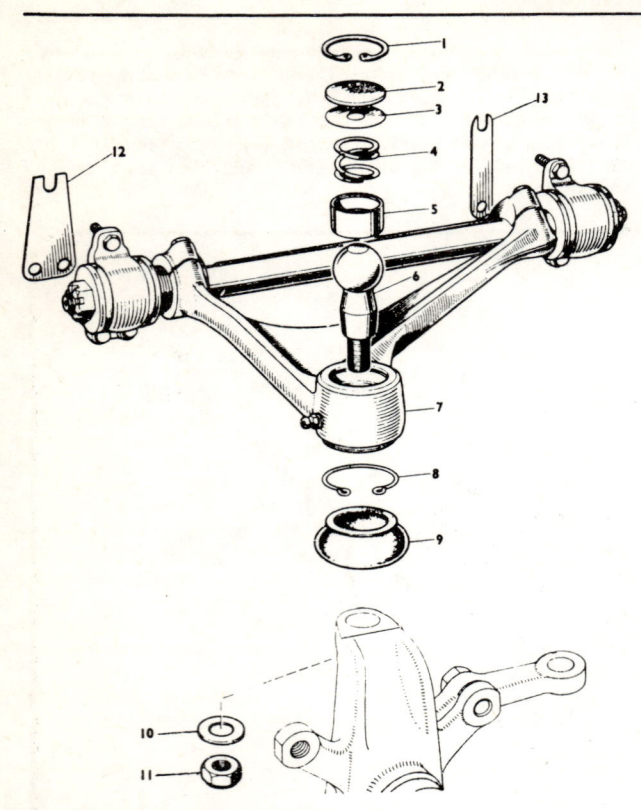

Fig. 11.4. The upper wishbone and ball pin

1 Circlip
2 Top cover
3 Shims
4 Socket spring
5 Ball pin socket
6 Ball pin
7 Upper wishbone
8 Circlip
9 Rubber gaiter
10 Washer
11 Nut
12 Camber shims (front carrier bracket)
13 Camber shims (rear carrier bracket)

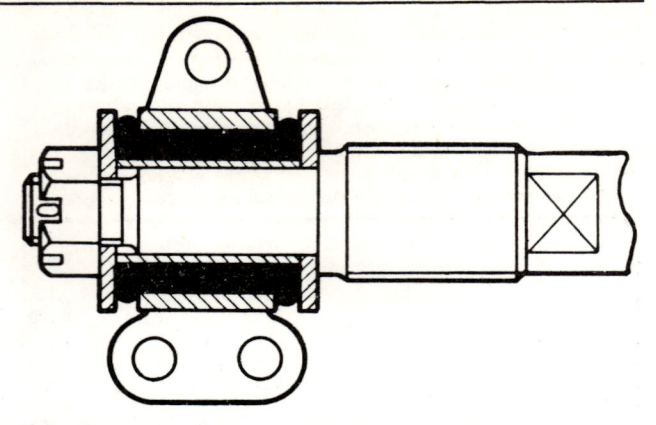

Fig. 11.5. Section through one of the upper wishbone rubber/steel bushed mounting brackets

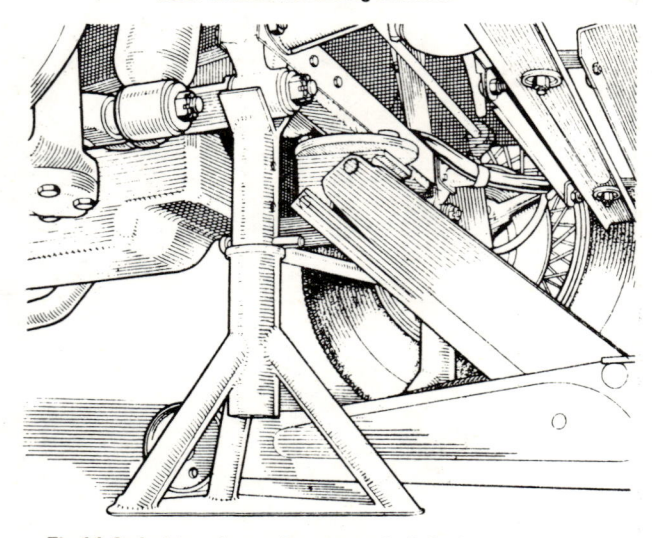

Fig. 11.6. Jacking the car for removal of the lower wishbone.

4 Front suspension lower wishbone - removal and refitting

1 Apply the handbrake and chock the rear wheels.
2 Slacken, but do not remove the hub cap to the front wheel of the side being worked on. The hub caps are marked "RIGHT (OFF) SIDE" and "LEFT (NEAR) SIDE" and the direction of rotation for removal is clockwise for the right hand side and anti-clockwise for the left hand side.
3 If both sides are to be worked on, make up a block of hardwood to fit the cross tube section as shown in Fig.11.6 to spread the load for the jack. If only one side is to be worked on at a time, the jack can be placed at the front end of the lower wishbone fulcrum shaft at the point where the stand is shown in position in the figure. Jacking on the cross tube will mean that you will first have to remove the cable harness from the tube by undoing the clips and clearing the harness out of the way for the hardwood block to be positioned.
4 Raise the car and when you are satisfied that it is safely supported, remove the road wheel(s).
5 Disconnect the hydraulic brake pipe from the frame connection and immediately blank off the pipe and the union to prevent ingress of dirt.
6 Take out, and discard, the split pin from the nut securing the steering tie-rod ball joint. Remove the nut and then separate the ball joint by tapping with a hammer on the side face of the steering arm adjacent to the ball pin.
7 Disconnect the upper wishbone ball joint in the manner described in Section 3 and, if the upper wishbone is being left in position, raise it to its full extent and tie it up to the frame, out of the way.
8 Remove the self locking nut of the lower wishbone ball joint and seperate the joint by tapping on the manner already described.
9 The axle carrier complete with the brake caliper and disc can now be removed.
10 Place a jack under the lower suspension arm and take the weight of the car but do not lift the car off the stands or jack already in position.
11 Remove the self locking nut securing the anti-roll bar to the lower suspension arm.
12 Remove the split pin and nuts securing the hydraulic damper to the frame and the wishbone. Remove the upper mounting bolt and take out the damper.
13 Lower and remove the jack under the lower suspension arm.
14 Remove the two bolts and their lockwashers which hold the torsion bar rear adjuster lever to the frame, and slide the lever forward until it is clear of the torsion bar splines.
15 Remove the locking bolt from the front mounting of the torsion bar and slide the torsion bar rearwards until the front splines are clear of the wishbone, and now withdraw it towards the front of the car.
16 Remove the two bolts and washers securing the rear fulcrum shaft carrier to the chassis frame.
17 Take out the four nuts and bolts holding the front carrier bracket to the chassis frame.
18 Take out, and discard the split pin locking the nuts of the lower wishbone shaft. Remove the nuts and withdraw the

brackets and their rubber bushes (Fig.11.7); If the bushes are to be replaced they can be drifted or pressed out of the brackets.
19 Refitting of the lower wishbone assembly is the reverse of the above sequence but the torsion bar will have to be reset as described in Section 9 and the brake system must be bled as described in Chapter 9. **Do not** tighten the slotted nuts at each end of the fulcrum shaft until the complete front suspension assembly has been fitted and the full weight of the car is on the suspension, failure to take this precaution will result in undue torsional loading of the rubber bushes and may lead to premature failure.

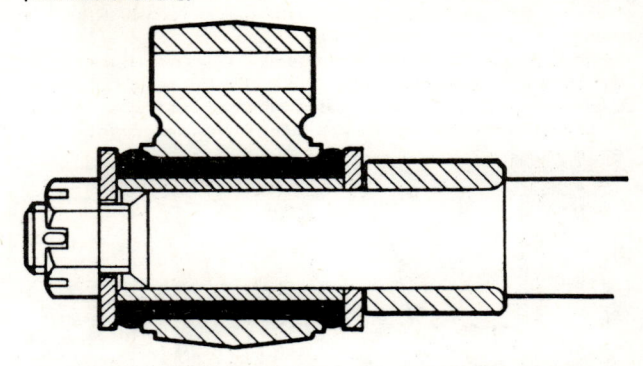

Fig. 11.7. Section through one of the lower wishbone rubber/steel bushed mounting brackets

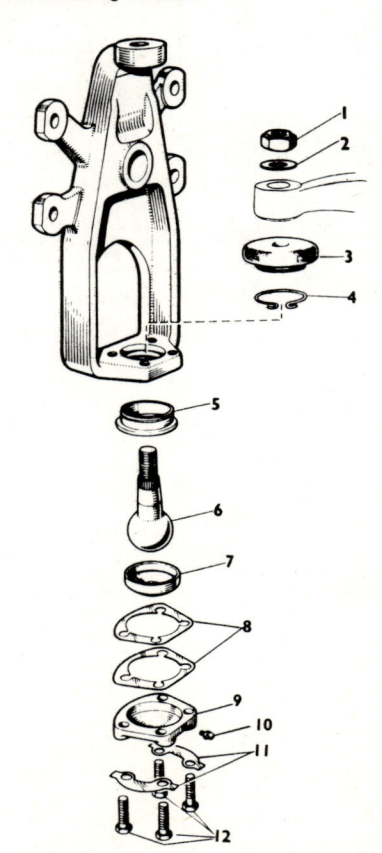

Fig. 11.8. Stub axle carrier and lower wishbone ball joint

1	Nut	7	Socket
2	Washer	8	Shims
3	Rubber gaiter	9	Ball pin cap
4	Circlip	10	Grease nipple
5	Spigot	11	Tab washers
6	Ball pin	12	Setscrews

5 Stub axle carrier and lower wishbone ball joint - removal and refitting

1 Follow the instructions given in Section 4 paragraph 1 to 9 for removal of the stub axle carrier from the car.
2 Remove the stub axle securing nut (Fig.11.8) and separate the front hub assembly from the carrier by tapping the carrier sharply with a ghammer to break the retaining taper.
3 The first stage in removing the ball joint, illustrated in Fig.11.8, is to remove the wire clip (4) and take off the rubber gaiter.
4 Open up the tab washers (11) and remove the four setscrews (12) which hold the ball pin cap to the stub axle carrier.
5 Remove the ball pin cap (9) and take out the shims (8) followed by the ball pin socket (7) and the ball pin (6).
6 Reassembly of the ball joint is a reversal of the dismantling sequence, but the joint has to be adjusted to obtain the correct clearance of 0.004" to 0.006" (0.10 to 0.15 mm). This is done by removing the shims one by one until, with the ball cap fully tightened, the ball pin is tight in its socket. When this stage is reached, fit shims to the value of 0.004" to 0.006" (0.10 to 0.15 mm) and in this condition it should be possible to move the shank of the ball pin by hand. Shims are available in thicknesses of 0.002" (0.05 mm) and 0.004" (0.10 mm).
Note: Shims must not be removed to make up for excessive wear in the ball pin and sockets which, if worn to that extent, should be replaced.
7 Follow the procedure, as applicable, given in Section 4 for refitting the stub axle carrier to the car.

6 Front wheel hubs - removal and refitting

1 Jack up the car and when you are satisfied that it is safely supported, remove the road wheel.
2 Disconnect the flexible hydraulic brake pipe from the frame connection and immediately blank off the pipe and the union to prevent the ingress of dirt.
3 Break the locking wire to the brake caliper mounting bolts and remove the two bolts taking note of the number and position of the shims fitted between the caliper and the mounting plate. Remove the caliper.
4 Refer to Fig.11.9.
5 Remove, and discard, the split pin (2). You will find that holes are provided in the side of the hub to give access for removal of the pin.
6 Unscrew and remove the slotted nut (1) and take off the washer (3) from the end of the stub axle shaft.
7 The wheel hub can now be withdrawn by hand.
8 If the hub is to be dismantled, first remove the oil seal (8). The inner races of the taper roller bearings can now be withdrawn.
9 We advise that the outer races are left in position unless the bearing is faulty in which case the outer race can be drifted out from the hub.
10 The brake disc is secured to the hub by five bolts and self locking nuts on removal of which, the disc can be tapped off the hub.
11 Reassemble the hub by first pressing the outer bearings (4 and 7) into position.
12 Refit the brake disc.
13 Make sure that the assembly is free from grease and then assemble the taper roller bearings followed by the oil seal (8).
14 Offer the assembly to the stub axle, place the plain washer in position and then secure with the slotted nut.
15 The next step is to check and adjust for correct end float of the bearings. The end float should be 0.003" to 0.005" (0.07 to 0.13 mm), it is important that the end float does not exceed 0.005" as this may cause the brakes to drag and not function properly, on the other hand, if end float is too small bearing failure may result.
16 The end float can be measured with a dial test indicator

228   Chapter 11/Suspension and steering

mounted so that the indicator button is against the hub. Tap the hub fully home and take a reading, now move the hub outwards and record the movement. Tighten or slacken the slotted nut as required until the correct amount of end float is recorded.
17 If you do not have a dial test indicator, the end float can be adjusted by tightening the slotted nut until there is no end float that is when the hub feels 'sticky' when rotated. Now slacken the nut between one and two flats depending on the position of the split pin hole relative to the slots in the nut.
18 In each case, fit a new split pin and open out the ends.
19 Using a grease gun, applied to the grease nipple in the hub, pump in grease until it appears past the outer hub bearing which can be observed through the bore of the splined hub.
20 Refit the brake caliper making sure that the correct clearances are maintained between the inner faces of the caliper and the disc as described in Chapter 9.
21 Reconnect the flexible brake pipe and bleed the system as described in Chapter 9.

7 Front hydraulic dampers - removal and refitting

1 The dampers are of the sealed type with no provision for adjustment or topping-up so if faults develop a replacement damper must be fitted.
2 The damper attachment points are shown in Fig.11.10.
3 Jack up the car under the lower wishbone close to the damper lower mounting but make sure that the jack is so positioned as not to interfere with access to the mounting.
4 Raise the car and when you are satisfied that it is safely supported, remove the road wheel.
5 Remove the split pin and nut from the top and bottom mounting bolts.
6 Withdraw the top mounting bolt and then pull the damper clear of the bottmo mounting and remove from the car.
7 If you use a jacking point other than the lower wishbone, support the outer end of the lower wishbone before removing the damper because, if the suspension unit is allowed to drop more than the normal rebound position, the top ball joint may "neck" in its housing and become damaged.

8 With the damper removed from the car, check it for correct action by holding it in its normal vertical position, either in a vice or you may be able to hold the lower mounting on the ground with your foot, and apply an up and down force on the shroud. There should be no lost motion throughout the travel of the shroud, if there is, make several short strokes (not exceeding halfway) on the shroud followed by extending the damper to its full extent two or three times. If this does not cure the trouble, replace the damper.
9 When a replacement damper is brought into use, carry out the operation of bleeding as described in paragraph 8 and then keep the damper in its normal upright position until you fit it to the car.
10 Refitting is the reverse of the removal procedure but do not tighten the slotted mounting nuts until the full weight of the car is on the ground otherwise undue torsional loading will be put on the rubber bushes and early failure of them may result.

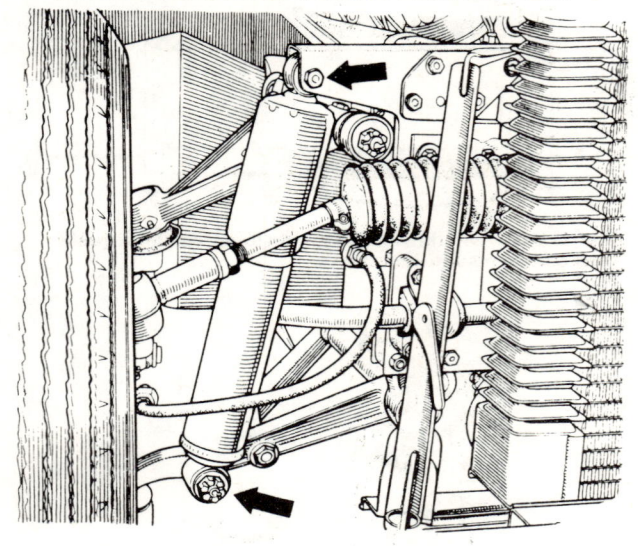

Fig. 11.10. Front hydraulic damper attachment points

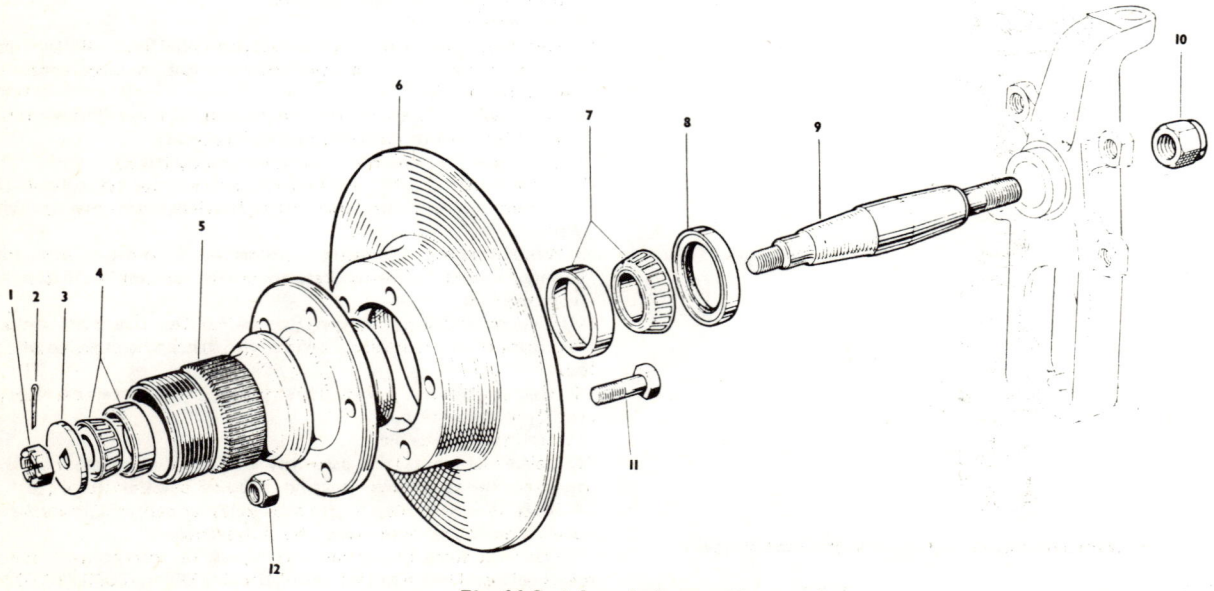

Fig. 11.9. A front hub assembly

1 Nut
2 Split pin
3 "D" washer
4 Outer bearing
5 Wheel hub
6 Brake disc
7 Inner bearing
8 Oil seal
9 Stub axle
10 Stub axle securing nut
11 Brake disc securing bolt
12 Nut

Chapter 11/Suspension and steering

8 Anti-roll bar - removal, renewing the link arm bushes and refitting

1 The anti-roll bar is illustrated in Fig.11.3.
2 Remove the four bolts, nuts and washers from the anti-roll bar support brackets on the chassis member. Take out the two distance pieces.
3 Remove the self locking nuts from the bolts attaching the link arms to the lower wishbone. Remove the anti-roll bar from the car.
4 Separate the link arm from the bar by removing the self locking nut and washer and then withdrawing the bolt.
5 You will find that the rubber bushes (50 in Fig.11.3) are split to enable them to be removed from the anti-roll bar. This means that they can be replaced without completely removing the bar from the car. All that is necessary is to remove the two brackets (51 in Fig.11.3), spring the bar outwards on the link arm, remove the old rubber bushes and fit the new ones.
6 If the link arm bushes are perished or worn they should be replaced by drifting or pressing out from the upper or lower eyes. Now press the new bush into the eye until it protrudes from each end of the eye by an equal amount. You will find that fitting will be made easier if you lubricate the bush with a solution of 12 parts water to 1 part of liquid soap.
7 Refitting is the reverse of the removal sequence but do not tighten any of the attachment bolts of the bar or link arm until the full weight of the car is taken on the suspension otherwise, as in other cases, you will place undue torsional loading on the rubber bushes which, as a result, may fail prematurely.

9 Torsion bar - adjustment

1 The car should be full of petrol, oil and water but if you are low on petrol, add additional weight - allowing approximately 80 lbs (36.0 kg) to compensate for 10 gallons of petrol.
2 Adjust the tyres to the correct inflation pressures.
3 Place the car on a perfectly level surface and, with the wheels in the straight ahead position, roll the car forward three lengths.
4 Now for 3.8 litre cars, refer to Fig.11.11 and check measurement "A" which, if the torsion bar is correctly adjusted, should be 8¾" ± ¼" (22.2 ± 0.64 mm). But for 4.2 "E" Type models, refer to Fig.11.12, obtain dimensions "A" and "B" and then subtract "B" from "A" to obtain dimension "C" which should be, for 4.2 "E" Type FHC and Open Sports cars, 3½" ± ¼" (88.9 ± 6.35 mm) and for 2 + 2 cars 3¾" ± ¼" (95.25 ± 6.35 mm)
5 If a check on both sides of the car indicates that one or both of the torsion bars require adjustment, refer to Fig.11.13 and make up a setting gauge to the general dimensions shown but with hole centre distances as follows:

3.8 litre models	17.13/16" (45.24 cm) as shown
4.2 "E" Type FHC and Open Sports	17.13/16" (45.24 cm)
4.2 "E" Type 2 + 2	18¼" (46.36 cm)

Commencing at the following chassis numbers, larger diameter torsion bars are fitted and hole centre distances for the gauge must be:

4.2 "E" Type FHC
Chassis 1E.35382 (LHD)) 17¾" (45.1 cm)
4.2 "E" Type Open Sports)
Chassis 1E.17532 (LHD))

4.2 "E" Type 2 + 2
Chassis 1E.50875 (RHD)) 18" (45.7 cm)
1E.77407 (LHD))

All air conditioned models are fitted with the larger diameter torsion bar and the hole centre distance for the 4.2 "E" Type FHC and Open Sports with this equipment is 17¾" as quoted above but the dimension for the 4.2 "E" Type 2 + 2 with air conditioning is 18.1/8" (48.87 cm).

6 Jack up the car and place stands under the lower wishbone fulcrum support brackets. **Do not** place the jack or stand under the forward frame tubes.
7 Remove the road wheels.
8 Refer to Section 3 and carry out those operations necessary to disconnect the upper wishbone ball joint from the stub axle carrier. Disconnect the steering tie rod ball joint from the stub axle carrier as described in Section 4 and then disconnect the anti-roll bar as described in Section 8.
9 Place a jack under the lower wishbone adjacent to the damper mountings and raise the wishbone, but do not lift the car off the stands or other supports already in use.
10 Take out the split pins and slacken the nuts holding the lower wishbone rubber mountings.
11 Refer to Section 7 and remove the damper, store it in its normal upright position.
12 Remove the two bolts and nuts which secure the torsion bar rear adjuster lever to the frame.
13 Now fit the appropriate setting gauge you made to the damper mounting points so as to position the lower wishbone. Adjust on the jack to allow the gauge to be fitted and, when the gauge is in position, remove the jack.
14 The two holes in the torsion bar adjuster lever and the corresponding holes in the frame, because the torsion bar requires adjustment, will now be misaligned.
15 Note which way the lever requires to be rotated. Mark the position of the lever on the shaft, remove the lever by sliding it off the splines, turn it in the direction required and locate on fresh splines to bring the holes into alignment with those in the frame. If the holes cannot be aligned by adjustment on the lever, remove the locking bolt and slide the torsion bar out of the front splines. As the torsion bar has 25 splines at the rear and 24 at the front, it can be used as its own vernier.
16 Turn the bar in the direction required and engage fresh splines. If the position of the lever is now correct, refit the rear bolts and nuts, and also the front locking nut, and fully tighten down.
17 Place the jack under the lower wishbone, take the weight and remove the setting gauge.
18 Locate the damper on the lower mounting, raise the jack until the upper retaining bolt will pass through the bracket and the damper eye. Fit the securing nuts but do not tighten at this stage.
19 Refit the top wishbone and steering tie rod ball joints and then refit the anti-roll bar, but not tighten the nuts.
20 Refit the road wheels and lower the car.
21 Tighten the damper securing nuts, the lower wishbone fulcrum shafts nuts and the nuts securing the anti-roll bar. It is best not to lock the nuts with split pins as yet.
22 Move the car at least three lengths and recheck the standing height which should now be correct.
23 If no further adjustment is required, lock the nuts, as applicable with new split pins.

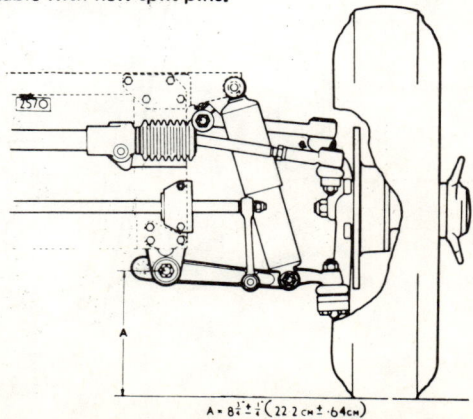

Fig. 11.11. Check of standing height, 3.8 litre models

Fig. 11.12. Check of standing height, 4.2 litre models

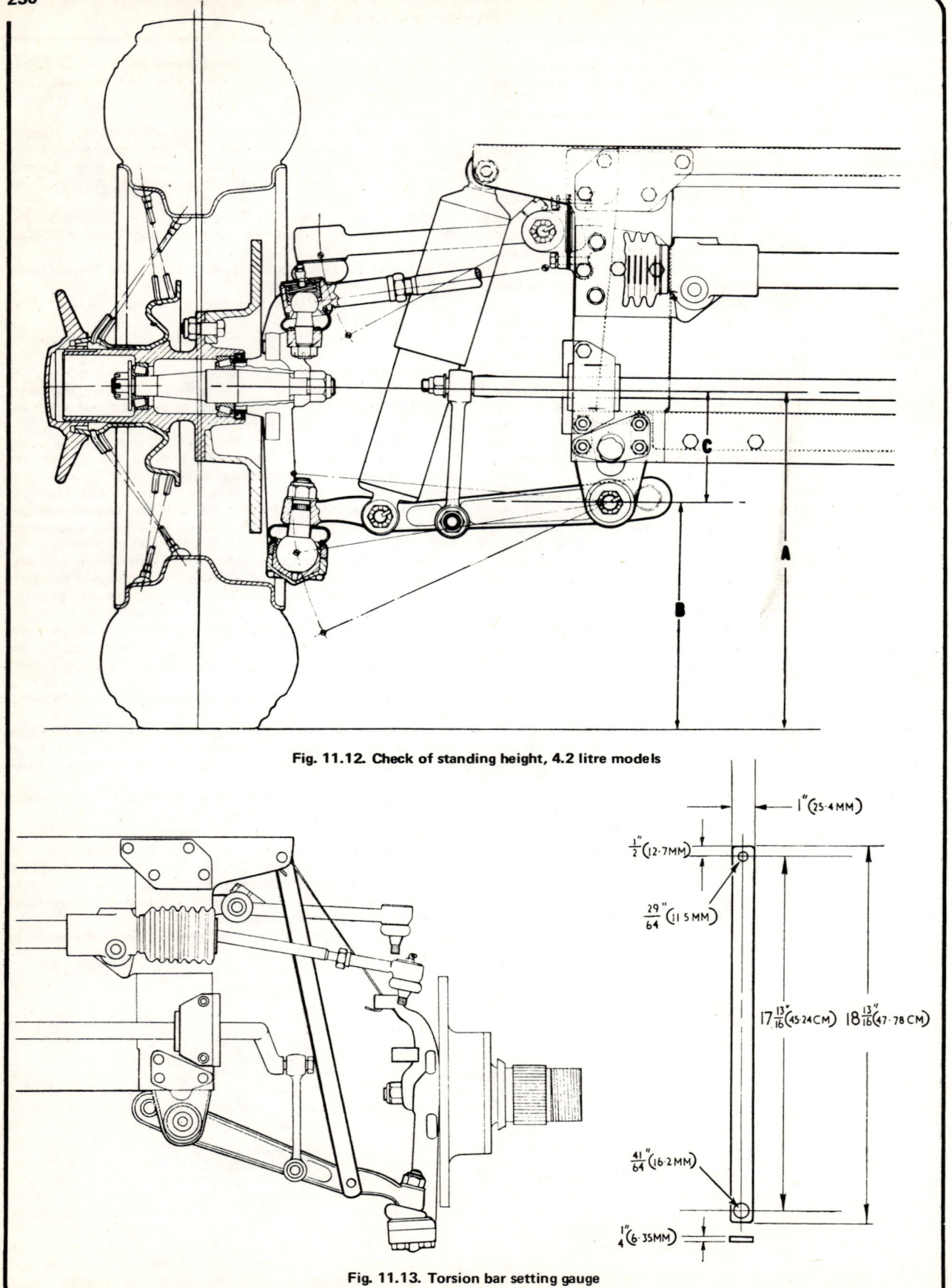

Fig. 11.13. Torsion bar setting gauge

Chapter 11/Suspension and steering

10 Castor angle - adjustment

Check and adjustment of front wheel castor angle is not a difficult task, but it does call for accurate angular measurement, so, if you do not have, and cannot borrow the necessary equipment, it is a job which must be left to your local Jaguar agent.

1 Special links are required to hold the front, and the rear, suspension in the mid-laden position and these can be made up to the dimensions given in Fig.11.14 for the front and Fig.11.35 for the rear suspensions.

2 Fit the front links over the top and bottom hydraulic damper mountings and use the setting links for the rear suspension in the manner described in Section 19.

3 Use the castor angle gauge to check the castor angle which should be $2° \pm ½°$ positive. Check both wheels and the angle for each wheel must not vary more than $½°$.

4 If adjustment is required, remove the split pins and slacken the nuts at each end of the upper wishbone fulcrum shaft. Release the wishbone clamping bolts and turn the fulcrum shaft by means of a spanner on the two flats provided on the shaft as illustrated in Fig.11.15.

5 To increase the positive castor angle, turn the shaft anti-clockwise as viewed from the front of the car and, to decrease the angle, turn the shaft clockwise. Tighten the fulcrum shaft clamp bolts when adjustment is completed.

6 Make sure that the full weight of the car is on the suspension before you tighten the slotted nuts on the fulcrum shaft.

7 If you had to make any adjustment to the castor angle, the alignment of the front wheels should now be checked and reset in the manner described in Section 31.

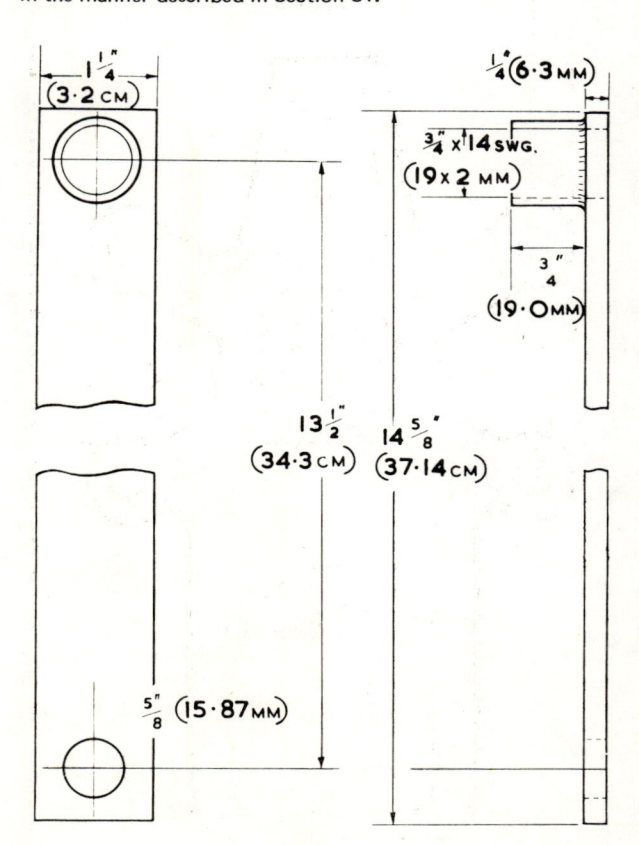

Fig. 11.14. Front suspension setting links

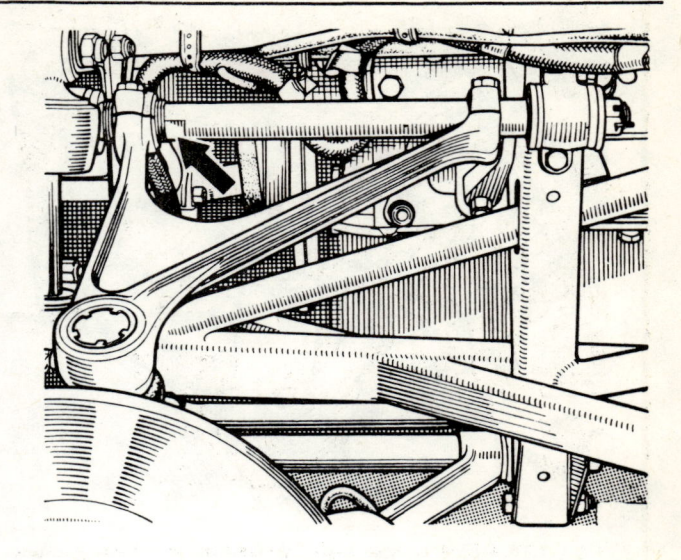

Fig. 11.15. Adjust the castor angle by rotating the shaft arrowed

11 Camber angle - adjustment

This is in the case of checking the castor angle, is a task which also calls for specialised equipment and should also be left to your Jaguar agent if this equipment is not available.

1 Check and adjust the tyres to the correct pressures.

2 Fit the special links (Section 10 paragraph 1) to bring the front and rear of the car to the mid-laden position.

3 Line up the front wheel being checked parallel to the centre line of the car.

4 Check the camber angle in one position and then rotate the wheel through $180°$ and recheck. The camber angle should be $¼° \pm ½°$ positive. Repeat for the opposite wheel, the angle for each wheel must not vary more than $½°$.

5 If adjustments are required slacken the top bolts to the upper wishbone fulcrum brackets (Fig. 11.4) and remove the bottom bolts. Add or subtract shims as required, to an equal amount behind the front and rear brackets. (Fig.11.16) You will find that the shims are slotted at the top and can be slid in behind the top bolts.

6 Inserting shims will increase the positive camber angle whilst removing shims will decrease it. Do make sure that you have an equal number of shims at the front and rear of the brackets otherwise the castor angle will be affected. It will be helpful if you remember that 1/16" (1.6 mm) of shimming will alter the camber by about $¼°$.

7 If any adjustment to the camber angle is made, the alignment of the front wheels must be checked and reset as described in Section 30.

12 Front suspension components - accidental damage

The dimensioned drawings at Figs 11.17, 18,19 and 20 are provided to enable you to assess **dimensional serviceability** of a component following accidental damage to your car. The component concerned should be removed from the car in the manner described in the appropriate section, cleaned and checked dimensionally against the drawing, at the same time look for fractures and other deterioration. **Under no circumstances** should any item found to be dimensionally incorrect be adjusted to bring it within the tolerance limits; discard it and fit a new item. The possibility of a component, although dimensionally correct, having suffered other damage must be appreciated, each case must be treated on its merits but our advice is that if you are at all doubtful as to the fitness of an item-discard it.

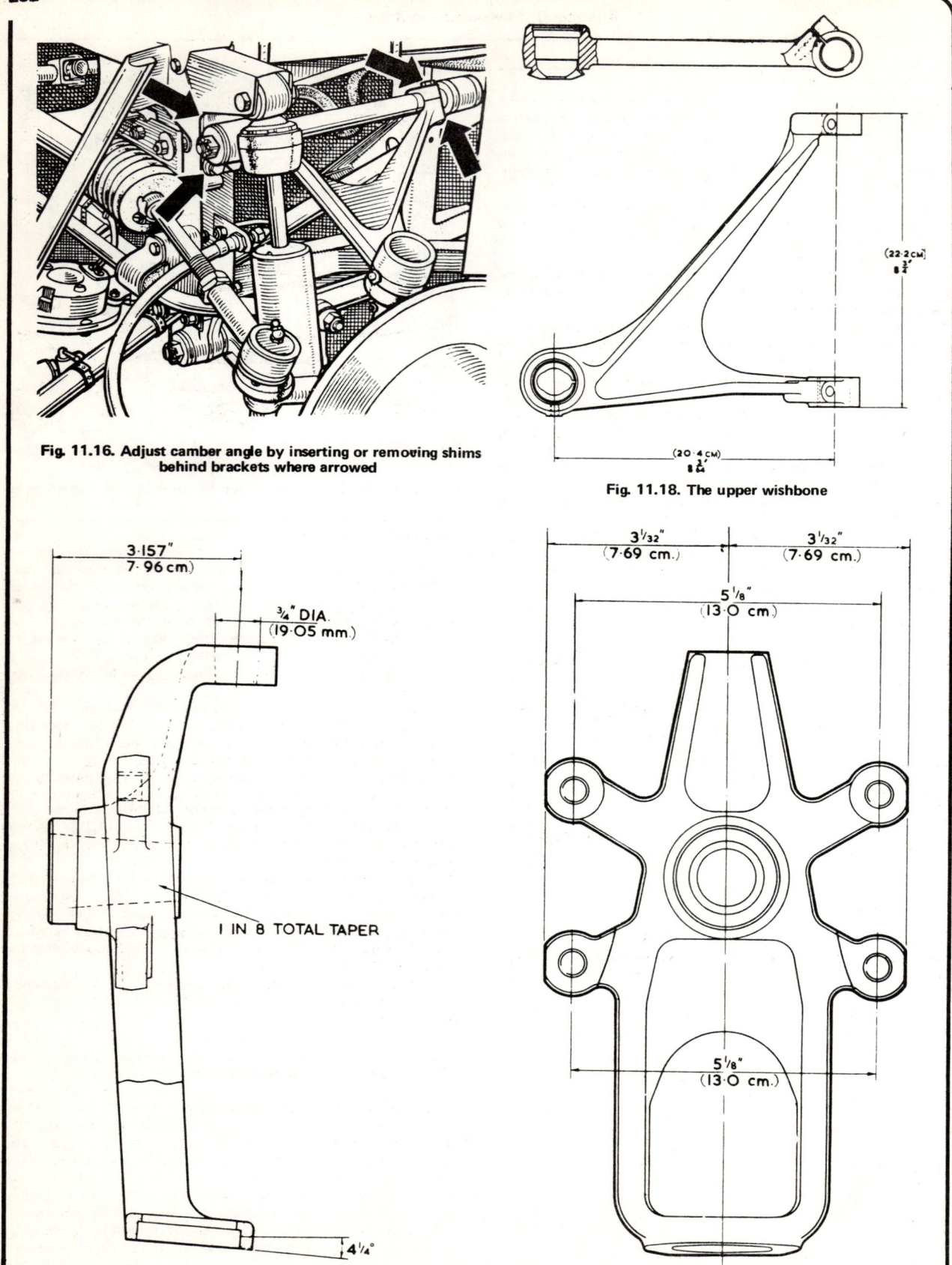

Fig. 11.16. Adjust camber angle by inserting or removing shims behind brackets where arrowed

Fig. 11.18. The upper wishbone

Fig. 11.17. The stub axle carrier

Chapter 11/Suspension and steering

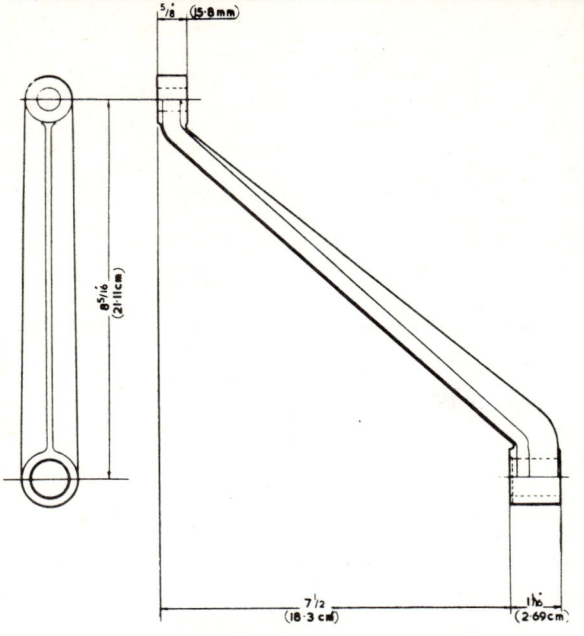

Fig. 11.19. The lower wishbone lever - rear

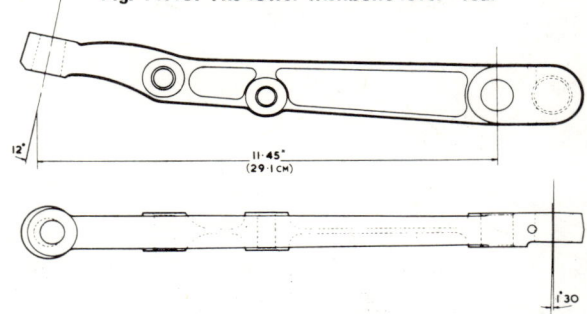

Fig. 11.20. The lower wishbone lever - front

13 Rear suspension - removal and refitting

1 The first job is to position the car over a pit or on a ramp to give access for removal of the exhaust tail pipes.
2 The tail pipes can be removed by slackening the clamps at the silencers and then removing the nuts, bolts and washers which secure the pipes to the centre mounting under the rear of the body. In practice you will probably find that the pipes are so rusted and burnt into position as to be immovable without damaging them or the silencers. In this case, remove the exhaust system as a complete unit: first disconnect the downpipes at the exhaust manifold and then disconnect the tailpipes from the centre mounting underneath the rear body. Now, remove the nuts from the four bolts holding the silencers to the brackets, have an assistant to take the weight of the exhaust system and then lever the rubber mounting brackets, with the bolts, clear of the silencers. The exhaust system can now be removed from the car. Collect the bolts from the brackets.
3 An exploded view of the rear suspension assembly is given in Fig.11.21 to which reference should be made during the following operations.
4 Detach the radius arms at the front end.
5 Place a stout piece of wood approximately 9" x 9" x 1" (22.8 cm x 22.8 cm x 25.4 mm) beneath the rear suspension tie plate and hold in position with a jack.
6 Raise the jack to a sufficient height to allow two chassis stands (of equal heights) to be placed under the body forward of the radius arm mounting posts. Place blocks of wood between the chassis stands and the body to avoid damage. Leave the jack in position.
7 Remove the rear road wheels.
8 Remove the two self locking nuts and bolts which secure the anti-roll bar links to the roll bar.
9 Disconnect the flexible hydraulic brake pipe at the connection on the body. Immediately blank off the pipe and the union to prevent the ingress of dirt.
10 Make sure that the handbrake is off, and then remove the split pin, washer and clevis pin securing the handbrake cable to the actuating levers on the suspension cross beam.
11 Slacken the locknut and screw the handbrake outer cable out of the adjuster block.
12 Remove the four bolts and self locking nuts holding the mounting rubbers at the front of the cross beam to the body frame. Take careful note of the number, and location, of the packing shims between the mounting rubbers and the bodyframe.
13 Remove the six self locking nuts and the four bolts securing the rear mounting rubbers to the cross beam.
14 Remove the four self locking nuts and bolts securing the propeller shaft to the pinion flange. Push the shaft forward clear of the pinion flange and tie it up to avoid strain being put on the front universal joint needle rollers.
15 Lower the rear suspension unit on the jack and withdraw it from under the car as shown in Fig.11.22.
16 Before refitting the rear suspension unit, check all mounting rubbers for deterioration and replace if necessary, ensure that the rubbers are refitted so that the cut-away flange is towards the suspension unit as illustrated in Fig.11.23.
17 Refitting is the reverse of the removal procedure but do not tighten the radius arm securing nuts, if they were removed, until the full weight of the car is on the suspension.
18 Finally bleed the braking system and adjust the handbrake as described in Chapter 9.

14 Road spring and hydraulic damper assembly - removal and refitting

1 The road spring and hydraulic damper assembly can be removed and replaced with the rear suspension assembly in position. The damper mounting points are shown in Fig.11.24.
2 Remove the self locking nuts and washer securing the dampers to the wishbone.
3 Support the wishbone on a jack to relieve the load on the dampers and then drift out the mounting shaft as illustrated in Fig.11.25 which, for the purpose of clarity, shows the suspension unit on the bench.
4 Remove the self locking nut and the bolt securing the dampers to the cross beam. Withdraw the damper and road spring assembly.
5 Refitting is the reverse of the above sequence.

15 Road spring and hydraulic damper assembly - dismantling and reassembly

1 The telescopic hydraulic dampers are of the sealed type with no provision for adjustment or topping-up, therefore, a replacement must be fitted if a damper is found to be unserviceable. In this event it is advisable to replace both dampers in order to retain balanced operation.
2 Remove the road spring and hydraulic damper as described in Section 14.
3 Using a suitable press, compress the road spring until the split collet can be removed from under the spring retaining pad. Fig.11.26 shows this being done but in this case Tool No J11 with adaptor SL 14 is being used to compress the spring.
4 Carefully release pressure on the spring and withdraw the damper.
5 On early models an aluminium pad was fitted to each end of the spring; the pad fitted to the shrouded end of the damper being recessed to receive the shroud.

6 Check the hydraulic damper by holding it in an upright position with the shroud uppermost and make several short strokes if there is any lost motion which cannot be eradicated by movement of the shroud, the damper is unserviceable.

7 Before a new damper is brought into use, hold it in its normal upright position with the shroud uppermost and then make several short strokes, not extending more than halfway, until there is no lost motion. Finish by extending the damper to its full length once or twice. After this bleeding operation, store the damper in its normal upright position until it is required for fitment to the car.

8 The first step in reassembly is to compress the road spring sufficiently for the hydraulic damper to be passed through the spring and for the spring pad and split collet to be placed in position. Ensure that the pad and split collet are seating correctly and then slowly release pressure on the spring. However, on those early models fitted with aluminium pads, first fit the recessed pad to the shrouded end of the damper, compress the spring, pass the damper through the spring and fit the other pad and split collet. Make sure that the collet is seated correctly and release pressure on the spring.

9 Refit the hydraulic damper and road spring assembly to the car in the reverse sequence to the removal procedure.

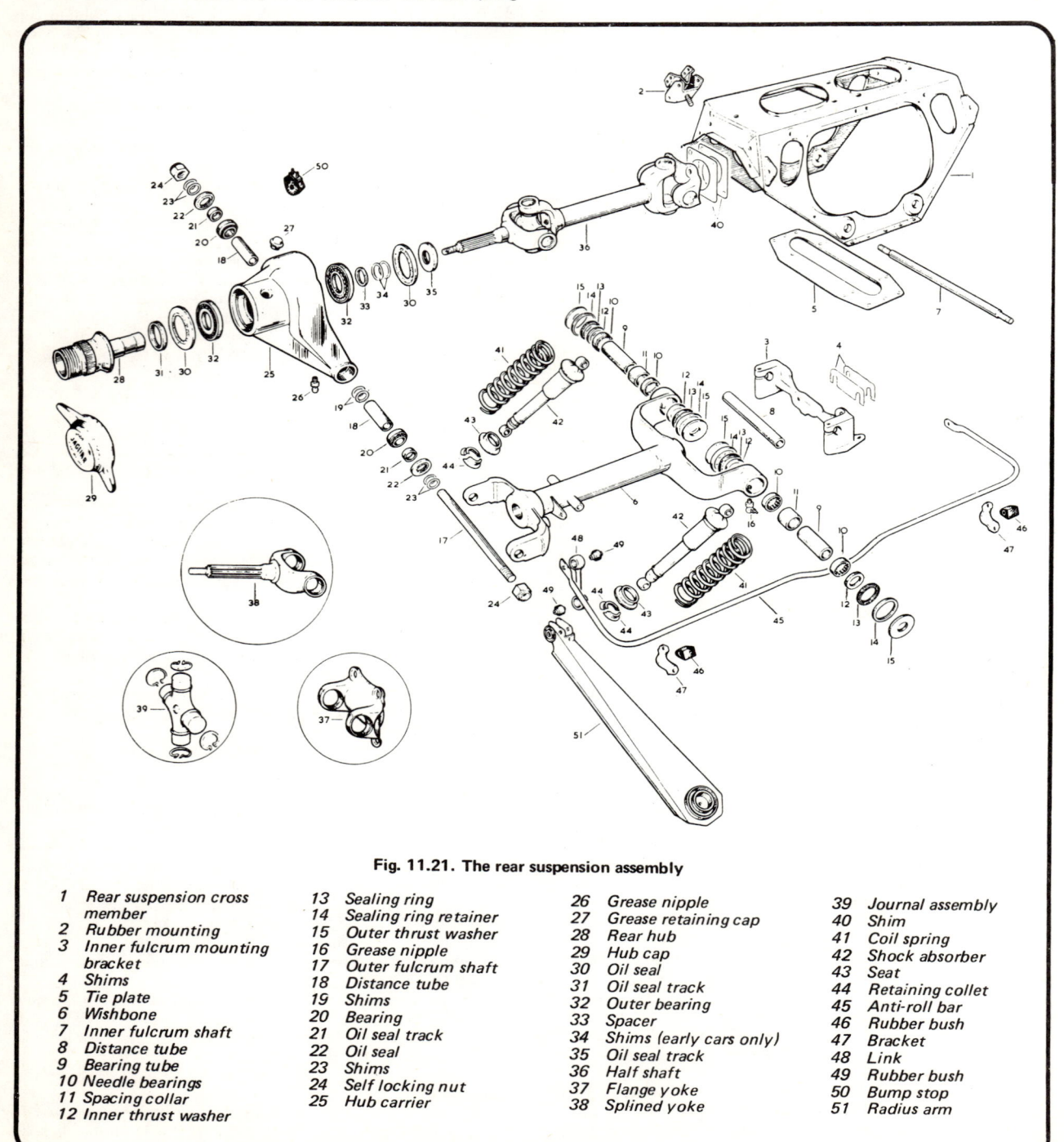

Fig. 11.21. The rear suspension assembly

1 Rear suspension cross member
2 Rubber mounting
3 Inner fulcrum mounting bracket
4 Shims
5 Tie plate
6 Wishbone
7 Inner fulcrum shaft
8 Distance tube
9 Bearing tube
10 Needle bearings
11 Spacing collar
12 Inner thrust washer
13 Sealing ring
14 Sealing ring retainer
15 Outer thrust washer
16 Grease nipple
17 Outer fulcrum shaft
18 Distance tube
19 Shims
20 Bearing
21 Oil seal track
22 Oil seal
23 Shims
24 Self locking nut
25 Hub carrier
26 Grease nipple
27 Grease retaining cap
28 Rear hub
29 Hub cap
30 Oil seal
31 Oil seal track
32 Outer bearing
33 Spacer
34 Shims (early cars only)
35 Oil seal track
36 Half shaft
37 Flange yoke
38 Splined yoke
39 Journal assembly
40 Shim
41 Coil spring
42 Shock absorber
43 Seat
44 Retaining collet
45 Anti-roll bar
46 Rubber bush
47 Bracket
48 Link
49 Rubber bush
50 Bump stop
51 Radius arm

235

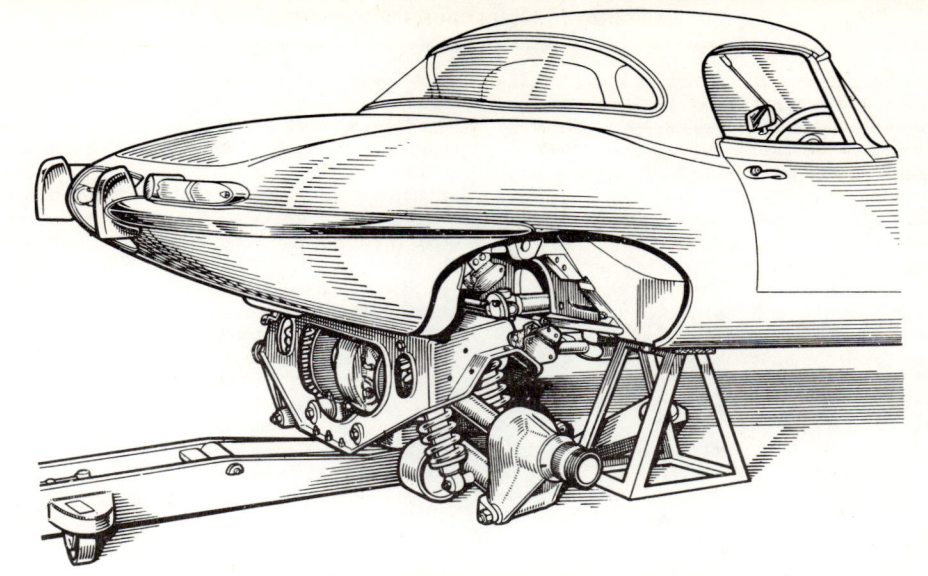

Fig. 11.22. Removing the rear suspension assembly from the car

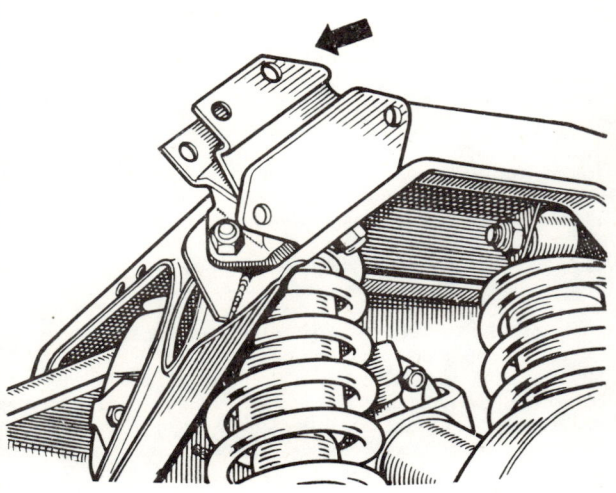

Fig. 11.23. Showing the correct position of the mounting rubber

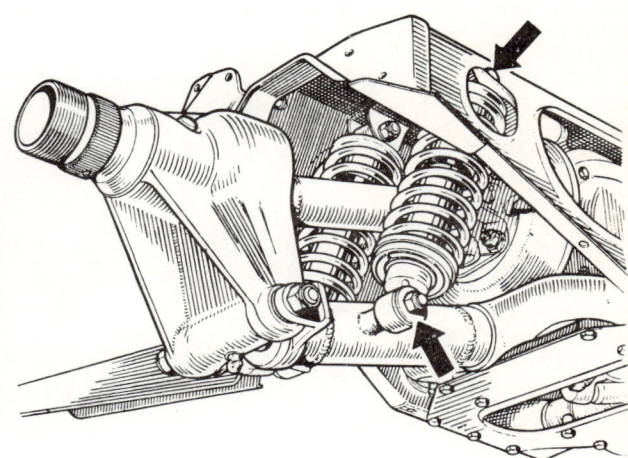

Fig. 11.24. Hydraulic damper mounting points

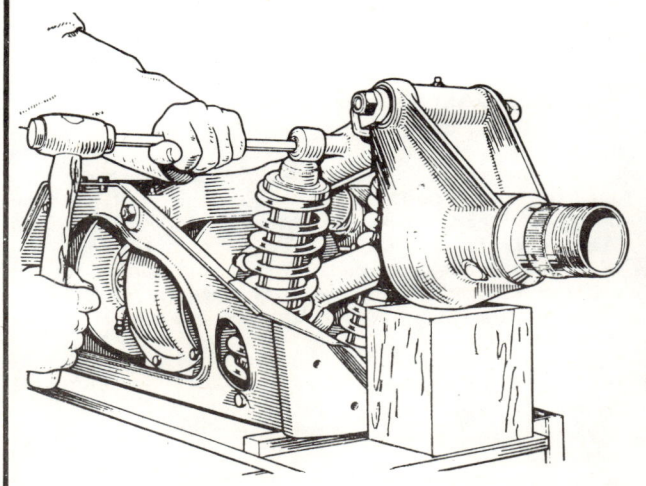

Fig. 11.25. Drifting out the damper mounting shaft

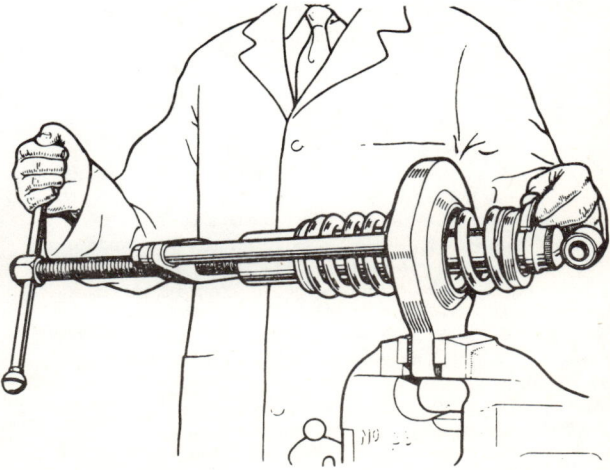

Fig. 11.26. Removing a rear road spring

16 Radius arm - removal and refitting

1 Raise the car and remove the appropriate rear road wheel.
2 Unscrew the two self locking nuts securing the safety strap to the body floor.
3 Remove the radius arm securing bolt and spring washer and then remove the safety strap.
4 Withdraw the radius arm from the mounting post on the body.
5 Remove the self locking nut and bolt securing the anti-roll bar to the radius arm.
6 Place a jack under the wishbone to take the load of the coil springs.
7 Remove one of the self locking nuts securing the hub bearing assembly fulcrum shaft to the wishbone.
8 Drift out the fulcrum shaft but be careful to hold the drift central on the shaft so that the thread is not damaged.
9 Position the wishbone so that the self locking nut and bolt securing the radius arm can be removed. Remove the radius arm.
10 Examine the radius arm mounting rubbers for deterioration, if they show signs of wear or perishing they can be pressed out. When fitting new rubbers, they should be pressed into position until there is an equal amount showing on each side of the radius arm. When replacing the large radius arm body mounting rubber, the two holes should be in the longitudinal position in the radius arm as shown in Fig.11.27.
11 Refitment of the radius arm is the reverse of the removal sequence but follow the instructions given in Section 17 when refitting the fulcrum shaft.

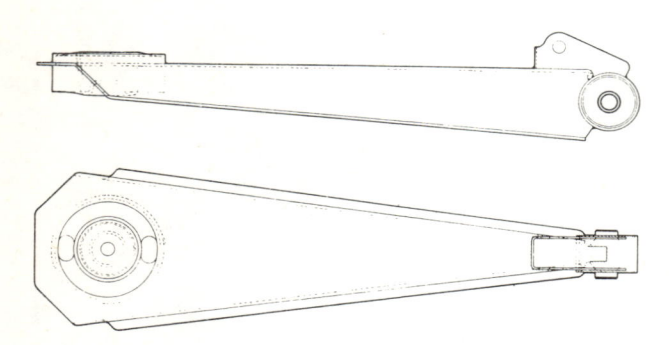

Fig. 11.27. Showing the position of the mounting rubbers in the radius arm

17 Wishbone - removal and refitting

1 Follow the instructions given in Section 13 for removal of the rear suspension assembly.
2 Remove the road spring and hydraulic damper assembly from the appropriate wishbone, this work is described in Section 14.
3 Invert the suspension and remove the six self locking nuts and bolts securing the tie plate to the cross beam, these are shown in Fig.11.28.
4 Remove the eight self locking nuts and bolts, Fig.11.29, securing the tie plate to the inner fulcrum wishbone mounting brackets. Remove the tie plate.
5 Remove one of the self locking nuts securing the hub bearing assembly fulcrum shaft to the wishbone and drift out the fulcrum shaft as illustrated in Fig.11.30, ensure that the drift is ket central on the shaft to avoid damage to the thread.
6 Seperate the hub carrier from the wishbone but look for shims which may be fitted between the wishbone and the hub assembly. Take note of the number and position of any shims as it is essential that the exact amount are replaced in the correct location.
7 Take precautions to prevent the hub carrier oil seal tracks becoming displaced. A piece of sticky tape is ideal to hold them in position.
8 Remove the self locking nut from the bolt securing the radius arm to the wishbone. Withdraw the bolt and remove the radius arm from the wishbone. Note that the bolt is a special type with a thin head.
9 Remove the self locking nut securing the wishbone inner fulcrum shaft to the cross beam mounting bracket.
10 Drift the inner fulcrum shaft out of the wishbone and the mounting bracket taking care, as with the outer fulcrum shaft, not to damage the thread. Take very careful note of the number and position of the various distance and bearing tubes, the inner and outer thrust washers, retainers and sealing rings which make up the fulcrum shaft assembly. These are illustrated, in part, in Fig.11.31.
11 Withdraw the wishbone assembly.
12 There is no need to remove the needle rollers unless these are to be replaced in which case they can be tapped out of the wishbone with a suitable drift.
13 Start reassembly by fitting the needle rollers (if they were removed) to the larger fork of the wishbone lever. Press one outer cage into position with the engraving on the cage facing outwards. Insert the spacing tube and then press in the other roller cage. Repeat for the other fork.
14 Insert the bearing tubes.
15 Smear the four outer thrust washers, the inner thrust washers, the oil seals and oil seal retainers with grease and place them in position on the wishbone (see Fig.11.31).
16 Offer up the wishbone to the mounting bracket making sure that the radius arm mounting bracket is facing forward.
17 Align the holes and all the washers and spacers ready to accept the fulcrum shaft. Because of the multiplicity of components you will find it difficult to keep them in line and so avoid damaging them when fitting the fulcrum shaft, therefore, it is usual to use a dummy shaft in the first place, indeed there is a special tool No J14 available for this purpose. When all items are aligned, tap the dummy shaft into position.
18 Grease the fulcrum shaft and then gently tap it through the cross beam, the wishbone and inner fulcrum mounting bracket as illustrated in Fig.11.32. Keep the dummy shaft in contact with the fulcrum shaft as it is being tapped home by excerting a slight pressure on the end of the dummy shaft, this will reduce the tendency for it to be knocked out of position and allow a spacer or thrust washer to be displaced.
19 When the fulcrum shaft is in position, tighten the two self locking nuts to a torque of 55 lbs f ft (7.60 kg f m).
20 Refit the tie plate in the reverse manner to the removal procedure.
21 Refit the radius arm to the wishbone using the special thin headed bolt.
22 Now remove the tape (or whatever was used) holding the oil seal tracks in position and offer up the wishbone to the hub assembly.
23 Place the sleeves, oil seals, spacers, retaining washers and shims (if fitted) in position, line up all the components and then fit a dummy shaft followed by the fulcrum shaft as described in paragraph 17 and 18.
24 Now check the clearance between the hub carrier and the wishbone fork using feeler gauges as shown in Fig.11.33. If necessary, fit shims between the hub carrier and the wishbone to centralise the carrier.
25 Replace the nuts on the fulcrum shaft and tighten to a torque of 55 lbs f ft (7.60 kg f m).
26 Refit the hydraulic dampers (Section 14).
27 Refit the rear suspension assembly to the car as described in Section 13.
28 Check, and adjust as necessary, the rear suspension camber angle as described in Section 19.
29 Finally, lubricate the fulcrum shafts in the manner described in "Routine Maintenance" at the beginning of this Manual.

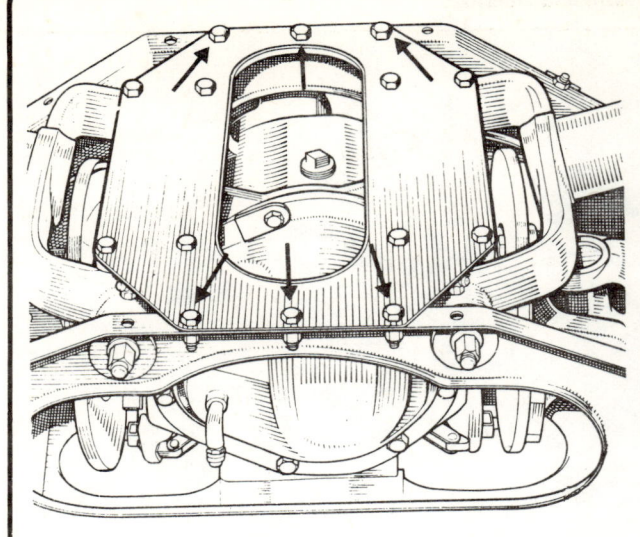

Fig. 11.28. Showing the six bolts which secure the tie plate to the cross beam

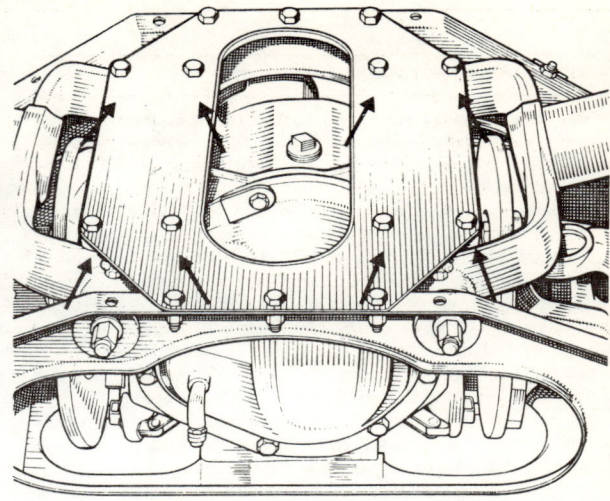

Fig. 11.29. Showing the eight bolts which secure the tie plate to the inner fulcrum mounting bracket

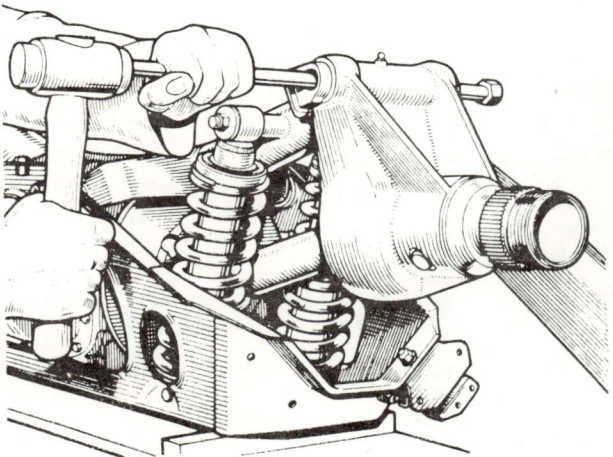

Fig. 11.30. Removing the outer fulcrum shaft

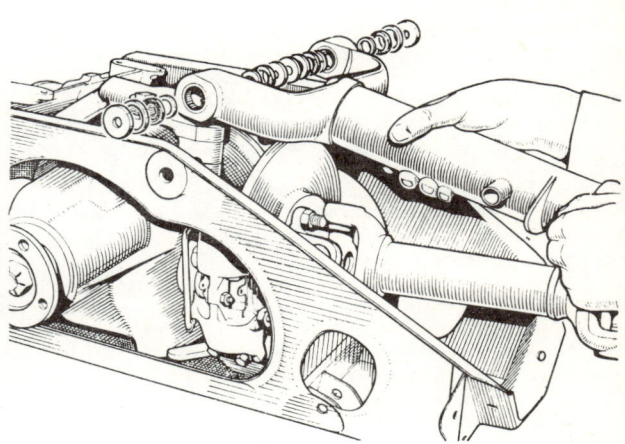

Fig. 11.31. The wishbone inner fork and components

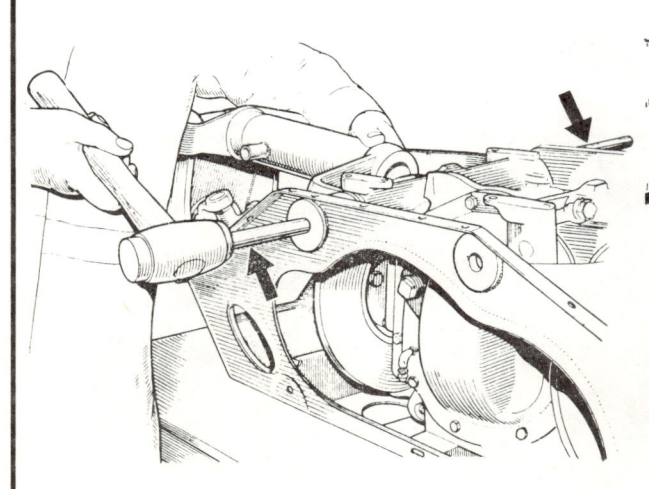

Fig. 11.32. Fitting the inner fulcrum shaft, dummy shaft arrowed

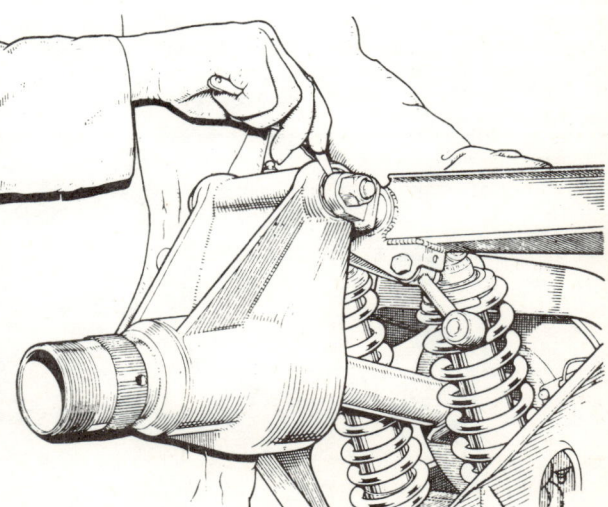

Fig. 11.33. Checking clearance between the hub carrier oil seal tracks and the wishbone fork

Chapter 11/Suspension and steering

18 Wishbone outer pivot bearing - adjustment

1 When new bearings are fitted to the outer wishbone pivot they will have to be adjusted by fitting shims between the two fulcrum shaft spacer tubes but this work may also become necessary in service.
2 The correct bearing preload is 0.000" to 0.002" (0.00 to 0.05 mm) and shims, to bring about this condition, are available in sizes of 0.004" (0.101 mm) and 0.007" (0.17 mm) thick and 1.125" (28.67 mm) in diameter.
3 The first step is to make up a holder consisting of a piece of steel plate about 7" x 4" x 3/8" (17.7 cm x 10.1 cm x 9.5 mm) in size drilled and tapped to receive the outer fulcrum shaft.
4 Next follow the instructions in Section 17 and remove the fulcrum shaft mounting the wishbone to the hub carrier.
5 Take out the split pin and remove the nut holding the hub carrier to the half shaft assembly, remove the hub carrier and hub and collect the inner and outer oil seals and seating rings.
6 Now place the steel plate in a vice and screw the fulcrum shaft into the plate. Slide an oil seal track onto the shaft.
7 Place the hub carrier assembly into position on the shaft minus the oil seals and with an excess of shims, of a known quantity, between the spacers.
8 Fit an inner wishbone fork outer thrust washer on the fulcrum shaft so that it abuts the oil seal track and then fill the remaining space on the shaft with washers and secure with the nut.
9 Tighten the nut to a torque of 5 lb f ft (7.60 kg f m).
10 Press the hub carrier towards the steel plate and at the same time apply a twisting motion to settle the rollers on the bearing surface.
11 Maintain a steady pressure against the hub carrier and, using a feeler gauge as illustrated in Fig.11.34, measure the amount of clearance between the large diameter washer and the machined face of the hub carrier.
12 Now pull the hub carrier towards the large diameter washer and again apply a twisting motion to settle the bearings. Check the clearance between the large diameter washer and the machined surface of the hub carrier in this position.
13 Subtract the second measurement from the first and this will give the amount of end float present in the bearings. Now remove sufficient shims to give a preload of 0.000" to 0.002" (0.00 to 0.05 mm).
14 For example, the **mean** preload is 0.001". The end float you measured is, 0.010" (0.25 mm), therefore, 0.010" + 0.001" = 0.011" (0.25 mm + 0.02 mm = 0.27 mm) to be removed to give the correct preload.
15 Refit the hub carrier as described in Chapter 8 and reconnect the outer wishbone as described in Section 17 but do not forget to measure, and shim as necessary, to centralise the hub carrier in the wishbone so as to prevent the wishbone forkends from closing inwards.

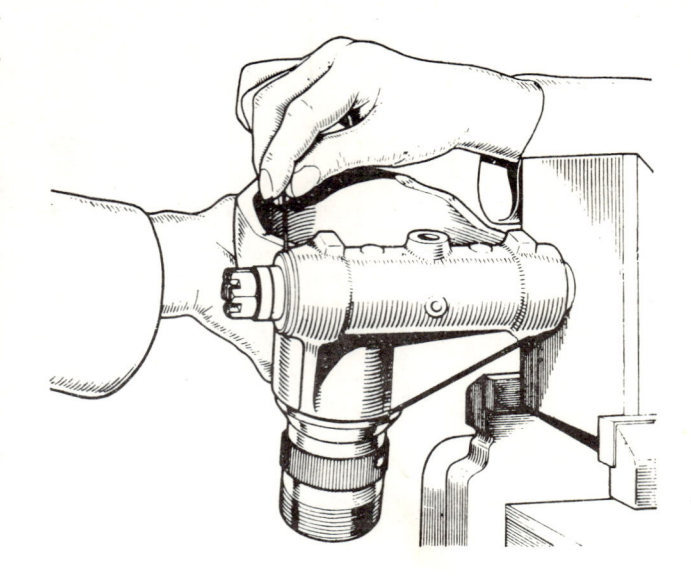

Fig. 11.34. Checking bearing end float

19 Rear wheel camber angle - adjustment

Check and adjustment of the rear wheel camber angle is not a different task but, as in the case of the front wheels, it is one which calls for specialised equipment and should, therefore be left to your local Jaguar agent if you do not have, and cannot borrow, the required gauge.
1 Make up two setting links to the dimensions shown in Fig.11.35. These items are required to lock the suspension in the mid-laden position.
2 Bring all tyres to the correct inflation pressures and then position the car on a flat surface.
3 Now fit the setting links at each side of the car as shown in Fig.11.35 by hooking one end in the lower hole of the rear mounting; then, have an assistant to depress the body of the car until the other end can be slid over the hub carrier fulcrum nut.
4 Remove the self locking nut securing the forward road spring and damper to the wishbone mounting pin at each side of the car. Drift out each pin so that the road spring and damper assembly is free at the wishbone.
5 Now check the camber of each rear wheel as illustrated in Fig.11.36 - the correct camber is -¾° ± ¼°. If the reading is incorrect you will have to add or subtract shims between the half

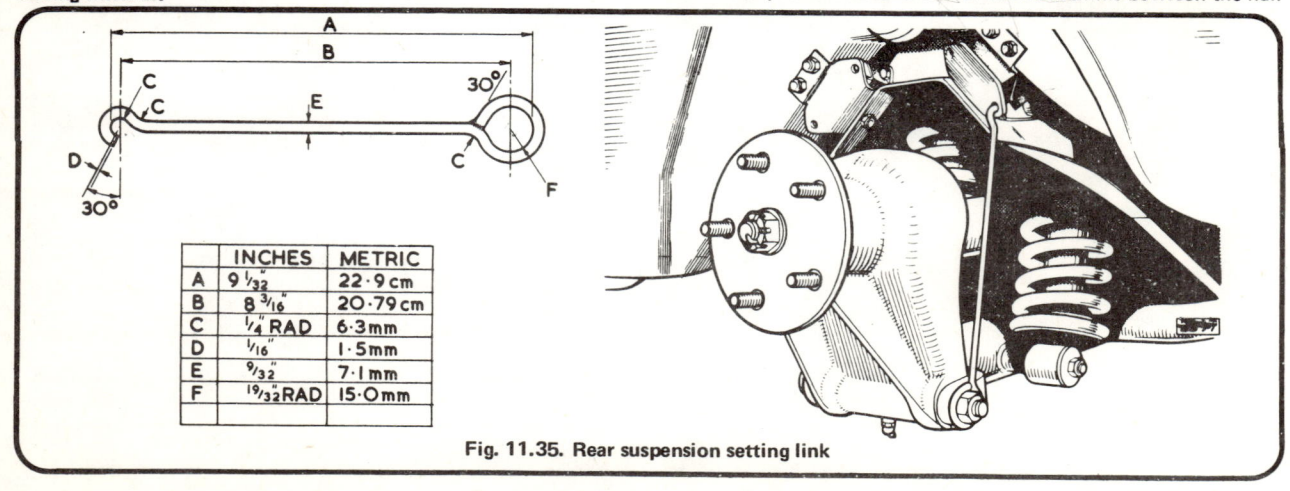

	INCHES	METRIC
A	9 1/32"	22·9 cm
B	8 3/16"	20·79 cm
C	1/4" RAD	6·3 mm
D	1/16"	1·5 mm
E	9/32"	7·1 mm
F	19/32" RAD	15·0 mm

Fig. 11.35. Rear suspension setting link

Chapter 11/Suspension and steering

shaft and the brake disc as shown in Fig.11.37. Shims are available in a thickness of 0.020" (0.5 mm) and one of these shims will alter the camber angle by approximately ¼°.
6 To fit the shims, jack up the car and remove the appropriate road wheel.
7 Remove the self locking nut and bolt securing the top of the road spring and damper assembly to the cross beam and remove the assembly.
8 Unscrew the four self locking nuts securing the half shaft to the brake disc and pull the half shaft away from the disc sufficiently to clear the mounting studs.
9 Remove or add shims as necessary.
10 Refit items in the reverse order to the above, lower the car and recheck the camber angle.
11 If the angle is now correct, reconnect the road spring and damper assembly to the wishbone and finally remove the setting links.

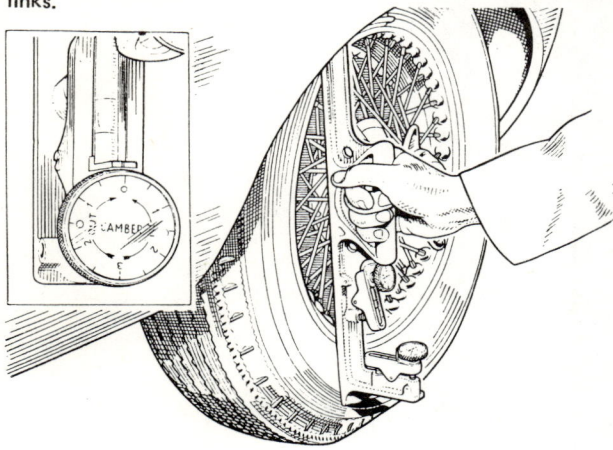

Fig. 11.36. Checking the camber angle of a rear wheel

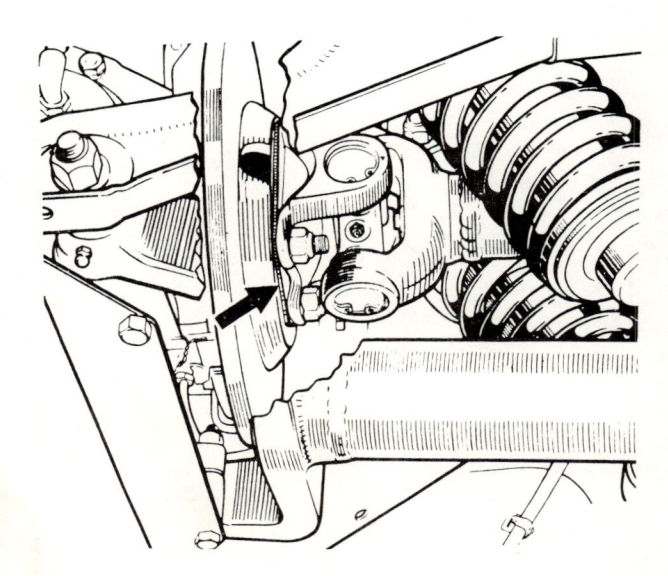

Fig. 11.37. Add or subtract shims where arrowed to adjust the camber angle of the rear wheels

20 Steering wheel - removal and refitting

1 Disconnect the battery.
2 Remove the three grub screws in the steering wheel hub and withdraw the horn push assembly.
3 Slacken the steering column adjuster nut.
4 Slacken and remove the locknut followed by the hexagon nut securing the steering wheel to the inner column shaft. Extract the flat washer located under the hexagon nut.
5 Note the position of the steering wheel in relation to the road wheels, excert a sudden pressure behind the steering wheel and withdraw it from the splines of the inner column.
6 Collect the two halves of the split cone which will now be uncovered around the inner column.
7 When refitting the steering wheel, first fully extend the sliding portion of the inner column and lock it in that position by screwing up on the steering column adjuster nut.
8 Fit the two halves of the split cone but make sure that the narrowest part of the cone is towards the top of the column. You will find it helpful if a small quantity of grease is put in the groove in the inner column to retain the cone when fitting the steering wheel.
9 Slide the steering wheel onto the splines - positioning it in relation to the front wheels as before removal. Make sure that the wheel is correctly aligned with the splines and keep it square when pushing it down the splines. Do not force it: if it is tight, remove it, investigate and rectify the cause.
10 After pushing the steering wheel fully home on the split cone, fit the flat washer and the hexagon nut and fully tighten down.
11 Fit the locknut and secure oy using a ring or a box spanner but do not overtighten the nut.
12 Refit the horn push with the Jaguar motif correctly positioned.
13 Reconnect the battery and check the horn for correct operation.

21 Upper steering column (conventional type) - removal and refitting

1 Disconnect the battery.
2 Identify and then remove the wires leading from the flasher switch at the multi connector located behind the side facia panel.
3 Disconnect the horn push cable from the connector at the lower end of the upper column tube.
4 Remove the nut and spring washer from the pinch bolt retaining the upper column to the lower steering column universal joint. Remove the bolt.
5 Remove the two nuts, bolts and washers securing the column to the lower support bracket.
6 Remove the nut, bolt, washer and spacer tube securing the column to the upper bracket which is located behind the side facia panel.
7 Take note of the position of the steering column in relation to the front road wheels and then withdraw the upper steering column from the lower column universal joint and remove from the car.
8 Refitting is the reverse of the removal sequence but before engaging the splines in the universal joint of the lower column, make sure that the steering wheel is in the same relationship to the front road wheels as before removal.

22 Upper steering column (conventional type) - dismantling and reassembly

1 A sectioned view of the upper steering column and an exploded view of the complete steering column assembly will be found at Figs.11.38 and 11.39 respectively.
2 Follow the instructions given in Section 21 for the removal

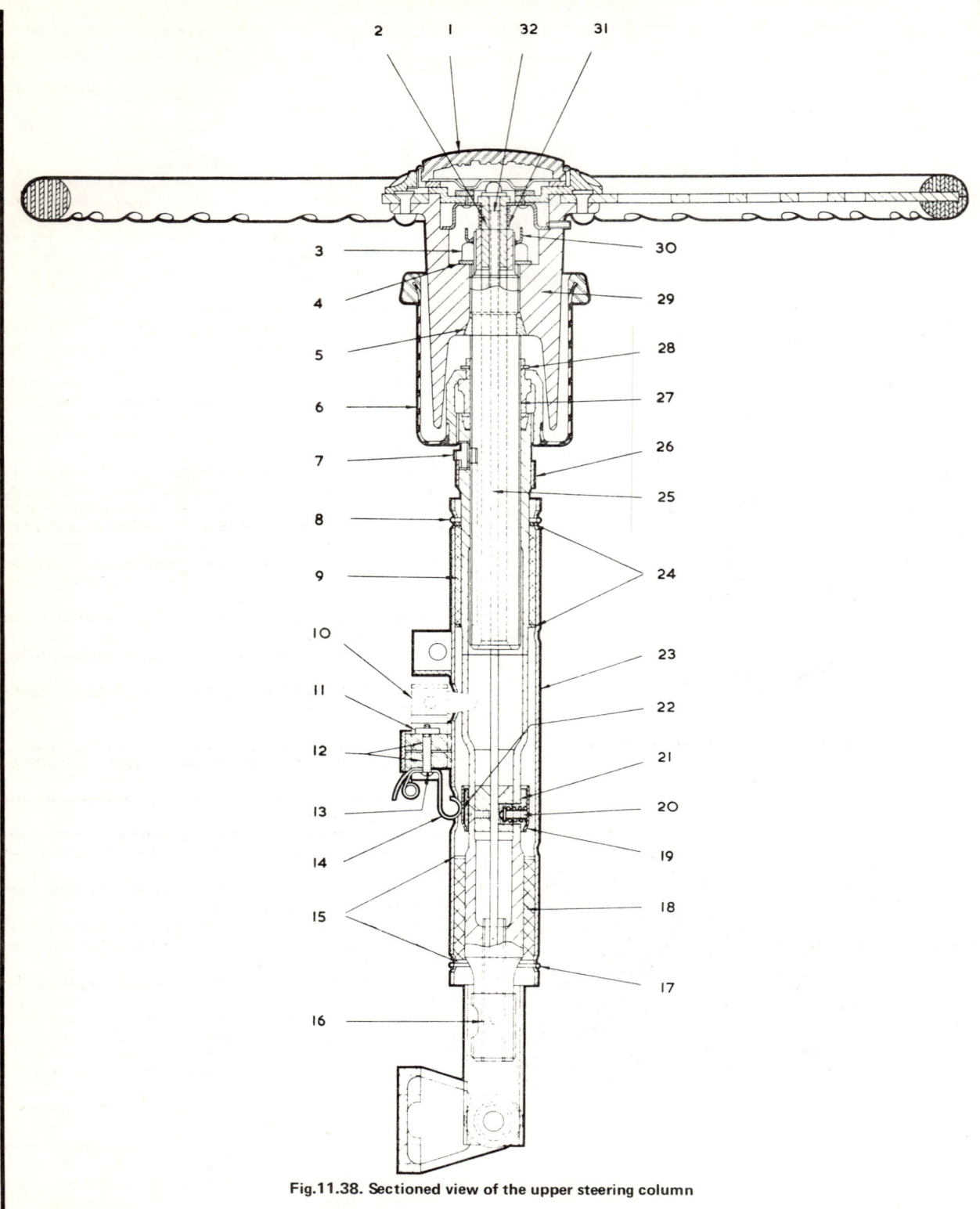

Fig.11.38. Sectioned view of the upper steering column

1	Horn button	9	Felt bush	17	Spring ring	25 Inner column (male)
2	Insulating bush	10	Earth contact	18	Felt bush	26 Control striker
3	Steering wheel nut	11	Contact securing nut	19	Slip ring	27 Split collet
4	Washer	12	Contact holder	20	Contact nipple	28 Circlip
5	Split cone	13	Contact securing bolt	21	Rotor - bottom half	29 Wheel hub assembly
6	Steering wheel locknut	14	Slip ring contact	22	Rotor - top half	30 Locknut
7	Stop button	15	Washer	23	Outer tube assembly	31 Horn switch contact spring
8	Spring ring	16	Insulating sleeve	24	Washer	32 Horn switch contact rod

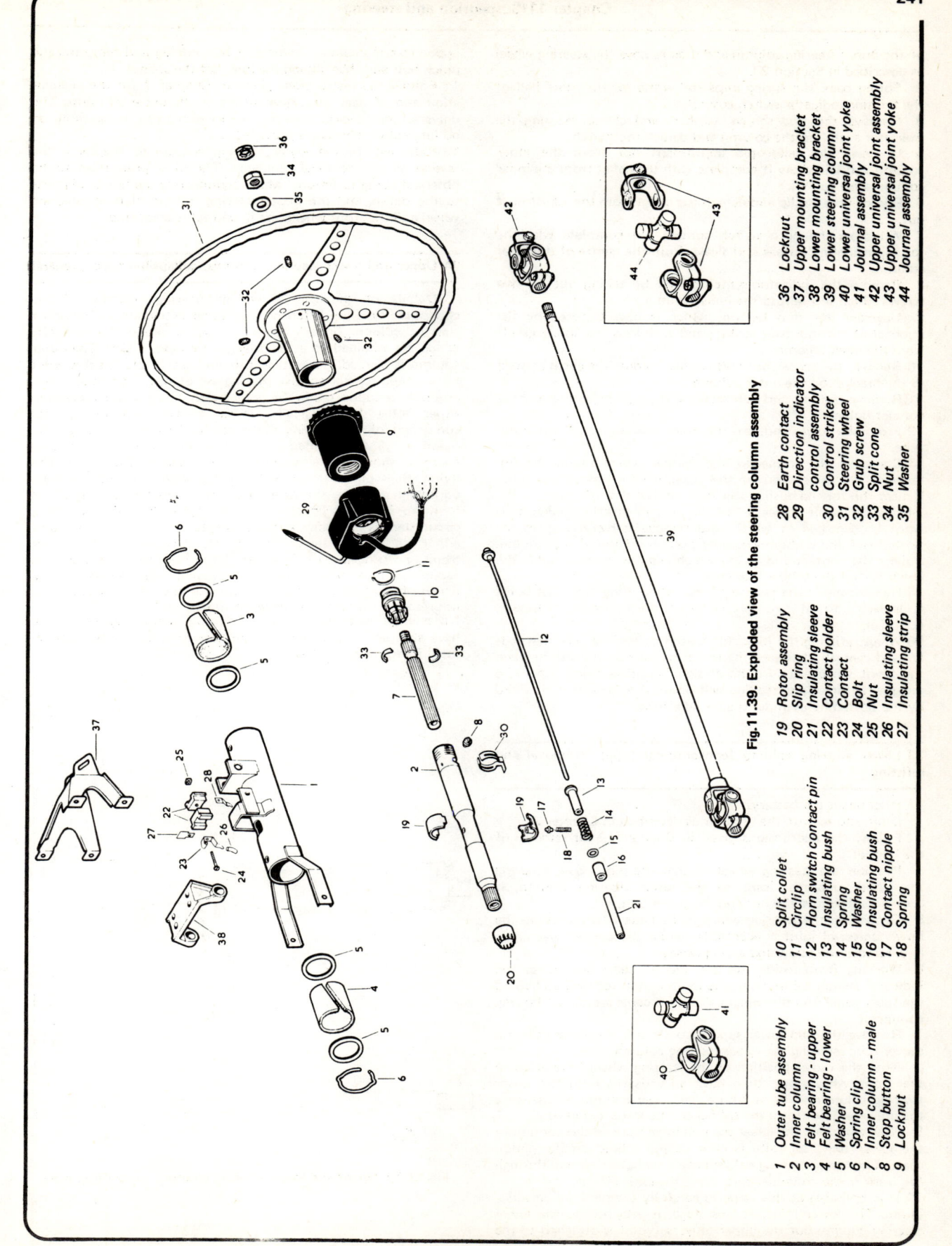

Fig.11.39. Exploded view of the steering column assembly

1 Outer tube assembly	10 Split collet	19 Rotor assembly	28 Earth contact
2 Inner column	11 Circlip	20 Slip ring	29 Direction indicator control assembly
3 Felt bearing - upper	12 Horn switch contact pin	21 Insulating sleeve	30 Control striker
4 Felt bearing - lower	13 Insulating bush	22 Contact holder	31 Steering wheel
5 Washer	14 Spring	23 Contact	32 Grub screw
6 Spring clip	15 Washer	24 Bolt	33 Split cone
7 Inner column - male	16 Insulating bush	25 Nut	34 Nut
8 Stop button	17 Contact nipple	26 Insulating sleeve	35 Washer
9 Locknut	18 Spring	27 Insulating strip	36 Locknut
			37 Upper mounting bracket
			38 Lower mounting bracket
			39 Lower steering column
			40 Lower universal joint yoke
			41 Journal assembly
			42 Upper universal joint yoke
			43 Upper universal joint yoke
			44 Journal assembly

of the upper steering column and then remove the steering wheel as described in Section 20.
3 Spring back the spring clips and withdraw the inner half of the flashing indicator switch cover.
4 Remove the two screws, washers and clamp securing the indicator switch to the column and detach the switch.
5 Unscrew the telescopic adjustment nut from the inner column and withdraw it complete with the collet from the inner column shaft splines.
6 Remove the circlip and detach the collet from the adjustment nut.
7 Withdraw the horn switch contact pin complete with the spring, insulating bushes and sleeve from the centre of the inner column.
8 Remove the indicator switch striker by taking out the two screws which secure it to the inner column.
9 Unscrew the stop button, which is now exposed by the removal of the indicator striker, and withdraw the splined shaft from the inner column.
10 Remove the screw, nut and washer holding the earth contact to the bracket on the outer column.
11 Remove the nut and bolt securing the slip ring contact to the contact holders.
12 From the lower end, tap the inner column out of the outer tube.
13 Compress the springclip and washer which retain the felt bush in the upper end of the column and remove the bush. Repeat this for the bush at the lower end of the column.
14 It is advisable to leave the slip ring in position unless it is worn or damaged in which case remove it by prising up the slotted end and withdrawing over the splined end of the column. Collect the contact and spring which will now be exposed in the lower half of the rubber insulator.
15 Reassembly is the reverse of the dismantling sequence but it is advisable to fit new felt bushes if their condition appears doubtful.
16 When refitting the indicator switch striker, be sure that it is located centrally between the two trip levers on the switch when the front wheels are in the straight ahead position otherwise unequal automatic cancelling will occur. Adjustment is provided by the slotted holes under the fixing screws.

23 Lower steering column (conventional type) - removal and refitting

1 Disconnect the battery.
2 Drain and remove the radiator as described in Chapter 2.
3 Follow the instructions given in Chapter 12 for removal of the bonnet.
4 Position the steering wheel so that the pinch bolt, securing the upper universal joint to the lower steering column, is accessible. Remove the nut and the pinch bolt.
5 Reposition the steering wheel so that the Allen screw in the lower universal joint is accessible. Using the correct size Allen key, remove the screw and spring washer.
6 Working from inside the car, identify and disconnect the indicator switch cables from the multi-connector located behind the facia panel and disconnect the horn cable from the slip ring connector.
7 Remove the two lower support bolts and the upper bolt and spacer tube securing the upper steering column.
8 Note the radial position of the steering wheel in relation to the front wheels and then have an assistant hold the lower steering column whilst you raise the upper column to disengage the universal joint from the splines on the lower column.
9 Now withdraw the lower column in an upward direction until the lower universal joitn is clear of the splines on the pinion shaft. The lower steering column can now be withdrawn through the front frame crossmembers above the anti-roll bar.
10 It is advisable at this stage to carefully examine the universal joints. The lower joint, at this stage, is attached to the lower steering column but the upper universal joint is attached to the upper steering column, remove it by undoing and removing the pinch bolt and then sliding the joint off the splines.
11 Follow the instructions given in Chapter 7 for the examination and, if necessary, dismantling of the universal joints. The universal joint bearings are prepacked with grease at assembly, so no lubrication points are provided.
12 Refitment of the lower steering column to the car is the reverse of the removal sequence. The only precaution to be observed being to ensure, when reconnecting the universal joints to the pinion and the upper steering column, that the steering wheel and the front wheels are in correct relationship.

24 Upper and lower steering columns, collapsible type - general

Details of the cars to which steering columns of the collapsible type are fitted will be found in Section 1. The upper steering column (inner) comprises two seperate sliding shafts which are retained to a fixed length by nylon plugs. The outer column is pierced in a lattice from and is protected at this point by a gaiter to seal against the ingress of dirt. The inner shaft assembly is supported in the outer column by two pre-lubricated taper roller bearings. The lower steering column is also comprised of two sliding shafts retained to a fixed length by nylon plugs. A sectioned view of the upper and lower steering columns showing the nylon plugs is given in Fig.11.40. The nylon plugs of the upper and lower steering columns will shear on impact allowing the steering columns and the steering wheel to move forward. Thus, it will be appreciated that **under no circumstances** must an excessive longitudinal force be applied to either the upper or lower steering columns when removing them from, or fitting them to, the car. This means that if difficulty is experienced in disengaging or engaging the splines of the universal joints with those on either of the columns you must not resort to the use of a mallet or similar tool.
No attempt must be made to repair any nylon plug which may have sheared, or partially sheared, due to impact. **New** replacement items **must** be fitted.

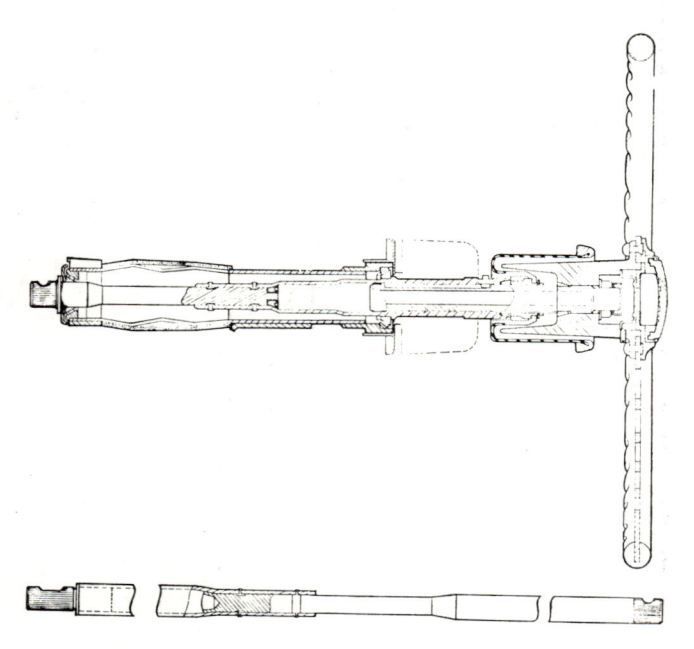

Fig.11.40. Upper and lower steering columns, collapsible type

Chapter 11/Suspension and steering

25 Upper steering column (collapsible type) - removal and refitting

1 Disconnect the battery.
2 Remove the under scuttle casing above the steering column by taking out the self tapping screws.
3 Withdraw the ignition key, unscrew the ring nut and detach the ignition switch from the mounting bracket on the steering column. This does not apply to those cars fitted with air-conditioning equipment where the switch is mounted on a bracket attached to the evaporator unit.
4 From this point onwards, follow the instructions given in Section 21 for the conventional type of upper steering column but take note of the warnings in Section 24 concerning the avoidance of the use of force when seperating the column from the universal joint splines.
5 Refitment is the reverse of the removal procedure but tighten the universal joint pinch bolts to a torque of 16-18 lb f ft.

26 Upper steering column (collapsible type) - dismantling and reassembly

Dismantling is confined to removal of the steering column adjuster locknut, the splined shaft and the direction indicator switch as detailed in Section 22.

27 Lower steering column (collapsible type) - removal and refitting

1 Remove the upper steering column from the car as described in Section 25.
2 Remove the nut, lockwasher and the bolt securing the column to the lower universal joint.
3 Take note of the warning in Section 24 concerning the avoidence of the use of force and then seperate the column from the lower universal joint and withdraw the column rearwards through the grommet.
4 At this stage it is advisable to carry out an examination of the unversal joints. Remove them from the upper column and the pinion splines and examine, and replace if necessary, in the manner described in Chapter 7.
5 Refitment of the lower steering column is the reverse of the removal procedure but tighten the pinch bolts to a torque of 16-18 lb f ft (2.2 - 2.5 kg f m).

28 Rack and pinion housing - removal and refitting

1 Drain and remove the radiator as described in Chapter 2.
2 Take out the split pin and remove the nut from each steering tie-rod ball joint. Tap the side face of the steering arm smartly to break the grip of the taper and separate the tie rod ends from the steering arms.
3 Turn the steering wheel to bring the Allen screw in the lower steering column lower universal joint to an accessible position. Using the correct size Allen key, remove the screw and spring washer.
4 From the pinion side, remove the two inner self locking nuts and the central bolt with the attached self locking nut securing the assembly to the rubber/steel bonded mounting.
5 Remove the top and bottom outer self locking nuts and withdraw the bolt, note the two spacer tubes fitted between the mounting bracket and the frame.
6 Repeat for the other side of the assembly but note that on the two outer fixings the spacer tubes are replaced by two adjuster locknuts.
7 Withdraw the rack and pinion assembly but, for those cars with collapsible steering columns, do not use excessive force when separating the pinion from the universal joint.
8 Refitting is the reverse of the removal procedure but be sure that the spacer tubes on the pinion side of the assembly are refitted and the nuts fully tightened. The mounting on the opposite side of the track must be adjusted as follows:
9 Slacken off the four nuts on the two outer bolts and fully tighten the two inner and the single central fixing.
10 Tighten the self locking nuts securing the two outer bolts until the flat washers under the bolt heads can be rotated by your fingers. Hold the nuts in this position and then screw up the locknuts and lock.
11 After final reassembly, check and adjust the alignment of the front wheels as described in Section 31.

29 Rack and pinion assembly - dismantling, examination and reassembly

1 An exploded view of the rack and pinion housing assembly is given at Fig.11.41.
2 Follow the instructions given in Section 29 for removal of the assembly from the car.
3 Slacken the locknuts locking the ball joints to the steering tie rods and unscrew the ball joints to remove them.
4 Slacken the clip and break the wire securing the bellows to the rack housing tie rods and remove the bellows.
5 Open the tabs of the washer locking the nut to each tie-rod ball housing. Slacken the locknut and then unscrew the ball housing to remove the tie-rod, the socket and the socket spring.
6 Remove the three set screws which hold the pinion shaft oil seal retainer to the housing. Remove the retainer followed by the "O" ring, the attachment plate and the thrust plate.
7 Withdraw the pinion shaft and bearings.
8 Remove the circlip (17 in Fig.11.41) and withdraw the cover, the belleville washer, shims and plunger (note that a coil spring is used instead of a belleville washer in 4.2 litre cars).
9 Withdraw the rack from the housing.
10 Thoroughly clean all parts and examine them for wear or damage. The safe course is to exchange any item if you are at all doubtful as to its serviceability. Pay particular attention to the tie-rod ball seats and housings.
11 The two outer ball joint assemblies cannot be dismantled so they must be replaced by new items if they show signs of wear.
12 The bush at each end of the rack tubing must be replaced if worn. It can be drifted out from the opposite end of the housing by means of a long drift. Press the new bush into the housing using a shouldered, and polished, mandrel of the same diameter as the shaft until the visible end of the bearing is flush with the rack tubing. Before fitting the bush, allow it to soak, in clean engine oil, for about 24 hours to allow the pores of the bush to be filled with lubricant. The bush must not be opened out after fitting.
13 Carefully check the condition of the bellows and the outer ball joint rubber seals and replace them if they are damaged or show signs of deterioration.
14 Start reassembly by applying a generous coating of grease to the rack and then insert the rack into the housing.
15 Grease the pinion shaft and then reassemble it to the housing followed by the thrust plate, the attachment plate, the "O" ring and the retainer. Secure in position with the three set screws (or self locking nuts for 4.2 litre cars) and fully tighten down.
16 Fit the steering damper plunger, the cover and the circlip to the housing but leave out the shims and belleville washer (or coil spring as the case may be) at this stage.
17 Now refer to Fig.11.42 and attach a suitable clock gauge to the housing as shown so that the plunger is in contact with the cover. Apply a downward pressure to the cover to ensure that the pinion is fully engaged with the rack at its lowest point. Set the gauge to zero in this condition.
18 Next apply an upward pressure to the rack to eliminate all end float in the plunger and in this condition the cover should be in contact with the circlip. Note the reading on the gauge.
19 Now measure the thickness of the belleville washer, if applicable, and subtract this figure from the reading obtained on the gauge. Select shims, which are available in thicknesses of

0.004" and 0.010", to give a final end float of 0.006" to 0.010" (0.15 to 0.25 mm).

20 Remove the clock gauge, fit the selected shims, the belleville washer and coil spring, the cover and then secure the assembly by the circlip.

21 Grease and then refit the tie rod assemblies in the reverse order to that in which you removed them and tighten the ball housing until no end float is felt in the tie-rod. When this condition is achieved screw down lightly on the locknut but do not lock with the tab washer as yet.

22 Fit the tie-rod ball joints.

23 Attach a spring balance to the outer ball joints as shown in Fig.11.43 and adjust the tie-rod ball housing until the tie rod will just move under a load of 7 lbs (3.18 kg).

24 Tighten down fully on the locknut to lock the ball housing and secure by closing over the tab washer.

25 Remove the tie rod ball joints and the locknuts.

26 Now refit the bellows and secure the large end to the groove in the housing using soft iron locking wire. Tighten the clip securing the small end to the tie-rod.

27 Refit the tie-rod ball joint locknuts and then screw on each ball joint an equal number of turns and lock.

28 Using a grease gun, pump a little grease into the rack and into the tie-rod ball joints but do not lubricate the ball joints to the extent where grease escapes from the rubber seal.

29 As removal of the tie-rod ball joints will have upset the setting for alignment of the front wheels, this must be checked and adjusted after the rack and pinion assembly is refitted to the car.

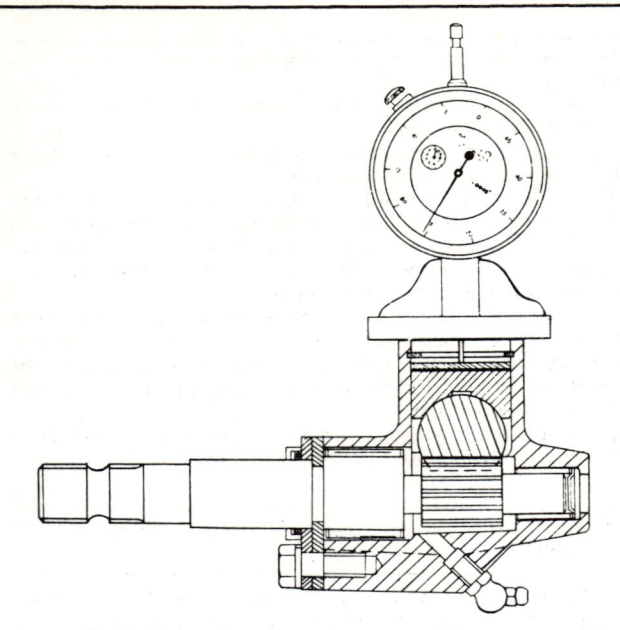

Fig.11.42. Checking endfloat of the cover plunger

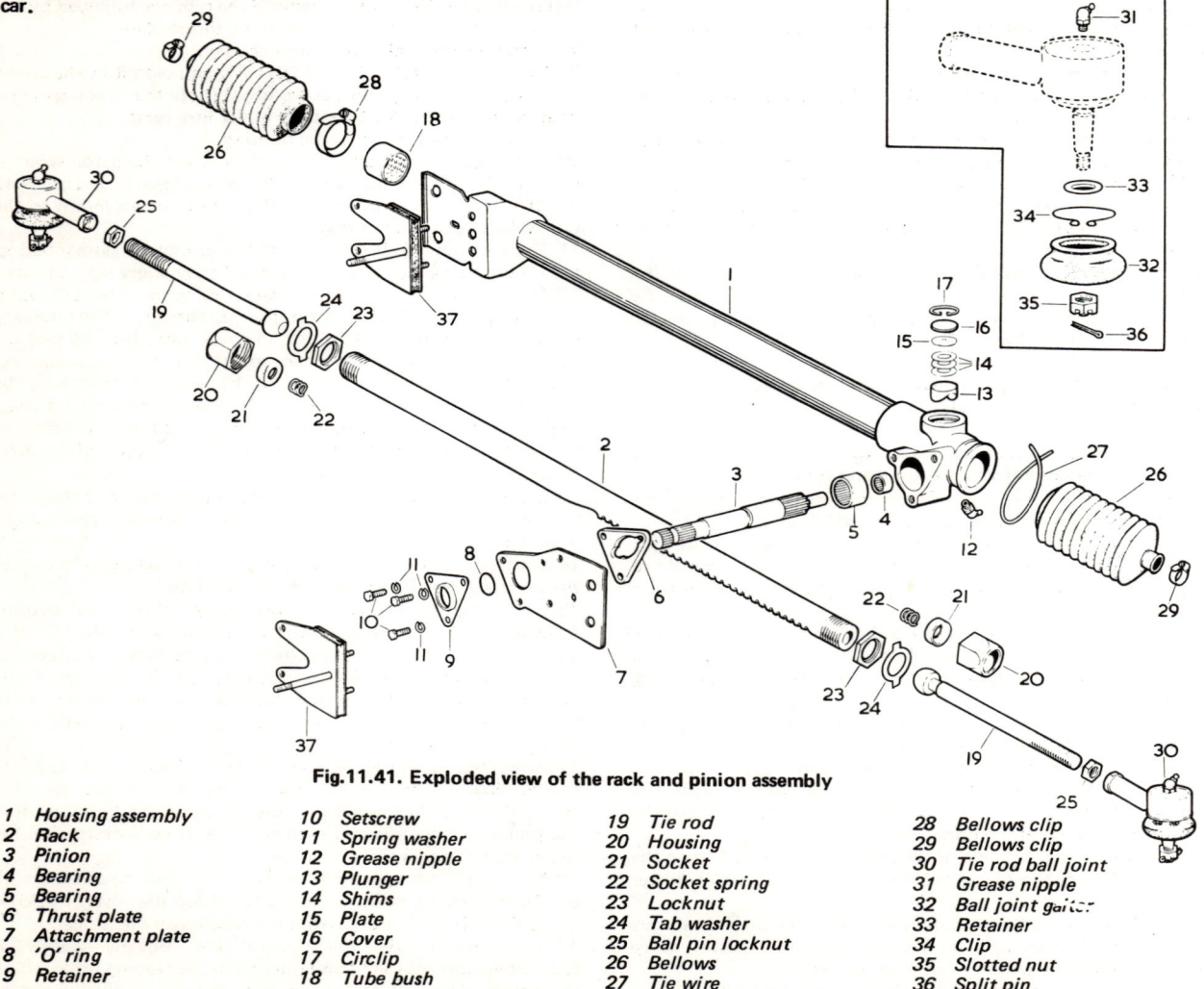

Fig.11.41. Exploded view of the rack and pinion assembly

1	Housing assembly	10	Setscrew	19	Tie rod	28	Bellows clip
2	Rack	11	Spring washer	20	Housing	29	Bellows clip
3	Pinion	12	Grease nipple	21	Socket	30	Tie rod ball joint
4	Bearing	13	Plunger	22	Socket spring	31	Grease nipple
5	Bearing	14	Shims	23	Locknut	32	Ball joint gaiter
6	Thrust plate	15	Plate	24	Tab washer	33	Retainer
7	Attachment plate	16	Cover	25	Ball pin locknut	34	Clip
8	'O' ring	17	Circlip	26	Bellows	35	Slotted nut
9	Retainer	18	Tube bush	27	Tie wire	36	Split pin

Chapter 11/Suspension and steering

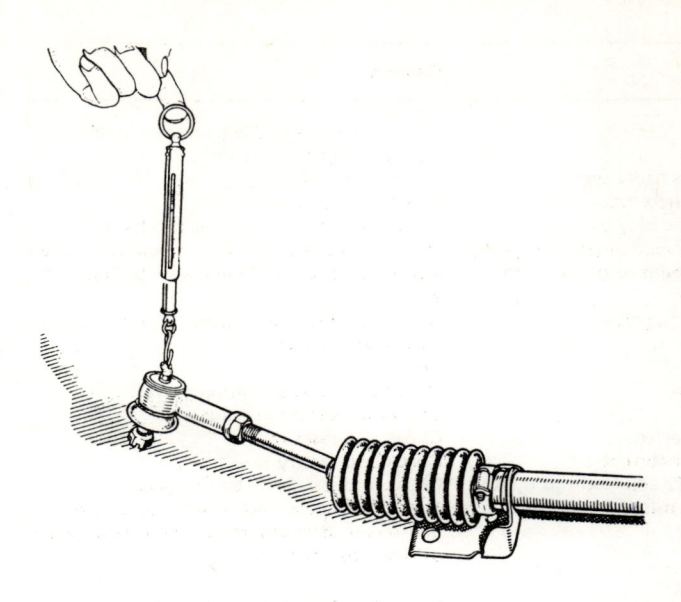

Fig. 11.43. Checking adjustment of the ball housing

30 Steering arm - removal and refitting

1 Raise the car by placing a jack under the lower wishbone fulcrum bearing bracket. Make provision for the car to be safely supported and then remove the road wheel.
2 Take out the split pin and then remove the nut from the steering tie-rod ball joint.
3 Remove the joint from the steering arm, in which it is a taper fit, by tapping smartly a number of times on the side face of the steering arm.
4 Remove the self locking nut and then take out the bolt and spring washer securing the steering arm to the stub axle carrier.
5 Remove the self locking nut securing the stub axle shaft to the carrier and withdraw the steering arm.
6 Refitting is the reverse of the above sequence.

31 Front wheel alignment

This is a task which calls for specialised equipment and thus, if this equipment is not available to you, is a job which should be left to your local Jaguar agent.
1 Check that the car is full of petrol, oil and water. If not, add weight to compensate for, say, a low level of petrol on the basis of 10 gallons of petrol weighing approximately 80 lbs (36.0 kg).
2 Bring tyres to the correct inflation pressures.
3 Place the car on a level surface with the front wheels in the straight ahead position.
4 Using a suitable track setting gauge, check the alignment of the front wheels which should be 1/16" to 1/8" (1.6 to 3.2 mm) toe-in measured at the wheel rims.
5 Recheck the alignment after the car has been pushed forward so that the wheels have turned through 180°.
6 If adjustment is required, first slacken the clip securing the bellows to each steering tie rod. Now slacken the nuts locking each steering tie rod ball joint and turn each tie rod by **equal amounts** in the appropriate direction until the alignment of the wheels is correct. Ensure that the bellows do not twist when turning the tie rods.
7 Tighten the ball joint locknuts and recheck the wheel alignment.
8 If alignment is satisfactory, ensure that the bellows are not twisted and then tighten down on the clips to secure them to the tie rods.

32 Accidental damage

The drawing of Fig. 11.44 is provided to enable you to assess **dimensional** serviceability of a steering arm following accidental damage to your car. We stress that the drawing is confined to dimensional serviceability only and that this is no guarantee that an item is fit for use following an accident. **Under no circumstances** should a steering arm be reworked to bring it to the dimensions quoted.

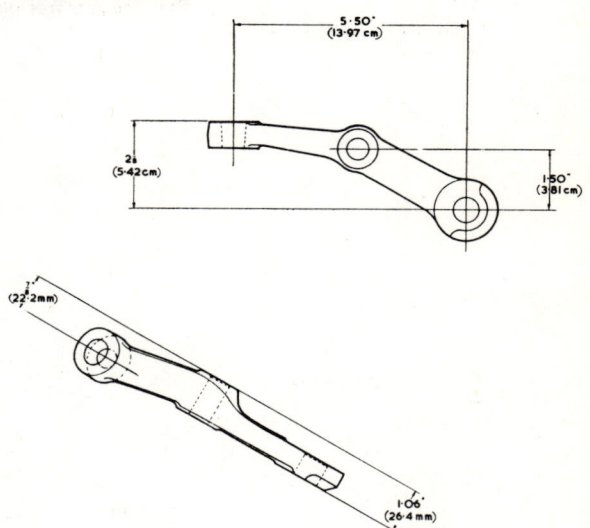

Fig. 11.44. The steering arm

33 Fault diagnosis

Symptom	Reason/s	Remedy
Steering feels vague, car wanders and floats at speed	Tyre pressures uneven	Check pressures and adjust as necessary.
	Shock absorbers worn	Test, and replace if worn.
	Steering gear ball joints badly worn	Fit new ball joints
	Suspension geometry incorrect	Check and rectify.
	Steering mechanism free play excessive	Adjust or overhaul steering mechanism.
	Front suspension and rear suspension pick-up points out of alignment or badly worn	Normally caused by poor repair work after a serious accident. Extensive rebuilding necessary.
	Front suspension lacking grease	Check condition and grease or replace worn parts and re-grease.
Stiff and heavy steering	Tyre pressures too low	Check pressures and inflate tyres.
	Lack of lubrication	Carry out correct maintenance.
	Front wheel toe-in incorrect	Check and reset toe-in.
	Suspension geometry incorrect	Check and rectify.
	Steering gear incorrectly adjusted too tightly	Check and re-adjust steering gear.
	Steering column badly misaligned	Determine cause and rectify (usually due to bad repair after severe accident damage and difficult to correct)
Wheel wobble and vibration	Wheel nuts loose	Check and tighten as necessary.
	Front wheels and tyres out of balance	Balance wheels and tyres and add weights as necessary.
	Steering ball joints badly worn	Replace steering gear ball joints.
	Hub bearings badly worn	Remove and fit new hub bearings.
	Steering gear free play excessive	Adjust and overhaul steering gear.

Chapter 12 Bodywork

Contents

Air conditioning refrigeration equipment - general ... 33	Hardtop (detachable) - removal and refitting ... 26
Body repairs - major ... 5	Heater - removal and refitting ... 31
Body repairs - minor ... 4	Heater water control tap - removal and refitting ... 32
Bonnet centre section - removal and refitting ... 18	Luggage compartment lid and hinges - removal and refitting ... 22
Bonnet - general repair ... 16	Maintenance - body and underframe ... 2
Bonnet - removal and refitting ... 15	Maintenance - upholstery and carpets ... 3
Bonnet side panel - removal and refitting ... 17	No draught ventilator (fixed head Coupe') - removal and refitting ... 14
Bonnet under panel - removal and refitting ... 19	Rear window glass (fixed head Coupe') - removal and refitting ... 25
Bumper (front) - removal and refitting ... 20	Tyre inflation pressures ... 35
Bumpers (rear) - removal and refitting ... 21	Tyre replacement and wheel interchanging ... 37
Door lock mechanism - removal and refitting ... 13	Wheel alignment and tyre wear ... 36
Door rattles - tracing and rectification ... 7	Wheels and tyres - general ... 34
Doors and hinges - removal and refitting ... 9	Window regulator - removal and refitting ... 11
Door trim casing - removal and refitting ... 8	Windscreen (fixed head Coupe') - removal and refitting ... 24
Door window glass - removal and refitting ... 10	Windscreen (open 2 seater) - removal and refitting ... 23
Door window outer seal - removal and refitting ... 12	Wing mirrors - guide lines for fitting ... 39
Exhaust system - removal and refitting ... 30	Wire spoke wheels - repair and adjustment ... 38
Facia panel (side) - removal and refitting ... 27	Wood trim panels - renovation ... 6
Facia panel (top) - removal and refitting ... 29	
General description ... 1	
Glovebox - removal and refitting ... 28	

1 General description

The range of Jaguar 'E' Type cars covered by this Manual have stressed steel bodies of monocoque construction made up of panels as illustrated at Figs. 12.1 and 12.2 for the Open 2 seater and Fixed Head Coupe models respectively. The Open 2 seater is provided with a folding hood, mounted on a special frame. The hood, when in the folded position is completely concealed by a detachable cover. A fibreglass detachable hardtop is available as an optional extra for the Open 2 seater model.

The bonnet of all models, illustrated at Fig. 12.3, is counter-balanced and is, in effect, a forward opening front section giving excellent accessibility to the engine compartment. The bonnet is hinged on the front subframe (Fig. 12.4) which is of square section steel tubing and also carries the power unit and the front suspension.

The interior features two bucket seats at the front which are adjustable for reach and rake with, in the case of 2 + 2 models, seating for two passengers at the rear and on this model the rear seat squab can be folded forward if desired and so extend the boot floor.

2 Maintenance - body and underframe

1 The condition of the bodywork is of considerable importance as it is on the visual condition of this that the re-sale value of the car will mainly depend. It is much more difficult, and costly, to repair neglected bodywork than it is to renew mechanical assemblies. Attention to the hidden portions of the body such as the wheel arches and the underframe is as important, if not more so, than periodic cleaning and polishing of the paintwork.
2 At frequent intervals, especially during the winter months, raise the car and thoroughly hose to remove all mud and road dirt from the wheel arches and projections on the underframe. Pay particular attention to the front part of the front wheel arches, the valance beneath the radiator, the space between the petrol tank and the body and the wheel covers themselves. These, amongst others, are all places where mud will collect and cause corrosion especially if impregnated with salt.
3 The insides of the doors are fairly well protected but nevertheless take off the door panels at least once a year, to make sure that the drain holes are clear and that there is no corrosion.
4 Once a year, preferably in the summer, it is advisable to have the underside of the body steam cleaned. This will remove all traces of dirt and oil so that the underside can be examined for rust, damaged pipes or electrical wiring.
5 The wheel arches should be given particular attention to ensure that the underseal has not been damaged by stones thrown up from the wheels. If damage is found, clean down to the bare metal using a wire brush and then paint on a rust inhibitor or, if preferred, red lead and finally recover the area with underseal.

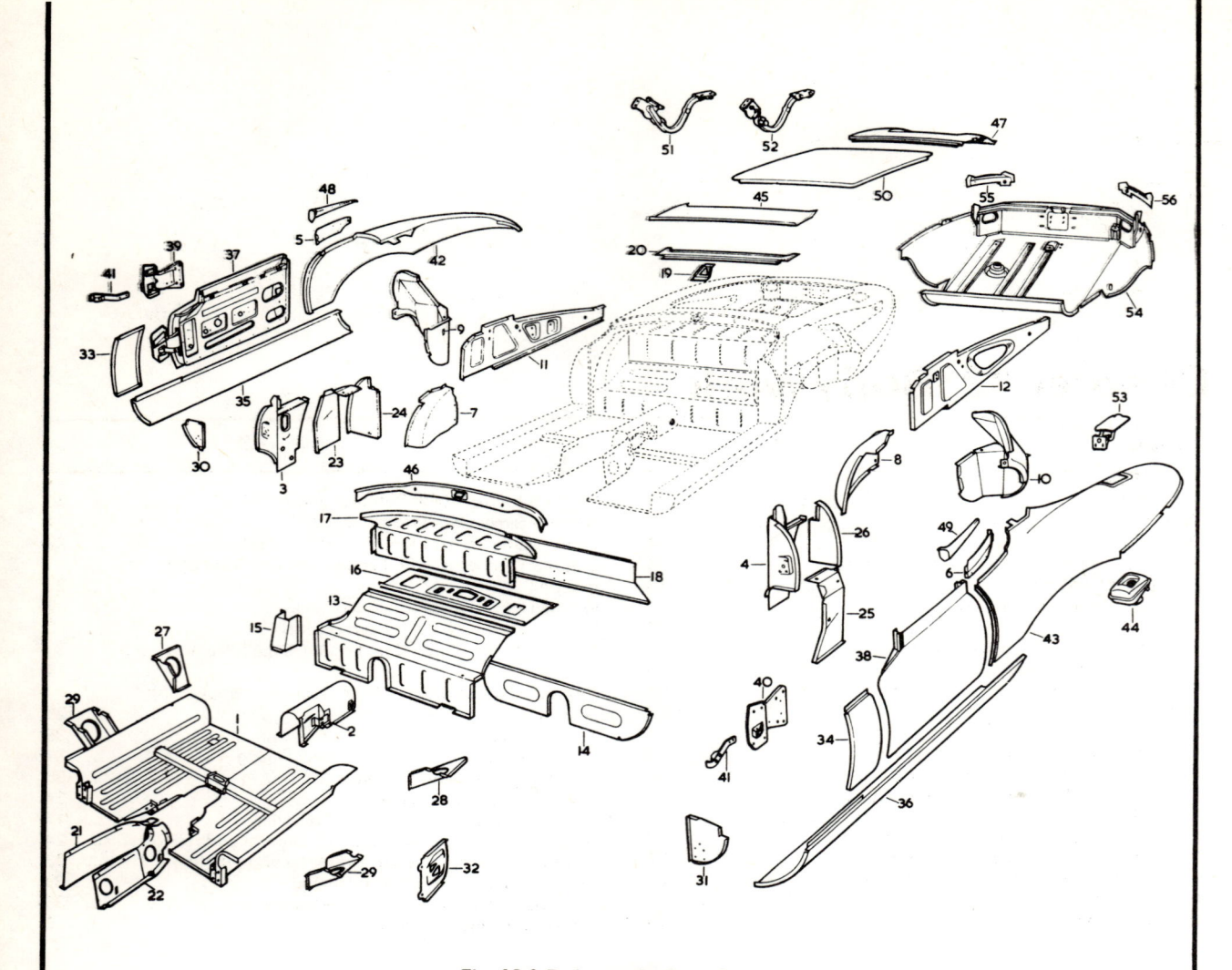

Fig. 12.1 Body panels, Open 2 seater

1 Floor assembly
2 Tunnel assembly
3 Shut pillar (right hand side)
4 Shut pillar (left hand side)
5 Support panel (right hand rear quarter)
6 Support panel (left hand rear quarter)
7 Wheel arch panel (right hand forward)
8 Wheel arch panel (left hand forward)
9 Wheel arch panel (right hand rear)
10 Wheel arch panel (left hand rear)
11 Valance (behind right hand wheel arch)
12 Valance (behind left hand wheel, arch)
13 Floor panel (rear)
14 Crossmember (rear floor)
15 Stiffener bracket (sides of rear crossmember)
16 Top panel (above rear floor)
17 Rear bulkhead panel assembly
18 Panel assembly (front of spare wheel compartment)
19 Shield (interior light)
20 Panel (reinforcing tonneau)
21 Gearbox panel (right hand)
22 Gearbox panel (left hand)
23 Reinforcement panel (right hand shut pillar)
24 Closing panel (right hand shut pillar)
25 Reinforcement panel (left hand shut pillar)
26 Closing panel (left hand shut pillar)
27 Reinforcement panel right hand sill, rear)
28 Reinforcement panel (left hand sill, rear)
29 Reinforcement panel (left and right hand sill, front)
31 Closing panel (left hand sill, front)
32 Reinforcement panel (left hand dash)
33 Exterior panel (right hand dash)
34 Exterior panel (left hand dash)
35 Sill outer panel (right hand)
36 Sill outer panel (left hand)
37 Door shell (right hand)
38 Door shell (left hand)
39 Hinge (right hand)
40 Hinge (left hand)
41 Check arm (both doors)
42 Rear wing panel (right hand)
43 Rear wing panel (left hand)
44 Petrol filler box
45 Tonneau top panel
46 Support panel (tonneau top panel)
47 Tonneau rear panel
48 Top quarter panel (right hand)
49 Top quarter panel (left hand)
50 Boot lid shell
51 Boot lid hinge (right hand)
52 Boot lid hinge (left hand)
53 Petrol filler box lid
54 Lower rear panel
55 Filler panel (right hand stop/tail lamp)
56 Filler panel (left hand stop/tail lamp)

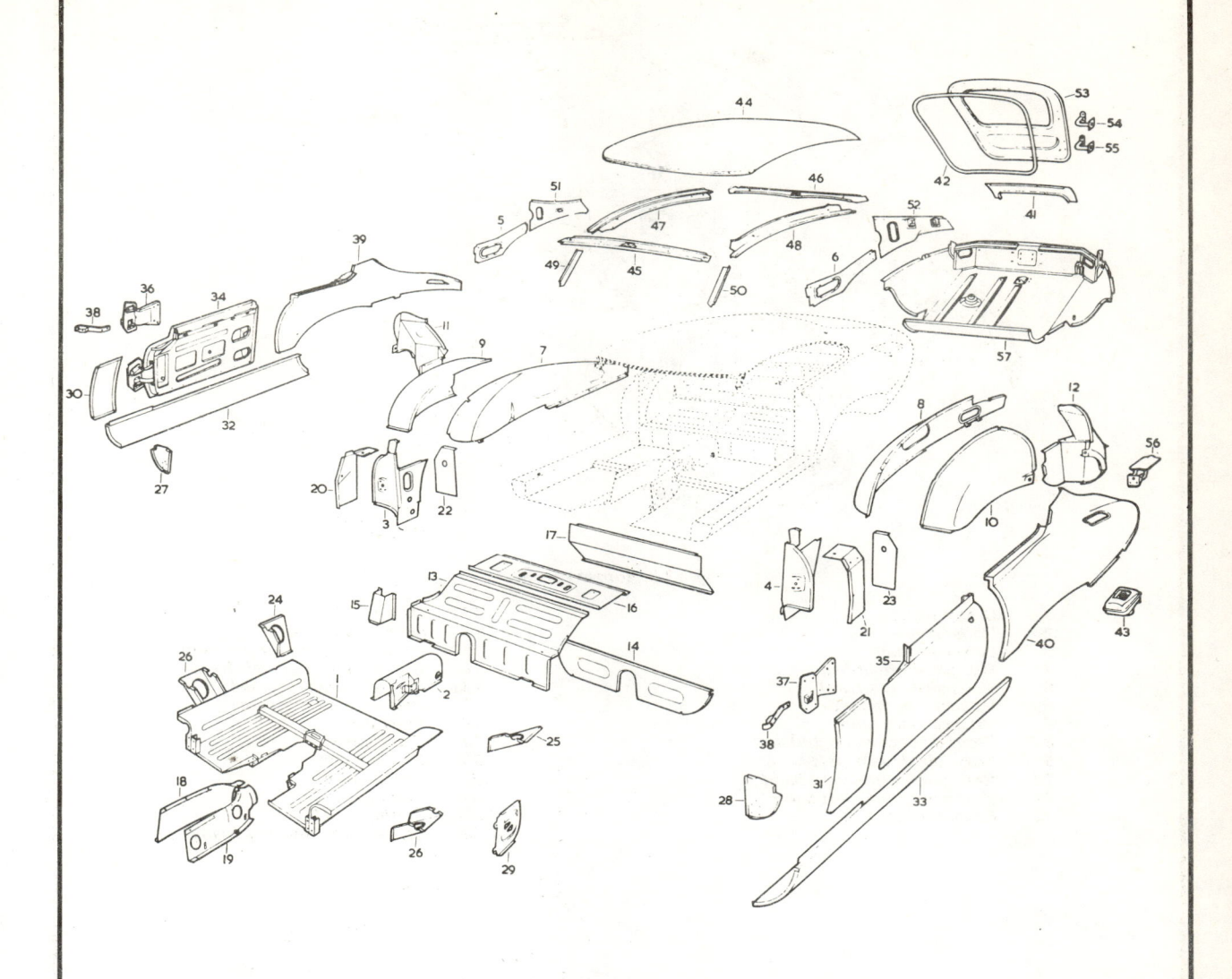

Fig.12.2. Body panels, Fixed Head Coupe

1. Floor assembly
2. Tunnel assembly
3. Shut pillar (right hand side)
4. Shut pillar (left hand side)
5. Support panel (right hand)
6. Support panel (left-hand)
7. Wheel arch (right-hand inner)
8. Wheel arch (left-hand inner)
9. Wheel arch (right-hand outer)
10. Wheel arch (left-hand outer)
11. Wheel arch (right-hand rear)
12. Wheel arch (left-hand rear)
13. Floor panel (rear)
14. Cross member (rear floor)
15. Stiffener bracket
16. Top panel assembly
17. Rear bulkhead panel assembly
18. Gearbox panel (right-hand)
19. Gearbox panel (left-hand)
20. Reinforcement panel (right hand shut pillar)
21. Reinforcement panel (left hand shut pillar)
22. Closing panel (right hand shut pillar)
23. Closing panel (left-hand shut pillar)
24. Panel (right-hand sill, rear)
25. Panel (left-hand sill, rear)
26. Panel (left and right-hand sill, front)
27. Closing panel (right-hand sill, front)
28. Closing panel (left-hand sill, front)
29. Panel (left-hand dash)
30. Exterior panel (right-hand dash side)
31. Exterior panel (left-hand dash side)
32. Sill outer panel (right-hand)
33. Sill outer panel (left-hand)
34. Door shell (right-hand)
35. Door shell (left-hand)
36. Hinge assembly (right-hand door)
37. Hinge assembly (left-hand door)
38. Check arm (left and right hand doors)
39. Rear wing panel (right hand)
40. Rear wing panel (left hand)
41. Tail panel
42. Gutter (boot lid aperture)
43. Petrol filler box
44. Roof panel
45. Windscreen header panel
46. Reinforcement rail (rear)
47. Cantrail panel (right-hand)
48. Cantrail panel (left-hand)
49. Bead extension (right-hand cantrail)
50. Bead extension (left-hand cantrail)
51. Support panel (right-hand)
52. Support panel (left-hand)
53. Boot lid shell
54. Upper hinge (boot lid)
55. Lower hinge (boot lid)
56. Petrol filler box lid
57. Lower panel (rear)

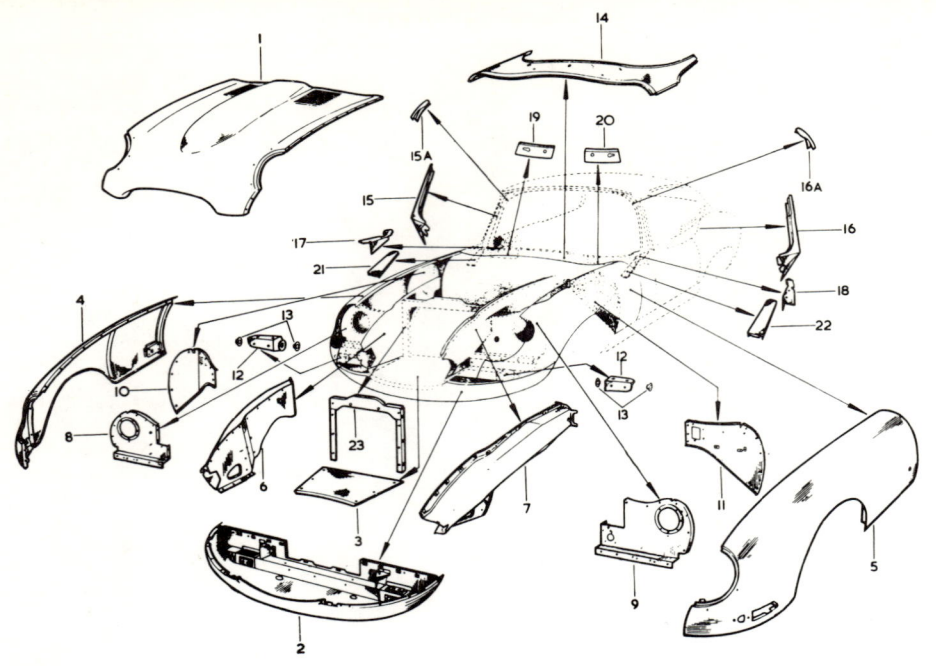

Fig. 12.3. Bonnet panels

1	Centre section	13	Nylon bush
2	Under panel	14	Scuttle top panel
3	Lower air duct	15	Windscreen pillar - right-hand
4	Side panel - right-hand	16	Windscreen pillar - left-hand
5	Side panel - left-hand	15a	Reinforcement channel
6	Valance - right-hand	16a	Reinforcement channel
7	Valance - left-hand	17	Filler panel
8	Front diaphragm - right-hand	18	Filler panel
9	Front diaphragm - left-hand	19	Corner panel
10	Rear diaphragm - right-hand	20	Corner panel
11	Rear diaphragm - left-hand	21	Closing panel
12	Bonnet hinge	22	Closing panel
		23	Stoneguard mounting frame

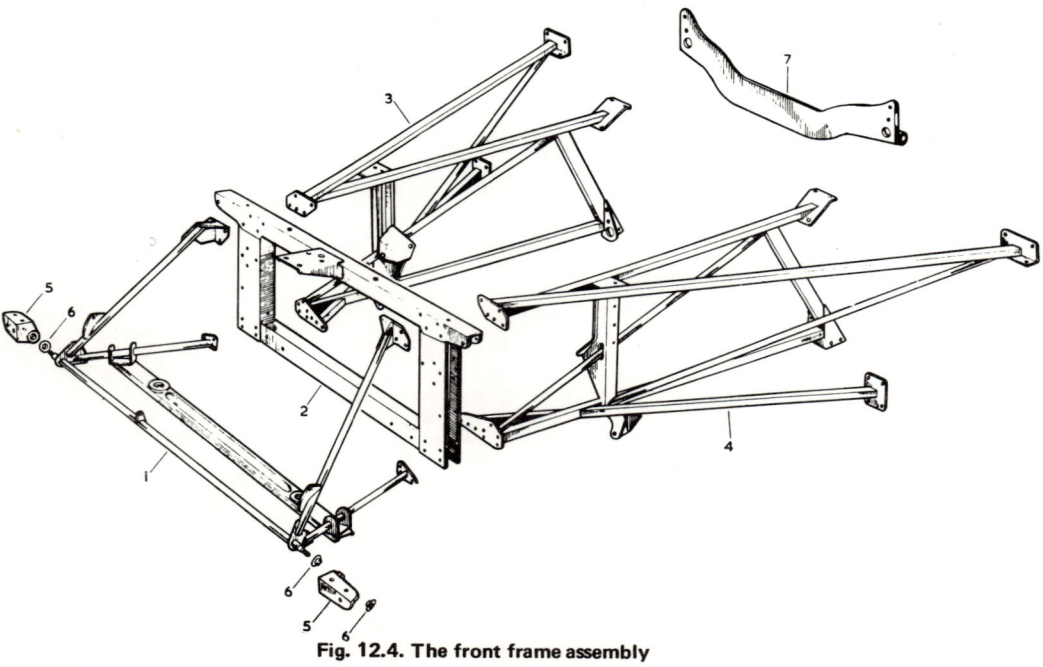

Fig. 12.4. The front frame assembly

1	Front sub frame assembly	5	Bonnet hinge bracket
2	Front cross member assembly	6	Nylon bush
3	Right-hand side member assembly	7	Torsion bar anchor bracket reaction plate
4	Left-hand side member assembly		

Chapter 12/Bodywork

3 Maintenance - upholstery and carpets

Remove the carpets and thoroughly vacuum clean the interior of the car at frequent intervals at the same time checking that dampness, to the extent of causing corrosion of the floor, is not present.

2 Use the vacuum cleaner on the seats and backrests to remove grit from the partitions as may cause chafing and breakage of the stitching.

3 Vacuum clean the carpets and if they are dirty they can be cleaned with one of the many proprietary cleaners which are now on the market.

4 It is suggested that you use saddle soap on the leather upholstery. This will not only clean the leather but will also 'feed' it and help keep it supple.

4 Body repairs - minor

1 The rectification of major body damage should be left to a body repair specialist having the necessary equipment and 'know-how' to restore the car to the condition it deserves. However, there is no reason why you cannot sucessfully beat out, repair and respray minor damage yourself.

2 The most common defect arising in the bodywork is chipping of the paint by stones thrown up from other cars. These blemishes can be repaired by first treating the chip with a rust inhibitor and then applying 'touch-up' paint of the matching colour of your car. Allow the paint to dry. You will probably find that it has shrunk and left a depression in which case apply more paint, but not too much, and again allow to dry. Carry on applying paint until it stands slightly proud of the chip and then leave it for a couple of days to harden off. Now obtain a handy sized block of wood for rubbing down, cover the block evenly with cloth (flannel is ideal) and impregnate the cloth with metal polish. Rub down on the paint using a motion to conform to the particular shape of the panel, do not worry about damaging the surrounding area of old paint finish as this will be much harder than the new surface which you have applied and which will quickly rub down to give a perfect match.

3 Larger areas of damage will need a different approach to the above.

4 Dents can be knocked out from underneath but never use a metal hammer as this will bruise and distort the metal. Use a mallet or hide faced hammer when knocking out a dent and at the same time support the metal with a sandbag or wooden block; beat out the dent as best you can but under no circumstances beat it out to the extent of standing proud of the surrounding metal because in this case the raised portion of metal will have to be cut in order to get a flat finish.

5 Having beaten out the dent, clean down to the bare metal and, then, if there is no rust present, apply a filler such as Holts 'Cataloy' and smooth it over the area as evenly as possible. Do not mix the filler too thinly or it will tend to run; on the other hand a stiff mix will be difficult to apply and will harden off as you are working.

6 Allow time for the filler to harden (usually about 30 minutes) and then follow the instructions for rubbing down the type of filler you used. Fillers of the Holts 'Cataloy' type can first be roughly filed to shape and then carry on with glass paper followed by a medium and fine grade wet and dry paper to feather the patch in to the surrounding area.

7 When doing any rubbing down, use a block of wood packed with cloth to support the rubbing down paper. If you use your hand or fingers they will follow any irregularities and you will never finish up with a flat surface.

8 Protect surfaces as necessary from paint using masking tape, newspaper held on by tape etc., or even use grease for small areas, but do not allow any oil or grease to come into contact with the area you are going to paint. Now spray on an undercoat, allow it to harden and look for blemishes in your filling work, its is surprising how defects will now show up. Use more filler and rub down again as necessary.

9 When satisfied with the undercoat, rub down with a fine grade wet and dry paper and then, when the surface is clean and dry, apply the finishing coats of the matching colour of your car. The number of coats applied is a matter of choice but do not try to finish the job in one application, several very light coats are better than a couple of heavy applications which in any event will most likely result in 'runs'. Rub down between each coat and then when the final coat has had time to harden, polish the whole area with a fine cutting paste.

10 The final treatment for the repair of holes or cuts is the same as the foregoing but first a base has to be provided to hold the filler. This can be done by welding (and this may be the only answer if a large hole, the result of body rot due to rust, is to be repaired) or by using fibreglass as a patching agent.

11 Assuming you are going to use fibreglass to repair a hole resulting from rust, first cut away all other corroded metal and then clean down to the bare metal for a distance of about 1½" all round the hole and apply a rust inhibitor.

12 Knock down the edge of the hole so that it is below the surface. Apply the fibreglass patch in accordance with the instructions for the type being used and then, when it is dry and hard, treat as for a dent as already described in this Section.

13 Beat out a cut but leave the edges below the surface. Clean down to the bare metal and then carry on to patch and fill as described previously.

5 Body repairs - major

Repairs to rectify extensive damage to the body and especially the front sub-frame, following an accident must be carried out so that when the repair is completed the main mounting points for the engine, the front and rear suspensions etc; are in correct relation to each other, in both the horizontal and vertical planes. This is not work that you should do yourself, it is a task which requires specialised equipment and knowledge of the car and its construction and so we strongly recommend that you leave the work to a Jaguar agent.

If accidental damage is sustained at the front frame and the remainder of the subframe assembly, except for that part, appears to be undamaged you can replace that part of the assembly but after you have completed the work it will be money well spent to have the alignment checked professionally. **Under no circumstances** should you weld or braze replacement tubes into the front sub-frame assembly and neither should you apply heat in any form to straighten them.

Where body panels are badly damaged and no affect on roadworthiness of the car can be foreseen, they can be cut out and a replacement panel welded in their place. You may find it an advantage to use only part of a panel so that your joint can be made in a more accessible position, care will have to be taken, of course, when cutting the mating parts of the panel to ensure that a perfect match is obtained.

6 Wood trim panels - renovation

Where wooden trim panels are used, they are prone to damage through the varnish lifting due to the heat of the sun. We have found that a good method of renovation is to remove them as described later in this Chapter and then scrape off all the old varnish and rub down to the bare wood with a fine garnet paper. Apply a walnut polyurethane wood filler and when this is dry, rub it down to a fine finish. Finally apply two coats of a gloss or matt, polyurethane varnish, depending on your choice, rubbing down after each coat. You will find that this will give a durable surface which will last for a considerable time.

7 Door rattles - tracing and rectification

1 The most common cause of door rattles is a misaligned, loose or worn striker plate.
2 To remove the striker plate, take out the three screws illustrated at Fig. 12.5 and lift the plate away from the door pillar.
3 Position the new striker plate and assemble the three screws loosely to the adjustable tapping plate in the door pillar.
4 Correct positioning is now a matter of trial and error until the door can be closed easily and when no lifting or dropping of the door is apparent during opening or closing. Further adjustment may be required after road trial to eliminate any rattle still present.
5 Other causes of door rattles may be :
a) Loose door handles, window winder handles or door hinges.
b) Loose, worn or misaligned door lock components.
c) Loose or worn remote control mechanisms.
d) A combination of the above.

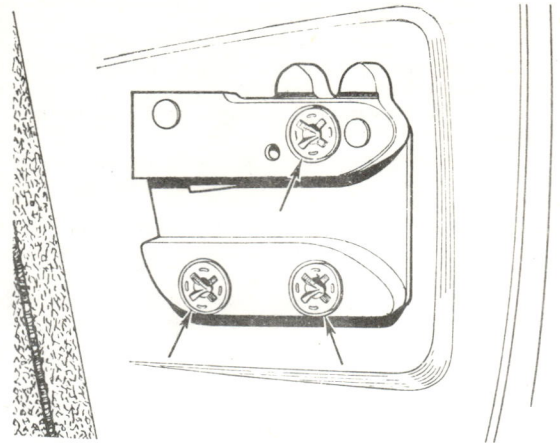

Fig. 12.5. The door striker plate securing screws

8 Door trim casing - removal and refitting

1 First remove the door handle and the window regulator handle by pressing inwards on the spring cap to expose the retaining pin. Drift out the pin and remove the handle. Try pushing inwards on the spring cap using your fingers and if you are not successful, use a screwdriver as illustrated in Fig. 12.6, but be careful not to leave unsightly scratches on the metal.
2 Remove the spring cap and escutcheon.
3 Insert a screwdriver under the top chrome strip and lever it away from the door casing starting at the hinge end of the door.
4 Take out the screws and remove the clips which held the chrome strip.
5 Now insert a thin screwdriver between the casing and the door frame and prise off the casing which is held by twenty-one spring clips ('F' in Fig. 12.7). If possible, leave the plastic cover 'G' in position. Work evenly round the door when prising off the casing, try not to distort it.
6 Refitting the casing is the reverse of the removal sequence.

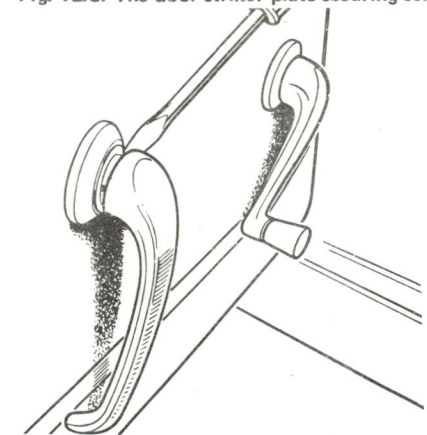

Fig. 12.6. Gaining access to the interior door lock handle retaining pin

9 Doors and hinges - removal and refitting

1 The location of the eight setscrews and two self tapping screws which secure the hinge assembly to the door and to the door pillar is shown in Fig. 12.8.
2 We think you will find it easier to remove the door from the car and then remove the hinge from the door if the hinge has to be renewed.
3 Mark the position of the hinge on the door pillar and then slightly slacken the four setscrews.
4 Have an assistant take the weight of the door and then remove the four setscrews from the door pillar and lift the door away from the car.
5 The first step in removing the hinge is to take off the door trim casing as described in Section 8.
6 Remove the four setscrews and their lockwashers followed by the two self tapping screws and then lift the hinge away from the door.
7 Refitting is the reverse of the removal procedure.
8 Position the hinge to the marks you made before removal and after final tightening of the setscrews check the door for ease of opening and closing. You may find that slight repositioning of the hinge, or adjustment on the striker plate (Section 7) is required.

10 Door window glass - removal and refitting

1 Refer to Section 7 and follow the instructions for removal of

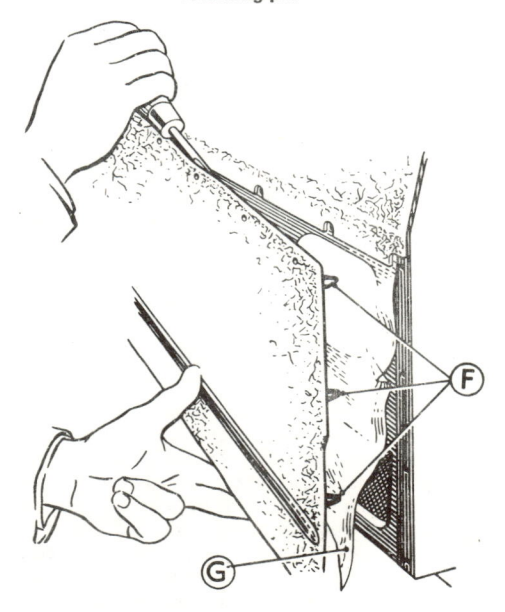

Fig. 12.7. Removing the door trim casing

F Retaining clips
G Plastic cover

Chapter 12/Bodywork

the door trim casing.
2 Remove the plastic sheet, 'G' in Fig. 12.7, which is stuck to the door frame.
3 Remove the closing strip from the top of the door frame by taking out the six screws and washers.
4 Wind the window down until you can get at the roller on the window regulator through the lower aperture in the door panel.
5 Now unscrew and remove the regulator stop pin arrowed in Fig. 12.9.
6 Place the window regulator handle over its spigot and raise the window until the regulator channel is above the door panel.
7 Ease the regulator slide from the channel and withdraw the glass.
8 If desired, the door window frame can now be removed.
9 Take out the three screws which secure the frame to the top of the door panel. Take note of the spacing shims fitted between the frame and the door panel for refitting in their original positions.
10 Remove the two nuts and washers securing the frame to the two brackets on the door lower panel.
11 Withdraw the frame.
12 The door window glass and frame are refitted in the reverse sequence to the foregoing.

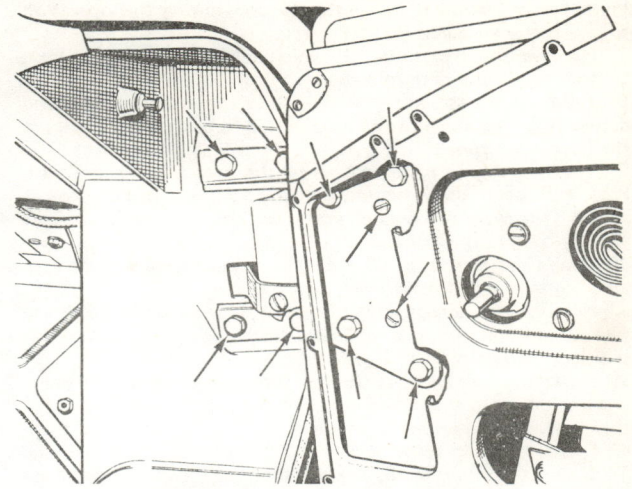

Fig. 12.8. Attachment points of the door hinge assembly

11 Window regulator - removal and refitting

1 Follow the instructions given in Sections 8 and 10 for the removal of the trim casing, the window glass and frame.
2 Fig. 12.10 shows the location of the four nuts and screws which secure the regulator mechanism to the door.
3 Remove the four nuts, and their lockwashers, which secure the regulator to the door frame.
4 Take out the four screws and their lockwashers, which secure the regulator spring to the door frame.
5 The regulator mechanism is now free and can be removed through the aperture at the bottom of the door panel.
6 Refitting is the reverse of the removal.

12 Door window outer seal - removal and refitting

1 Follow the instructions given in Sections 8 and 10 for the removal of the door trim casing and the window glass.
2 Remove the five screws and the outer seal retaining strip which hold the seal to the door panel.
3 Remove the seal.
4 The chrome finisher which is fitted to the top of the door panel on the Open 2 Seater can be taken off, when the seal is removed, by taking out the screw at each end of the finisher.
5 Refit the chrome finisher (if applicable) and the seal in the reverse sequence to the above.

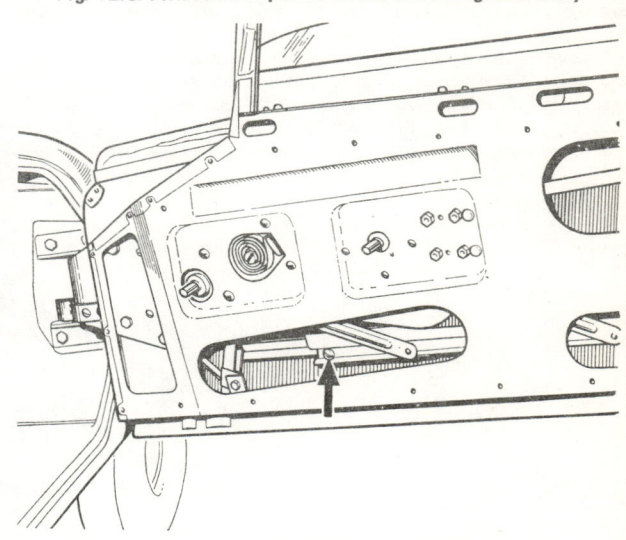

Fig. 12.9. Location of the window regulator stop pin

13 Door lock mechanism - removal and refitting

1 Refer to Fig. 12.11 which shows an exploded view of the door lock mechanism.
2 Follow the instructions given in Section 8 for the removal of the door trim casing.
3 The lock and the remote control units are joined by a connecting link and can be separated, for independent removal, by taking off the circlip and then lifting the link off the dowel '' in Fig. 12.11.
4 Release the spring 'I' holding the link 'J' to the dowel 'K'. You can get at this through the aperture in the inner door panel.
5 Take out the four screws in the front edge of the door (Fig. 12.12).
6 The lock can now be removed by pressing inwards and downwards and passing it around the window channel which is

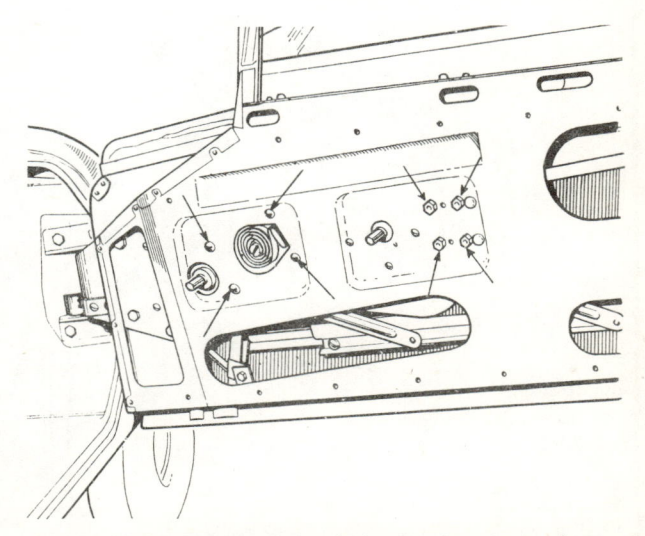

Fig. 12.10. Attachment of the window regulator and spring to the door frame

immediately behind it. Move the window during this operation to give better access to the lock.

7 Remove the two nuts 'M' which secure the outside push button handle and remove it.

8 Take out the screws 'N' to free the remote control unit and connecting link which can now be passed through the aperture in the inner door panel.

9 The first step in refitting is to get the lock unit into position. Pass it through the aperture in the inner door panel, work it round the window channel and position it until it projects slightly beyond the front edge of the door frame.

10 Secure the lock in position by fitting and tightening the four securing screws and shakeproof washers.

11 Now for the remote control unit. **This must be fitted in the locked position** and you will find that a new unit is supplied pinned as shown at 'P' in Fig. 12.11.

12 Insert the unit through the aperture in the inner door panel and position it so that the spindle and the pin 'P' project through their respective holes in the panel.

13 Loosely fit the securing screws 'N' and shakeproof washers.

14 Connect the link to the dowel 'H' on the lock operating lever and secure it with its circlip. Note that a waved washer goes between the lever and the link and that a plain washer is fitted under the circlip.

15 Now move the remote control unit towards the lock until the operating lever is in contact with the lock case and in this position tighten the securing screws 'N'. In cases where a new lock or remote control unit are being brought into use, you may have to elongate the holes in the inner door panel to achieve the foregoing positioning of the remote control unit.

16 Hold the push button handle in position on the door panel and, without depressing the push button, measure the clearance between the plunger 'Q' and the lock contactor 'R'. the clearance should be 1/32" (0.8 mm).

17 If adjustment is required, turn the operating lever 'S' to the unlocked positon so that depression of the push button moves the plunger through its housing. In this condition, release the locknut 'T' and then screw the plunger bolt 'Q' in or out as required. Retighten the locknut before releasing the push button.

18 Attach the link 'J' to the operating lever 'S' and retain it with its circlip.

19 Now turn the operating lever to the locked position so that the holes in the lever and plunger housings are in line.

The operating lever has to be held in this positon for the time being. This can be done by inserting a short length of 1/8" (3.2 mm) rod 'U' through the locating holes. Bend up one end of the rod to aid subsequent removal.

20 Manoeuvre the rod and the link 'J' through the handle aperture so that they hang down inside the door and now fit, and tighten, the fixing nuts 'M'.

21 Check that the remote control cam 'P' is pinned in the locked position; you will then find that any of the three holes in the bottom of the link 'J' can be aligned with the dowel 'K'. Press the link onto the dowel and it will be retained by the spring 'I'.

22 Withdraw the rod 'U' and the pin 'P'.

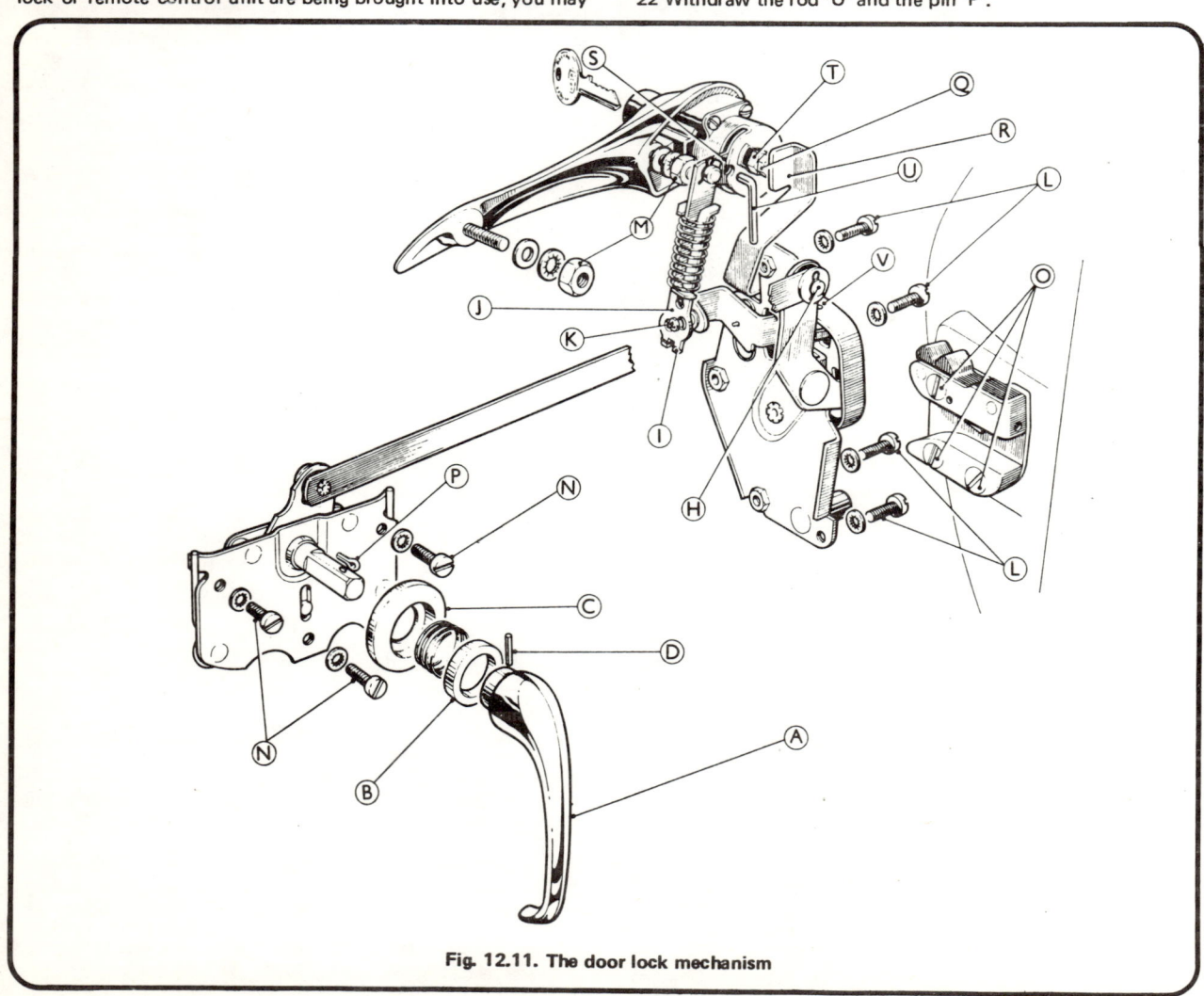

Fig. 12.11. The door lock mechanism

Chapter 12/Bodywork

23 Test the fit of the door and adjust the striker plate as may be necessary (Section 7).
24 Now test the door lock mechanism by fittin an inside handle vertically downwards, turn the handle forward to the locked position and it should automatically return to the central position when released.
25 Next close the door whilst holding the push button in the fully depressed position, the door should remain locked although the push button can be fully depressed.
26 Unlock the door using the key, push button control should be restored and the key should automatically return to the vertical position.
27 If the tests are satisfactory, use a protective grease on all moving parts.
28 Refit the door trim casing (Section 8) and introduce a few drops of thin machine oil into the hole 'V' on the top of the lock case.

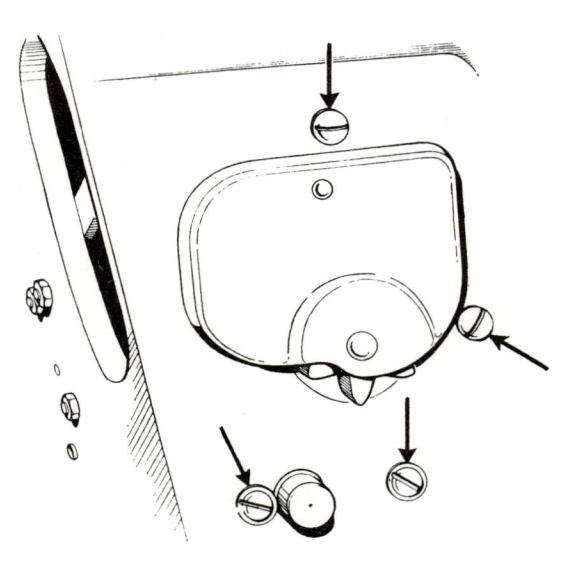

Fig. 12.12. The screws retaining the door lock

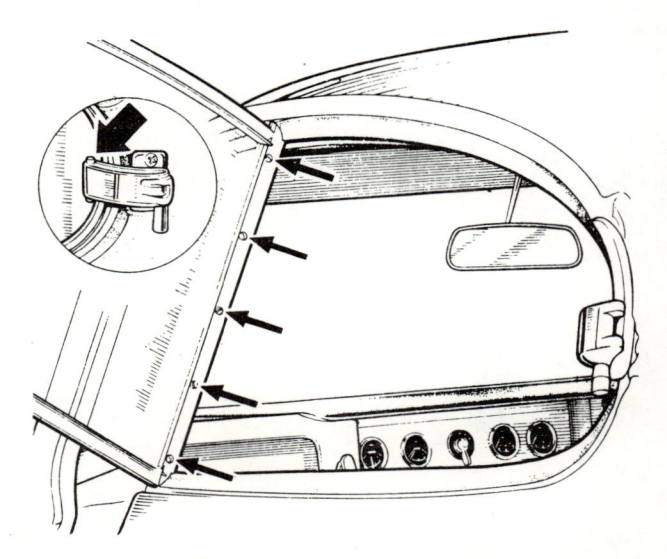

Fig. 12.13. Location of the screws securing the no draught ventilator hinge. Inset shows the catch retaining pin

14 No draught ventilator (Fixed Head Coupe) - removal and refitting

1 Take out the two screws which secure the catch arm bracket to the body. These are accessible from inside the car.
2 Take out the five screws shown in Fig. 12.13 which secure the hinge to the frame post and remove the ventilator.
3 Refitting is the reverse of the removal procedure.

15 Bonnet - removal and refitting

1 Before starting to remove the bonnet, make sure that you have adequate assistance to handle it. The bonnet is deceptively heavy and we found that two men were required to support it whilst a third carried out the removal procedure and then, to ensure that it was not dragged on the floor or otherwise damaged, four men were needed to carry it.
2 The first step is to open the bonnet using the 'T' handle provided for early cars or by turning the two small levers located on the right-hand and left-hand door hinge posts on later models.
3 Press the safety catch under the rear edge of the bonnet and raise it.
4 Disconnect the battery.
5 Disconnect the multi-pin socket, Fig. 12.14, at the left hand side of the bonnet.
6 Mark the position of the hinges on the bonnet to assist in correct location when refitting.
7 Close the bonnet slightly and place some wooden blocks on the ground so that the bumper will come up against them when the bonnet is raised again. This will help support the bonnet and will make removal of the hinges easier.
8 Have assistants each side of the bonnet to support it.
9 Remove the two self locking nuts and washers which secure the bonnet hinges to the front sub frame mounting pin, these are shown in Fig. 12.15.
10 Now take out the two pivot pins and nuts which secure the helper spring mechanism to the subframe, Fig. 12.16.
11 Slacken the four setscrews and washers securing either the right-hand or left-hand hinge to the bonnet. Make sure that your assistants have got the weight of the bonnet and then remove the four setscrews and the hinge, taking careful note of the number and location of the packing pieces between the hinge and the bonnet.
12 It is now possible to slide the bonnet off the mounting pin of the hinge still in position on the bonnet.
13 Take the bonnet away from the car and store it in a safe place where it will not be damaged.
14 Refitting is the reverse of the removal procedure but some adjustment may be required to ensure locking of the bonnet.
15 Adjustment on early models is achieved by packing pieces inserted uner the bonnet lock plate which is attached by two screws to the body, add or remove packing pieces until the lock pawl retains the bonnet firmly when locked.
16 Adjustment on later models is provided by means of rubber buffers which are attached to adjustable spigot pins. Release the spigot pin locknut and turn the pin until the lock pawl retains the bonnet firmly when locked and then tighten down on the locknut.

16 Bonnet - general repair

An exploded view of the bonnet is given in Fig. 12.3. It is composed essentially of certain main components each of which si replaceable if damaged.
Commencing at the following chassis numbers, the bonnet and front wing assemblies have been modified and individual parts are not, therefore, interchangeable with earlier models.

	R.H. Drive	L.H. Drive
4.2 litre Open 2 Seater	1E.1479	1E.12580
Fixed Head Coupe	1E.21228	1E.32632

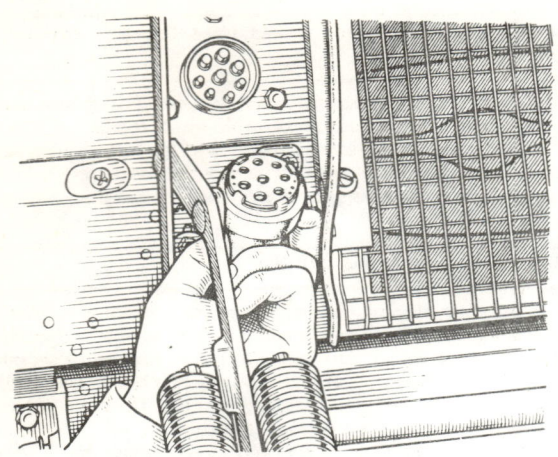

Fig. 12.14. The multi-pin socket connection

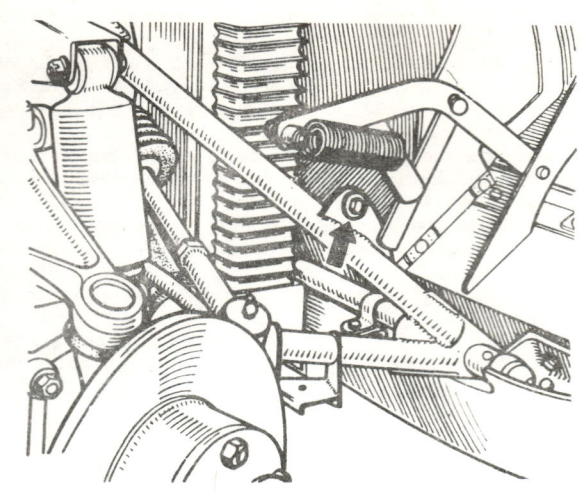

Fig. 12.16. The bonnet spring mechanism pivot points

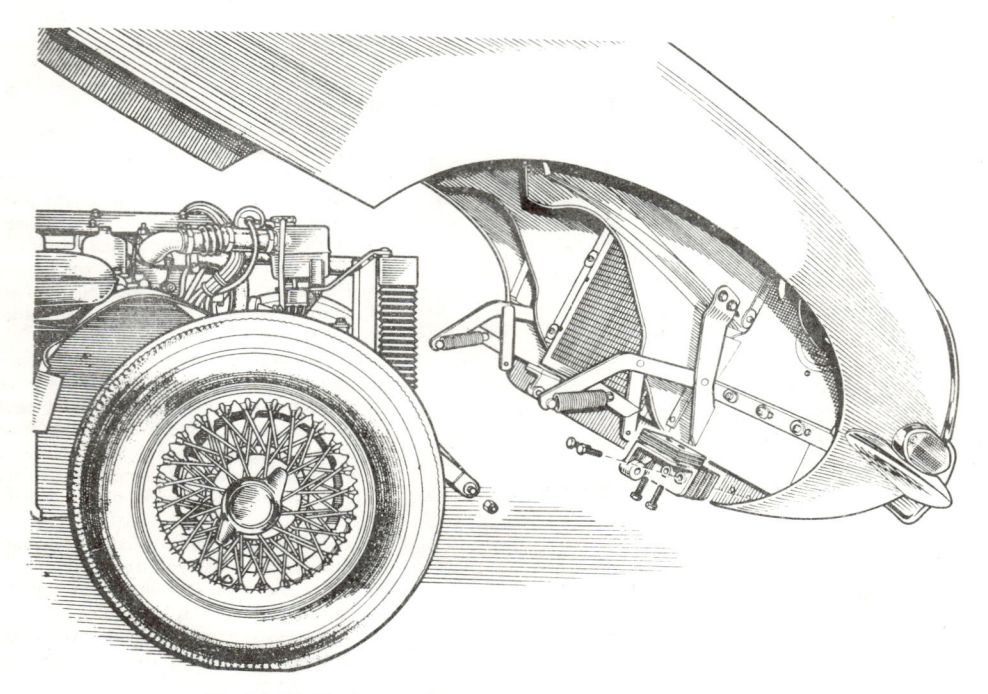

Fig. 12.15. The bonnet hinge mounting points

17 Bonnet side panel - removal and refitting

1 Remove the bonnet as described in Section 15
2 Remove the front bumper as described in Section 20.
3 Take out the three fixing screws and remove the side flasher lamp after disconnecting the two cables from the snap connectors in the headlamp nacelle.
4 Remove the five bolts, nuts, plain and lock washers which secure the side panel to the centre panel.
5 Now remove the four nuts and bolts with their plain and lock washers which secure the side panel to the headlamp mounting diaphragm.
6 Now take off the nine nuts, bolts and washers attaching the side panel to the centre panel along the crown line of the side panel.
7 Next remove the five nuts, bolts and washers which hold the side panel to the engine valance panel followed by the four nuts bolts and washers attaching the side panel to the rear diaphragm.

8 Open the brass tabs of the two chromium beading strips and remove the strips. You will find that there are nine clips on the long and two clips on the smaller strip.
9 Take out the four setscrews and washers which attach the side panel to the centre panel at the rear.
10 The side panel can now be removed.
11 Refitting is the reverse of the removal sequence but you must take care to ensure that the edges of the centre section and side panels are flush when you bolt them together otherwise the chrome strip will be prevented from fitting neatly to the bonnet. Do not forget to refit the brass clips for the chrome strips when replacing the bolts to the centre section and the side panels.
12 Replace the chrome strips and then bend over the clips to secure them. If a clip is inadvertently broken a replacement should be obtained and iftted otherwise the chrome strip may not seat correctly.
13 Finally, coat all under wing joints with a good quality sealing compound.

Chapter 12/Bodywork

18 Bonnet centre section - removal and refitting

1 Remove the bonnet from the car as described in Section 15.
2 Refer to Chapter 10 and follow the instructions for removal of the headlamp covers and ducts.
3 Working from the rear of the front bumper extension, remove the two hexagon headed setscrews which secure the Motif Bar.
4 Remove both front bumpers as described in Section 20.
5 Take out the eight cross headed screws and the two nuts and bolts which secure the stone guard to the bonnet. Withdraw the stone guard from the bottom and collect the felt sealing strip at its top edge.
6 Working on one side of the bonnet, take out the cross headed drive screw and washers which attach the centre section to the valance.
7 Take out the three screws from the rear diaphragm.
8 Move to the headlamp nacelle and remove the two nuts, bolts and washers from the vertical flange holding the side panel to the centre section and now remove the two nuts, bolts and washers which secure the centre section to the under panel.
9 Straighten the brass tabs of the chrome beading strips and remove the strips.
10 Remove the nine nuts, bolts and washers which secure the centre panel to the side panel along the crown line.
11 Take out the four setscrews and washers and remove the closing plate at the rear between the centre section and the side panel.
12 Repeat operations 6 to 11 for the other side of the bonnet.
13 Reassembly is the reverse of the foregoing but make sure that the edge lines of the centre section and the side panels are flush when you bolt them together, otherwise the chrome strips will be prevented from fitting neatly to the bonnet. Do not forget to fit the clips for the strips when replacing the bolts to the centre section and the side panels. If a clip is inadvertently broken, a replacement should be obtained and fitted, otherwise the chrome strip may not seat correctly.
14 When you refit the stone guard, make sure that the felt sealing strip is in good condition, fit a new strip if necessary.
15 Finally, coat all under wing joints with a good quality sealing compound.

19 Bonnet under panel - removal and refitting

1 Remove the bonnet from the car as described in Section 15.
2 Refer to Chapter 10 and follow the instructions for the removal of both headlamp covers and ducts.
3 Working from the rear of the front bumper extension, remove the two hexagon headed screws which secure the motif bar and take off the bar.
4 Remove both front bumpers as described in Section 20.
5 Take out the eight cross headed screws which secure the stone guard to the bonnet. Withdraw the stone guard from the bottom. Collect the felt sealing strip and discard it if it is in poor condition.
6 Working on one side of the bonnet; remove the nuts, bolts and washers, located in the headlamp nacelle, which secure the under panel to the centre section and side panel.
7 Repeat for the other side of the bonnet.
8 Remove the four setscrews and washers securing each bottom hinge bracket, remove the bracket and take careful note of the number of spacer shims fitted under the brackets.
9 Mark the position of the bonnet spring brackets and then take out the four setscrews and remove the brackets.
10 Remove the five cross headed screws which secure the under panel to each headlamp mounting diaphragm and then take out the screws holding the under panel to the valance at each side.
11 Take out the three cross headed screws and the two nuts, bolts and washers attaching the under panel to the orifice lower panel and remove the lower panel.
12 Refitting is the reverse of the above, but as a final task, coat all under wing joints with a good quality sealing compound.

20 Front bumper - removal and refitting

1 The mounting points for the front bumpers of 3.8 litre and early 4.2 litre cars are indicated in Fig. 12.17. However, commencing from the chassis numbers quoted in Section 16 the front bumpers of later model 4.2 litre cars have been modified to allow access from the outside of the car; their removal and refitting is, therefore, a straightforward operation consisting merely of the removal of two setscrews per bumper.
2 For early models, remove the glass headlamp cover by taking out the six screws which hold the cover retaining ring to the wing. Remove the ring and rubber seal. We have experienced difficulty in removing the ring securing screws and it is suggested that you first treat each screw with penetrating oil before trying to remove it.
3 Remove the headlamp glass cover and then take out the three setscrews which secure the headlamp duct to the diaphragm panel, the duct can now be withdrawn forward through the nacelle.
4 Remove the motif bar by taking out the two hexagon headed setscrews which are accessible from the rear of the bumper extension. A 7/16" A.F. socket wrench of the ratchet type is the best tool for this job.
5 Remove the two setscrews, located in the wing nacelle, which secure the bumper to the wing.
6 Detach the bumper and collect the sealing strip.
7 Take out the two setscrews and detach the curved extension from the bumper.
8 Remove the nut, plain washer and lockwasher which hold each over-rider to the bumper and withdraw the over-riders. Collect the sealing strip between them and the bumper.
9 Refitting is the reverse of the above sequence but watch that the various sealing strips are seated properly.

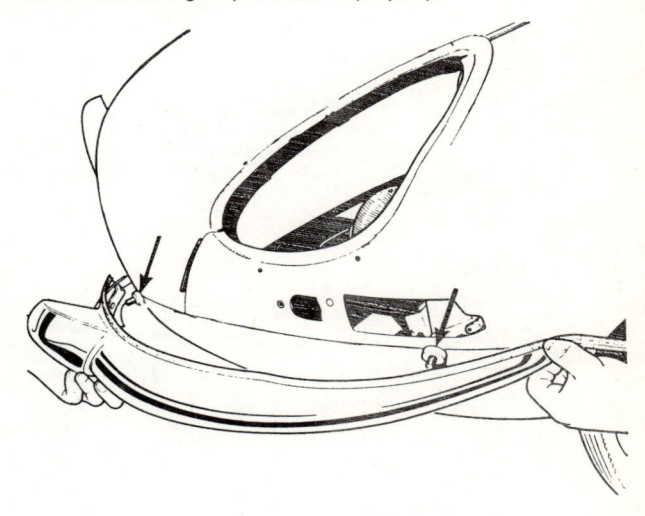

Fig. 12.17. The front bumper attachment points (left hand side illustrated)

21 Rear bumpers - removal and refitting

1 From the chassis numbers quoted below, the mounting points of the rear bumpers of 4.2 litre models are accessible from outside the car; their removal and refitting is, therefore a straightforward operation.

	RH Drive	LH Drive
4.2 'E' Type Open 2 Seater	1E.1413	1E.11741
4.2 'E' Type Fixed Head Coupe	1E.21000	1E.32010

2 The mounting points for the rear bumper of models earlier than the above are shown at Fig. 12.18.
3 Remove the section of the boot floor over the spare wheel and then take out the spare wheel.
4 To remove the right-hand bumper, take out the three chrome screws and remove the side trim casing.
5 The three bumper retaining screws are now accessible. The forward screw is located within the wheel arch and the other two are accessible from within the boot interior.
6 Remove the bumper retaining screws and withdraw the right-hand bumper and collect the sealing strip.
7 To remove the left hand bumper, first remove the spare wheel.
8 Then disconnect the battery.
9 Take out the countersunk screws and remove the floor board covering the petrol tank.
10 Identify and then disconnect the two cables from the petrol tank gauge unit.
11 Remove the cover from the junction block, identify and then disconnect the petrol pump cables.
12 Disconnect the pipe from the petrol tank, cover the unions and then tie up the pipe to prevent fuel from syphoning out.
13 Take out the three drive screws and remove the side trim casing.
14 Slacken the clips and remove the petrol filler hose.
15 Follow the instructions given in Chapter 3 for the removal of the petrol tank.
16 The left hand bumper retaining screws are now accessible. The forward screw is located within the wheel arch and the other two are within the boot interior. Remove the screws and withdraw the bumper collecting the sealing strip.
17 The over-riders can be seperated from the bumpers after removing the securing nut, plain washers and lockwashers.
18 Refitting is the reverse of the above sequence but make sure that the various parts of sealing strips are correctly seated.

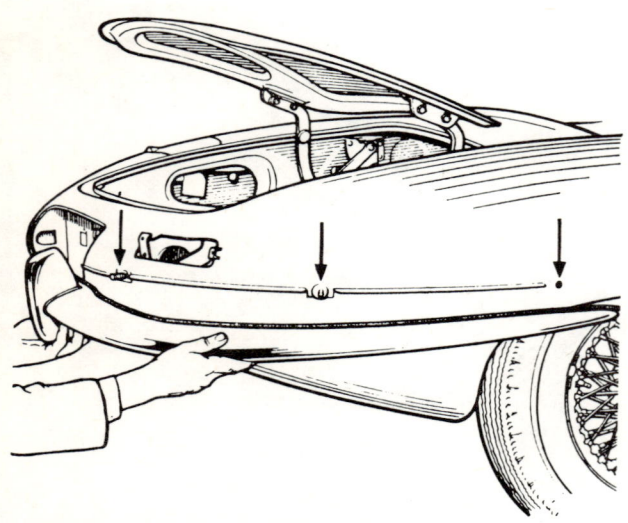

Fig. 12.18. Rear bumper attachment points (early models)

22 Luggage compartment lid and hinges - removal, refitting and adjustment.

1 Raise the luggage compartment lid and, if applicable, secure it in the open position with the stay.
2 Mark the position of the hinges on the lid and then remove the four setscrews, plain washers and lockwashers and remove the lid.
3 Mark the position of the hinges on the body and then remove the four setscrews, nuts and plain washers and remove the hinges.
4 Refitting is the reverse of the above sequence but adjustment of the lock may be required.
5 Refer to Fig. 12.19 which shows the location of the screws for adjustment of the luggage compartment lid striker on Open 2 seater cars. Slacken the four setscrews and move the striker (the fixing holes are elongated) until the lock operates correctly and the lid does not rattle. Fully tighten the screws after correct adjustment is obtained.
6 Fig. 12.20 shows the location of the screws for adjustment of the luggage compartment lid striker fitted to Fixed Head Coupe models. Slacken the two cross headed screws and the two nuts securing the striker to the lid, move the striker (the fixing holes are elongated) until the lock operates correctly and the lid does not rattle. If adjustment on the striker is insufficient, the lock itself can be moved through the four slotted holes in the body.

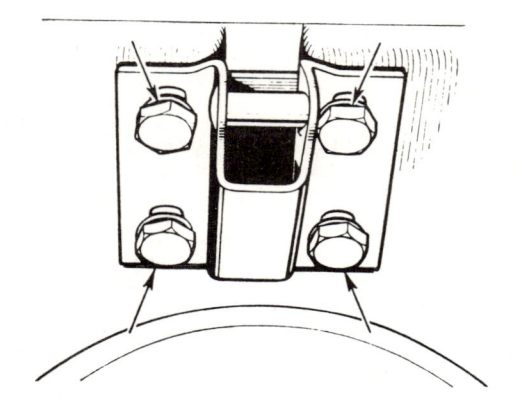

Fig. 12.19. Adjustment of the luggage compartment lid striker — Open 2 seater

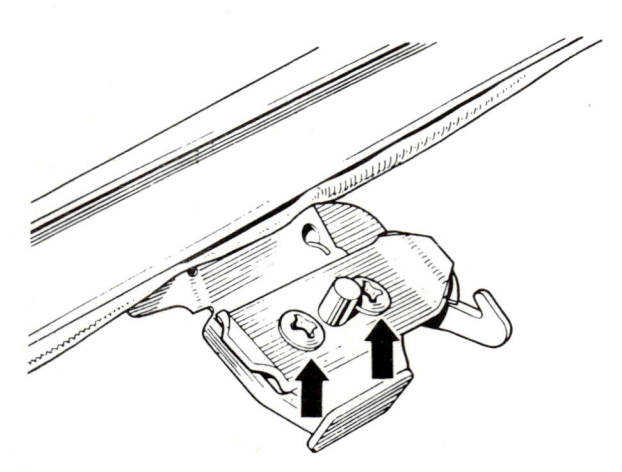

Fig. 12.20. Adjustment of the luggage compartment lid striker — Fixed head coupe

23 Windscreen (Open 2 seater) - removal and refitting

Replacement of a windscreen is one of those jobs you would be well advised to leave to a professional. It is not a very difficult task but, if you fail to seat it correctly, breakage or rain getting past the rubber seal at high speeds, may result.

Chapter 12/Bodywork

1 Generally, the following removal procedure is applicable, if the screen is being removed as a complete item, or if the remains are to be removed after breakage.
2 Take out the two slotted setscrews and remove the windscreen stay from the bracket attached centrally to the top screen frame.
3 Remove the two cross headed screws (note, for refitting, that they have different sized heads and detach the chrome screen pillar cappings from the screen pillars.
4 Carefully pull the screen pillar trim welts away from the flange on the pillars.
5 Two pop rivets securing the chrome finisher to each screen pillar will now be exposed. These will have to be drilled out using a No. 35 drill following this, the finisher can be prised away from the screen rubber.
6 Prise the chrome finisher from the bottom of the windscreen rubber.
7 Now extract one end of the screen rubber insert and withdraw it completely.
8 Run a thin bladed tool around the windscreen to break the seal between the rubber and the windscreen aperture flange. If the same windscreen is to be replaced, it is advisable to use a wooden tool for this task otherwise the screen will be scratched.
9 Now strike the glass with the flat of the hand from inside the car as shown in Fig. 12.21. Start in one corner and work from top to bottom all round the screen until it can be removed.
10 Remove the windscreen top frame in a similar manner to that described in paragraph 8. Do not use undue force when removing the frame.
11 Now work all round the windscreen flange and remove all trace of old sealer.
12 Examine the screen rubber for cuts and discard it if there are any signs of damage. It is a worthwhile proposition to replace the rubber as a matter of course.
13 If the windscreen broke for no apparent reason, examine the aperture flange for bumps or irregularities in the metal which, if present, should be filed down. Faults of this nature are possible but, due to the high standards maintained during manufacture, improbable.
14 The first step in refitting is to attach the rubber to the windscreen aperture with the flat side of the rubber to the rear.
15 There are two special tools available to assist in seating the screen in the rubber seal. These are illustrated in Fig. 12.22, and although they are not indispensable they do make the job much easier and so we suggest you try to borrow or hire them if the occasion arises.
16 Insert the screen into the rubber along the bottom edge first, you will have no success in fitting one end and then trying to get the other end in place. Now use Tool 'A' to lift the rubber over the glass (Fig. 12.23).
17 When the screen is in position, use Tool 'B' to fit the rubber sealing strip (Fig. 12.24) with the wide rounded edge towards the outside.
18 Now apply the nozzle of a pressure gun (Fig. 12.25) filled with sealing compound between the metal body flange and rubber to fill the gap all round with compound. Use a copper or soft non-ferrous nozzle in order to avoid scratching the screen.
19 Clean away all excess compound when you have filled the joint. Use a rag soaked in white spirit.
20 Fit the chrome strip on top of the windscreen rubber after coating the inside of the strip with Bostik 1251 and allowing it to become tacky. You may have to bend the strip to suit the contour.
21 Place the chrome strip on the rubber over the sealing strip and lip the rubber over the strip using Tool 'A'.
22 Refit the windscreen top frame using a length of new sealing strip. It may be an advantage to lubricate the sealing strip and glass with a liquid soap solution.
23 Apply Bostik 1251 to the inside face of the screen pillar finisher in the area where it comes into contact with the screen rubber. Allow the Bostik to become tacky and then place the finisher on the rubber and lip the the rubber over the finisher using Tool 'A'.
24 Secure the finisher to the screen pillar with two pop rivets.
25 Refit the screen pillar cappings to the screen pillars. Make sure that the flat countersunk screw is fitted to the inside face of the screen pillar capping and that the raised screw is fitted to the top face.

Fig. 12.21. Removing the windscreen

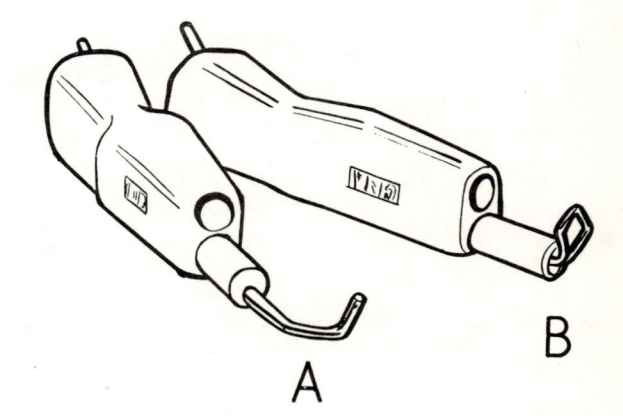

Fig. 12.22. Tools used when re-fitting the windscreen

Fig. 12.23. Using Tool "A" to lift the rubber over the glass

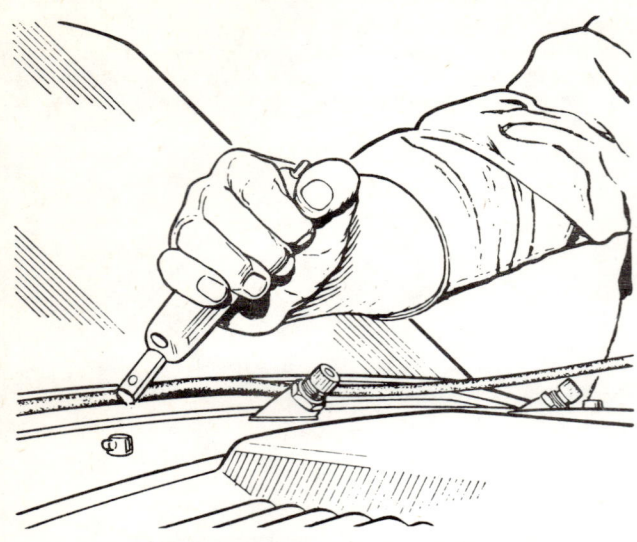

Fig. 12.24. Using Tool "B" to fit the sealing strip

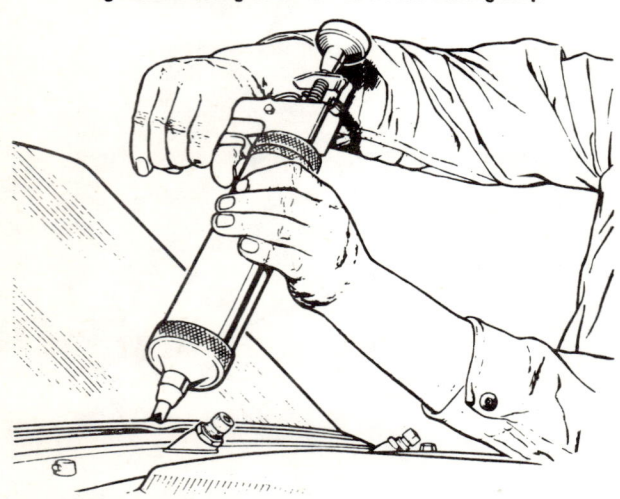

Fig. 12.25. Using a pressure gun to inject sealing compound between the surround and the glass

24 Windscreen (Fixed Head Coupe) - removal and refitting

This, as in the case of the Open 2 seater, is a job which should, preferably, be left to a professional.
1 Prise off the chrome screen pillar finishers from the rubber and then remove the upper and lower finishers in the same manner.
2 Extract one end of the rubber insert and withdraw it completely.
3 Now carry on as described in Section 23 paragraphs 8 to 13 for the removal of the windscreen and cleaning up of the surround.
4 The method of refitting and sealing the windscreen is similar to that described for the Open 2 Seater at Section 23.
5 When the screen is in position and sealed and all excess sealing compound has been removed, coat the inside of the upper chrome strip with Bostik 1251 and after waiting for a period to allow it to become tacky fit the strip on top of the windscreen rubber and use a hook, or Tool 'A' to lip the rubber over the finisher. It is permissible to bend the strip to fit the contour if necessary.
6 Repeat the above with the lower chrome strip and then the screen pillar finishers which should overlap the upper and lower finishers at the two ends.

25 Rear window glass (Fixed Head Coupe) - removal and refitting

1 Prise off the chrome finisher strip from the outside of the rubber.
2 Extract one end of the rubber insert and withdraw it.
3 Run a thin bladed tool around the glass (Fig. 12.26) to break the seal between the rubber and the glass aperture flange.
4 Have an assistant to strike the glass with the flat of his hand from inside the car starting in one top corner and working to the bottom. Work all round the glass in this manner and be prepared, outside, to catch the glass when it is released.
5 Removal of old sealer from the glass flange, examination of the rubber and refitting of the glass is similar to the instructions given in Section 23 for fitting the windscreen.
6 Coat the inside of the chrome strip with Bostik 1251, allow it to become tacky and then fit the strip over the rubber and, using Tool 'A', lip the rubber over the finisher strip.
7 The rear light on the detachable hard top can be changed in the same manner as described above.

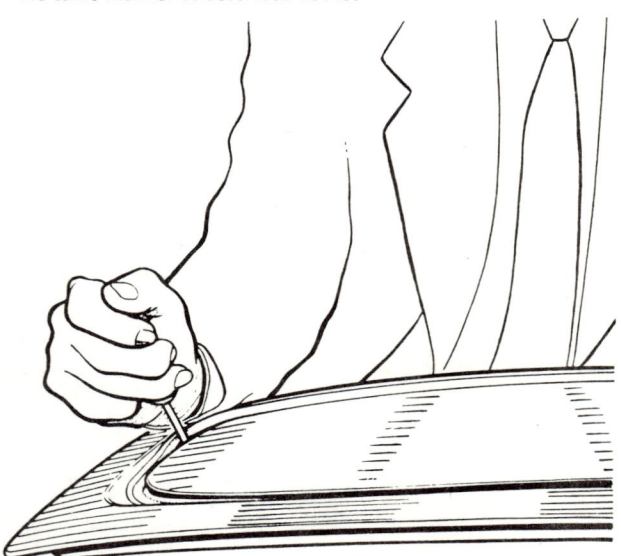

Fig. 12.26. Removing the rear window glass — Fixed head coupe and hardtop

26 Detachable Hardtop - removal and refitting

1 Lower the windows.
2 Pull down on the levers of the three fasteners which retain the hardtop to the windscreen surround (Fig. 12.27).
3 Remove the retaining bolts at each side as shown in Fig. 12.28.
4 Lift off the hardtop. You can probably do this by yourself but it is advisable to have assistance, if available.
5 The hardtop is refitted in the reverse manner to the above but do ensure that the fasteners over the windscreen are in good condition.

27 Side facia panel - removal and refitting

1 Disconnect the battery.
2 Unscrew the two bezels each securing the speedometer trip and the clock control cables to the under scuttle casings.
3 Take out the drive screws and withdraw the under scuttle casings from the retaining clips.
4 Withdraw the ignition, fuel warning and headlamp beam

Chapter 12/Bodywork

warning lights from the rear of the speedometer.
5 Unscrew the union nut and withdraw the speedometer cable from the rear of the speedometer.
6 Take out the nut and bolt securing the upper steering column to its top support bracket and collect the distance tube from between the bracket side flanges.
7 Take out the upper steering column lower mounting nuts and bolts.
8 Support, but lower the steering column slightly to give acces to the flasher switch cables at the multi snap connector behind the facia panel. Disconnect the cables.
9 Lower the steering column and allow the wheel to rest on the drivers seat.
10 Remove the two thumb screws which secure the centre instrument panel and allow the panel to hinge down into the horizontal position.
11 Remove the three slotted setscrew (Fig. 12.29) which hold the side facia panel to the instrument panel bracket.
12 Remove the nut and bolt attaching the mixture control bracket to the centre panel support bracket.
13 Remove the chrome ring nut securing the headlamp dipper switch to the facia and withdraw the switch lever through the panel.
14 You will find there are two nuts and washers behind the facia panel securing it to a bracket adjacent to the door hinge post. Remove them and the panel can then be withdrawn forwards.
15 Slacken the setscrew securing the mixture control inner cable to the control lever and disengage the cable.
16 Remove the mixture control warning light bulb holder and disconnect the two cables from the warning light switch.
17 Withdraw the two instrument illumination bulb holders from the speedometer.
18 Withdraw the two illumination bulb holders from the revolution counter and then disconnect the two cables from the Lucar tags.
19 Withdraw the flasher indicator warning light bulb holders from the indicator light unit.
20 Disconnect the two cables from the brake fluid warning light.
21 Disconnect the cable to the clock.
22 The side facia panel can now be removed from the car.
23 Refitting the panel is the reverse of the above. Points to watch are correct adjustment of the mixture control cable (see Chapter 3) and movement of the heater controls and correct connection of the electrical cables; refer to the wiring diagram given in Chapter 10 if you are in any doubt.

28 Glovebox - removal and refitting

1 The glovebox attachment points (right-hand drive models) are shown in Fig. 12.30. The attachment points for the left-hand drive cars are similar but are reversed.
2 Disconnect the battery.
3 Take out the screws securing the under scuttle casing and withdraw the casing from the retaining clips.
4 Remove the thumb screws securing the centre instrument panel and allow the panel to hinge down to the horizontal position.
5 Remove the grab handle by taking out two setscrews from the hidden face of the glovebox and then, on early models, one setscrew from the base of the screen pillar; this screw will be exposed after lifting the draught rubber and pulling away the trim welt. On later models the setscrew is located at the base of the pillar.
6 Take out the three slotted setscrews which retain the glovebox to the support bracket of the centre instrument panel.
7 Remove the two nuts and lockwashers at the rear of the glovebox which secure it to a bracket on the body adjacent to the door hinge post.
8 On right-hand drive models, disconnect the heater controls.
9 On left-hand drive models, disconnect the mixture control warning light and switch.
10 The glovebox can now be withdrawn from the car.

11 Refitting is the reverse of the above. Make sure that full movement of the heater control is obtained and that the mixture control is correctly set.

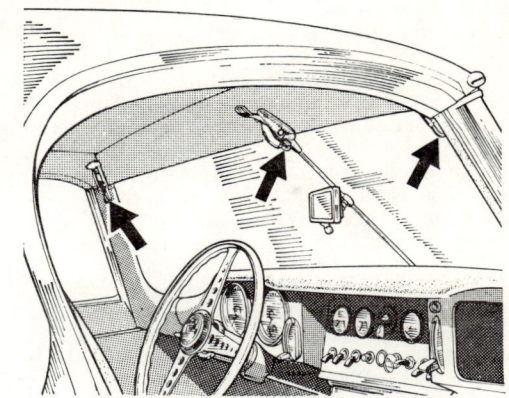

Fig. 12.27. The Hardtop retaining fasteners

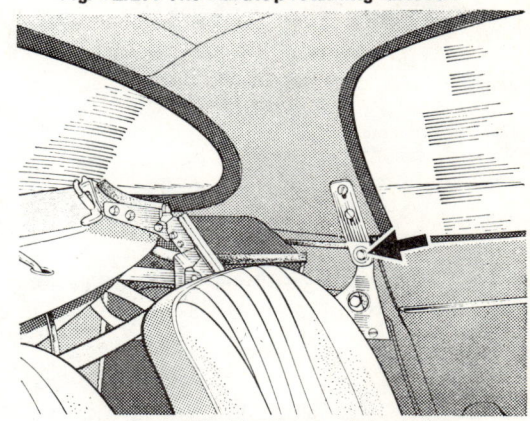

Fig. 12.28. The Hardtop retaining bolts

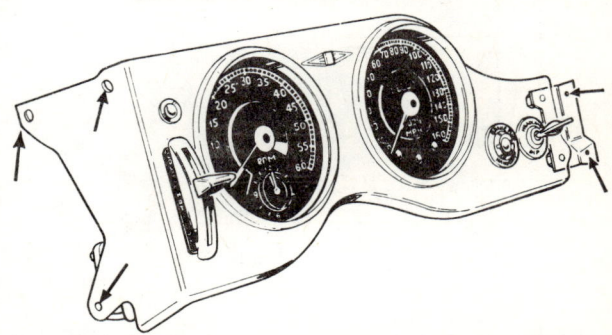

Fig. 12.29. The side facia panel attachment points (right-hand drive illustrated)

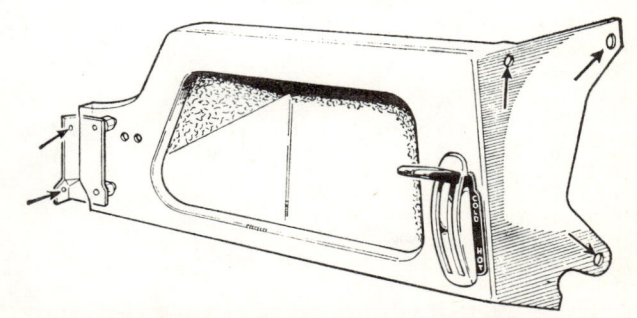

Fig. 12.30. The glovebox attachment points (right-hand drive illustrated)

29 Top facia panel - removal and refitting

1 The top facia attachment points are shown in Fig. 12.31
2 Disconnect the battery.
3 Take out the drive screws and remove the under scuttle casings.
4 Remove the central console panel by taking out the four large round headed setscrews which attach the panel to the body bracket. If a radio is fitted, you will have to disconnect the aerial and input cables.
5 Take out the thumb screws of the centre instrument panel and allow the panel to hinge down to a horizontal position.
6 Remove the two centre nuts attaching the top panel to the brackets on the body and then take off the two outer securing nuts.
7 Pull off the three demister conduit pipes from the rubber elbow connections on the bulkhead below the instrument panel.
8 Identify and then disconnect the two cables to the map light.
9 The top facia panel can now be removed complete with the demister nozzles and pipes.
10 Refit the panel in the reverse sequence to the above using the slotted holes in the brackets to adjust the forward edge of the panel to the screen frame.

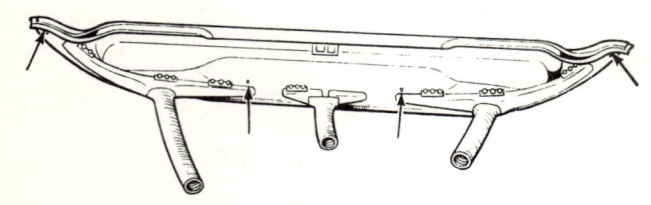

Fig. 12.31. The top facia panel attachment points

30 Exhaust system - removal and refitting

An exploded view of the exhaust system is given at Fig. 12.32. The attachment points for the two downpipes are their connections to the exhaust manifold on the engine and then by means of clips to stub pipes at the front of the silencers. The silencers, which are 'handed' are attached through four rubber mountings to brackets on the body underframe. The tail pipe assembly attaches by clips to stub pipes at the rear of the silencers and are then secured to the body of the car by a mounting positioned centrally between the two detachable mufflers as shown in Fig. 12.33.

It is our experience that the various joints of any exhaust system are usually in a semi-'welded' condition especially if the system has been in use for a long period and thus their separation, without damage, is difficult. For that reason we advise removal of the exhaust system as a complete assembly if removal is for purposes other than replacement. A full description on the method of removal of the complete exhaust system will be found in Chapter 1 under the Section dealing with removal of the engine. However, the procedure for removing particular items of the exhaust system is as follows, this procedure can, of course, be short circuited as desired.
1 Slacken the clips securing the tail pipes to the rear of the silencer assembly.
2 Remove the nut, bolt and washer securing the two mufflers.
3 Remove the nut, bolt and washer which secure the mufflers to the central mounting.
4 Work the tail pipes out of the silencer stub pipes and withdraw them complete with the mufflers.
5 Release the clips securing the silencers to the two downpipes.
6 Remove the nuts from each of the four rubber mountings of

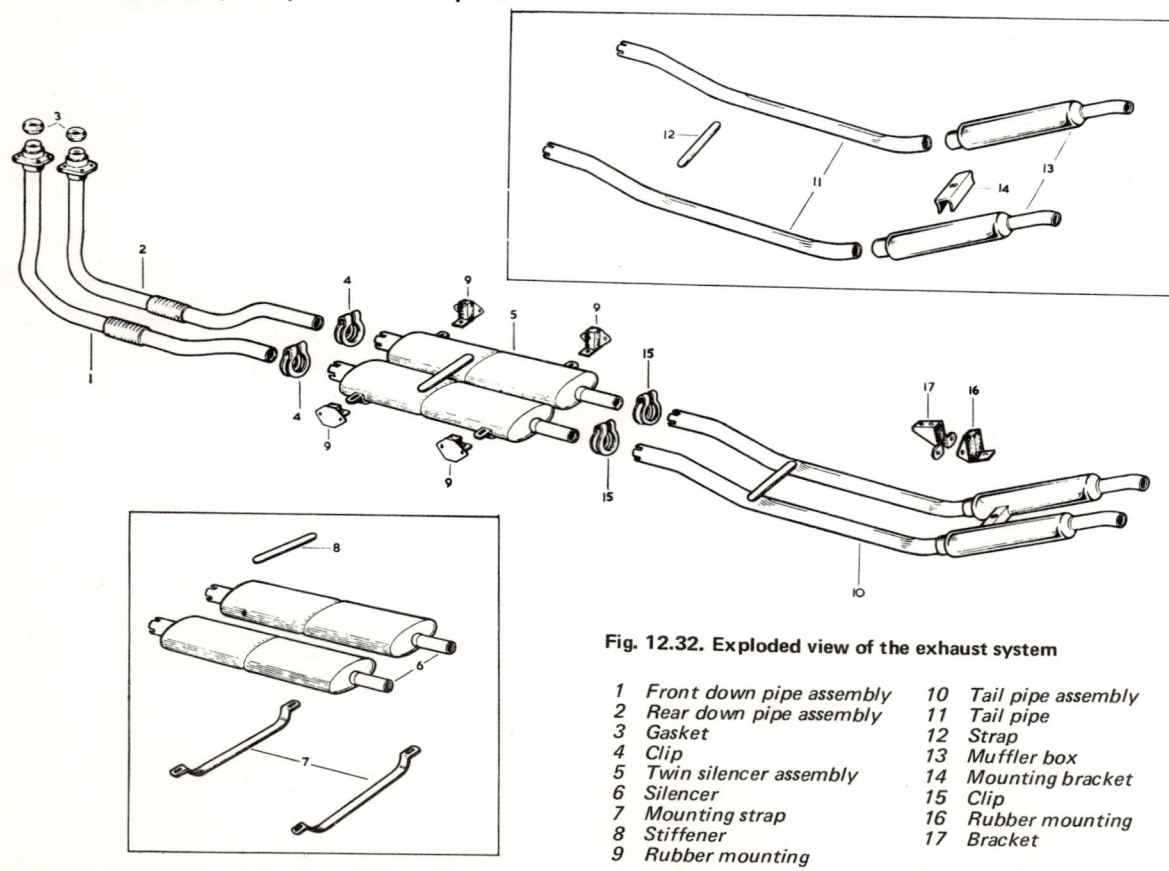

Fig. 12.32. Exploded view of the exhaust system

1 Front down pipe assembly
2 Rear down pipe assembly
3 Gasket
4 Clip
5 Twin silencer assembly
6 Silencer
7 Mounting strap
8 Stiffener
9 Rubber mounting
10 Tail pipe assembly
11 Tail pipe
12 Strap
13 Muffler box
14 Mounting bracket
15 Clip
16 Rubber mounting
17 Bracket

Chapter 12/Bodywork

the silencer assembly. Leave the bolts in position.
7 Take the weight of the silencers and then, using a screwdriver, lever on the rubber mountings to disengage the bolts.
8 Work the silencers off the downpipes.
9 Remove the four nuts and washers securing each downpipe to the exhaust manifold. As described in Chapter 1, you may find it easier to do this by working from underneath the car using a socket wrench and a long extension.
10 Separate the downpipes from the exhaust manifold and remove them from below the car. Collect the sealing rings located between the exhaust manifold and each downpipe.
11 The silencer rubber mounting brackets can be removed by taking off the nut and washer securing the bracket to the body. The muffler mounting bracket is removed by unscrewing the two nuts and bolts which secure it to the body.
12 Refit the exhaust system in the reverse sequence to the above but fit new sealing rings between the downpipes and the exhaust manifold.

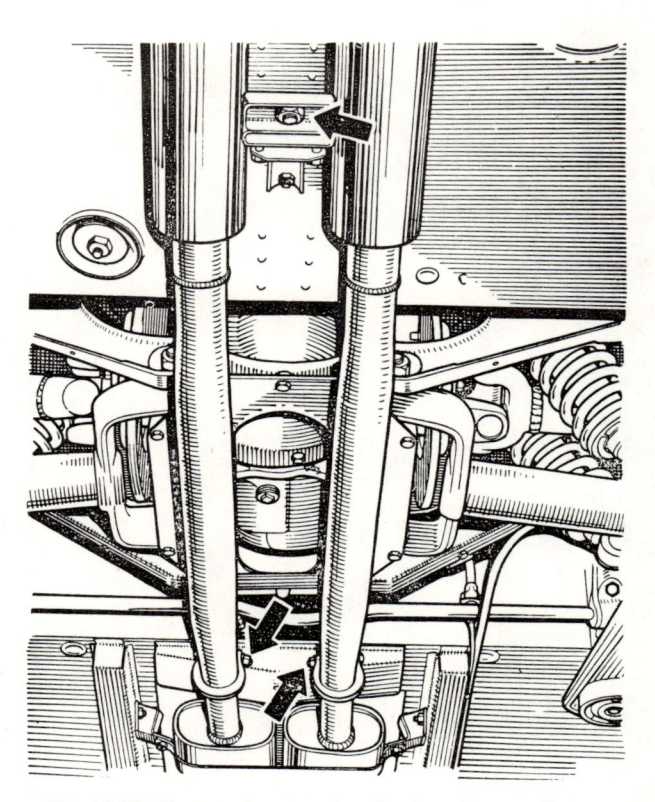

Fig. 12.33. The attachment points for the exhaust tail pipes

31 Heater - removal and refitting

See Fig. 12.34
1 Follow the instructions given in Chapter 2 for draining the cooling system.
2 Disconnect the battery.
3 Slacken the two jubilee clips which secure the heater hoses to the body of the heater.
4 Slacken the pinch bolt clamping the heater control flap cable to the lever.
5 Slacken the pinch bolt securing the conduit casing to the heater body and remove the cable.
6 Identify and then disconnect the three cables for the fan.
7 Remove the four bolts with their plain and serrated washers which secure the heater to the scuttle, and then remove the two screws which attach the heater body bracket to the subframe.
8 The heater can now be removed.

9 When refitting the heater, which is a straightforward reversal of the above removal sequence, make sure that the rubber seal between the heater body outlet and the car aperture is in the correct position.
10 When the heater is secured in position, move the heater flap lever on the heater into the fully forward position, as shown 'A' in fig. 12.35, and pass the control cable through the attachment points. Tighten the two pinch bolts to secure the control cable and casing.

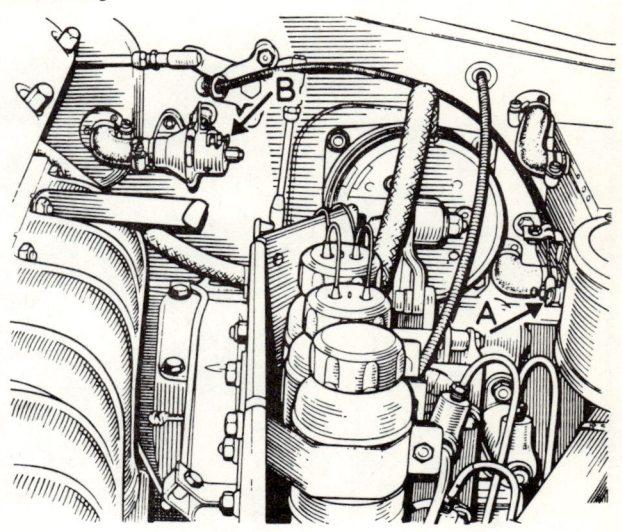

Fig. 12.35. Setting the heater operating levers

32 Heater water control tap - removal and refitting

1 Follow the instructions given in Chapter 2 for draining the cooling system.
2 Slacken the pinch bolts securing the water tap control cable and casing.
3 Slacken the clip securing the rubber hose to the water tap.
4 Remove the bolts which secure the water tap to the mounting block.
5 Remove the tap and collect the rubber seal.
6 Refitting is the reverse of the above but you must make sure that the sealing rubber is seated correctly in the machined faces of the water tap and the distance piece.
7 Before connecting the operating cable, move the lever on the tap fully forward to the 'HOT' position ('B' in Fig. 12.35) and place the heater control on the facia in the fully 'ON' position. Fit the control cable and casing in place and tighten the pinch bolts.

33 Air conditioning refrigeration equipment - general

Air conditioning equipment is available as a factory installed optional extra and is fitted in conjunction with the standard heating and ventilating equipment.
A basic knowledge of refrigeration systems and special tools is required before any servicing operations can be attempted. It is **dangerous** for any unqualified person to disconnect, or attempt to disconnect, or remove, any part of the air conditioning system. This is because the system is pressurised with a refrigerant which is a halogenated hydrocarbon (dichlorodifluoromethane!) and blindness or other serious injury may occur if the refrigerant makes contact with the eyes.
In view of the foregoing, we are deliberately confining information on servicing procedures to those given in Routine Maintenance at the beginning of this Manual. Consequently, if

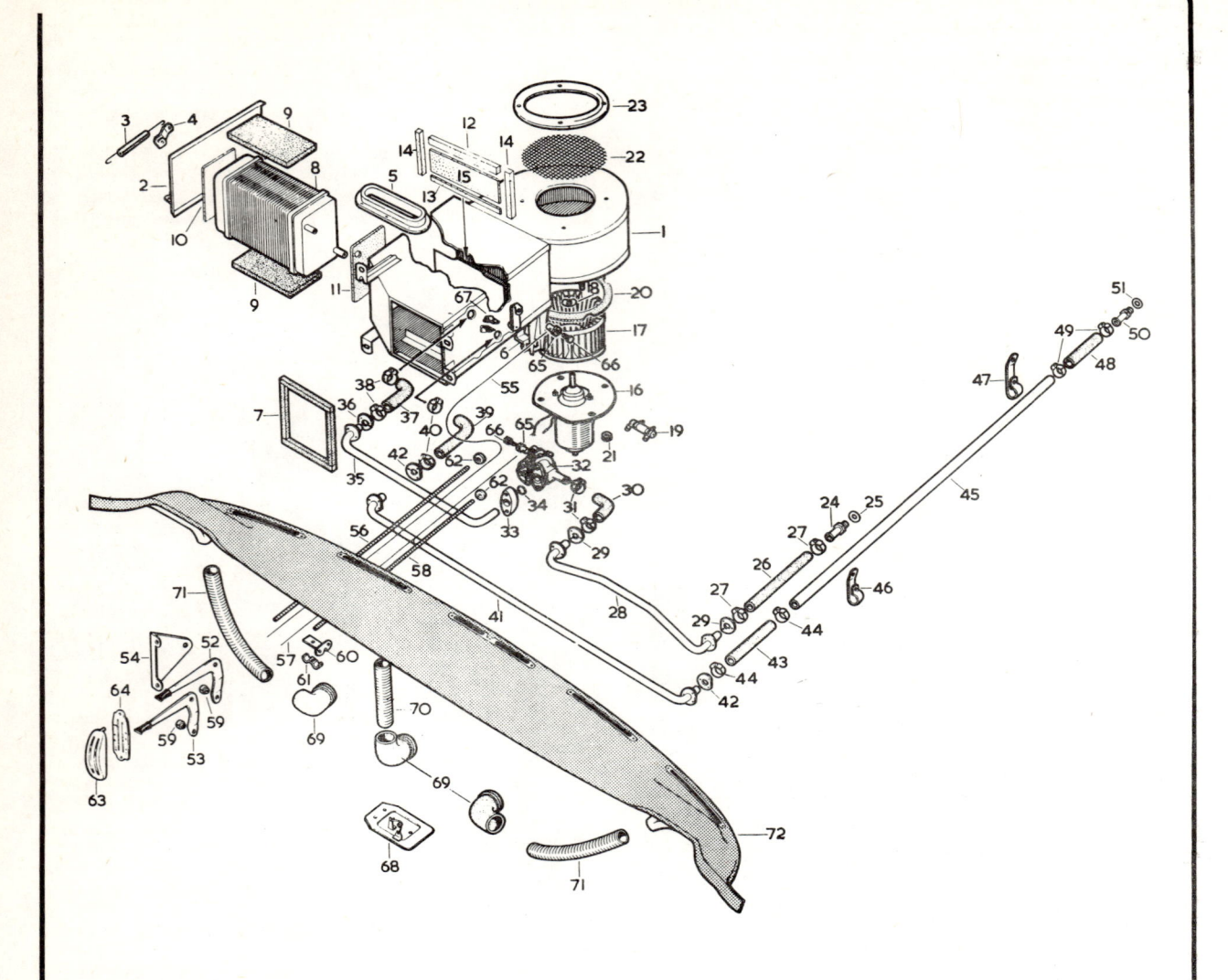

Fig. 12.34. Exploded view of the heater components

1 Heater case
2 Side panel
3 Spring
4 Flap lever
5 Air release duct
6 Mounting bracket
7 Seal between heater case and dash
8 Water radiator for heater
9 Felt seal
10 Seal
11 Seal
12 Seal on air control flap
13 Seal on heater case
14 Seal on air control flap
15 Air release duct seal
16 Fan motor
17 Fan
18 Spire nut
19 Electrical resistance
20 Sealing ring
21 Grommet
22 Wire mesh
23 Wire mesh securing ring
24 Manifold heater pipe adaptor
25 Copper washer
26 Water hose
27 Hose clip
28 Water feed pipe
29 Feed pipe flange
30 Water hose elbow
31 Hose clip
32 Water control tap
33 Control tap mounting block
34 Sealing ring
35 Feed pipe from water control tap
36 Feed pipe securing flange
37 Water hose elbow
38 Hose clip
39 Water hose elbow
40 Hose clip
41 Water return pipe
42 Securing flange
43 Water hose
44 Hose clip
45 Water return pipe
46 Return pipe mounting clip
47 Return pipe mounting clip
48 Water hose
49 Hose clips
50 Water pump adaptor
51 Copper washer
52 Air flap control lever
53 Water control tap lever
54 Control lever support bracket
55 Air flap control cable
56 Conduit for air flap control cable
57 Water tap control cable
58 Conduit for water tap control cable
59 Control cable retaining clip
60 Cable abutment clamp bracket
61 Abutment clamp
62 Grommet
63 Control lever escutcheon
64 Plate
65 Inner control cable trunnions
66 Setscrew
67 Abutment clamp
68 Heater doors
69 Rubber elbow
70 Demister hose
71 Demister hose
72 Screen rail

Chapter 12/Bodywork

work on your car, not necessarily concerned with the air conditioning equipment, requires removal of any part of the air conditioning equipment, in order to give access to engine or other components, we strongly recommend that you have the system de-pressurised and the concerned item of air conditioning equipment removed professionally.

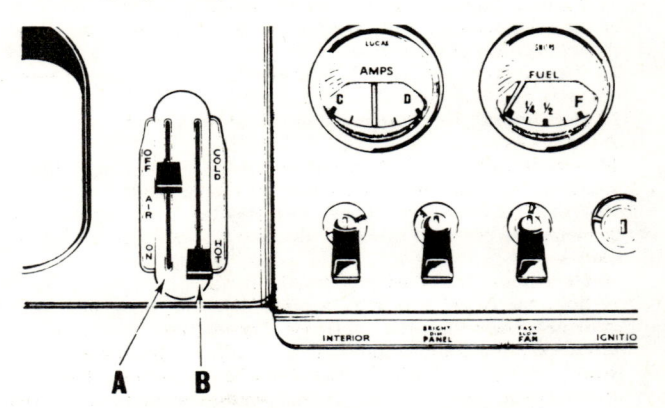

The air conditioning control panel is shown in Fig 12.36. Turning switch 'A' from the 'OFF' position starts the air flow from the supply outlet and also engages the compressor clutch. This switch has three blower speed positions 'LOW' -'MEDIUM' - 'HIGH' as indicated by the appropriate letters.

Switch 'B' increases the length of time that the compressor stays on and in its furthest clockwise position gives the coldest air flow.

To start the air conditioning system, turn switch 'B' to the required temperature setting; turn switch 'A' to the 'OFF' position. This, besides switching off the blower, will also disengage the compressor clutch.

If, in use, the volume of air decreases it is possible that the coil in the evaporator is freezing up. This can be corrected by turning switch 'B' counter-clockwise until the condition improves. If the car interior becomes too cold even when the air volume is on 'LOW', turn switch 'B' until the desired temperature is reached. In order to help in maintaining the system in good working order, it is recommended that the air conditioning system is operated for short periods each week during the winter or cool periods.

An indication of the refrigerant level is given by a sight glass attached to the receiver/drier unit at the left-hand side of the engine compartment. If bubbles or foam appear whilst the compressor is running it is an indication that the system requires servicing and for this the car should be taken to an Authorised Dealer. The air conditioning unit **must always** be **switched off** after stopping the engine.

Fig. 12.36. The air conditioning control panel

A Air volume
B Air temperature

34 Wheels and tyres - general

All cars are fitted with wire spoke wheels each having 24 long and 48 short spokes. The rim section is specified as 5K and rim diameter is 15" (381 mm).

The manufacturers' recommendation for Dunlop tyres are as follows:

3.8 litre Type 640 x 15 Road speed R.S.5

Pressures	Front	Rear
Normal use up to maximum speeds of 130 mph (210 kph)	23 lbs sq.in. (1.62 kg/cm^2)	25 lbs sq.in. (1.76 kg/cm^2)
For sustained high speed and max: performance	30 lbs sq.in. (2.11 kg/cm^2)	35 lbs sq.in. (2.46 kg/cm^2)

4.2 'E' Type and 2 + 2 Type 640 x 15 R.S.5
Pressure details as above, or Type SP 41.HR 185 x 15

Pressures		
For speeds up to 125 mph (200 kph)	32 lbs sq.in. (2.25 kg/cm^2)	32 lbs sq.in. (2.25 kg/cm^2)
For speeds up to maximum	40 lbs sq.in. (2.81 kg/cm^2)	40 lbs sq.in. (2.81 kg/cm^2)

Note: The Dunlop SP 41.HR tyre is the high speed version of the SP 41 which must not be used on 'E' Type cars unless the maximum speed is restricted to 125 mph (200 kph).

4.2 litre 'E' Type Series 2 Type SP Sport 185 VR15

Pressures	Front and Rear
For speeds up to 125 mph (200 kph)	32 lbs sq.in. (2.25 kg/cm^2)
For speeds up to maximum	40 lbs sq.in. (2.81 kg/cm^2)

Your preference for make of tyre may differ to those given above, however, before fitting a different type of tyre, you are strongly advised to enquire closely into the pressures to be used and the maximum safe permitted speed for the tyre. Furthermore, when worn or damaged tyres are not being replaced all round, it is essential that only tyres and tubes having the same characteristics as those already in use are fitted.

Due to the high speed performance capabilities of the car, it is important that no attempt is made to repair damaged or punctured tyres. Any tyre which is suspect should be taken out of use immediately and submitted to the tyre manufacturer for examination and report.

All cars sold in the U.S.A. will have a tyre information panel moulded in the wall of the tyre, necessary to conform to U.S.A. Federal Regulations.

These panels give the following information:
Maximum vehicle load
Maximum tyre pressures

A 'Tyre Information Plate' attached to the glovebox lid states:
1—Vehicle capacity weight. 2—Designated seating capacity. 3—Designated seating distribution. 4—Recommended tyre pressures. 5—Recommended tyre size.

Special tyres sizes 650 x 15 are available for competition purposes for fitment to the rear wheels only. These tyres must be fitted to special wheels (Part No.C.18922) which have a wider rim section and revised spoking to maintain the normal clearance between the tyre and the wheel arch and at the same time to increase the rear track. The wheels are supplied only as spares and not as part of the equipment of a new car, they must not **in any circumstances** be used in the front wheel positions. It is not desirable that cars should be used under normal touring conditions when fitted with the special wheels and tyres because, although they give better handling qualities under racing conditions they do not have the same qualities for touring purposes as those tyres quoted previously; furthermore, the tyre walls are more liable to damage through 'kerbing'.

35 Tyre inflation pressures

It is important to maintain the tyre pressures at the correct figures as given in Section 34 as incorrect pressures will affect the steering, riding comfort and tyre wear.

The pressure should be checked when the tyre is cold and not when it has attained its normal running temperature when an increase in pressure due to increased tyre temperature is to be expected and is allowed for in the quoted pressure when cold. Always ensure that the caps are fitted to the valves as they not only prevent the ingress of dirt but also act as a secondary seal to the valve core.

Chapter 12/Bodywork

36 Wheel alignment and tyre wear

It is most important that correct alignment of the front wheels (see Chapter 11) is maintained as misalignment causes a tyre tread to be scrubbed off laterally because the natural direction of the wheel will differ from that of the car.

Misalignment of the front wheel is indicated by an upstanding sharp 'fin' thrown up on the edge of each patter rib. 'Fins' on the inside edges of the ribs, that is the edges nearest the car, indicate excessive 'toe-in' whilst those on the outside edges show that the wheels are 'toeing-out'.

Road camber affects the direction of the car as it imposes a side thrust and the car drifts towards the nearside and this is instinctively, although perhaps not knowingly, corrected by steering towards the centre of the road. This action results in a crabwise motion of the car as shown in exaggerated form in Fig. 12.37 which also illustrates why nearside tyres are very sensitive to too much 'toe-in' and offside tyres to 'toe-out'.

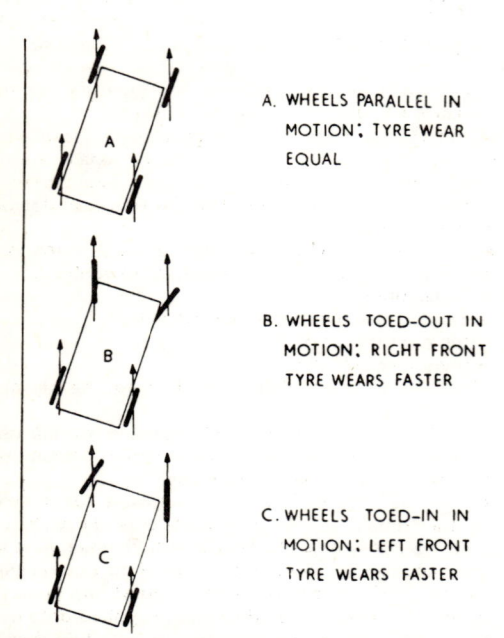

A. WHEELS PARALLEL IN MOTION: TYRE WEAR EQUAL

B. WHEELS TOED-OUT IN MOTION: RIGHT FRONT TYRE WEARS FASTER

C. WHEELS TOED-IN IN MOTION: LEFT FRONT TYRE WEARS FASTER

Fig. 12.37. Showing how road camber and front wheel misalignment affect tyre wear

37 Tyre replacement and wheel interchanging

It is common practice to interchange wheels at regular intervals with the object of getting even wear on the tyres. However, changing wheels with part worn tyres from the rear to the front positions can cause very adverse changes in steering characteristics. It is our experience that this can result in 'wander' and vagueness in 'feel' of the steering as is usually associated with misalignment of the front wheels. We recommend, therefore, as do Jaguar Cars Ltd, that wheels with part worn tyres are not transferred from the rear to the front positions, when the time comes to change the rear tyres, and these will invariably wear quicker than the front, fit these wheels, with the new tyres, on the front and transfer those from the front, which should still have a useful tyre life left, to the rear. If the tyre on the spare wheel is new this can, of course, be brought into use in the front position and one of the rear wheels with a new tyre then held as the spare.

38 Wire spoke wheels - repair and adjustment

Wire spoke wheels should be examined at regular intervals for damage and looseness of spokes. Either fault should be rectified without delay as it is likely that the wheel will be out of truth in a lateral or radial direction or a combination of both.

Spikes, 24 long and 48 short per wheel, with their securing nipples are available as spares and although their fitment present no very great problem you may find difficulty in trueing the wheel as a free runing trueing stand, and a certain amount of expertise are called for. However, if you wish to repair the wheel yourself, and have no access to a trueing stand and provided there is no play in the bearings of the front wheel hub on your car you may consider using the hub as a trueing stand, but it will make for awkward working.

The following paragraphs assume that the wheel is being completely rebuilt.
1 Place the wheel centre and the rim on a flat surface with the valve hole upwards in the 6 o'clock position.
2 Refer to Fig. 12.38
3 Starting at the valve hole fit one A,B,C, and D spoke to produce the pattern shown in the drawing.
4 Once you have established the correct pattern, remove the A and B spokes.
5 Fit the nipple to the D spoke and screw it up finger tight. Leave the C spoke loose and without a nipple fitted.
6 Assemble all the D spokes and screw up the nipples finger tight.
7 Now insert all the C spokes through the hub shell but do not fit the nipples.
8 Attach all the B spokes and fit their nipples finger tight.
9 Repeat with all the A spokes
10 Attach the nipples finger tight to all the C spokes.
11 Tighten the two C spokes and the two D spokes on each side of the valve hole until the ends of the spokes are just below the slot in the head of the nipple.
12 Repeat with the two C and two D spokes diametrically opposite the valve hole.
13 Work round the wheel and tighten all the C and D spokes as above.
14 Repeat with all the A and B spokes until the end of these spokes are also just below the slot in the nipple heads.
15 Work round the wheel and tighten the nipples on diametrically opposed spokes until some resistance is felt on all spokes.
16 Now mount the wheel on a trueing stand.
17 Spin the wheel and, with a piece of chalk, mark any high spots near the wall of the rim flange. Tighten the A and B spokes and slacken the C and D spokes in the region of the marks.
Note: No spoke should be tightened to the extent that it is impossible to tighten it further (maximum normal torque is 60 lb f in (0.7 kg f m)). If a spoke is as tight as it will go, all the other spokes must be slackened.
18 Carry on until all lateral errors are corrected.
19 Radial errors must now be corrected and this is done by spinning the wheel and marking the high spots on the horizontal tyre seat.
20 Tighten all the spokes in the region of the marks, or, if those spokes are on the limit of tightness, slacken all the others.
21 Now refer to Fig. 12.39 and check the 'dish' of the wheel. This is the lateral dimension from the inner face of the flanges of the wheel centre to the inner edge of the wheel rim and should be 3.7/16" ± 1/16" (8.73 mm ± 1.58 mm).
22 If the 'dish' is in excess of the above dimension, tighten all A and B spokes and slacken all C and D by a similar amount. Conversely, if the dimension is less that that quoted, slacken all A and B and tighten all C and D spokes by a similar amount.
23 If correction for 'dish' has been made, it will be necessary to repeat the lateral and radial trueing until the wheel is not more than 0.060" (1.5 mm) out of truth in either direction.
24 If the trueing operation has been completed properly, all the spokes should be uniformly tensioned and to a reasonably high

Chapter 12/Bodywork

degree. A correctly tensioned spoke should emit a high pitched note when lightly tapped with a hammer. If a nipple spanner of the torque recording type is used, the reading, as stated above, should be in the region of 60 lb f in (0.0 kg f m) for a properly tightened spoke.

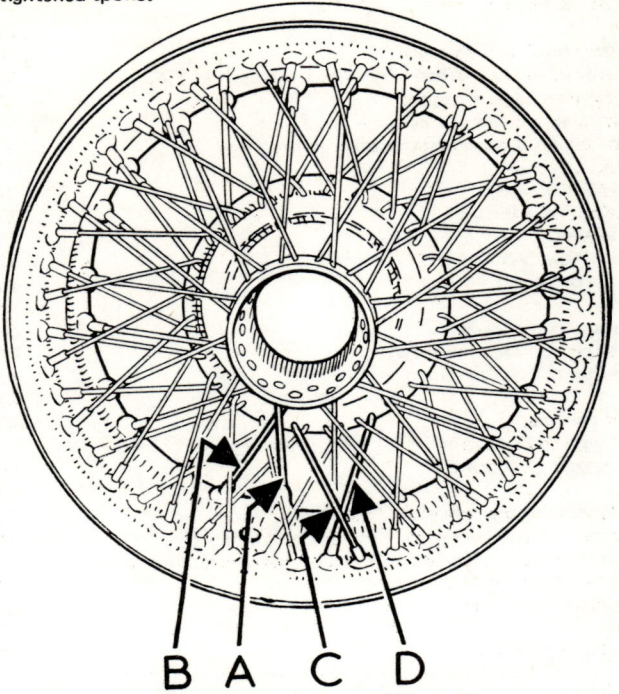

Fig. 12.38. Showing the spoking arrangement

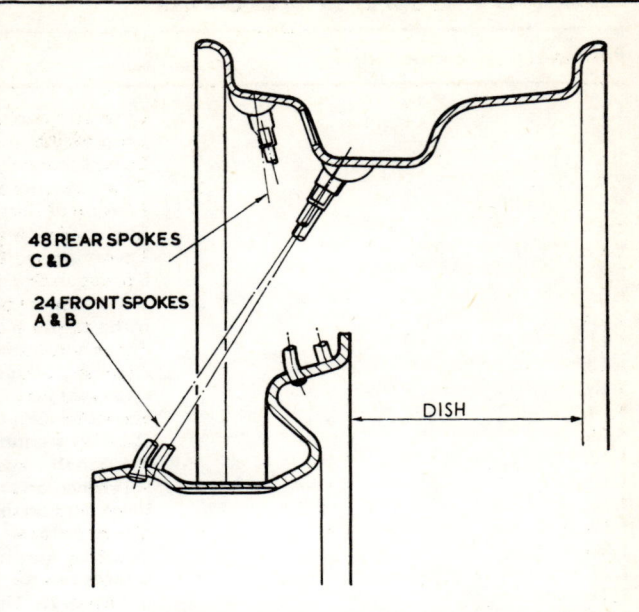

Fig. 12.39. Measurement of "dish" and showing location of the spokes

39 Wing mirrors - guidelines fitting

Wing mirrors are supplied as standard equipment on cars exported to certain Continental countries and can be supplied as optional extras on new cars ex-works.

There are certain factors to be taken in consideration if you decide to fit wing mirrors to you car yourself. Firstly the type of mirror; this, of course, is a matter of individual choice but from aesthetic aspects we feel that the over-riding factor should be to choose a pair of mirrors which will suit the lines of the car. The mirrors supplied as optional extras are Magnatex M2VC/6C (Jaguar Part No. C.19909) and it may be worthwhile visiting your local Dealer and examining these in the first place.

Having obtained the mirrors, the next question is where to fit them, how, will be self evident. They need to be fitted in a position where they can be used to obtain a good rearward view without your having to move your head or body when you are in your normal driving position. Here the services of an assistant are required, seat yourself in the car as for driving and have your assistant place the mirror in various positions until you are suited. Do not forget that horizontal and vertical adjustment will be possible after fitment. Mark the position chosen with a pencil or something which will not scratch the finish of the car because the next step is to examine the body below the spot chosen and ensure that drilling of the body or fitment of the mirror will not interfere with an existing fitment. The position chosen may well be on the bonnet in which case one of the bonnet under panels may have to be removed (see Section 16), on the other hand if it is to go on the body just forward of the screen pillar access may be gained by removal of scuttle under casings and side panels (see Section 27 et. seq). If the latter position is chosen you must look out for possible fouling of electrical cables.

The next task is to drill the body using a suitably sized drill to suit the shank of the mirror post. First, lightly mark the centre with a centre punch and then use a hand brace to drill a small pilot hole to ensure that the larger drill does not run off centre and scratch the paintwork. Having drilled the fixing hole, it is a good idea to treat the bare edge with paint or other preservative to guard against subsequent rusting. When fitting the mirror and we suggest the use of a rubber washer - if one is not supplied; to act as a seal between the plain washer and fixing nut on the underside of the body. Having got one mirror positioned, measure off from some datum point and fix the mirror on the other side of the car in a matching position. Both mirrors can now be adjusted to your individual requirement.

List of illustrations

Fig. No.		Page
1.1.	Cross sectional view of engine	23
1.2.	Longitudinal view of the engine	24
1.3.	Sump drain plug	25
1.4.	Balance marks on the clutch and flywheel	32
1.5.	Position of distributor drive shaft offset when No. 6 piston is at TDC	32
1.6.	Exploded view of cylinder head assembly	35
1.7.	Loosening and tightening sequence of cylinder head nuts	36
1.8.	Corresponding numbers on the bearing cap and cylinder head	37
1.9.	Valve support block	37
1.10.	Valve tappet and adjusting pad	37
1.11.	Piston and connecting rod assembly	39
1.12.	Crankshaft thrust washers	41
1.13.	Exploded view of cylinder block assembly	42
1.14.	Exploded view of the oil pump	43
1.15.	Oil filter assembly	44
1.16.	Crankshaft damper and components	45
1.17.	Identification marks on compression rings	47
1.18.	Identification marks on the piston crown	47
1.19.	E type pistons	47
1.20.	Checking the piston ring gap	48
1.21.	Dimensions for fitting a spark plug insert	50
1.22.	3.8 litre) Timing gear arrangement	50
1.23.	4.2 litre)	50
1.24.	Exploded view of the timing gear assembly (3.8 litre)	51
1.25.	Oil pump checking clearance between inner and outer rotors	51
1.26.	Oil pump measuring the clearance between the outer rotor and pump body	51
1.27.	Oil pump checking the end float of the rotors	52
1.28.	The bottom timing chain tensioner	52
1.29.	Air cleaner assembly	53
1.30.	Fitting the bottom half of the rear oil seal	54
1.31.	Corresponding marks on the main bearing caps and crankcase	54
1.32.	Assembly of big-end bearing shells and caps	55
1.33.	View of engine showing cylinder numbers and firing order	55
1.34.	Position of the bottom timing chain tensioner	56
1.35.	Suggested tool for compressing valve springs	57
1.36.	Positioning of camshaft for assembly	58
1.37.	Valve timing gauge in position	59
1.38.	Valve timing diagram	60
1.39.	Serrated plate for adjustment of top timing chain	60
1.40.	TDC marks	60
1.41.	Camshaft sprocket assembly	61
1.42.	Engine breather assembly	61
1.43.	Engine stabiliser	62
2.1.	Radiator drain plug	68
2.2.	Cylinder block drain plug	68
2.3.	Bolts securing the duct shield	70
2.4.	Attachment of radiator to sub-frame	70
2.5.	Attachment of cowl to radiator	71
2.6.	Header tank mounting points	71
2.7.	Fan motor thermostatic switch	71
2.8.	Exploded view thermostat and housing 3.8 litre cars	72
2.9.	Thermostat and housing 4.2 litre cars	72
2.10.	Dynamo/alternator mounting points	72
2.11.	Exploded view of the water pump) 3.8 litre cars	73
2.12.	) 4.2 litre cars	73
2.13.	Sectional view of the water pump	74
2.14.	Withdrawing the fan hub from the spindle	74
2.15.	Removing impeller from spindle	75
2.16.	Fitting impeller to spindle	75
3.1.	The fuel system (Lucas 2 FP fuel pump	77
3.2.	Lucas 2 FP fuel pump	78
3.3.	Removing pump from tank	78
3.4.	Fuel pump relief valve adjusting screw	79
3.5.	AUF 301 fuel pump	80
3.6.	Fuel pump location	80
3.7.	Exploded view of AUF 301 fuel pump	81
3.8.	Attaching pedestal to coil housing	83

List of illustrations

Fig. No.		Page
3.9.	Rocker and contact clearance	83
3.10.	Setting the diaphragm	83
3.11.	Sectional view of SU HD8 carburettor	85
3.12.	SU carburettor controls	85
3.13.	Exploded view of HD8 carburettor	86
3.14.	Checking the float level setting	88
3.15.	Centring the jet	88
3.16.	Location of needle in piston	88
3.17.	HD8 carburettor adjustment	89
3.18.	Piston lifting pin	90
3.19.	Balancing SU carburettors by car	90
3.20.	Stromberg 175 CD 2 SE carburettor	90
3.21.	Choke limiting device	91
3.22.	Temperature compensator	92
3.23.	Throttle by-pass valve	92
3.24.	Stromberg carburettors - removal and refitting	93
3.25.	Exploded view - Stromberg carburettor	94
3.26.	Stromberg carburettor) Dismantling	95
3.27.	) Assembling	95
3.28	Checking the height of the float	96
3.29.	Checking needle for distortion	96
3.30.	Carburettor) slow idle adjustment	97
3.31.	) fast idle adjustment	97
3.32.	Adjustment of fast idle screw	97
3.33.	Layout of carburettors and linkage	98
3.34.	Fuel evaporation loss control system	98
3.35.	Carbon canister removal and refitting	98
3.36.	Air delivery pump	98
3.37.	Air delivery pump belt adjustment	99
3.38.	Air rail removal and refitting	99
3.39.	Air duct removal and refitting	99
3.40.	Air cleaner and filter element - removal and refitting	100
3.41.	Temperature sender unit	100
3.42.	Gulp valve	100
3.43.	Check valve	101
3.44.	Secondary throttle housing) Removal and refitting	101
3.45.	) Dismantling	102
3.46.	) Re-assembling	102
3.47.	Inlet manifold	102
3.48.	Thermostatic vacuum switch	102
4.1.	DMBZ 6A distributor) Checking contact breaker gap	105
4.2.	22D6 distributor)	105
4.3.	Distributor lubrication points	106
4.4.	DMBZ 6A distributor - exploded view	107
4.5.	Ignition timing scale crankshaft damper	107
4.6.	Spark plug maintenance and electrode conditions	110
5.1.	10 A6 G type clutch) Sectional view	113
5.2.	) Exploded view	114
5.3.	Clutch operating system	114
5.4.	Laycock) 10" diaphragm spring clutch - exploded view	115
5.5.	Borg & Beck)	115
5.6.	Clutch fluid reservoir	116
5.7.	Bleed screw location	116
5.8.	Adjustment for clutch pedal free travel	117
5.9.	Balance marks clutch and flywheel	117
5.10.	Removing adjusting nuts	118
5.11.	Press for removal of adjusting nuts	118
5.12.	Disengaging the strut from the lever	118
5.13.	Collapsing the centre sleeve	118
5.14.	Showing maximum permissible reduction	118
5.15.	Setting the release levers	119
5.16.	Check for distortion of lever pressing	120
5.17.	Special tool for compressing diaphragm spring	120
5.18.	Assembly of fulcrum ring to cover pressing	121
5.19.	Clutch and base plate of tool inverted	121
5.20.	Positioning the tool for refitting	121
5.21.	Rivet securely with a hand punch	121
5.22.	Special tool SSC 805	121
5.23.	Staking the centre sleeve to the release plate	121

List of illustrations

Fig. No.		Page
5.24.	Clutch master cylinder - sectional view	123
5.25.	Clutch slave cylinder	124
5.26.	Setting dimension - Hydrostatic slave cylinder	125
6.1.	EB/JS gearbox casing and top cover - Exploded view	129
6.2.	EB/JS gearbox) Exploded view of internal components	130
6.3.	EJ - KE and KJS series gearboxes)	
6.4A.	Automatic transmission unit - Exploded view	132
6.4B.	Converter housing transmission mounting - Exploded view	134
6.5.	Automatic transmission controls	135
6.6.	Top cover removal showing layout of mainshaft gears	136
6.7.	Dummy countershaft	137
6.8.	Removing the constant pinion shaft and rear bearing	137
6.9.	Removing the mainshaft	137
6.10.	Depressing the 3rd speed thrust washer locking plunger	137
6.11.	Oil pump	137
6.12.	Checking layshaft end float	138
6.13.	Showing the holes through which the thrust washer locking plungers are depressed	138
6.14.	Alignment of the relieved tooth and stop pin	138
6.15.	Checking the 2nd speed gear for freedom	139
6.16.	Assembly of the operating sleeve to the inner synchronising sleeve	139
6.17.	The relieved tooth must be in line with the ball and plunger holes	139
6.18.	Location of the operating sleeve on the mainshaft	140
6.19.	Checking operation of the interlock plunger	140
6.20.	Checking 4th (top gear) interlock plunger and assembly	140
6.21.	Fitting the dummy countershaft	140
6.22.	Identification grooves 3rd/top synchro assembly	141
6.23.	Assembly of the synchro hub and operating sleeve	142
6.24.	Showing the relative position of the detent ball, plunger and thrust member	142
6.25.	Assembling the synchro hub to the sleeve	142
6.26.	Fitting the springs, plungers and thrust members	142
6.27.	Compressing the springs	142
6.28.	Checking layshaft end float	142
6.29.	Transmission pressure 'take off' point	144
6.30.	Kickdown cable adjustment	144
6.31.	Manual selector linkage adjustment	145
6.32.	Adjusting the front band	145
6.33.	Access for adjustment of the rear band	146
7.1.	Early type propeller shaft assembly - Exploded view	150
7.2.	Access hole in gearbox cowl for lubrication - Front universal joint and sliding spline	151
7.3.	Sliding joint assembly	151
7.4.	Removing a universal joint bearing	152
7.5.	Withdrawing the bearing	152
7.6.	Tapping out a bearing	152
7.7.	Separating the yokes	153
7.8.	Replacing a gasket	153
8.1.	Axle unit) Sectional view	155
8.2.	) Exploded view	156
8.3.	Thornton 'Powr-Lok' differential - Exploded view	156
8.4.	Differential carrier mounting bolts	157
8.5.	Removing the cross beam for axle unit	157
8.6.	Withdrawing a rear hub and carrier	157
8.7.	Withdrawing a halfshaft	157
8.8.	Align split pin hole and access hole	157
8.9.	Removing a bearing	158
8.10.	Withdrawing) a universal joint bearing	158
8.11.	Tapping out)	158
8.12.	Separating the universal joint yokes	158
8.13.	Replacing a gasket retainer	158
8.14.	Pressing a hub from the carrier	159
8.15.	A hub removed from its carrier	159
8.16.	Pressing in the hub inner bearing inner race using a special collar of Tool J15	159
8.17.	Checking end float of hub bearing	160
8.18.	Drive shaft assembly - Exploded view	160
8.19.	Removing a drive shaft	160
8.20.	Withdrawing the pinion inner bearing outer race	161
8.21.	Withdrawing the pinion inner bearing inner race	161

List of illustrations

Fig. No.		Page
8.22.	Mating marks on differential case	162
8.23.	Withdrawing a differential bearing	162
8.24.	Tightening the differential casing bolts with drive shaft in position	162
8.25.	Fitting the pinion inner bearing outer race	163
8.26.	Typical markings on the ground face of the pinion	163
8.27.	Pinion setting distances	164
8.28.	Checking pinion cone setting	164
8.29.	Differential bearing cap markings	164
8.30.	Checking run of the crown wheel	164
8.31.	Checking backlash and position of drive shafts	165
8.32.	Tooth contact indication	166
8.33.	Fitting the pinion oil seal	165
9.1.	Disc brakes	168
9.2.	Self adjusting mechanism) Early type	169
9.3.	) Later type	169
9.4.	Brake fluid reservoirs) Right hand drive	169
9.5.	) Left hand drive	169
9.6.	Brake bleed screw	170
9.7.	Bleeding the brakes	170
9.8.	Friction pad assembly	170
9.9.	Resetting a piston	170
9.10.	Front brake caliper	171
9.11.	Rear brake assembly - exploded view) Early 3.8 litre models	172
9.12.	) Late 3.8 and 4.2 litre models	172
9.13.	Location of shims fitted between the caliper and adjuster plate	173
9.14.	Flexible pipe connection	173
9.15.	Brake master cylinder - Sectional view 3.8 litre models	175
9.16.	Dunlop bellows type vacuum servo unit 3.8 litre models	175
9.17.	Lining the bosses on the retainer sleeve with the recesses on the bellows	176
9.18.	Lining up the bosses on the bellows with the recesses on the mounting plate and valve housing	176
9.19.	Vacuum reservoir and check valve	177
9.20.	Checking the servo with a vacuum gauge	177
9.21.	Dual-line servo braking system) Early 4.2 litre cars	178
9.22.	) Later 4.2 litre cars	178
9.23.	Remote servo and slave cylinder	180
9.24.	Master cylinder and re-action valve	182
9.25.	Removing the brake/clutch pedal box assembly	183
9.26.	Brake controls 3.8 litre models	184
9.27.	Handbrake actuating mechanism	185
9.28.	Self adjusting handbrake mechanism	185
9.29.	Preloading the handbrake caliper return spring	186
9.30.	Handbrake adjustment non-self adjusting type	187
10.1.	Dynamo model) C42	192
10.2.	) C45 PVS 6 (End plate removed)	193
10.3.	Testing brush spring tension	194
10.4.	Undercutting the commutator insulators	194
10.5.	Correctly and incorrectly cut insulators	194
10.6.	Fitting a bearing bush	194
10.7.	Drive end bracket bearing assembly	194
10.8.	Dynamo lubrication	195
10.9.	Control box) RB310	195
10.10.	) RB340	196
10.11.	Circuit diagram of RB340 control box	197
10.12.	Lucas 11 AC alternator - Exploded view	197
10.13.	Diodes and connections in the slip ring end cover	197
10.14.	Slip ring end cover - Exploded view	198
10.15.	Detach the connections from the base of the control unit	199
10.16.	Checking the rotor coil using a battery and ammeter	199
10.17.	Circuit diagram 4TR control unit	200
10.18.	Potentiometer adjuster	201
10.19.	Fuse unit model 4 FJ	201
10.20.	Location of flasher unit	201
10.21.	Flasher unit circuit diagram	201
10.22.	Location of horn relay	202
10.23.	WT 618 horn (dome removed)	202
10.24.	Lucas horn type 9H	203
10.25.	Headlamp removal (non sealed beam type)	203

List of illustrations

Fig. No.		Page
10.26.	Headlamp adjustment screws (non sealed beam type)	203
10.27.	Sealed beam headlamp	204
10.28.	Removing a side lamp bulb	204
10.29.	Headlamp setting screws sealed beam type	204
10.30.	Rear brake/rear flasher bulb removal	204
10.31.	Starter motor model M45G	204
10.32.	Internal connections of the M45 C starter motor	205
10.33.	Pre-engaged starter motor model M45G	205
10.34.	Checking the brush gear	206
10.35.	Checking the brush spring tension	206
10.36.	Commutator end bracket brush connections	206
10.37.	Starter drive assembly	206
10.38.	Setting the pinion movement	207
10.39.	Starter switch	207
10.40.	Roller clutch drive components	209
10.41.	Windscreen washer assembly	209
10.42.	DLS windscreen wiper motor and linkage	210
10.43.	Windscreen wiper parking adjustment screw (Early models)	210
10.44.	Windscreen parking adjustment (Later models)	210
10.45.	Spindle housing assembly	211
10.46.	Wiring connections - windscreen wiper switch to motor	211
10.47.	Electric clock - later models	212
10.48.	Changing the clock battery	212
10.49.	Clock adjustment (early models)	212
10.50.	Rocker type switches (locking tabs)	213
10.51.	Lucas 47A combined ignition and starter switch	213
10.52.	Tachometer plug and socket assembly	213
10.53.	Checking the inner cable for kinks	214
10.54.	Protrusion of inner flex from outer cable	215
10.55.	Checking concentricity of the inner flex	215
10.56.	Location of code number on the speedometer face	215
10.57.	Cardboard counter to check rear axle and gearing	215
10.58.	Location of TPM register	215
10.59.	Apparent source and type of noise	216
10.60.	Combined wiring diagram of fuel tank contents, water and temperature gauges with voltage regulator	216
10.61.	Circuit diagram - fuel contents gauge	217
10.62.	Circuit engine oil pressure gauge	217
10.63.	Circuit water temperature gauge	217
10.64.	Layout of wiring harness	218
10.65.	Wiring diagram) 3.8 litre models	219
10.66.	) 4.2 litre models	220
10.67.	) 4.2 litre models	221
11.1.	Front suspension assembly	224
11.2.	Rear suspension assembly (one side)	224
11.3.	Front suspension assembly (Exploded view)	225
11.4.	Upper wishbone and ball pin	226
11.5.	Section through a upper wishbone rubber/steel bushed mounting bracket	226
11.6.	Jacking the car for removal of the lower wishbone	226
11.7.	Section through a lower wishbone rubber/steel bushed mounting bracket	227
11.8.	Stub axle carrier and lower wishbone ball joint	227
11.9.	Front hub assembly	228
11.10.	Front hydraulic damper attachment points	228
11.11.	Check of standing height) 3.8 litre models	229
11.12.	) 4.2 litre models	230
11.13.	Torsion bar setting gauge	230
11.14.	Front suspension setting links	231
11.15.	Castor angle adjustment	231
11.16.	Camber angle adjustment	232
11.17.	Stub axle carrier	232
11.18.	Upper wishbone	232
11.19.	Lower wishbone lever) Rear	233
11.20.	) Front	233
11.21.	Rear suspension assembly	234
11.22.	Removal of rear suspension assembly from car	235
11.23.	The correct position of mounting rubber	235
11.24.	Hydraulic damper mounting points	235
11.25.	Driving out a damper mounting shaft	235
11.26.	Removing a rear road spring	235

List of illustrations

Fig. No.		Page
11.27.	Position of mounting rubbers on the radius arm	236
11.28.	The six bolts which secure the tie plate to the cross beam	237
11.29.	The eight bolts which secure the tee plate to the inner fulcrum mounting bracket	237
11.30.	Removing the outer fulcrum shaft	237
11.31.	Wishbone inner fork and components	237
11.32.	Fitting the inner fulcrum shaft	237
11.33.	Checking clearance between the hub carrier oil seal tracks and the wishbone fork	237
11.34.	Checking bearing end float	238
11.35.	Rear suspension setting link	238
11.36.	Checking the camber angle of a rear wheel	239
11.37.	Camber angle adjustment of rear wheels	239
11.38.	Upper steering column (sectional view)	240
11.39.	Steering column assembly (Exploded view)	241
11.40.	Upper and lower steering columns	242
11.41.	Rack and pinion assembly	244
11.42.	Checking end float of the cover plunger	244
11.43.	Checking adjustment of the ball housing	245
11.44.	Steering arm	245
12.1.	Body panels) Open two seater	248
12.2.	) Fixed head coupe	249
12.3.	Bonnet panels	250
12.4.	Front frame assembly	250
12.5.	Door striker plate screws	252
12.6.	Access to door lock handle retaining pin	252
12.7.	Removing door trim casing	252
12.8.	Attachment points - door hinge assembly	253
12.9.	Location of window regulator stop pin	253
12.10.	Attachment of window regulator	253
12.11.	Door lock) Mechanism	254
12.12.	) Retaining screws	255
12.13.	No draught ventilator hinge retaining screws	255
12.14.	Multi-pin socket connection	256
12.15.	Bonnet hinge mounting points	256
12.16.	Bonnet spring mechanism points	256
12.17.	Front) Bumper attachment points	257
12.18.	Rear)	258
12.19.	Adjustment of luggage compartment lid striker (Open two seater)	258
12.20.	(Fixed head coupe)	258
12.21.	Windscreen removal	259
12.22.	Tools for windscreen fitment	259
12.23.	Tool 'A' to lift rubber over glass	259
12.24.	Tool 'B' to fit sealing strip	260
12.25.	Pressure gun to inject sealing compound	260
12.26.	Removal rear window glass	260
12.27.	Hardtop retaining (Fasteners	261
12.28.	(Retaining bolts	261
12.29.	Side facia panel attachment points	261
12.30.	Glovebox attachment points	261
12.31.	Top facia panel attachment points	262
12.32.	Exhaust system - Exploded view	262
12.33.	Exhaust tail pipes - attachment points	263
12.34.	Heater components - Exploded view	263
12.35.	Setting the heater operating levers	263
12.36.	Air conditioning control panel	265
12.37.	Showing hoe road camber and front wheel misalignment effect tyre wear	266
12.38.	Wheel spoke arrangement	267
12.39.	Dish measurement and spoke locations	267

Metric conversion tables

Inches	Decimals	Millimetres	Millimetres to Inches		Inches to Millimetres	
			mm	Inches	Inches	mm
1/64	0.015625	0.3969	0.01	0.00039	0.001	0.0254
1/32	0.03125	0.7937	0.02	0.00079	0.002	0.0508
3/64	0.046875	1.1906	0.03	0.00118	0.003	0.0762
1/16	0.0625	1.5875	0.04	0.00157	0.004	0.1016
5/64	0.078125	1.9844	0.05	0.00197	0.005	0.1270
3/32	0.09375	2.3812	0.06	0.00236	0.006	0.1524
7/64	0.109375	2.7781	0.07	0.00276	0.007	0.1778
1/8	0.125	3.1750	0.08	0.00315	0.008	0.2032
9/64	0.140625	3.5719	0.09	0.00354	0.009	0.2286
5/32	0.15625	3.9687	0.1	0.00394	0.01	0.254
11/64	0.171875	4.3656	0.2	0.00787	0.02	0.508
3/16	0.1875	4.7625	0.3	0.01181	0.03	0.762
13/64	0.203125	5.1594	0.4	0.01575	0.04	1.016
7/32	0.21875	5.5562	0.5	0.01969	0.05	1.270
15/64	0.234375	5.9531	0.6	0.02362	0.06	1.524
1/4	0.25	6.3500	0.7	0.02756	0.07	1.778
17/64	0.265625	6.7469	0.8	0.03150	0.08	2.032
9/32	0.28125	7.1437	0.9	0.03543	0.09	2.286
19/64	0.296875	7.5406	1	0.03937	0.1	2.54
5/16	0.3125	7.9375	2	0.07874	0.2	5.08
21/64	0.328125	8.3344	3	0.11811	0.3	7.62
11/32	0.34375	8.7312	4	0.15748	0.4	10.16
23/64	0.359375	9.1281	5	0.19685	0.5	12.70
3/8	0.375	9.5250	6	0.23622	0.6	15.24
25/64	0.390625	9.9219	7	0.27559	0.7	17.78
13/32	0.40625	10.3187	8	0.31496	0.8	20.32
27/64	0.421875	10.7156	9	0.35433	0.9	22.86
7/16	0.4375	11.1125	10	0.39370	1	25.4
29/64	0.453125	11.5094	11	0.43307	2	50.8
15/32	0.46875	11.9062	12	0.47244	3	76.2
31/64	0.484375	12.3031	13	0.51181	4	101.6
1/2	0.5	12.7000	14	0.55118	5	127.0
33/64	0.515625	13.0969	15	0.59055	6	152.4
17/32	0.53125	13.4937	16	0.62992	7	177.8
35/64	0.546875	13.8906	17	0.66929	8	203.2
9/16	0.5625	14.2875	18	0.70866	9	228.6
37/64	0.578125	14.6844	19	0.74803	10	254.0
19/32	0.59375	15.0812	20	0.78740	11	279.4
39/64	0.609375	15.4781	21	0.82677	12	304.8
5/8	0.625	15.8750	22	0.86614	13	330.2
41/64	0.640625	16.2719	23	0.90551	14	355.6
21/32	0.65625	16.6687	24	0.94488	15	381.0
43/64	0.671875	17.0656	25	0.98425	16	406.4
11/16	0.6875	17.4625	26	1.02362	17	431.8
45/64	0.703125	17.8594	27	1.06299	18	457.2
23/32	0.71875	18.2562	28	1.10236	19	482.6
47/64	0.734375	18.6531	29	1.14173	20	508.0
3/4	0.75	19.0500	30	1.18110	21	533.4
49/64	0.765625	19.4469	31	1.22047	22	558.8
25/32	0.78125	19.8437	32	1.25984	23	584.2
51/64	0.796875	20.2406	33	1.29921	24	609.6
13/16	0.8125	20.6375	34	1.33858	25	635.0
53/64	0.828125	21.0344	35	1.37795	26	660.4
27/32	0.84375	21.4312	36	1.41732	27	685.8
55/64	0.859375	21.8281	37	1.4567	28	711.2
7/8	0.875	22.2250	38	1.4961	29	736.6
57/64	0.890625	22.6219	39	1.5354	30	762.0
29/32	0.90625	23.0187	40	1.5748	31	787.4
59/64	0.921875	23.4156	41	1.6142	32	812.8
15/16	0.9375	23.8125	42	1.6535	33	838.2
61/64	0.953125	24.2094	43	1.6929	34	863.6
31/32	0.96875	24.6062	44	1.7323	35	889.0
63/64	0.984375	25.0031	45	1.7717	36	914.4

Metric conversion tables

1 Imperial gallon = 8 Imp pints = 1.16 US gallons = 277.42 cu in = 4.5459 litres

1 US gallon = 4 US quarts = 0.862 Imp gallon = 231 cu in = 3.785 litres

1 Litre = 0.2199 Imp gallon = 0.2642 US gallon = 61.0253 cu in = 1000 cc

Miles to Kilometres		Kilometres to Miles	
1	1.61	1	0.62
2	3.22	2	1.24
3	4.83	3	1.86
4	6.44	4	2.49
5	8.05	5	3.11
6	9.66	6	3.73
7	11.27	7	4.35
8	12.88	8	4.97
9	14.48	9	5.59
10	16.09	10	6.21
20	32.19	20	12.43
30	48.28	30	18.64
40	64.37	40	24.85
50	80.47	50	31.07
60	96.56	60	37.28
70	112.65	70	43.50
80	128.75	80	49.71
90	144.84	90	55.92
100	160.93	100	62.14

lb f ft to Kg f m		Kg f m to lb f ft		lb f/in^2 : Kg f/cm^2		Kg f/cm^2 : lb f/in^2	
1	0.138	1	7.233	1	0.07	1	14.22
2	0.276	2	14.466	2	0.14	2	28.50
3	0.414	3	21.699	3	0.21	3	42.67
4	0.553	4	28.932	4	0.28	4	56.89
5	0.691	5	36.165	5	0.35	5	71.12
6	0.829	6	43.398	6	0.42	6	85.34
7	0.967	7	50.631	7	0.49	7	99.56
8	1.106	8	57.864	8	0.56	8	113.79
9	1.244	9	65.097	9	0.63	9	128.00
10	1.382	10	72.330	10	0.70	10	142.23
20	2.765	20	144.660	20	1.41	20	284.47
30	4.147	30	216.990	30	2.11	30	426.70

Use of English

As this book has been written in England, it uses the appropriate English component names, phrases, and spelling. Some of these differ from those used in America. Normally, these cause no difficulty, but to make sure, a glossary is printed below. In ordering spare parts remember the parts list will probably use these words:

Glossary

English	American
Allen screw	Hexagon socket screw
Anti-roll bar	Stabiliser or Sway bar
Bonnet (engine cover)	Hood
Boot (luggage compartment)	Trunk
Bottom gear	1st gear
Bulk head	Firewall
Clearance	Lash
Crown wheel	Ring gear (of differential)
Catch	Latch
Camfollower (or tappet)	Valve lifter
Drop arm (steering box)	Pitman arm
Drop arm shaft	Pitman shaft
Dynamo	Generator (DC)
Damper	Shock absorber
Earth (electrical)	Ground
Free play	Lash
Free wheel	Coast
Gudgeon Pin	Piston pin or wrist pin
Gearchange	Shift
Gearbox	Transmission
Hood	Soft top
Hard top	Hard top
Leading shoe (of brake)	Primary shoe
Lay shaft (in gearbox)	Counter shaft
Mudguard or wing	Fender
Motorway	Freeway
Petrol	Gas
Reverse	Back-up
Split cotter (as in valve spring cap)	Lock (for valve spring retainer)
Split pin	Cotter pin
Sump	Oil pan
Silencer	Muffler
Self-locking nut	Pawl nut
Steering arm	Spindle arm
Saloon	Sedan
Side light	Parking light
Tappet	Valve lifter
Tab washer	Tang; lock
Top gear	High
Transmission	Whole drive line from clutch to axle shaft
Trailing shoe (of brake)	Secondary shoe
Track rod (of steering)	Tie rod (or connecting rod)
Warning light	Tell Tale
Windscreen	Windshield

Miscellaneous points

An "Oil seal" is fitted to components lubricated by grease!

A "Damper" is a "Shock absorber": it damps out bouncing, and absorbs shocks of bump impact. Both names are correct, and both are used haphazardly.

Note that British drum brakes are different from the Bendix type that is common in America, so different descriptive names result. The shoe end furthest from the hydraulic wheel cylinder is on a pivot; interconnection between the shoes as on Bendix brakes is most uncommon. Therefore the phrase "Primary" or "Secondary" shoe does not apply. A shoe is said to be Leading or Trailing. A "Leading" shoe is one on which a point on the drum, as it rotates forward, reaches the shoe at the end worked by the hydraulic cylinder before the anchor end. The opposite is a trailing shoe, and this one has no self servo from the wrapping effect of the rotating drum.

Index

A

Air cleaner - 99
Anti-freeze - 74
Anti-roll bar - 229
Air conditioning refrigeration equipment - 263
Alternator - 198

B

Battery:
    Charging - 192
    Electrolyte replenishment - 192
    Maintenance - 191
    Removal and replacement - 191
Big-end bearings - examination - 44
Bleeding the hydraulic system - 168
Bodywork and underframe:
    Air conditioning refrigeration equipment - 263
    Bonnet - 255
    Bumpers - front and rear - 257
    Doors - 252
    Exhaust system - 262
    Hardtop - 260
    Heater - 263
    Maintenance: Body & underframe - 247
                            Upholstery and carpets - 251
    Windscreen - 258
    Wood trim panels - 251
Braking system:
    Bleeding - 168
    Brake/clutch pedal box - 181
    Brake overhaul - general - 173
    Front and rear disc brakes - 173
    Handbrake - 184
    Lockheed dual line servo system - 177
    Master cylinders - 174
    Remote servo and slave cylinder - 179
    Vacuum servo unit - 176
Bumpers - front and rear - 257

C

Camshaft - removal - 36
Carbon canister - 97
Carburettors:
    Dismantling and reassembly - 85
    Examination and repair - 87
    Float chamber flooding - 87
    Float needle sticking - 87
    Jet centring - 88
    Needle replacement - 88
    Piston sticking - 87
    Stromberg - 90
    SU types - 84

Carpets - 251
Clock - electric - 212
Clutch:
    Bleeding - 113
    Dismantling - 117
    Examination - 118
    Fork and release bearings - 123
    Judder - 126
    Master cylinder - 124
    Pedal - free travel - 116
    Squeal - 125
    Slip - 125
    Spin - 126
Coil - 104
Condenser - 106
Connecting rods - 48
Contact breaker points - 105
Cooling system:
    Anti-freeze - 74
    Draining - 68
    Filling - 69
    Flushing - 68
    Radiator - 69
    Thermostat - 71
    Water pump - 72
Crankshaft:
    Examination and renovation - 43
    Removal - 40
    Replacement - 54

D

Decarbonisation - 49
Diagrams - wiring - 219-221
Differential - 161
Disc brakes - 173
Distributor:
    Adjusting points - 105
    Dismantling, inspection, reassembly - 106
    Replacing points - 105
    Testing condenser - 106
Doors - 252
Door trim - 252
Drive shafts - 160
Dynamo - 192

E

Electrical system:
    Alternator - 198
    Battery - 191
    Dynamo - 192

Index

Electric clock - 212
Flasher unit - 201
Fuse units - 201
Horns - 202
Instrument panel - 212
Lamps - 203
Output control unit - 200
Pre-engaged starter - 206
Revolution counter - 213
Speedometer - 214
Starter motor - 205
Switches - 212
Traffic hazard warning device - 202
Voltage regulator - 195
Windscreen washer - 209
Windscreen wiper - 210
Wiring diagrams - 219-221
Engine:
   Big-end and main bearings - 44
   Camshaft removal - 36
   Connecting rods - 48
   Crankshaft - 43
   Cylinder block - 46
   Cylinder head - reassembly - 57
   Dismantling - 30
   Distributor drive - removal - 40
   Examination - 43
   Gearbox and clutch housing refitting - 61
   General dismantling - 30, 32
   Gudgeon pin removal - 39
   Mountings - 53
   Oil filter - removal - 41
   Oil pump - removal - 40
   Pistons and rings - 46
   Removal - 25-30
   Replacement - 61
   Tappets and valve adjusting pads - 49
   Timing gear - removal - 39
   Valve clearance adjustment - 58
   Valve springs - 48
   Valve timing - 59
   Water pump - 61
Exhaust system - 262
Exhaust emission control - 96

F

Fan - electric - 71
Fault diagnosis:
   Braking system - 188
   Clutch - 125
   Cooling system - 75
   Electrical system - 216
   Engine - 64
   Fuel system - 103
   Gearbox - manual - 143
          automatic - 147
   Ignition - 109
   Suspension and steering - 246
Flywheel:
   Examination and renovation - 45
   Refitting with clutch - 57
Fuel system:
   Air duct - 99
   Air rail - 99
   Carbon canister - 97
   Carburettors - 84-96
   Check valve - 101
   Exhaust emission control - 96
   Fuel gauge - tank unit - 84
   Fuel tank - 84
   Fuel pumps - 78-83
   Fuse units - 201

G

Glass:
   Windscreen - 259-260
   Rear window - 260
Gearbox:
   Manual:
   Dismantling - 128
   Examination - 137
   Reassembly - 138-141
   Removal and replacement - 128
   Automatic:
   Automatic shift speeds - 146
   Fluid level - 144
   Front band adjustment - 145
   Governor - examination - 146
   Manual linkage - 145
   Rear band adjustment - 145
   Throttle kickdown cable - 144
Gudgeon pins - 39

H

Handbrake - 184
Headlamps - 203
Heater - 263
High tension leads - 109
Hood (bonnet) - 255
Horns - 202
Hydraulic system bleeding:
   Brakes - 168
   Clutch - 113

I

Ignition system:
   Coil - 104
   Condenser - 106
   Contact breaker points - 105
   Distributor - 106
   Timing - 107
   Spark plugs and leads - 109

K

Kickdown cable/throttle - 144

L

Leads - H&T and spark plugs - 109
Lighting system:
   Headlamp   )
   Side lamp/front flasher   )
   Rear brake/rear flasher   ) 203
   Number plate   )
Locks - door - 253
Lubrication chart - 16
Lubricants and fluids - approved list - 17

M

Main bearings - 44
Maintenance - routine - 8
Major operations:
   Engine in car - 25
   Engine removed - 25
Mats - carpets - 251
Metric conversion tables - 272 - 273
Mountings - Engine - 53

O

Oil pump.
   Removal - 40
   Renovation - 52
   Reassembly - 56
Oils and greases (Castrol) - 18
Ordering spare parts - 7

Index

P

Pedal box (brake/clutch) - 181
Pistons:
    Examination and renovation - 47
    Reassembly - 54
Propeller shafts:
    Removal and replacement - 150
    Sliding joint - 151
    Universal joints - 151
Pump:
    Fuel pump - 78-83
    Oil - 52
    Water - 72

R

Radiator:
    Anti-freeze - 74
    Draining - 68
    Flushing - 68
    Removal, cleaning and refitting - 69
Rings - piston:
    Examination and renovation - 47
    Replacement - 55
Routine maintenance - 8

S

Shock absorbers:
    Front - 228
    Rear - 233
Spark plugs - 109
Specifications - technical:
    Braking system - 167
    Clutch - 112
    Cooling system - 67
    Electrical system - 189
    Engine - 19
    Fuel system - 76
    Gearbox - 127
    Ignition system - 104
    Rear axle - 154

Suspension and steering - 222
Starter motor - 205
Steering:
    Camber angle - 231
    Castor angle - 231
    Front wheel alignment - 245
    Lower steering columns - 243
    Rack and pinion - 243
    Upper steering columns - 243
    Wishbone - 236
Sump:
    Cleaning and examination - 52
    Refitting - 57
Switches - 212

T

Tank - fuel - 84
Tappets - 49
Technical specifications: See specifications
Timing - ignition - 107
Transmission - see Gearbox
Torsion bar - front - 229
Tyres and wheels - 265

U

Universal joints - 151
Upholstery - 251

V

Valves - removal - 37
    timing - 59

W

Water pump - refitting - 72
Wheels - wire - 265
Windscreen:
    Glass - 259
    Washer - 209
    Wiper - 210
Wiring diagrams - 219-221
Wishbones - 236

Printed by
J. H. HAYNES & Co. Ltd
Sparkford Yeovil Somerset
ENGLAND